Speech Recognition and Understanding

Recent Advances, Trends and Applications

NATO ASI Series

Advanced Science Institutes Series

A series presenting the results of activities sponsored by the NATO Science Committee, which aims at the dissemination of advanced scientific and technological knowledge, with a view to strengthening links between scientific communities.

The Series is published by an international board of publishers in conjunction with the NATO Scientific Affairs Division

A Life Sciences B Physics	Plenum Publishing Corporation London and New York
C Mathematical and Physical Sciences D Behavioural and Social Sciences E Applied Sciences	Kluwer Academic Publishers Dordrecht, Boston and London
F Computer and Systems Sciences G Ecological Sciences H Cell Biology I Global Environmental Change	Springer-Verlag Berlin Heidelberg New York London Paris Tokyo Hong Kong Barcelona Budapest

NATO-PCO DATABASE

The electronic index to the NATO ASI Series provides full bibliographical references (with keywords and/or abstracts) to more than 30000 contributions from international scientists published in all sections of the NATO ASI Series. Access to the NATO-PCO DATABASE compiled by the NATO Publication Coordination Office is possible in two ways:

– via online FILE 128 (NATO-PCO DATABASE) hosted by ESRIN, Via Galileo Galilei, I-00044 Frascati, Italy.

– via **CD-ROM** "NATO-PCO DATABASE" with user-friendly retrieval software in English, French and German (© WTV GmbH and DATAWARE Technologies Inc. 1989).

The CD-ROM can be ordered through any member of the Board of Publishers or through NATO-PCO, Overijse, Belgium.

Series F: Computer and Systems Sciences Vol. 75

Speech Recognition and Understanding

Recent Advances, Trends and Applications

Edited by

Pietro Laface

Dipartimento di Automatica e Informatica
Politecnico di Torino
Corso Duca degli Abruzzi 24, 10129 Torino, Italy

Renato De Mori

School of Computer Science
3480 University St., Montreal, Quebec H3A 2A7, Canada

Springer-Verlag
Berlin Heidelberg New York London Paris Tokyo
Hong Kong Barcelona Budapest
Published in cooperation with NATO Scientific Affairs Division

Proceedings of the NATO Advanced "Study Institute on Speech Recognition and Understanding. Recent Advances, Trends and Applications" held in Cetraro, Italy, July 1–13, 1990.

ISBN-13: 978-3-642-76628-2 e-ISBN-13: 978-3-642-76626-8
DOI: 10.1007/978-3-642-76626-8

Softcover reprint of the hardcover 1st edition 1992
Typesetting: camera ready by authors
45/3140-543210 – Printed on acid-free-paper

Preface

The book collects the contributions to the NATO Advanced Study Institute on "Speech Recognition and Understanding: Recent Advances, Trends and Applications", held in Cetraro, Italy, during the first two weeks of July 1990.

This Institute focused on three topics that are considered of particular interest and rich of innovation by researchers in the fields of speech recognition and understanding: Advances in Hidden Markov modeling, connectionist approaches to speech and language modeling, and linguistic processing including language and dialogue modeling.

The purpose of any ASI is that of encouraging scientific communications between researchers of NATO countries through advanced tutorials and presentations: excellent tutorials were offered by invited speakers that present in this book 15 papers which summarize or detail the topics covered in their lectures. The lectures were complemented by discussions, panel sections and by the presentation of related works carried on by some of the attending researchers: these presentations have been collected in 42 short contributions to the Proceedings.

This volume, that the reader can find useful for an overview, although incomplete, of the state of the art in speech understanding, is divided into 6 Parts.

The first Part covers **Recent Advances in Hidden Markov Models (HMM)** and contains 2 invited and 12 contributed papers. Tutorial contributions are presented by L. R. Rabiner on strengths and limitations of HMM techniques for speech recognition systems, and by R. Schwartz on the techniques for speaker adaptation. The tutorials on this field have been complemented by a number of contributions from the participants reporting results obtained, in different countries, for isolated word recognizers developed with the HMM paradigm.

The second Part, **Continuous Speech Recognition Systems**, contains 2 invited and 10 contributed papers. The first invited paper, by C. H. Lee, describes a large vocabulary continuous speech recognition system based on continuous densities HMM and the methods used for selecting and modeling its set of speech units. The second invited paper, by K.F. Lee, reprinted with permission by the IEEE Transactions on ASSP summarizes the work done at Carnegie Mellon University on context-dependent unit modeling for speaker-independent continuous speech recognition. Several contributed papers either present other continuous speech recognition systems or focus on the relevant issue of system efficiency.

Part 3, **Connectionist Models of Speech**, deals with different approaches to speech and language modeling using emerging technologies based on Connectionist Models. It contains 4 invited and 4 contributed papers. Historical and mathematical background are presented together with biological considerations. In particular a model inspired by biology is proposed by J.P. Haton, the relations between artificial neural networks and HMMs are discussed by J. Bridle, while an interesting recursive network for state-based system modeling is presented by F. Fallside, finally A. Waibel reports on advanced connectionist architectures and system performances in a paper reprinted with permission by the IEICE. Most of the contributions from the participants are related to phoneme and phonetic feature recognition.

Part 4, **Stochastic Models for Language and Dialogue**, that contains 3 invited and 3 contributed papers, focuses on stochastic language modeling. An introduction to stochastic context-free grammars and their application to Automatic Speech Recognition is illustrated in the paper by H. Ney, while an in depth discussion of the algorithms for training these grammars is presented by F. Jelinek. In the third invited paper, R. De Mori reviews the models used for oral dialogues and proposes some ideas on how these models can be made stochastic ones. Contributions refer to a framework for the integration of syntactic, semantic and pragmatic information, and to methods for language modeling and adaptation.

In Part 5, **Understanding and Dialog Systems** that contains 4 invited and 2 contributed papers, three systems for speech understanding and dialogue are described that are characterized by a knowledge-based approach: the EVAR system by H. Niemann and the Susy system by E. Giachin, and the Voyager system along with a probabilistic syntactic parser by S. Seneff. A syllable knowledge-based approach is proposed in the two contributed papers.

Part 6, **Acoustic-Phonetic Coding and Segmentation** contains 10 contributed papers, that deals with the problems related to automatic segmentation and modeling of acoustic-phonetic units: from data base management to automatic generation of structural models of sublexic units.

The editors would like to thank the attendees of this NATO ASI for their scientific value and kindness that made the scenic beauty of Cetraro even more stimulating for discussions and intellectual efforts.

Thanks are also due to the contributors for their efforts to provide very good papers, to the publisher for preparing this volume and to the NATO Office of Scientific Affairs in Brussels, to the International Association for Pattern Recognition, to the Fondazione "Ugo Bordoni" in Rome, to the Istituto per la Ricerca Scientifica e Tecnologica in Trento, and to Politecnico di Torino for their sponsorship, financial, and organization support of the ASI.

It was also a pleasure to see a temporary reunion of former members of the Speech Group at the Politecnico di Torino. This book is dedicated to the memory of Professor Rinaldo Sartori who put this group together in the seventies.

P. Laface and R. De Mori
Torino and Montreal

Summer 1991

Table of Contents

2 Continuous Speech Recognition Systems 131

Invited papers

Contributed papers

Contributed papers

5 Understanding and Dialogue Systems 401

Invited papers

Contributed papers

6 Speech Analysis, Coding and Segmentation 493

Contributed papers

Part 1

Recent Results on Hidden Markov Models

Hidden Markov Models for Speech Recognition – Strengths and Limitations

L. R. Rabiner and B. H. Juang

Speech Research Department, AT&T Bell Laboratories, 600 Mountain Avenue, Murray Hill, NJ 07974

Abstract: The use of hidden Markov models for speech recognition has become predominant for the last several years, as evidenced by the number of published papers and talks at major speech conferences. The reasons why this method has become so popular are the inherent statistical (mathematically precise) framework, the ease and availability of training algorithms for estimating the parameters of the models from finite training sets of speech data, the flexibility of the resulting recognition system where one can easily change the size, type, or architecture of the models to suit particular words, sounds etc., and the ease of implementation of the overall recognition system. However, although hidden Markov model technology has brought speech recognition system performance to new high levels for a variety of applications, there remain some fundamental areas where aspects of the theory are either inadequate for speech, or for which the assumptions that are made do not apply. Examples of such areas range from the fundamental modeling assumption, i.e. that a maximum likelihood estimate of the model parameters provides the best system performance, to issues involved with inadequate training data which leads to the concepts of parameter tying across states, deleted interpolation and other smoothing methods, etc. Other aspects of the basic hidden Markov modeling methodology which are still not well understood include; ways of integrating new features (e.g. prosodic versus spectral features) into the framework in a consistent and meaningful way; the way to properly model sound durations (both within a state and across states of a model); the way to properly use the information in state transitions; and finally the way in which models can be split or clustered as warranted by the training data. It is the purpose of this paper to examine each of these strengths and limitations and discuss how they affect overall performance of a typical speech recognition system.

NATO ASI Series, Vol. F 75
Speech Recognition and Understanding. Recent Advances
Edited by P. Laface and R. De Mori

1. Introduction

The use of hidden Markov models (HMM's) for speech recognition has become increasingly popular in the past few years [4,11,15,18,21,31,35]. The widespread popularity of the HMM framework can be attributed to its simple algorithmic structure that is straightforward to implement and to its clear performance superiority over alternative recognition structures.

The strength of the hidden Markov model method lies in two major areas; namely the mathematical framework (theory) and the implementational structure (practice). Therefore, in this paper we first reexamine the strengths of the HMM methodology and next we discuss some limitations of the HMM method that potentially affect speech recognition performance for various implementations of the system.

2. Hidden Markov Modeling of Speech

Speech is essentially a non-stationary signal. When we speak, our articulatory apparatus modulates the air pressure and flow to produce an audible sequence of sounds. While the spectral content of any particular sound may include frequencies up to several thousand Hertz, our articulatory configuration changes on the order of about 10 times per second. Speech modeling thus involves the analysis of the short time spectral properties of individual sounds, and characterization of the long time changes in the articulatory configuration leading to different speech sounds.

The short time properties of an individual sound can be conveniently represented by a spectral measurement vector obtained via such standard methods as measurement of the discrete (fast) Fourier transform (FFT), the linear prediction method (LPC) and autoregressive/moving average (ARMA) models [1,2,10,27,28]. We generally call this short time spectral vector an observation. Observations are typically measured once every 10-30 msec.

There are many ways to characterize the temporal sequence of sounds, i.e. running speech, as represented by a sequence of spectral observations. The most direct way is to register the spectral sequence directly, i.e. without further modeling. Another way is to model the temporal sequence of spectra in terms of a Markov chain to describe the way one sound changes to another. As will be explained later, the two modeling phases, namely the short time spectra of individual sounds and the probabilistic description of the sound changes, can be described in a mathematically consistent framework, thereby offering analytic solutions to speech problems.

For historical reasons, the direct representation of a sequence of speech spectra, without explicit probabilistic modeling, is called a "deterministic" template. If we denote the spectral vector at time t by $\mathbf{O}_t$ and the observed spectral sequence lasts from $t=1$ to $t=T$, such a deterministic sequence is then represented by $\{\mathbf{O}_t\}_{t=1}^T = (\mathbf{O}_1, \mathbf{O}_2, ..., \mathbf{O}_T)$. In contrast, when an explicit probabilistic structure is imposed on the representation of the evolutional sequence, the resultant model is often classified as "stochastic".

Based on the above discussion, we are now ready to formally define a hidden Markov model for speech recognition. Consider a first-order N-state Markov chain as illustrated for $N=3$ in Fig. 1.

HIDDEN MARKOV MODEL

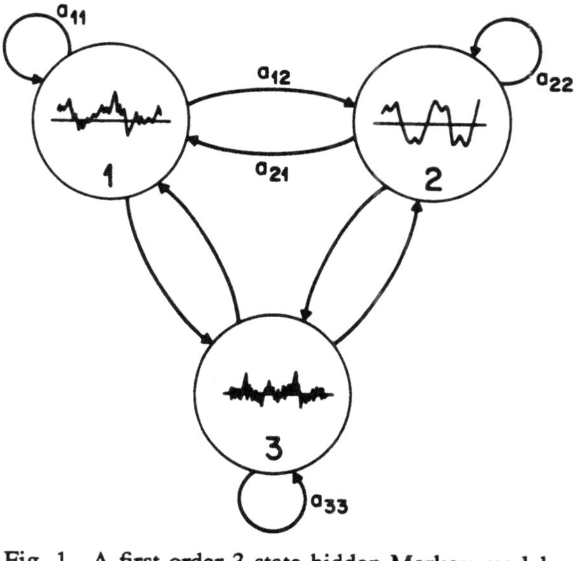

Fig. 1. A first-order 3-state hidden Markov model.

The system can be described as being in one of the N distinct states, $S_1, S_2, ..., S_N$, at any discrete time instant t. We use the state variable, q_t, as the state of the system at discrete time t. The Markov chain is then described by a state transition probability matrix $A = [a_{ij}]$ where

$$a_{ij} = P(q_t = S_j | q_{t-1} = S_i), \quad 1 \leq i, \ \ j \leq N \tag{1}$$

Note that in (1) we have assumed homogeneity of the Markov chain so that the transition probabilities do not depend on time. These transition probabilities satisfy the axiomatic constraints that $a_{ij} \geq 0$ and $\sum_{j=1}^{N} a_{ij} = 1$ for all i. Assume that at $t=0$, the state of the system, q_0, is specified by an initial state probability vector $\pi' = [\pi_1, \pi_2 \ldots \pi_N]$ where $'$ denotes matrix transpose. Then, for any state sequence $\mathbf{q} = (q_0, q_1, q_2, \ldots, q_T)$, where $q_t \in \{S_1, S_2, \ldots, S_N\}$, the probability of $\mathbf{q}$ being generated by the Markov chain is

$$P(\mathbf{q}|A, \pi) = \pi_{q_0} \; a_{q_0 q_1} \; a_{q_1 q_2} \cdots a_{q_{T-1} q_T} \tag{2}$$

Suppose now that the state sequence $\mathbf{q}$ cannot be readily observed. Instead, we envision each observation $\mathbf{O}_t$ as being produced with the system in state q_t (where $q_t \in \{S_1, S_2, \ldots, S_N\}$). We assume the production of $\mathbf{O}_t$ in each possible state S_i, $i = 1, 2, \ldots, N$ is stochastic, and is characterized by a set of observation probability measures $B = \{b_i(\mathbf{O}_t)\}_{i=1}^N$ where

$$b_i(\mathbf{O}_t) = P(\mathbf{O}_t | q_t = S_i) \; . \tag{3}$$

If the state sequence $\mathbf{q}$ that led to the observation sequence $\mathbf{O} = (\mathbf{O}_1, \mathbf{O}_2, \ldots, \mathbf{O}_T)$ is known, the probability of $\mathbf{O}$ being generated by the system is assumed to be

$$P(\mathbf{O}|\mathbf{q}, B) = b_{q_1}(\mathbf{O}_1) \; b_{q_2}(\mathbf{O}_2) \cdots b_{q_T}(\mathbf{O}_T) \; . \tag{4}$$

Therefore, the joint probability of $\mathbf{O}$ and $\mathbf{q}$ being produced by the system is simply the product of (2) and (4), written as

$$P(\mathbf{O}, \mathbf{q}|\pi, A, B) = \pi_{q_0} \prod_{t=1}^{T} a_{q_{t-1} q_t} \; b_{q_t}(\mathbf{O}_t) \tag{5}$$

It then follows that the stochastic process, represented by the observation sequence $\mathbf{O}$, is characterized by

$$P(\mathbf{O}|\pi, A, B) = \sum_{\mathbf{q}} P(\mathbf{O}, \mathbf{q}|\pi, A, B)$$

$$= \sum_{\mathbf{q}} \pi_{q_0} \prod_{t=1}^{T} a_{q_{t-1} q_t} \; b_{q_t}(\mathbf{O}_t) \tag{6}$$

which describes the probability of $\mathbf{O}$ being produced by the system without assuming the knowledge of the state sequence in which it was generated. The triple (π, A, B) is called a hidden Markov model and (6) is the probability measure defined by the model. The model parameter set is often denoted as $\lambda = (\pi, A, B)$.

In the development of the hidden Markov model methodology, the following problems due to Ferguson [12] are of particular interest. First, given the observation sequence $\mathbf{O}$, and a model λ, how do we efficiently evaluate the probability of $\mathbf{O}$ being produced by the source model λ, i.e. $P(\mathbf{O}|\lambda)$? Second, given the observation $\mathbf{O}$, how do we solve the inverse problem of estimating the parameters in λ? Although (6) defines the probability measure associated with model λ, the knowledge of the most likely state sequence $\mathbf{q}$ that led to the observation $\mathbf{O}$ is desirable in many applications. The third problem then is how do we deduce from $\mathbf{O}$ the most likely state sequence $\mathbf{q}$ in a meaningful manner? According to convention [12] we call these three problems 1) the evaluation problem, 2) the estimation problem, and 3) the decoding problem, respectively. In the following sections we describe several conventional solutions to these three standard problems.

2.1 The Evaluation Problem

The main concern in the evaluation problem is computational efficiency. Without complexity constraints, one can simply evaluate $P(\mathbf{O}|\lambda)$ directly from the definition of (6). Since the summation in (6) involves N^{T+1} possible $\mathbf{q}$ sequences, the total computational requirements are on the order of $2T \cdot N^{T+1}$ operations. The need to compute (6) without the exponential growth of computation, as a function of the sequence length T, is the first challenge for implementation of the HMM technique. Fortunately, using the well-known forward-backward procedure [6], this exorbitant computational requirement of the direct summation can be easily alleviated. By visualizing the Markov chain as a trellis structure, a forward induction procedure allows evaluation of the probability $P(\mathbf{O}|\lambda)$ to be carried out with only a computational requirement linear in the sequence length T, and quadratic in the number of states, N. Such a tremendous reduction in computation makes the HMM method attractive and viable for speech recognizer designs because the evaluation problem can be viewed as one of scoring how well an unknown observation sequence (corresponding to the speech to be recognized) matches a given model (or sequence of models) source, thus providing a mechanism for classification.

2.2 The Estimation Problem

Given an observation sequence (or a set of sequences) $\mathbf{O}$, the estimation problem involves finding the "right" model parameter values that specify a model most likely to produce the given sequence. In speech recognition, this is often called "training" and the given sequence, based on which we obtain the model parameters, is called the training sequence, even though the formulation here is statistical.

In solving the estimation problem, we often follow the method of maximum likelihood; that is, we choose model λ such that $P(\mathbf{O}|\lambda)$, as defined by (6), is maximized for the given training sequence $\mathbf{O}$. The Baum-Welch algorithm [7] (often blended with the forward-backward algorithm because of its interpretation as an extension of the forward induction procedure to the evaluation problem) cleverly accomplishes this maximization objective in a two step procedure. Based on an existing model λ' (possibly obtained randomly), the first step transforms the objective function $P(\mathbf{O}|\lambda)$ into a new function $Q(\lambda', \lambda)$ that essentially measures a divergence between the initial model λ' and an updated model λ. The Q function is defined, for the simplest case, as

$$Q(\lambda', \lambda) = \sum_{\mathbf{q}} P(\mathbf{O}, \mathbf{q}|\lambda') \log P(\mathbf{O}, \mathbf{q}|\lambda) \tag{7}$$

where $P(\mathbf{O}, \mathbf{q}|\lambda)$ is given in (5). Because $Q(\lambda', \lambda) \geq Q(\lambda', \lambda')$ implies $P(\mathbf{O}|\lambda) \geq P(\mathbf{O}|\lambda')$, we can then simply maximize the function $Q(\lambda', \lambda)$ over λ to improve λ' in the sense of increasing the likelihood $P(\mathbf{O}|\lambda)$. This maximization of the Q function over λ is the second step of the algorithm. The algorithm continues by replacing λ' with λ and repeating the two steps until some stopping criterion is met. The algorithm is of a general hill-climbing type and is only guaranteed to produce fixed point solutions, although in practice the lack of global optimality does not seem to cause serious problems in recognition performance [30].

2.3 The Decoding Problem

As noted above, we often are interested in uncovering the most likely state sequence that led to the observation sequence $\mathbf{O}$. Although the probability measure defined by a hidden Markov model, by definition, does not explicitly involve the state sequence, it is important, in many applications, to have the knowledge of the most likely state sequence for several reasons. As an example, if we use the states of a word model to represent the sounds in the word, it may be desirable to know the correspondence between the speech segments and the sounds of the word, as the duration of the individual speech segments provides useful information for speech recognition purposes.

As with the second problem, there are several ways to define the decoding objective. The most useful choice is to work on the entire state sequence $\mathbf{q}$ by maximizing $P(\mathbf{q}|\mathbf{O}, \lambda)$ for three reasons: 1) it is optimal for the unknown observation $\mathbf{O}$ in the MAP sense; 2) speech utterances are usually not prohibitively long so as to require locally (rather than globally) optimal decoding; and 3) it is possible to formulate the maximization of $P(\mathbf{q}|\mathbf{O}, \lambda)$ in a sequential manner so as to be solved by dynamic programming methods such as the Viterbi algorithm [13].

Maximization of $P(\mathbf{q}|\mathbf{O}, \lambda)$ is equivalent to maximization of $P(\mathbf{q}, \mathbf{O}|\lambda)$ because $P(\mathbf{O}|\lambda)$ is not involved in the optimization process. From (5), we see that

$$P(q_1, q_2, \ldots, q_t, \quad O_1, O_2, \ldots, O_t | \lambda)$$

$$= P(q_1, q_2, \ldots, q_{t-1}, \quad O_1, O_2, \ldots, O_{t-1} | \lambda) \cdot a_{q_{t-1} q_t} \, b_{q_t}(O_t) \,. \tag{8}$$

If we define

$$\delta_t(i) \triangleq \max_{q_1, q_2, \ldots, q_{t-1}} P(q_1, q_2, \ldots, q_t = S_i, \quad O_1, O_2, \ldots, O_t | \lambda) \,, \tag{9}$$

then the following recursion is true

$$\delta_{t+1}(j) = [\max_i \, \delta_t(i) \, a_{ij}] \; b_j(O_{t+1}) \,. \tag{10}$$

The optimal state sequence is thus the one that leads to $\delta_T(q_T)$. This recursion is in a form suitable for the application of the Viterbi algorithm.

2.4 Speech Recognition Using Hidden Markov Models

The typical use of hidden Markov models in speech recognition is not very different from the traditional pattern matching paradigm. Successful application of HMM methods usually involves the following steps: 1) define a set of L sound classes for modeling, such as phonemes or words; call the sound classes $V = \{v_1, v_2, \ldots, v_L\}$; 2) for each class, collect a sizeable set (the training set) of labeled utterances that are known to be in the class; 3) based on each training set, solve the estimation problem to obtain a "best" model λ_i for each class v_i, $i = 1, 2, \ldots, L$; 4) during recognition, evaluate $P(O|\lambda_i)$, $i = 1, 2, \ldots, L$, for the unknown utterance O and identify the speech that produced O as class v_j if

$$P(O|\lambda_j) = \max_{1 \le i \le L} P(O|\lambda_i) \,. \tag{11}$$

Since the detailed characteristics of how to implement an HMM recognizer are not essential to this paper, we will omit them here.

3. Strengths of the Method of Hidden Markov Models

As mentioned in the Introduction, the strengths of the hidden Markov model method lie in two broad areas: its mathematical framework and its implementational structure.

3.1 The Consistent Statistical Framework of the HMM Methodology

The foundation of the hidden Markov model methodology is built on the well established field of statistics and probability theory. That is to say, the development of the methodology follows a tractable mathematical structure that can be examined and studied analytically.

The basic theoretical strength of the hidden Markov model is that it combines modeling of stationary stochastic processes (for the *short time* spectra) and the temporal relationship among the processes (via a Markov chain) together in a well defined probability space. The measure of such a probability space is defined by (6). This combination allows us to study these two separate aspects of modeling a dynamic process (like speech) using one consistent framework.

In addition, this combination of short time static characterization of the spectrum within a state, and the dynamics of change across states, is rather elegant because the measure of (6) can be decomposed simply into a summation of the joint probability of $\mathbf{O}$, the observation, and $\mathbf{q}$, the state sequence, as defined by (5). The decomposition permits independent study and analysis of the behavior of the short time processes and the long term characteristic transitions. Since decoding and recognition are our main concerns, this also provides an intermediate level of decision that can be used to choose among alternate configurations of the models for the recognition task. This kind of flexibility with consistency is particularly useful for converting a time-varying signal such as speech, without clear anchor points that mark each sound change, into a sequence of (sound) codes.

3.2 The Training Algorithm for HMM's

Another attractive feature of hidden Markov models comes from the fact that it is relatively easy and straightforward to train a model from a given set of labeled training data (one or more sequences of observations).

When the maximum likelihood criterion is chosen as the estimation objective, i.e. maximization of $P(\mathbf{O}|\lambda)$ over λ, the well known Baum-Welch algorithm is an iterative hill climbing procedure that leads to, at least, a fixed point solution. If we choose the state-optimized (or decoded) likelihood defined by

$$L_\lambda(\bar{\mathbf{q}}) = \max_{\mathbf{q}} P(\mathbf{O}, \mathbf{q} | \lambda) \tag{12}$$

where

$$\bar{\mathbf{q}} = \arg\max_{\mathbf{q}} P(\mathbf{O}, \mathbf{q} | \lambda) \tag{13}$$

as the optimization criterion, the segmental k-means algorithm [19], which is an extended version of the Viterbi training/segmentation algorithm [4], can be conveniently used to accomplish the parameter training task.

The segmental k-means algorithm, as can be seen from the objective function of (12), involves two optimization steps, namely, the segmentation step and the optimization step. In the segmentation step, we find a state sequence $\bar{\mathbf{q}}$ such that (12) is obtained for a given model λ and an observation sequence $\mathbf{O}$. Then, given a state sequence $\bar{\mathbf{q}}$ and the observation $\mathbf{O}$, the optimization step finds a new set of model parameters $\bar{\lambda}$ so as to maximize (12), i.e.

$$\bar{\lambda} = \arg\max_{\lambda} \{\max_{\mathbf{q}} P(\mathbf{O}, \mathbf{q} | \lambda)\} . \tag{14}$$

Equation (14) can be rewritten as

$$\bar{\lambda} = \arg\max_{\lambda} \{\max_{\mathbf{q}} [\log P(\mathbf{O} | \mathbf{q}, \lambda) + \log P(\mathbf{q} | \lambda)]\} . \tag{15}$$

Note that $\max_{\lambda} [\log P(\mathbf{O} | \bar{\mathbf{q}}, \lambda) + \log P(\bar{\mathbf{q}} | \lambda)]$ consists of two terms that can be separately optimized since $\log P(\bar{\mathbf{q}} | \lambda)$ is a function of only A, the state transition probability matrix, and $\log P(\mathbf{O} | \bar{\mathbf{q}}, \lambda)$ is a function of only B, the family of (intrastate) observation distributions. (We neglect the initial state probability for simplicity in presentation.) This separate optimization is the main distinction between the Baum-Welch algorithm and the segmental k-means algorithm.

These two training algorithms (Baum-Welch and segmental k-means) both result in well formulated and well behaved solutions. The segmental k-means algorithm, however, due to the separate optimization of the components of the model parameter sets, leads to a more straightforward (simpler with less computation and numerical difficulties) implementation.

The ease of HMM training also extends to the choice of observation distributions. It is known [25] that these algorithms can accommodate observation densities that are; 1) strictly log-concave densities; 2) elliptically symmetric densities; 3) mixtures of distributions of the above two categories; and 4) discrete distributions. These choices of observation distribution in each state of the model allow accurate modeling of virtually unlimited types of data.

3.3 Modeling Flexibility

The flexibility of the basic HMM is manifested in three aspects of the model, namely: model topology, observation distributions, and decoding hierarchy.

Many topological structures for HMM's have been studied for speech modeling. For modeling isolated utterances, (i.e. whole words or phrases) we often use left-to-right models [5, 35] since the utterance begins and ends at well identified time instants (except in the case of very noisy or corrupted speech) and the sequential behavior of the speech is well represented by a sequential HMM. For other speech modeling tasks, the use of ergodic models is often more appropriate. The choice of topological configuration and the number of states in the model often is a reflection of the *a priori* knowledge of the particular speech source to be modeled and is not in any way related to the mathematical tractability or implementational considerations.

Previously we pointed out that the range of observation distributions that can be accommodated by well developed training algorithms is rather large. There are no real analytical problems that make the use of any of this rather rich class of distributions impractical. Since speech has been shown to display quite irregular probability distributions both in waveform and spectral parameters, one indeed needs the freedom to choose an appropriate distribution model that fits the observation well and yet is easy to obtain.

In modeling spectral observations, we have found the use of mixture densities beneficial. With $f_i(\cdot)$ denoting the kernel density function, the mixture density assumes the form

$$b(\mathbf{O}) = \sum_{i=1}^{M} c_i f_i(\mathbf{O}) \tag{16}$$

where c_i is the mixture component weight, $\sum_{i=1}^{M} c_i = 1$, and M is the number of mixture components. This mixture distribution function is used to characterize the distribution of the observations in each state. By varying the number of mixture components, M, it can be shown that it is possible to approximate densities of virtually any shape (unimodal, multimodal, heavy-tail, etc.).

The density of (16) can be modified to take the form

$$b(\mathbf{O}) = \sum_{i} c_i I(\mathbf{O}, i) \tag{17}$$

where $I(\mathbf{O}, i)$ is the indicator function

$$I(\mathbf{O}, i) = \begin{cases} 1 & , \ i = \arg\max_j f_j(\mathbf{O}) \\ 0 & , \ \text{otherwise} \ . \end{cases} \tag{18}$$

The density function $b(\mathbf{O})$ then becomes a discrete distribution in (17) and the resulting model is often called a discrete HMM. The mixture density thus allows us to bridge the gap between parametric and non-parametric distribution models.

Although the above discussion focuses on the choice of observation distributions within individual states of the model, one can apply these same ideas to the modeling of the *entire observation space* (lumped from all the classes to be modeled), as if it were a single state HMM source, before further modeling for classification. This flexibility of configuration gives rise to the so-called vector quantizer hidden Markov models (VQ-HMM) [35], semi-continuous HMM [16], or continuous HMM [31], all of which share the mathematical foundation described above.

3.4 Ease of Implementation

Two areas of concern in the implementation of any algorithm are the potential for numerical difficulties and the computational complexity. The hidden Markov model is no exception.

The potential numerical difficulties in implementing HMM systems come from the fact that the terms in the HMM probability measure of (5) and (6) are multiplicative. A direct outcome of the multiplicative chain is the need for excessive dynamic range in numerical values in order to prevent underflow or overflow problems in digital implementations.

There are two reasonable ways of avoiding such numerical problems: numerical scaling and interpolation. The scaling algorithm, well documented in [23] alleviates the dynamic range problem by normalizing the forward and backward probabilities at each time instance before they cause overflow or underflow. The scaling algorithm is naturally blended in the forward-backward procedure which is the core of the solution to the evaluation problem and the estimation problem above. However, normalization alone does not entirely solve the numeric problems which result from insufficient data support. Insufficient data support can cause spurious singularities in the model parameter space. One may resort to parameter smoothing and interpolation to alleviate such numerical singularity problems. Although the topic of model parameter smoothing and interpolation is still one of the open issues in HMM theory, some trivial measures such as setting a numeric floor to prevent singularity are often found beneficial and are straightforward to implement.

4. Hidden Markov Model Limitations

The basic theory of hidden Markov modeling has been developed over the last two decades. As explained above, the theory includes the definition of the probability measure, the evaluation algorithm, the decoding algorithm, several parameter estimation procedures which are readily interpreted in terms of signal modeling, and a rather rigorous treatment of the HMM measure space. However, when applied to speech recognition, there are still some limitations which potentially affect speech recognizer performance.

4.1 Parameter Estimation Criteria

The original HMM parameter estimation was formulated as an inverse problem: given an observation sequence $\mathbf{O}$, and an assumed source model, estimate the (source) parameter set λ which maximizes the probability that $\mathbf{O}$ was produced by the source. The method of maximum likelihood (ML), which seeks to maximize $P(\mathbf{O}|\lambda)$, is optimal according to this criterion. The Baum-Welch reestimation algorithm, is a convenient, straightforwardly implementable solution to the maximum likelihood HMM estimation problem.

The method of maximum likelihood, however, need not be optimal in terms of minimizing classification error rate in recognition tasks in which the observation $\mathbf{O}$ is said to be produced by one of the many (say L) source classes, $\{C_i\}_{i=1}^{L}$. This is the classic problem in isolated and connected word recognition tasks. To achieve the minimum classification error rate, the classical Bayes rule requires that

$$C^*(\mathbf{O}) = C_i \quad \text{if} \quad C_i = \arg\max_j P(C_j|\mathbf{O}) \tag{19}$$

where $\mathbf{O}$ is the unknown observation, which is to be classified into (recognized as) one of the L classes, $C^*(\cdot)$ denotes the decoded class of $\mathbf{O}$, and $P(C_j|\mathbf{O})$ is the (true) a posteriori probability of C_j given the observation $\mathbf{O}$. The decision rule of (19) is the well known MAP decoder. Since

$$P(C_j|\mathbf{O}) = \frac{P(\mathbf{O}|C_j)\, P(C_j)}{P(\mathbf{O})} \tag{20}$$

and since $P(\mathbf{O})$ is a constant in this decoding context, the decision rule of (19) is often written as

$$C^*(\mathbf{O}) = C_i \quad \text{if} \quad C_i = \arg\max_j P(\mathbf{O}|C_j)\, P(C_j) \tag{21}$$

in terms of the class prior $P(C_j)$ and the conditional probability $P(\mathbf{O}|C_j)$. This factorization, i.e. writing the joint probability as a product of an *a priori* probability and a conditional probability, is particularly appropriate in speech recognition as will be explained shortly.

It is clear that the difficulty in minimizing misclassification rate stems from the fact that both the prior distribution $P(C_i)$ and the conditional distribution $P(\mathbf{O}|C_i)$ are generally unknown and have to be estimated from a given, finite, training set. In order to obtain reliable estimates of $P(C_i)$ and $P(\mathbf{O}|C_i)$, we generally need a sufficient size training set (i.e. large enough to adequately sample all relevant class interactions). When the vocabulary is large, this is difficult if not impossible to achieve. For example, suppose the vocabulary has 10,000 words, each of which represents a class. If we assume 10 occurrences each, on average, are needed for reliable estimates of both $P(C_i)$ and $P(\mathbf{O}|C_i)$, this amounts to a total of $10{,}000 \times 10 = 10^5$ word utterances. (This is equivalent to ~14 hours of speech, assuming 2 words per second.) Therefore, some other strategy is required to estimate $P(C_j)$ and $P(\mathbf{O}|C_j)$ reliably from a smaller size training set. One possibility is to choose a set of classes to represent, instead of words, a reduced set of sub-word units, e.g. phonemes. This greatly reduces the requirements on the amount of training data since the number of sub-word classes essentially does not grow with the size of vocabulary. A consequence of using sub-word unit classes to represent the basic set of speech sounds is that an estimate of class probability, $P(C_i)$, can be obtained directly from a lexical description of the words, independent of the *spoken* training set. Furthermore, we can estimate $P(\mathbf{O}|C_i)$ for words from each of the sub-word units in the lexical entry for the words. This type of decomposition, namely breaking large sound classes like words and phrases into smaller ones like sub-word units, leads to one particular problem in HMM speech recognition: i.e., given an independent (and possibly incorrect) estimate of word probability, $P^*(C_i)$, how do we estimate $P^*(\mathbf{O}|C_i)$ such that the Bayes' minimum error rate is attained? Note that in this discussion, the association between the training data $\mathbf{O}$ and the (subword) class is assumed to be known *a priori* (often as a result of hand labeling). We call this case the complete label case. Typical examples of complete label systems include most isolated word and connected word tasks, and the case of hand-segmented and hand-labeled continuous speech recognition.

The decomposition described above (namely from ASCII text to words or sub-word units), while circumventing some of the training problems, leads to other problems, e.g. the need for a detailed segmentation and labeling of the speech to correspond to the basic speech classes. For large vocabulary continuous speech recognition, where the training data is extensive, this labor intensive task cannot realistically be accomplished. Instead, we often have to rely on only partial knowledge of the data. For example, we generally know or assume we know the phoneme sequence of the words in the string, but not the direct relationship between each phoneme and the corresponding segment of speech. In these cases, we need some other criterion to guide the parameter estimation. We call this case the incomplete label case and typical examples are the problems of estimating models of sub-word speech units from continuous speech and those of words in connected word tasks without prior word segmentation.

It is important to note that if the chosen class model (the conditional as well as the *a priori* probability) is indeed the correct one, the method of maximum likelihood will lead to the asymptotically best recognition performance (in terms of highest correct classification rate) while allowing classes to be added or removed from the system specification without the need for complete retraining of all class reference models. These assumptions, however, are rarely true in practice. The above examples attempt to address these unrealistic assumptions from the viewpoint of modeling for recognition, aiming at the minimum Bayes' error rate. The idea is to either obtain extremely accurate reference models or allow compensation for incorrect *a priori* probabilities in the overall modeling. Another alternative, called *corrective training*, tries to minimize the recognition error rate directly by identifying, during training, sources that lead to recognition errors (or near errors) and applying some type of correction rule to the parameter estimation to reduce the probability of these errors or near-misses. This type of corrective training procedure intentionally seeks model parameter estimates which give sub-optimal likelihood scores in order to reduce recognition error rate on the training set. Unlike the above cases, corrective training uses the HMM as a form of *discriminant function* and is not concerned about modeling accuracy per se but more with recognition errors as they appear during training.

We now provide a more complete discussion of the complete label case and the incomplete label case.

The Complete Label Case

In the complete label case, the association between the training data $\mathbf{O}$ and the class C_i is known *a priori* during training. This is a typical situation in classical pattern classification theory and all the concerns about supervised learning apply. However, what is unique in the current speech recognition problem is the use of a prescribed class prior model $P(C_i)$.

A detailed account of the issues involved in the complete label case is given in [29]. Consider the set of L HMM class models, $\Lambda = \{\lambda_i\}_{i=1}^{L}$, which are used to model the classes $\{C_i\}_{i=1}^{L}$ respectively. The complete set of models, Λ, defines a probability measure

$$P_\Lambda(\mathbf{O}) = \sum_i P_\Lambda(\mathbf{O}\,|\,C_i)\, P_\Lambda(C_i)\ . \tag{22}$$

The notation of (22) is slightly different from that used earlier because of the fact that we are explicitly using both classes and associated models simultaneously for classification purposes. Assume the prior $P_\Lambda^*(C_i) = P_0(C_i)$ is given or obtained independent of the *spoken* training set $\{\mathbf{O}^{(i)}\}$ where $\mathbf{O}^{(i)}$ are the portion of the training data that are labeled as C_i. For this case, Nadas *et al.* [29] proposed the use of a slightly different training measure, namely the conditional maximum likelihood (CMLE) estimator, obtained by

$$\Lambda_{\mathrm{CMLE}}^* = \arg\max_\Lambda \prod_i P_\Lambda(C_i\,|\,\mathbf{O}^{(i)})\ . \tag{23}$$

The motivation for choosing this estimator is that the assumed model, $P_\Lambda(\mathbf{O}^{(i)}\,|\,C_i)$ *or* $P(\mathbf{O}^{(i)}\,|\,\lambda_i)$, *may be inappropriate for* $P(\mathbf{O}^{(i)}\,|\,C_i)$, so that the asymptotic optimality of the standard ML estimator is no longer guaranteed. Furthermore, the prior model $P_0(C_i)$ could be incorrect (poor estimate etc.), so that the true MAP decoder of (19) is virtually impossible to implement.

When the class *a priori* probability $P_0(C_i)$ is obtained independent of $\{\mathbf{O}\}$ and is not part of the model Λ to be optimized, then

$$\begin{aligned} \Lambda_{\mathrm{CMLE}}^* &= \arg\max_\Lambda \prod_i P_\Lambda\ (C_i\,|\,\mathbf{O}^{(i)}) \\ &= \arg\max_\Lambda \sum_i \log \frac{P_\Lambda(\mathbf{O}^{(i)}, C_i)}{P_\Lambda(\mathbf{O}^{(i)})\, P_0(C_i)} \\ &= \Lambda_{\mathrm{MMI}}^* \end{aligned} \tag{24}$$

which is the well known maximum mutual information (MMI) estimator. Based on a simple example, it was concluded in [29] that if $P_0(C_i)$ is not the *true* prior distribution (because of bad assumptions or estimation errors), the MLE will implement a suboptimal decoder while the CMLE of (23) will lead to the correct MAP decoding result (asymptotically) because of the compensation built into the estimate of $P(\mathbf{O}\,|\,C_i)$.

Although the criterion of conditional maximum likelihood or maximum mutual information (for training) is attractive in terms of compensation problems associated with

MAP decoding, some important concerns remain unanswered. The most immediate concern in using this criterion is the lack of a convenient and robust algorithm to obtain the estimate $P^*_{\mathrm{CMLE}}(\mathbf{O}\,|\,C_i)$. In many practical situations, the procedure for obtaining the solution is very complicated, particularly when HMM's are involved and the observation distribution is not of a discrete type. Another problem is that the conditional maximum likelihood estimate has a larger variance than the maximum likelihood estimate. This therefore undermines the potential gain in offsetting the sensitivity due to the independently obtained $P_0(C_i)$ possibly being inaccurate when the decoder based on finite training data is used on testing data not included in the training set. While the motivation to have a criterion specifically aimed at minimizing the recognition error is a good one, the goal has not yet been achieved by CMLE methods.

The Incomplete Label Case

The case of incomplete labeling arises because of: 1) practical difficulties in labeling and segmenting any large continuous speech database; and 2) the inherent ambiguity among different sound classes in terms of both class definition (i.e. inherent similarities in sound classes) and time uncertainty as realized in speech signals (i.e. it is not clear that exact segmentation boundaries between adjacent sounds universally exist). For the case of decomposing an isolated word into a prescribed phoneme sequence, we usually have a lexical description of the word in terms of the phoneme sequence (as described in a dictionary), and the spoken version of the word without any explicit time segmentation into corresponding phonemes. Under these conditions (which are typical for speech recognition), training of the prescribed sub-word unit models is rather difficult due to the lack of a definitive labelling relating sub-word classes to specific intervals of speech. After all, if we do not know for sure that a training token $\mathbf{O}$ is in sound class C_i, the likelihood function $P(\mathbf{O}\,|\,C_i)$ cannot be defined, not to mention optimized.

There are several ways to handle the problem of incomplete labeling based on the idea of embedded decoding. One way is to retain the constraints of the known class sequence (in the previous example, the phoneme sequence) and solve for the "optimal" set of class models sequentially. Another alternative is to solve for the models of the sound classes simultaneously with the decoding into classes.

Consider first the case where we have partial knowledge; that is, a given training token $\mathbf{O}$ is known to correspond to a sequence of u class labels $h = (h_1, h_2, \ldots, h_u)$ (as determined from dictionary lookup of the words realized by $\mathbf{O}$) where $h_j \in \{C_i\}_{i=1}^{L}$. The goal is to obtain the L models $\Lambda = \{\lambda_i\}_{i=1}^{L}$ corresponding to the L sound classes

$C = \{C_i\}_{i=1}^{L}$, using the number of segments u, and the class labels h as hard constraints in the decoding. The procedure begins by assuming a uniform segmentation $\mathbf{X} = (x_1, x_2, ..., x_u)$ of $\mathbf{O}$ into u speech intervals with the ith interval, x_i, corresponding to sound class h_i. Based on this initial segmentation, the likelihood functions are defined and individual class models are obtained by maximum likelihood (via the forward-backward procedure). For example, if $h_i = C_j$, then the segment x_i is used to define a likelihood function $P(x_i | C_j)$ for maximization. As a result, a set of sound unit models are created. With the new set of unit models, we further refine the segmentation of $\mathbf{O}$ into $\mathbf{X}$ (again assuming exactly u segments) by optimally decoding $\mathbf{O}$ using the Viterbi algorithm. This leads to an improved segmentation of $\mathbf{O}$ which can then be used to give a refined set of sound models. This process is iterated until a reasonably stationary segmentation of $\mathbf{O}$ into intervals $\mathbf{X}$ is obtained.

This constrained decoding approach to the incomplete label case has been used in explicit acoustic modeling of phonemic classes [20] with good success. There are, however, some theoretical shortcomings to the method. One problem is that the segmentation/decoding results will be different for different numbers of segments, u, in the given string. Thus even the simple expedient of having multiple dictionary definitions of words can lead to inconsistent segmentation into sound classes. Since the criterion for decoding is conditioned on the "sub-optimal" segmentation (due to the constraint of a designated number of segments), the overall model parameter estimation does not obey any global optimality principle similar to that of maximum likelihood.

Corrective Training

As explained earlier, the minimum Bayes' risk or error rate is the theoretical recognizer performance bound, conditioned on the exact knowledge of both the prior and the conditional distributions. When both distributions are not known exactly, and the classifier needs to be designed based on a finite training set, there are several ways to try to reduce the error rate. One method is based on the theoretical link between discriminant analysis and distribution estimation. The idea here is to design a classifier (discriminant function) such that the minimum classification error rate is attained on the training set. In particular, we wish to design a classifier that uses estimates of $P(C_i)$ and $P(\mathbf{O} | C_i)$, and that achieves a minimum error rate for the training set in the same way a discriminant function is designed. The reason for using the HMM statistical framework in $P(\mathbf{O} | \lambda_i)$ for modeling $P(\mathbf{O} | C_i)$, as opposed to other discriminant functions, is to exploit the strengths of the HMM's, namely – consistency, flexibility and computational ease, as well as its ability to generalize classifier performance to independent (open) data sets. The generalization capability of HMM's, as discriminant functions, is somewhat beyond the scope of this article and will not be discussed here. In the following, we focus on the issue of treating the estimation of the distributions $P(C_i)$ and $P(\mathbf{O}|\lambda_i)$ as a discriminant function design, to attain the minimum error rate.

Bahl *et al.* [3] were the first to propose an error-correction strategy, which they named corrective training, to specifically deal with the misclassification problem. Their training algorithm was motivated by analogy with an error-correction training procedure for linear classifiers. In their proposed method, the observation distribution is of a discrete type, $B = [b_{ij}]$, where b_{ij} is the probability of observing a vector quantization code index (acoustic label) j when the HMM source is in state i. Each b_{ij} is obtained via the forward-backward algorithm as the weighted frequency of occurrence of the code index [35]. The corrective training algorithm of [3] works as follows. First, use a labeled training set to estimate the parameters of the HMM's $\Lambda = \{\lambda_i\}$ with the forward-backward algorithm. For each utterance $\mathbf{O}$, labeled as C_k, for example, evaluate $P(\mathbf{O}|\lambda_k)$ for the correct class C_k and $P(\mathbf{O}|\lambda_l)$ for each incorrect class C_l. (The evaluation of likelihood for the incorrect classes need not be exhaustive.) For every utterance where $\log P(\mathbf{O}|\lambda_l) > \log P(\mathbf{O}|\lambda_k) - \delta$, where δ is a prescribed threshold, modify λ_k and λ_l according to the following mechanism: 1) Apply the forward-backward algorithm to obtain estimates b'_{ij} and b''_{ij}, using the labeled utterance $\mathbf{O}$ only, for the correct class C_k and incorrect class C_l, respectively; 2) Modify the original b_{ij} in λ_k to $b_{ij} + \gamma b'_{ij}$ and the b_{ij} in λ_l to $b_{ij} - \gamma b''_{ij}$. When the state labels are tied for certain models, the above procedure is equivalent to replacing the original b_{ij} by $b_{ij} + \gamma(b'_{ij} - b''_{ij})$. The prescribed adaptation parameter, γ, controls the "rate of convergence" and the threshold, δ, defines the "near-miss" cases. This corrective training algorithm therefore focuses on those parts of the model that are most important for word discrimination, a clear difference from the maximum likelihood principle.

While Bahl *et al.* reported that the corrective training procedure worked better (in isolated word recognition tasks) than models obtained using the maximum mutual information or the conditional maximum likelihood criterion, the lack of a rigorous analytical foundation for the algorithm is one problem. Without a better theoretical understanding of the algorithm, the appeal of the method is primarily experimental.

4.2 Integration of Non-Spectral and Spectral Features

The use of HMM's in speech recognition has been mostly limited to the modeling of short time speech spectral information. That is, the observation **O** typically represents a smoothed representation of the spectrum of the speech signal at a given time, be it in terms of the traditional Fourier transform coefficients, linear prediction coefficients, or any other transformations of the estimated (power) spectrum. The spectral feature vector has proved extremely useful and has led to a wide variety of successful recognizer designs. This success can be attributed both to the range of spectral analysis techniques developed in the past three decades as well as to our understanding of the perceptual importance of the speech spectrum to the recognition of sounds. The success of spectral parameters for characterizing speech was further augmented by the introduction of the so-called delta-cepstrum [14] which attempts to model the differential speech spectrum much as the cepstrum models the instantaneous speech spectrum.

Besides spectral parameters, there are other speech features that are believed to contribute to the recognition and understanding of speech by humans. One such category of non-spectral speech features is prosody as is manifested on both the segmental and the supra-segmental level. Physical manifestations of prosody, in the speech signal, include signal energy (suitably normalized), differential energy, and signal pitch (fundamental frequency). There are at least two issues of concern in integrating non-spectral with spectral features in a statistical model of speech, namely do such features contribute to the performance of the statistical model for actual recognition tasks, and how should the features be integrated so as to be consistent with their physical manifestations in the production and perception of speech. The first issue is relatively easy to resolve based on experimental results. Several HMM-based recognition systems have incorporated log energy (and differential log energy) either directly into the feature vector, or as an additional feature whose probability (or likelihood) is factored into the likelihood calculation [32], with moderate success (i.e. higher recognition performance). However the level of performance improvement is considerably smaller than one might anticipate based on the importance of prosody in speech.

The problem with combining spectral and non-spectral features in a statistical framework is one of temporal rate of change. To attain adequate time resolution for the spectral parameters that characterize the varying vocal tract, we need to sample the spectral observations at a rate on the order of 100 times per second (10 msec frame update). The prosodic features of speech characterize stress and intonation and these occur at a syllabic rate or about 10 times per second. (Of course we can always oversample the features associated with the prosodic parameters to keep the rate the same as that of the spectral parameters – this in fact is what is currently done in most systems, e.g. [32]). Furthermore, the sequential characteristic change of different features may be too different to warrant a single, unified Markov chain. Thus one key question is how do we combine two feature sets with fundamentally different time scales and possibly different sequential characteristics so as to be able to perform optimum modeling and decoding.

A second question in combining fundamentally different (in nature) features concerns their statistical characterization. To be technically correct we need to know the joint distribution of the two feature sets. For statistically independent (or often just uncorrelated) feature sets we can represent the joint density as a product of the individual densities. However, in practice, there is usually some correlation between any pair of speech feature sets; hence some correction for the correlation is usually required. One proposed method of handling this problem is to perform a principal component analysis on the joint feature set before hidden Markov modeling is performed [8]. Although this alleviates the difficulties somewhat, it is not a totally satisfying solution since the resulting feature set usually has no straightforward physical significance. Furthermore the set of principal components is a function of the data and hence need not be optimal for unseen data (open set problem).

The key points raised here concern structural issues for combining traditional spectral features with other features such as pitch and signal energy. Although the features can be simply combined in traditional ways, there is probably a much better way of integrating these fundamentally different features into a consistent statistical structure which achieves the minimum decoding error.

4.3 Duration Modeling in HMM's

One of the inherent limitations of the HMM approach is its treatment of temporal duration – both within a model and within units composed of one or more models (e.g. a word made up from sub-word models). Inherently, within a state of an HMM, the probability distribution of state duration is exponential, of the form:

$$P_i(d) = (a_{ii})^{d-1}\ (1-a_{ii}) \tag{25}$$

$$= \text{probability of staying in state } S_i \text{ for exactly } d \text{ observations}$$

where a_{ii} is the self-loop state probability. This exponential duration model is inappropriate for almost any speech event.

There have been proposed several alternatives for implementing different state duration models. The most straightforward approach is to allow explicit state duration distributions, $P_{q_i}(d_i)$, to be associated with each state, S_i, using the concept of a semi-Markov chain [12, 36] in which state transitions do not occur at regular time intervals. More formally we assume that for a given state sequence $\mathbf{q}$ in which there are $r-1$ state transitions such that the states visited are $q_1, q_2, \ldots, q_r$ with associated state durations of $d_1, d_2, \ldots, d_r$ (frames), the joint probability of the observation sequence, $\mathbf{O}$, and the state sequence, $\mathbf{q}$, given the model, λ, becomes

$$
\begin{aligned}
P(\mathbf{O}, \mathbf{q} \mid \lambda) = \pi_{q_1} & P_{q_1}(d_1)\, P(\mathbf{O}_1, \mathbf{O}_2, \ldots, \mathbf{O}_{d_1} \mid q_1) \\
& a_{q_1 q_2} P_{q_2}(d_2)\, P(\mathbf{O}_{d_1+1} \ldots \mathbf{O}_{d_1+d_2} \mid q_2) \\
& \vdots \\
& a_{q_{r-1} q_r} P_{q_r}(d_r)\, P(\mathbf{O}_{d_1+d_2+\ldots+d_{r-1}+1} \ldots \mathbf{O}_T)\, q_r
\end{aligned}
\tag{26}
$$

where $P_{q_i}(*)$ are the state duration probabilities. The probability measure for such a Markov model is, accordingly:

$$
P(\mathbf{O} \mid \lambda) = \sum_{q} P(\mathbf{O}, \mathbf{q} \mid \lambda) \tag{27}
$$

Based on these definitions, modeling of the source, including the duration model $P_i(d_i)$, can be implemented using a hill-climbing reestimation procedure of the type used previously [22]. Typically, $P_i(d_i)$ is treated as a discrete distribution over the range $1 \le d_i \le D_{\max}$, where $D_{\max}$ represents the longest possible dwell in any state.

While the above formulation handles the duration model *simultaneously* with the Markovian transition and the local (state) distribution models, and can lead to analytical solutions, there are drawbacks with such a duration model for speech recognition. One such drawback is the greatly increased computational complexity due to the loss of regularity in transition timing. In the traditional HMM, transitions are allowed to occur at every time frame, since the model topology includes a return transition back to the original state. In the semi-Markov model, transitions do not occur at regular time (the return transition is part of the duration model $P_i(d_i)$) and this leads to a significantly more complicated lattice for decoding an input string. In [33], it was estimated that the semi-Markov model incurs a factor of 300 increase in computational complexity for a value of $D_{\max} = 25$. The much increased complication in decoding lattice often renders many search algorithms such as the beam search [26] and the stack algorithm [17] for handling large problems extremely difficult to implement. Another problem with the semi-Markov model is the large number of parameters (D_{max}) associated with each duration model that must be estimated in addition to the usual HMM parameters. Finally, it is not clear how accurate or how robust the estimated $P_i(d_i)$ needs to be in order to be beneficial in speech recognition.

One proposal [22] to alleviate some of these problems is to use a parametric state duration distribution model instead of the nonparametric ones used above. Several parametric distributions have been considered, including the Gaussian family (with constraints to avoid negative durations) and the Gamma family [22]. These families represent the most rigorous attempt to explicitly and analytically utilize the duration information in speech modeling and recognition to date. Other less involved attempts include modeling the state duration with a uniform distribution, requiring only the estimate of the duration range, minimum and maximum duration allowed in a particular state. This simple duration model has been applied with good success [26] for some tasks.

The major difficulty in modeling the durational information is that it is much more sparse than spectral information – i.e. there is only 1 duration per state. Hence we either alter the structure of the HMM (e.g. to a semi-Markov model) thereby losing many of the regularity properties of the original HMM formulation, or we seek alternative implementation structures, as in the case of prosodic information. For durational information a simple-minded approach is to treat the spectral modeling and the duration modeling as two separate, loosely connected, problems. Hence the regular HMM estimation (spectral) is performed on the given observation sequence $\mathbf{O}$. Then the best state sequence $\bar{\mathbf{q}}$, which maximizes $P(\mathbf{O}, \mathbf{q} | \lambda)$ is found using the Viterbi algorithm. Finally estimates of $P_i(d_i)$ are obtained based on the optimal state sequence $\bar{\mathbf{q}}$ by either the maximum likelihood method or from simple frequency of occurrence counts [21]. Often the duration d_i for state i is normalized by the overall duration T in order to account for the inherent variation in speaking rate. This approach is usually called the post-processor duration model because the standard decoding is performed first and the duration information is only available after the initial processing is finished. Although the postprocessor duration model has had some success, the questions of optimality of the estimate, robustness of the solution, and other criteria for successful utilization of duration information, especially as applied to speech recognition, remain unanswered.

4.4 Other Issues

Although we have discussed several of the most important open issues in HMM systems, especially as applied to speech recognition, there are a few more areas of concern worth noting. These include the issue of model clustering and model splitting for handling inhomogeneous data, and the assumption that observations within a state of a semi-Markov model are statistically independent. We briefly discuss each of these issues here.

Model Clustering and Splitting

One of the basic assumptions in statistical modeling is that the variability in the observations from an information source can be modeled by statistical distributions. For speech recognition, the source could be a single word, a sub-word unit like a phoneme, or a word sequence. Because of variability in the source production (e.g. accents, speed of talking), or the source processing (e.g. transmission distortion, noise), it is often expedient to consider using more than a single HMM to characterize the source. There are two motivations behind this multiple HMM approach. First, lumping all the variability together from inhomogeneous data sources leads to unnecessarily complex models, often yielding lower modeling accuracy. Second, some of the variability, or rather the inhomogeneity in the source data, may be known *a priori*, thus warranting separate modeling of the source data sets. Here, our main concern is the first case, i.e. automatic modeling of an inhomogeneous source with multiple HMM's, as the latter (manual) case is basically straightforward.

There exist a number of generalized clustering algorithms, such as the k-means clustering algorithm, the generalized Lloyd algorithm as is widely used in vector quantizer designs [24], or the greedy growing algorithm found in set partition or decision tree designs [9], all of which are suitable for the purpose of separating inconsistent training data so that each divided sub-group becomes more homogeneous and therefore is better modeled by a single HMM. The nearest neighbor rule required in these clustering algorithms is simply to assign an observation sequence $\mathbf{O}$ to cluster i if

$$P(\mathbf{O}|\lambda_i) = \max_j P(\mathbf{O}|\lambda_j)$$

where λ_j's denote the models of the clusters. Successful application of the model clustering algorithms to the speech recognition problem, using the straightforward maximum likelihood criterion, has been reported [34]. However, when other estimation criteria are used, the interaction between multiple HMM modeling and the Bayes minimum error classifier design remains an open question in need of further study.

An alternative to model clustering is to arbitrarily sub-divide a given speech source into a large number of sub-classes with specialized characteristics, and then consider a generalized procedure for model merging based on source likelihood considerations. By way of example, for large vocabulary speech recognition we often try to build specialized units (context sensitive) for recognition. For example, we could consider building units which are a function of the sound immediately preceding the unit (left-context) and the sound immediately following the unit (right-context). There are on the order of 10,000 such units in English. Many of the units are functionally almost identical. The problem is how to determine which pairs of units should be merged (so that the number of model units is manageable and the variance of the parameter estimate is reduced). To set ideas, consider two unit models, λ_a and λ_b, corresponding to training observation sets $\mathbf{O}_a$ and $\mathbf{O}_b$, and the merged model λ_{a+b}, corresponding to the merged observation sets $\{\mathbf{O}_a, \mathbf{O}_b\}$. We can then compute the change in entropy (i.e. loss of information) resulting from the merged model as:

$$\Delta H_{ab} = H_a + H_b - H_{a+b} \tag{28}$$

$$= -P(\mathbf{O}_a|\lambda_a) \log P(\mathbf{O}_a|\lambda_a) - P(\mathbf{O}_b|\lambda_b) \log P(\mathbf{O}_b|\lambda_b) + P(\{\mathbf{O}_a, \mathbf{O}_b\}|\lambda_{a+b}) \log P(\{\mathbf{O}_a, \mathbf{O}_b\}|\lambda_{a+b}) . \tag{29}$$

Whenever ΔH_{ab} is small enough, it means that the change in entropy resulting from merging the models will not affect system performance (at least on the training set) and the models can be merged. The question of how small is acceptable is dependent on specific applications. Other practical questions of a similar nature also remain. However this model merging technique has been used successfully by Lee [21] to create a generalized set of triphone models for large vocabulary speech recognition.

Parameter Significance and Statistical Independence

Consider an observation sequence $\mathbf{O} = \{\mathbf{O}_1, \mathbf{O}_2, ..., \mathbf{O}_T\}$. If the state sequence that led to the production of $\mathbf{O}$ is known, then the conditional probability of $\mathbf{O}$ given the source model and the state sequence is expressed in (4), where the product form implies statistical independence [37]. The partial HMM measure of (5) shows that the contribution of the underlying Markov chain is to multiply the result of (4), a result equivalent to assuming that the observations within a state are independent.

Although, in theory, the importance of the state transition coefficients, a_{ij}, is the same as that of the state observation density $b_j(\mathbf{O})$ (in training, a_{ij} affects the parameter estimate of $b_j(\cdot)$ and vice versa), in normal practice of speech recognition this is not usually the case. This paradox is due to the discrimination capability of the a_{ij}'s relative to that of the $b_j(\mathbf{O})$'s. It should be clear that since a_{ij} is the relative frequency of state transition from state i to state j, its dynamic range (especially in a left-to-right model) is severely constrained. For example, the self-loop coefficient, a_{ii}, which can be interpreted roughly as $(\bar{d}_i - 1)/\bar{d}_i$ where $\bar{d}_i$ is the average duration in state i, typically is constrained between 0.5 (for $\bar{d}_i = 2$) and 0.95 (for $\bar{d}_i = 20$); hence the dynamic range over all states and models is less than 2 to 1. Similarly, the jump coefficient, $a_{i,i+1} = 1 - a_{ii}$, is constrained to be between 0.05 and 0.5, a somewhat larger dynamic range. However, there is only 1 jump coefficient for every $\bar{d}_i$ self transitions, again effectively reducing the dynamic range of the product of the a_{ij} coefficients.

Consider now the $b_j(\mathbf{O})$ densities. These often involve the product of D local densities, each of which has almost unlimited dynamic range. Furthermore each of the state densities is highly localized in the acoustic space, thereby leading to the excellent classification/discrimination capability of HMM's.

When we combine the a_{ij}'s and the $b_j(\mathbf{O})$'s to give the probability distribution of the HMM, we find in practice that, for a left-to-right model, the a_{ij}'s can be neglected entirely (i.e. set all $a_{ii} = a_{i,i+1} = 0.5$) with no effect on recognition performance. This effect is due to the lack of discrimination capability of the a_{ij}'s, as described above. There seems to be no simple way of changing this situation – nor is it entirely clear that it is important to do so.

The above analysis points out the existence of the unbalanced numerical significance of the a's and the b's in the likelihood calculation of the HMM and should not be taken as to totally discredit the usefulness of the Markov chain contribution. In fact, the transition probability still plays an important role in parameter estimation. For some specific HMM's, e.g. large ergodic models with a large number of states, accurate calculation of the state transition probabilities is vital in obtaining a good model, although the trend of unbalanced significance between the state transition probability and the observation density remains.

5. Summary

In this paper we have reviewed the statistical method of hidden Markov models (HMM). We showed that the strengths of the method lie in the consistent statistical framework that is flexible and versatile, particularly for speech applications, and the case of implementation that makes the method practically attractive. We also pointed out some areas of the general HMM method that deserve more attention with the hope that increased understanding will lead to performance improvements for many applications. These areas include the modeling criteria, particularly the problem of minimum classification error, incorporation of new (non-spectral) features as well as prior linguistic knowledge, and the modeling of state durations and its use in speech recognition. With our current understanding, HMM systems have been shown capable of achieving recognition rates of upwards of 95% word accuracy in certain speaker independent tasks with vocabularies on the order of 1000 words. With further progress, it is not difficult to foresee HMM-based recognition systems that perform well enough so as to make the technology usable in everyday life.

Acknowledgement

The authors acknowledge the excellent comments and criticisms provided by Yariv Ephraim and C. H. Lee on earlier drafts of this manuscript.

References

1. Allen, J.B. and Rabiner, L.R.: "A unified theory of short-time spectrum analysis and synthesis," *Proc. IEEE*, vol. 65, no. 11, pp. 1558-1564, Nov. 1977.
2. Atal, B.S. and Hanauer, S.L.: "Speech analysis and synthesis by linear prediction of the speech wave," *J. Acoust. Soc. Am.*, vol. 50, no. 2, pp. 637-655, Aug. 1971.
3. Bahl, L.R., Brown, P.F., de Souza, P.V. and Mercer, R.L.: "A new algorithm for the estimation of hidden Markov model parameters," *Proc. ICASSP-88*, pp. 493-496, New York, Apr. 1988.
4. Bahl, L.R., Jelinek, F. and Mercer, R. L.: "A maximum likelihood approach to continuous speech recognition," *IEEE Trans. Pattern Analysis Machine Intell.*, vol. PAMI-5, pp. 179-190, 1983.
5. Bakis, R.: "Continuous speech word recognition via centisecond acoustic states," in *Proc. ASA Meeting* (Washington, DC), Apr. 1976.
6. Baum, L.E.: "An inequality and associated maximization technique in statistical estimation for probabilistic functions of Markov processes," *Inequalities*, vol. 3, pp. 1-8, 1972.
7. Baum, L.E., Petrie, T., Soules, G. and Weiss, N.: "A maximization technique occurring in the statistical analysis of probabilistic functions of Markov chains," *Ann. Math. Stat.*, vol. 41, no. 1, pp. 164-171, 1970.
8. Bocchieri, E.L. and Doddington, G.R.: "Frame-specific statistical features for speaker independent speech recognition," *IEEE Trans. Acoust. Speech Signal Processing*, vol. ASSP-34, no. 4, pp. 755-764, Aug. 1986.
9. Breiman, L. *et al.*: *Classification and regression trees*, Wadsworth, Monterey, CA 1984.
10. Cadzow, J.A.: "ARMA modeling of time series," *IEEE Trans. Pattern Anal. Machine Intell.*, vol. PAMI-4, Mar. 1982.
11. Chow, Y.L. *et al.*: "BYBLOS: The BBN continuous speech recognition system," in *Proc. ICASSP '87* (Dallas, TX), Paper 3.7.1, pp. 89-92, Apr. 1987.
12. Ferguson, J.D.: "Hidden Markov Analysis: An Introduction," in *Hidden Markov Models for Speech*, Institute for Defense Analysis, Princeton, NJ, Oct. 1980.
13. Forney, G.D.: "The Viterbi algorithm," *Proc. IEEE*, vol. 61, pp. 268-278, Mar. 1973.
14. Furui, S.: "Speaker independent isolated word recognition based on dynamics emphasized cepstrum," *Trans. IECE Japan*, vol. 69, no. 12, pp. 1310-1317, Dec. 1986.
15. Gupta, V.N., Lennig, M. and Mermelstein, P.: "Intergration of acoustic information in a large vocabulary word recognizer," in *Conf. Proc. IEEE Int. Conf. on Acoustics, Speech, and Signal Processing*, pp. 697-700, Apr. 1987.

16. Huang, X. and Jack, M.A.: "Unified techniques for vector quantization and hidden Markov models using semi-continuous models," *Proc. ICASSP-89*, pp. 639-642, Glassgow, Scotland, May 1989.
17. Jelinek, F.: "A fast sequential decoding algorithm using a stack," *IBM J. Res. Develop.*, vol. 13, pp. 675-685, Nov. 1969.
18. Jelinek, F.: "Continuous speech recognition by statistical methods," *Proc. IEEE*, vol. 64, pp. 532-536, Apr. 1976.
19. Juang, B.H. and Rabiner, L.R.: "The segmental k-means algorithm for estimating parameters of hidden Markov models," *IEEE Trans. Acoust. Speech Signal Processing*, vol. 38, no. 9, pp. 1639-1641, Sept. 1990.
20. Lee, C.H, Rabiner, L.R., Pieraccini, R. and Wilpon, J.G.: "Acoustic Modeling for Large Vocabulary Speech Recognition," Computer Speech and Language, vol. 4, no. 2, pp. 127-166, April 1990.
21. Lee, K.F.: "Automatic Speech Recognition, The Development of the SPHINX System," Kluwer Academic Publishers, Boston, 1989.
22. Levinson, S.E.: "Continuously variable duration hidden Markov models for automatic speech recognition," *Computer, Speech and Language*, vol. 1, no. 1, pp. 29-45, Mar. 1986.
23. Levinson, S.E., Rabiner, L.R. and Sondhi, M.M.: "An introduction to the application of the theory of probabilistic functions of a Markov process to automatic speech recognition," *Bell Syst. Tech. J.*, vol. 62, no. 4, pp. 1035-1074, Apr. 1983.
24. Linde, Y., Buzo, A. and Gray, R.M.: "An algorithm for vector quantizer design," *IEEE Trans. Comm.*, COM-28, pp. 84-95, Jan. 1980.
25. Liporace, L.A.: "Maximum likelihood estimation for multivariate observations of Markov sources," *IEEE Trans. Informat. Theory*, vol. IT-28, no. 5, pp. 729-734, 1982.
26. Lowerre, B. and Reddy, R.: "The HARPY speech understanding system," in *Trends in Speech Recognition*, W. Lea, Editor. Englewood Cliffs, NJ: Prentice-Hall, 1980, pp. 340-346.
27. Makhoul, J.: "Linear Prediction: A tutorial review," *Proc. IEEE*, vol. 63, pp. 561-580, 1975.
28. Markel, J.D. and Gray, A.H. Jr.: *Linear Prediction of Speech*, Springer-Verlag, New York, 1976.
29. Nadas, A., Nahamoo, D. and Picheny, M.A.: "On a model-robust training method for speech recognition," *IEEE Trans. Acoust. Speech Signal Processing*, vol. ASSP-36, no. 9, pp. 1432-1436, Sept. 1988.
30. Paul, D.B.: "Training of HMM recognizers by simulated annealing," *Proc. ICASSP-85*, pp. 13-16, Tampa, Florida, March 1985.
31. Poritz, A.B. and Richter, A.G.: "On hidden Markov models in isolated word recognition," in *Proc. ICASSP '86* (Tokyo, Japan), pp. 705-708, Apr. 1986.
32. Rabiner, L.R.: "On the application of energy contours to the recognition of connected word sequences," *AT&T Bell Labs Tech. J.*, 63, pp. 1981-95, Dec. 1984.
33. Rabiner, L.R.: "A tutorial on hidden Markov models and selected applications in speech recognition," *Proceedings of the IEEE*, vol. 77, no. 2, pp. 257-286, Feb. 1989.
34. Rabiner, L.R., Lee, C.H., Juang, B.H. and Wilpon, J.G.: "HMM clustering for connected word recognition." *Proc. ICASSP-89*, pp. 405-408, Glassgow, Scotland, May 1989.
35. Rabiner L.R., Levinson, S.E. and Sondhi, M.M.: "On the application of vector quantization and hidden Markov models to speaker-independent isolated word recognition," *Bell Syst. Tech. J.*, vol. 62, no. 4, pp. 1075-1105, Apr. 1983.
36. Russell, M.J. and Moore, R.K.: "Explicit modeling of state occupancy in hidden Markov models for automatic speech recognition," in *Proc. ICASSP '85* (Tampa, FL), pp. 5-8, Mar. 1985.
37. Wellenkens, C.: "Explicit time correlation in hidden Markov models for speech recognition," *Proc. ICASSP*, pp. 384-386, 1987.

Hidden Markov Models and Speaker Adaptation

Richard Schwartz and Francis Kubala

BBN Systems and Technologies, 10 Moulton St. Cambridge, MA, 02138, USA

Abstract: In this chapter we first review the use of Hidden Markov Models for Continuous Speech Recognition. Then we discuss techniques for adapting speech recognizers to the speech of new speakers. In the first section particular attention is paid to the need to incorporate basic speech knowledge within the model. We present the N-Best Paradigm, which provides a simple way of integrating speech recognition with natural language processing. We describe several algorithms for computing the N best sentence hypotheses, and show that one of them is both efficient and empirically accurate. We also introduce the Forward-Backward Search Algorithm, which reduces the computation needed for the N-Best search by an additional factor of 40. We review many different speaker adaptation techniques and distinguish among their salient features. We present a new more practical method for training a speaker-independent system. In particular, we show that it is not necessary to have training speech from a very large number of speakers. Finally, we described a method for adapting a speaker-independent model to a new speaker. This speaker adaptation method results in half the error rate of the unadapted speaker-independent system.

Keywords: continuous speech recognition, hidden Markov models, speaker adaptation, search algorithms, speaker-independent models, integration of speech and natural language.

1 Introduction

Over the years, several techniques have been proposed for speech recognition. The most successful speech recognition systems today are based on Hidden Markov Models (HMM) of speech. HMMs are an effective model for speech recognition because they provide a plausible model for speech and also have a sound mathematical basis, which makes automatic estimation of the model parameters possible. The ability to train acoustic models is perhaps the single factor that most accounts for the success of HMMs. In any real system we frequently need to reestimate models for changing conditions: new vocabularies, speakers, acoustic environments, speaking styles, etc. Also, since they use a simple probabilistic scoring mechanism, it is relatively straightforward to integrate knowledge of different types in a recognition and understanding system.

One of the major causes of variability in speech is the differences between the speech of different speakers. We know that the observable parameters of the speech signal vary significantly from speaker to speaker. As a result a Speaker-Dependent (SD) recognition system (a system whose parameters or rules have been designed for one speaker) will be much

NATO ASI Series, Vol. F 75
Speech Recognition and Understanding. Recent Advances
Edited by P. Laface and R. De Mori

less accurate when used by new speakers (Cross-Speaker mode). But SD training requires substantial amounts of training speech from each user of a system. (For example, to record 600 training sentences, or about 1/2 hour of speech we require 1-2 hours with a data collection program, depending on the speaker.) While this cost is small for many real applications, there are also applications - such as short telephone access, that cannot afford any significant SD training.

To alleviate this problem partially we can train the system on many speakers. We frequently call this a Speaker-Independent (SI) system. Typically we treat the speech from all the different speakers as if it all came from one speaker, and use the same techniques as used in the SD training scenario. During recognition we make believe that the new speaker is also the same speaker. Thus when we speak of a speech recognition system being SD or SI we generally are speaking of the training and testing paradigm and not the techniques. SI training will improve performance on new speakers somewhat at the expense of much worse performance than with SD training.

An area that has received much attention is Speaker Adaptation (SA). In contrast with the SD and SI training paradigms there are many different forms of SA. The common feature to all of them is that they start with speech models derived from one or more speakers and then, based on a much smaller amount of speech from a new speaker, they modify the models so that recognition performance on the new speaker is improved.

The remainder of this chapter contains three main sections. Section 2 discusses the use of HMMs. First, we briefly review the HMM speech recognition techniques currently in use for Continuous Speech Recognition (CSR). Then we review the major improvements to these basic techniques. Finally we describe the N-Best paradigm for integrating speech recognition with higher-level linguistic knowledge sources. We have found that this paradigm is computationally efficient, makes integration trivial, and empirically results in lower search error rate than more tightly coupled strategies. We present an approximate algorithm for finding the N most likely sentences that is quite efficient, and yet has been found to be quite accurate. Finally, we will describe the Forward-Backward Search Strategy, which is essentially a sentence-level fast match algorithm that can reduce the search computation by one or two orders of magnitude.

Section 3 of this chaper discusses speaker adaptation. First, we try to define terminology that can be used to characterize the many forms of SA. The features relate to the modes of training and to the characteristics of the algorithms used. Next we describe and characterize several representative techniques that have been used in the past, including those used in the BBN BYBLOS CSR system.

In Section 4 we consider the problem of speaker adaptation when we have available the speech from many speakers. We start by considering different ways of estimating speaker-independent models. In particular, we will demonstrate that, contrary to popular belief, we do not need a large number of speakers to train an SI system. Then we consider two different techniques for adapting from many reference speakers.

2 Continuous Speech Recognition Using HMMs

2.1 Modeling Speech with HMMs

Figure 1 illustrates the basic paradigm used in statistical pattern recognition. We call this the synthesis/recognition process. We assume that an underlying event e in E (the set of possible events) has occurred. We assume that the synthesis process consists of a structural process (e.g. a speech production model) that we understand, a random process that we can only measure, and a deterministic synthesis process that produces the observable signal. For recognition, we first invert the synthesis process to produce a feature vector. Then, we consider each possible event and choose the event that was most likely to result in the observed feature vector, given the structural process and the model of variability.

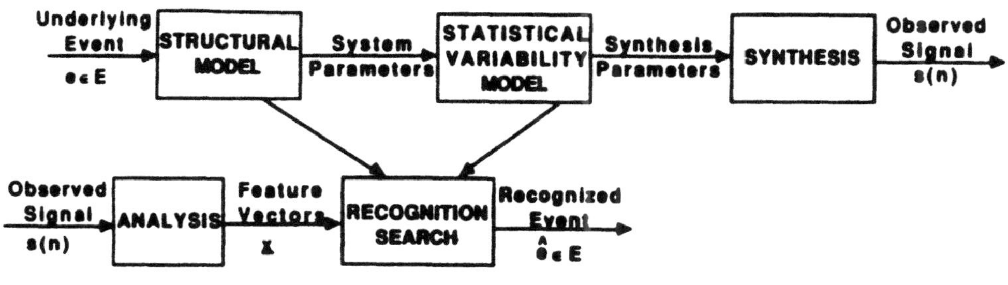

Figure 1. Synthesis/Recognition Model for Speech Recognition

Continuous speech recognition (CSR) is different from the typical statistical pattern recognition problem in several ways: 1) In addition to variability in the primary feature vector (the short-term spectral envelope), speech also exhibits variability in its timing. 2) Rather than recognizing a single isolated event, we must recognize a sequence of concatenated events of unknown duration. 3) We must integrate several diverse knowledge sources (KSs) in the search process. The best phoneme recognizers perform at 70% to 80% accuracy. And yet, for a useful system, we require 95% sentence accuracy. This requires that we integrate many different knowledge sources at several different levels (lexical, statistical language, syntax, semantics, pragmatics, discourse) in order to recognize the words with 99% accuracy and the phonemes with almost 99.8% accuracy. This says that even though our phonetic recognition models may not always have the highest score for the correct phoneme, they must never have a score of zero for the correct phoneme.

The Hidden Markov Model or Markov Random Process is a model that is appropriate for speech. Figure 2 shows a HMM that we can use to model the spectral sequence that results from one phoneme. We can imagine that the spectral sequence for a phoneme is generated by this HMM. At each state we have a probability density that reflects the variability in the spectrum. The alternate transitions allow for variability in timing. We can construct a model for a sequence of phones (any word or sentence) by simply concatenating the corresponding phonetic HMMs. And, in principle, we can construct an HMM for a language simply by making a network of phonetic HMMs. Thus the HMM is easily applied to CSR.

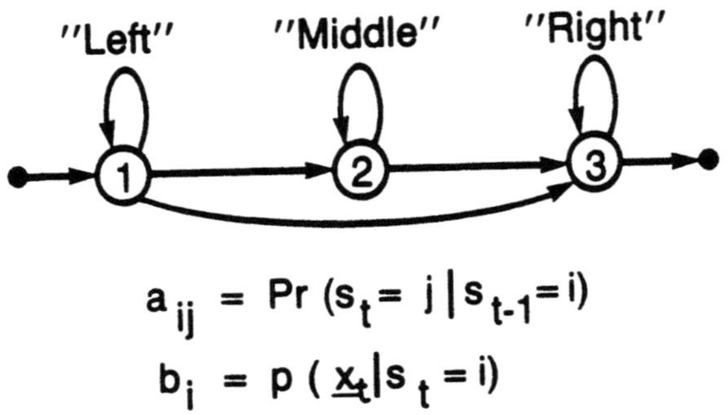

Figure 2. HMM of a Phoneme

2.2 Types of HMMs

There is considerable flexibility in the topology (shape) of the HMM and the type of output distribution used. For a phonetic model we use a left-to-right topology since a phoneme proceeds from beginning to end. Typically two types of output densities are used: continuous densities or discrete densities.

Continuous Densities

Continuous density systems often use a parametric model for the density. For example, each distribution may be modeled as a Gaussian or a Laplacian. If we want to avoid the assumption that the density for each state, s, can be modeled by a single parametric form, we can use a sum of densities. e.g.

$$p(\underline{x}|s) = \sum_i w_i N(\underline{u}_i, \Lambda_i); \quad \sum_i w_i = 1 \tag{1}$$

which is equivalent to a set of parallel states with transition probabilities w_i. Parametric densities have the advantage that the resulting distributions are automatically smoothed. That is, two vectors that are close together are assumed to have similar probability densities. When using continuous parametric models the maximum likelihood solution is often degenerate in that it entails setting one of the variances to zero. However this solution is clearly not useful. Therefore, one usually must avoid singularities by some arbitrary method - either by tying the covariances across many states, or by setting somewhat arbitrary lower bounds on the variance.

Discrete Densities

To use discrete densities we first divide the feature space into disjoint regions using a clustering algorithm [22]. Typically the regions are defined by a codebook of means. Next, we quantize each vector into one of the regions using Vector Quantization (VQ). Once we have determined the VQ index of a vector, we can look up the probability for the state using the vector index as an offset into a set of probabilities stored for that state. We can then estimate the probability of each region or bin. Discrete densities allow nonparametric densities with greatly reduced computation since the codebook is shared among all the states. As with continuous densities we must also be careful to avoid poor estimates with discrete densities. In particular, we must not allow any of the bin probabilities to be zero. We can accomplish this by setting a lower bound on the distribution or by interpolation [18] and smoothing methods.

Semi-Continuous Densities

Recently several researchers have used a compromise between discrete and continuous densities called semi-continuous densities [17] or tied-mixtures [6]. Each bin of the VQ codebook is modeled as a Gaussian, rather than a disjoint region with uniform probability density. Thus the regions overlap. (This is the probabilistic generalization of Fuzzy VQ [41]). The result is that the distributions are again continuous and somewhat smoothed, but we still have the computational advantage of the discrete densities since the Gaussian components are shared among all the states. Equation 2 below shows the form of the probability model for semi-continuous densities.

$$p(\underline{x}|s) = \sum_j p(\underline{x}|bin_j)\ Pr(bin_j|s) \tag{2}$$

where the $p(\underline{x}|bin_j)$ are shared across all states and can be thought of as the probability that the input vector was in bin_j. The $Pr(bin_j|s)$ are just the usual discrete densities. Alternatively as illustrated in Figure 3 one can think of the distribution as being a mixture of continuous densities consisting of N continuous densities in parallel with the transitions (mixture weights) being the same values as the "discrete bin probabilities". The discrete case can be thought of as assuming that the vector x has probability 1 of being in exactly one of the N bins.

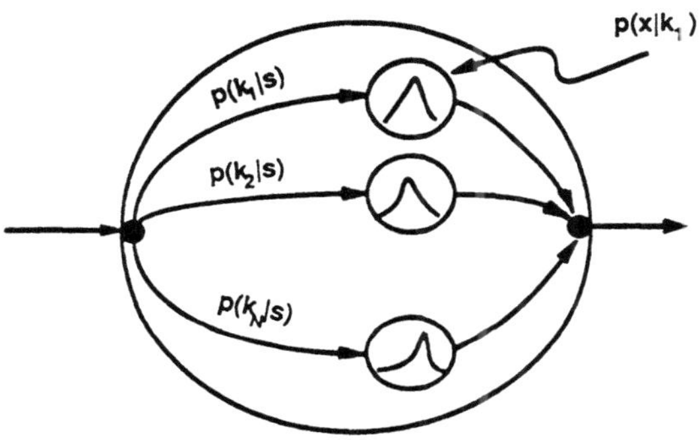

Figure 3. Semi-continuous/tied-mixture density of one state

Which form of the output distribution is best? At this time, despite strong intuitions and claims in either direction, there is no evidence of any significant differences. It is fairly well established that a single Gaussian distribution is not sufficient for one state of an HMM. The semi-continuous distribution appears to reduce the error relative to the discrete density system by about 10% to 20% under some conditions. These differences are often small compared to other system differences.

Training Methods: MLE vs MMI

The Forward-Backward or Baum-Welch [5] training algorithm attempts to find the HMM parameters for which the likelihood of all the training data is maximum. Thus we call this estimate the Maximum Likelihood Estimate (MLE). If the parametric model is correct and we have sufficient training data, this estimate will also yield the minimum error rate. However, we know that, in practice, the models we use are certainly not perfect, and we always have far too little training data for good parameter estimation.

Therefore, there has been much interest in recent years in training criteria that are directed more nearly at reducing error rate. Perhaps the most popular of these are Maximum Mutual Information (MMI) [2, 7] and corrective training [3, 20]. These algorithms start with the MLE estimate. Then, they perform recognition on the training data or held out data and try to modify the parameters of the HMM such that errors will be corrected. These algorithms are generally quite expensive compared to the much simpler ML estimates, especially for large vocabulary CSR. In practice they result in significant improvements over MLE only for small vocabulary problems, such as letters or digits, where one can carefully tune discriminant boundaries. If there is an improvement, this improvement often disappears when the model is improved.

2.3 Incorporating Speech Knowledge

In this section we point out that while the HMM model is a good one, it must be specialized for the speech recognition problem. The gain for this specialization is almost an order of magnitude reduction in the recognition error rate.

Common Corpus for all Results

The DARPA Spoken Language System (SLS) Program sponsors research in Continuous Speech Understanding at several research laboratories. In order to enable close comparison of results, all CSR experiments are carried out on a common corpus. Therefore when one research group reports an improvement, the others can assess whether the results are significant since they have direct experience with the same corpus. Currently, the common corpus is the 1000-Word Resource Management (RM) Corpus [26]. It is a simulation of a navy resource management domain. There are 2850 sentences which were generated from 950 sentence patterns. The speakers are classified as being from one of 8 broad dialects representative of American English. The corpus contains two segments: a speaker-dependent segment containing 12 speakers with 600 training sentences each (about 1/2 hour of speech) and 4 speakers with 2400 sentences each, and a speaker-independent segment containing 109 training speakers and 51 test speakers, who read 40 sentences each (about 2 minutes). 60% of the speakers in each segment of the corpus are male. Both the SD and SI segments are relatively balanced across the American dialects.

For evaluation purposes an artificial deterministic grammar was constructed that allows all bigrams allowed in a grammar made from the 950 sentence patterns. It is called the "Word-Pair" grammar. The test set perplexity of this word-pair grammar is 60. All results are reported in terms of the percent word error, defined as:

$$error = 100 \times \frac{substitutions + deletions + insertions}{total\ words}$$

Baseline Recognition Accuracy

If we construct 3-state phonetic models as described above and perform a speech recognition experiment using the SD portion of the corpus the word error rate is a disappointing 15%. Thus, merely using HMMs is not sufficient to achieve high accuracy. It is essential to incorporate as much information about speech in the phonetic model as possible, while also being careful about obtaining robust estimates of all of the system parameters. During the past five years there have been many modifications suggested for HMM recognition, but only a few of these have resulted in significant improvements on this corpus.

Context-Dependent Phonetic Models

First, we know that the acoustic realization of each phoneme is greatly affected by the adjacent phonemes due to coarticulation. We can model the coarticulation by estimating context-dependent sets of parameters for each phonetic HMM depending on the neighboring phones [30, 31, 32]. Sometimes we call these models triphones even though we really mean *triphone-dependent phonetic models*. To ensure that the estimates of the parameters are robust we interpolate the estimates of the context-dependent models with those that depend on less context. To get the full benefit of context-dependent models we must also model coarticulation that occurs between words, since phonemes on the ends of adjacent words affect each other also.

Derivative Features

Another important piece of knowledge is that spectral changes are important in human perception. The second major advance is the successful use of the time-derivatives of the spectral envelope parameters. Again, robust estimation is necessary for the new parameters to be useful. We measure the derivatives as the slope over several frames of speech [11]. If we used the derivative parameters jointly with the steady-state parameters, the dimensionality of the vector becomes too large to estimate the probability density robustly. Therefore, we assume that, conditioned on a state, the derivative parameters are independent of the steady-state parameters [15]. Thus, we use a separate VQ codebook for the derivative parameters. The joint probability of the steady-state vector, $\underline{x}_s$, and the derivative parameter vector, $\underline{x}_d$, is approximated by the product of the two separate probability densities.

$$p(\underline{x}_s, \underline{x}_d | s) \approx p(\underline{x}_s | s)\, p(\underline{x}_d | s) \tag{3}$$

Robust Estimation

Finally, as we add more and more parameters to our model, we have less and less data to estimate each parameter. We find that it is necessary to smooth these estimates in many ways. One method for smoothing discrete densities is to use a probabilistic smoothing matrix [34, 19]. Each entry in the matrix represents the probability that two of the VQ bins will cooccur in any probability density. As figure 4 illustrates, we can multiply the vector of probabilities associated with each state probability density by this matrix to get a new, more robust probability density.

A second method for smoothing is the semi-continuous density method mentioned earlier. We find that it is beneficial to use both smoothing methods simultaneously, since they have different properties.

Table 1 lists the improvements that result from each of these enhancements. The gains are listed in terms of the percent reduction in error and the (corresponding) factor by which

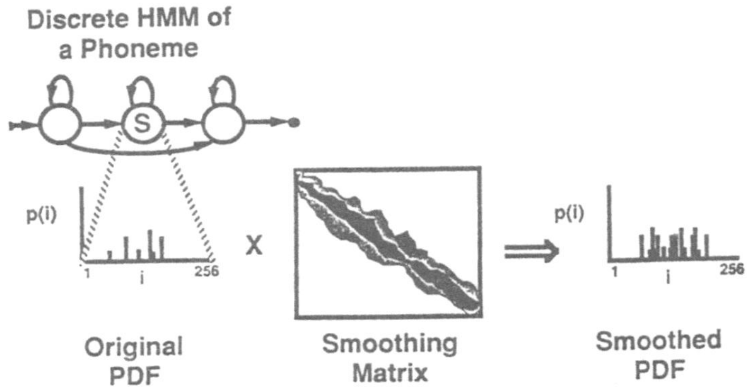

Figure 4. **Probabilistic smoothing. Each pdf is multiplied by the triphone model coccurrence** smoothing matrix

the error rate is divided. Amazingly, these improvements are completely independent, which means that the factors multiply. Thus, the final error rate today is 8.6 times lower than in the simple HMM system that uses context-independent phonetic models with steady-state cepstral parameters! As might be expected, the same factor of error reduction can be observed in the simple SI training scenario, since we use the same SD techniques.

Technique	% Error Reduction	Reduction Factor
Coarticulation Models		
(within words)	50	2
(between words)	30	1.5
Difference Parameters		
(1st derivative)	50	2
(2nd derivative)	10	1.1
Smoothing	25	1.3
Total Improvement	88%	8.6

Table 1. Improvements in SD Recognition. Word error was rate reduced from 15% to 1.7%

2.4 Integration of Speech Recognition and Natural Langugage

If our application for CSR is speech understanding, then we also have an additional Knowledge Source (KS). The linguistic understanding model of natural language can be used to aid in the recogition of the words. Therefore, when we perform our recognition search, we would really like to find the legal sentence that is most likely, given our acoustic and statistical language models. (One might correctly point out that the linguistic model might often reject the correct sentence, thereby reducing the accuracy. However, we assume here that we are not interested in recognizing the correct string of words if they can not be understood.) Most linguistic models of language are quite large and complex. Searching for the most likely word string requires consideration of a large number of sentence fragments. At the same time, the linguistic language models are usually not as powerful (in terms of language perplexity) as a simple statistical language model. Therefore, we have chosen, instead to use a multi-pass N-Best strategy.

The N-Best Paradigm

Figure 5 illustrates the N-Best Paradigm for the general case.

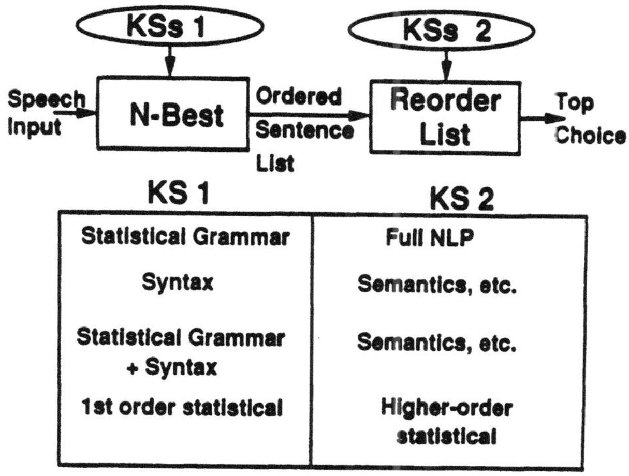

Figure 5. The N-Best Paradigm: The most efficient knowledge sources (KS1) are used first to produce a list of sentence hypotheses, which are then filtered and reordered using the remaining knowledge sources (KS2)

We use the most powerful and least expensive KSs first to produce not one but an ordered list of the N most likely sentences. Then in one or more successive passes, we use the remaining KSs to filter and reorder the list. Finally we choose the most likely sentence remaining. The idea of using a staged recognition process is not new [42]. The questions are: which knowledge sources are most efficiently used in which stage, and whether there is an efficient algorithm for finding the N-Best hypotheses. Presently we find it most efficient to use a bigram statistical language model together with the acoustic speech models to produce the N-Best list, and the remainder of the Natural Language (NL) models (syntax, semantics, pragmatics) to filter the list. As long as we can guarantee that the sentence we would choose in an exhaustive search is on the list, the N-Best paradigm is admissible. In practice, if all of the NL models only accept or reject a sentence, this means the list must be long enough so that it always includes a sentence hypothesis that would be accepted by NL. In practice we measure a much more conservative statistic - the percent of the time that the correct answer is on the list.

Efficient N-Best Search Algorithms

We have considered several algorithms for finding the N-Best hypotheses. An expensive but exact algorithm is given in [8, 35]. This algorithm computes the score of each sentence hypothesis independently. While this algorithm theoretically could require exponential time, in practice we find that we can simply keep up to N active sentence hypotheses for each HMM state, in which case the computation is linear in N. Much less costly approximations that require little more computation than the usual Viterbi search have also been proposed [40, 23]. These algorithms do not replicate any HMM states. Instead, they simply remember the alternate words that end at the same time. In [36, 37] we describe two approximate algorithms. One is a minor adaptation of the fast algorithm mentioned above. The second algorithm which is called the Word-Dependent N-Best algorithm is a compromise between the exact Sentence-Dependent search and the fast approximations. We perform separate computations for each active HMM state, depending on the previous word only. Thus, the computation is proportional to the number of plausible previous words. We typically allow up to 6 copies of each state. We find, empirically, that the Word-Dependent N-Best algorithm is just as accurate as the exact algorithm. Figure 6 shows the cumulative distribution of the rank of the correct answer for each of the three algorithms.

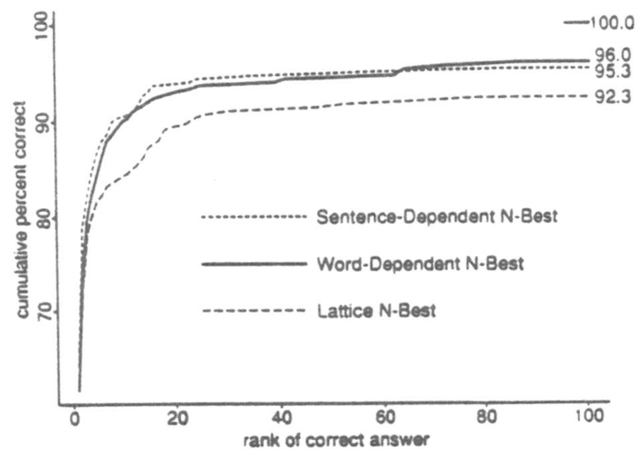

Figure 6. Comparison of N-Best Algorithms. The Word-Dependent algorithm is as accurate as the exact Sentence-Dependent algorithm

We plot the results up to 100 sentence hypotheses. (In this experiment, we used a statistical bigram class grammar with perplexity 100. Typically this grammar results in twice the error rate obtained with the Word-Pair Grammar.)

When the Word-Dependent N-Best algorithm is run allowing 6 preceding words, it will produce up to 6^L different sentence hypotheses where L is the number of words in the sentence. As can be seen in the figure, the correct sentence is included in the list 96% of the time, or missed 4% of the time. This error rate is about 1/10th of the error rate of the 1-Best answer. Thus the contribution to the error rate from keeping only 100 hypotheses is currently small compared to the overall error rate. If the secondary KSs become much more powerful then we must either make the N-Best list longer or include these KSs in the N-Best stage of the recognition.

2.5 Forward-Backward Search

Even with the fast algorithms described above, the computation required for finding the N-Best sequences is too large for a general purpose workstation. Therefore we used an additional multiple-pass strategy - the Forward-Backward Search Strategy [36, 38].

Most of the computation required by the time-synchronous beam search [21] involves considering the first few states of all words almost all the time. This is because any word that ends with a good score can be followed, with some probability by every other word. However, most of the states in most words have low enough probabilities so that the computation for those states is pruned out during the beam search. Thus, most words do not have high ending scores at any given frame. If we could look into the future of the sentence we could determine which of the words are plausible and avoid all the computation for any other words. Several fast match schemes have been developed for estimating the score for the next word [10, 4, 14, 1]. However, the score estimates from these algorithms suffer from being quite approximate. In addition, the further ahead these algorithms look, the longer the delay incurred. Finally, the fast match algorithms do not tell us whether the hypothesized word is grammatically consistent with the words that appear after it.

We have defined a whole-sentence 2-pass strategy called the Forward-Backward Search Algorithm, based on its mathematical similarity to the training algorithm of the same name. First we perform a normal 1-best search in the forward direction (which can be performed in real time if the hardware is fast enough), with no extra delay. In addition to the usual traceback information, we remember at each frame the ending score for each word that is above the beam threshold. This is actually the likelihood of the best sequence of words ending with that word at that time. When the end of the sentence is detected we perform a backward search looking for the N-Best hypotheses. Whenever we are about to start a word (in the backward direction) we first look up the best forward path score of the word. We multiply the backward path score by the saved forward path score. This product is the likelihood of the whole sentence given that the word sequence goes through the state from the right and ends with this word at this time. If the product (normalized apprpriately) is below the beam threshold, we do not even begin the computation for that word. The result is that fewer than 1% of the vocabulary is involved in the backward computation at any frame, and the backward pass is sped up by a factor of 40 with no additional observed pruning errors. Thus although the backward pass only begins after the sentence has been spoken, the delay is quite short.

The Forward-Backward Search is generally useful whenever there are expensive KSs to apply. For example, we could perform recognition in the forward direction using a simple bigram grammar and in the backward direction using a more expensive trigram or context-free grammar.

2.6 Complete Spoken Language System

We have implemented all of the search algorithms described above to produce an integrated Spoken Language System. The signal processing and vector quantization are performed on a Sky Challenger board, which has two TMS320C30 array processing chips. The forward 1-Best and backward N-Best speech recognition searches currently run on a SUN 4/330 workstation. The forward 1-Best pass can, with some approxiations, run in real time, which means that as soon as the system detects that the speaker has stopped speaking (say after 1/2 second of silence), the text of the 1-Best answer is sent to the Natural Language processor, which runs on another workstation. While the NL processor is parsing the first answer, the backward N-Best pass runs, which typically requires about a 1-2 second delay. Then, the texts of the N-Best answers are sent to the NL processor. The NL processor parses each answer in the list until it finds an acceptable one, and then prints the answer to the query. (Note that if any of the N-Best answers is the same as the 1-Best answer, the NL does not have to reparse it. Most often, the 1-Best answer is acceptable and is the same as the first of the N-Best answers, which means that the answer can be printed as soon as the N-Best answers come out.)

3 Speaker Adaptation

The main topic of this chapter is speaker adaptation. It is well known that the acoustic signal produced by different speakers is quite different. The differences are the result of varying vocal tract length and shape, voice characteristics, regional accent, and habit. And yet, we have the subjective impression that people can recognize speech from different speakers with little difficulty. (This author is not aware of any experiments that document the increase in human recognition error rate when the speaker is new, for instance when a child first hears speech from a different person than its immediate family. It could be that the error rate does, in fact, increase by a similar factor, but that the error rates under normal listening conditions are so low that measuring the increase would be difficult.) As an example of the effect of different speakers on a simple speech recogition system, if we test the BYBLOS system with 600 sentences of training (1/2 hour of speech) from one speaker (SD mode), the average word error rate is 1.7%, averaged over the 12 speakers in the DARPA SD segment. When the system is tested on another speaker of the same sex, without any adaptation, the error rate increases to 10%. If we test the system on a speaker of the opposite sex, the error rate increases to 77%. Thus, even for a different speaker of the same sex, the error rate increases by a factor of 6!

There has been considerable effort devoted to making a speaker-independent (SI) model by training the same system on the speech of 109 speakers, with a total of about 4000 sentences (3-1/2 hours of speech) [19]. In this case, the error rate is about 6% when tested on new speakers. Thus, the SI error rate is about 3 times the SD error rate with 7 times more training data from 109 speakers and 1/2 the error rate of a system trained on 2 speakers. A general alternative to the SD and SI training scenarios is the class of speaker-adaptive (SA) training scenarios. There are many such scenarios possible. In contrast to the SD and SI training scenarios, the techniques used for the SA scenarios are different. In one very economical scenario, we collect a substantial amount of speech from only one speaker (the reference speaker), and then we record 40 sentences (about 2 minutes of speech) of adaptation speech from a new speaker who wants to use the system (the target speaker). This two minutes of speech is used to normalize all of the speech models from the reference speaker so that they are appropriate for the target speaker. The economic advantage of this scenario is that to get started in a new domain we only need to record a large amount of speech from one speaker, which can be accomplished in a very short time. (We find that it takes 1 to 2 hours to record 1/2 hour of read speech from one speaker.)

3.1 Speaker Adaptation Terminology

In this subsection we define several features that can be used to characterize the techniques and training modes of a SA system. The issues often relate either to the amount of inconvenience felt by the user (which is determined by the scenario), or the technical capabilities of the algorithms used. The features we will discuss below are: batch vs incremental scenarios, supervised vs unsupervised training, training vs normalization algorithms, feature transformation vs model transformations, and deterministic vs probabilistic transformations.

Batch vs Incremental Adaptation

One can require that a user supply the adaptation speech before ever using the system (or at least before the adaptation takes place). We call this a *batch adaptation* scenario, since it takes place all at once. In this case, one would usually read a prepared list of sentences, so this scenario would also allow for *supervised* techniques (see below). The advantage of batch adaptation is that the data can be collected very quickly. After the few minutes needed to collect the data (and whatever computation is needed), the system will be ready to use with the adapted level of performance. The obvious disadvantage is that there is a delay before the user can use the system.

An alternative to batch adaptation is *incremental adaptation*, in which the system makes use of the speech that is collected while the user makes actual use of the system. This allows immediate use, and could possibly track dynamically any changes in the speaker or the environment. However, if the reason that we are using an adaptation technique is that the system is really not good enough without it, then this mode is not practical, since the system will behave poorly until it has enough adaptation speech. This means that the interactions will be slower than desired, to allow for corrections. The time needed to collect 40 confirmed sentences in this mode might be 20 frustrating minutes instead of 5 minutes of reading. Therefore, it would seem that incremental adaptation only makes sense after the system has already reached an acceptable level of performance.

Supervised Adaptation

It is easier (algorithmically) to use the adaptation speech if we know exactly which words it contains. This is because we can directly compare the way the target speaker produced a particular phoneme with the way the reference speaker would produce the same phoneme in the same context. Thus, the normalization that we would perform would be more precise. In addition, it becomes feasible to have the normalization be dependent on the model. (see below)

There are two simple ways to collect supervised adaptation data: 1) Have the speaker read prompted text in the batch adaptation mode, or 2) Require that the user confirm or correct all sentences spoken to the system. (Actually, we might reasonably assume that the user would normally need to correct any recognition mistakes in order to use the system properly anyway.) In the first case, we can also use *text-dependent* algorithms that require that the target speaker actually speak the same sentences as spoken by the reference speaker.

Incremental Training vs Normalization

It is important, from an algorithmic point of view to distinguish incremental training methods from global normalization methods. This should not be confused with incremental adaptation scenarios above which can use either incremental training or global normalization methods. We define as *incremental training*, any method that modifies only the parameters of the models observed in the adaptation speech. For example, we might modify the means of some distributions to be a weighted combination of the reference means and the observed target means.

$$\mu = a\, \mu_{target} + (1-a)\, \mu_{reference}$$

Incremental training requires a large amount of adaptation speech because it will take a long time before most of the models have occurred. However, eventually, when all of the parameters have changed sufficiently, the system should become speaker-dependent.

By *normalization* we mean that, based on the limited amount of adaptation speech, *all* of the models are modified. For example, shifting, rotating, and scaling the whole feature space would be a global normalization. In general, normalization methods requires less adaptation speech but are not necessarily guaranteed to converge to the SD model.

Feature Transformation vs Model Transformation

Another algorithmic distinction is whether the transformation could be applied to the input speech feature vectors, or whether they must be applied to the models. Transformations that could be applied directly to the features might require slightly more computation at the time of recognition, but they avoid the delay that occurs when all of the system parameters are normalized. The feature normalization could either be a linear transformation, e.g.

$$\underline{y}_t = A\underline{x}_t + \underline{b}$$

or it could be nonparametric. For example, the transformation could depend on the VQ region of the test speech vector.

If, on the other hand, we transform the models, we have much more flexibility. We can, of course simulate any feature transformation, but we can also allow each model to have its own normalization. For example, we might want the transformation to depend on the phoneme in order to be able to model dialects or accents, in which only a few phonemes have changed. We find this effect particularly powerful when adapting the system to a nonnative speaker of the language.

Deterministic vs Probabilistic Transformation

Finally we can ask whether the speaker transformation maps points in the feature space on a one-to-one basis, which we call a *deterministic mapping*, as illustrated in Figure 7a. While a deterministic mapping is simpler, we don't believe that it can fully model the differences between speakers.

A *probabilistic mapping*, on the other hand maps each point in one space to a distribution in the other as illustrated in Figure 7b. Thus, it is more flexible. (The probabilistic mapping is capable of representing a deterministic mapping in the degenerate case.) The probabilistic mapping combines a probabilistic model of speech with a probabilistic model of the speaker. This is also necessary to account for a lack of knowledge about the target speaker.

Summary

There are several data representations used in the speech recognition process: parameter vectors, VQ indices, probability densities. The different methods of speaker adaptation modify the recognition system at different stages. For example, we might perform a parametric

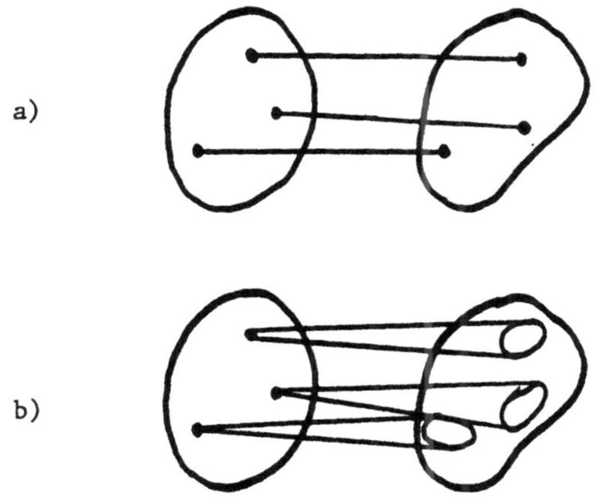

Figure 7. a) deterministic mapping between the spectral spaces of two speakers. b) probabilistic mapping between the two spaces

normalization on the parameter vectors, a nonparametric normalization on the VQ indices, and a probabilistic model-dependent normalization on the probability densities of the models themselves.

3.2 Survey of Adaptation Methods

In this section we survey several specific methods of speaker adaptation. The survey is by no means intended to be complete, but rather a sampling of the different types of adaptation.

Formant Range Normalization

In 1967 Gerstman [13] proposed a formant normalization that would explain the differences between the formant space of different speakers. He used the Peterson-Barney vowel data [25], which consisted of vowel formants and pitch that had been manually extracted from spectrograms. The data contained two readings of each vowel. He first found the minimum and maximum values of the first and second formants from the first reading of */iau/*, and then normalized the formant frequencies of the second reading based on these extrema.

$$fi' = \frac{fi - fi_{min}}{fi_{max} - fi_{min}}; \quad i = 1, 2, 3$$

The result was that he was able to recognize 97% of the confirmed vowels in the data. This method clearly requires a batch, supervised scenario, and the techniques would be characterized as a deterministic feature normalization.

Speaker-Independent Normalization by Pitch

In 1971 Schwartz [29] proposed a self-normalization formula based on the pitch frequency. Listening experiments showed that subjects' perception of the formant frequencies was affected by the pitch. This makes sense since pitch is highly correlated to vocal tract length. Based on these experiments the formant frequencies of each vowel sample in the Peterson-Barney data were normalized by a nonlinear function of the pitch

$$fi' = \frac{fi}{g(pitch)}; \quad i = 1, 2, 3$$

The result was that 91% of the vowels from men, women, and children were recognized correctly without any prior knowledge of the speaker. This method does not fit into most of the categories, since it does not require prior adaptation speech. However, it is a deterministic feature normalization. While the performance of this self-normalization was higher than any other attempts at speaker-independent recognition of this data, the fact that it is lower than the performance achieved by Gerstman shows that one frequency is not sufficient to characterize a speaker.

VQ Codebook Mapping

Many speaker adaptation algorithms are based on either changing the VQ codebook or mapping VQ codebooks from one person to another. Shikano [39] proposed a nonparametric method of normalization based on mapping a VQ codebook from one speaker to another. The training scenario is batch and text-dependent. The method is a deterministic feature normalization. The method is summarized below.

1. Record the same text for the reference and target speakers.
2. Time-align the cepstral sequences using dynamic time warping (DTW).
3. Collect all target frames that align to each reference VQ index.
4. Shift each reference codebook mean by the corresponding mean difference.
5. Iterate if necessary.
6. Use adapted codebook to quantize new test speech from the target speaker.

There were several variations on this algorithm. The basic idea was very powerful because it was nonparametric. That is, any deterministic mapping of the cepstral space was possible.

VQ-Based Speech Normalization

Rigoll [27] also proposed a VQ-based normalization, except that instead of transforming the test speech, he normalized all of the training speech into the space of the target speaker. First, he built a discrete HMM that expressed the likely reference VQ labels for each target VQ label. Then he relabeled all of the reference speech and retrained the entire HMM system on the normalized speech.

Re-Estimation of Mixture Means

Like Shikano, Rtischev [28] also adapted the codebook means, but did it using the Forward-Backward algorithm within an HMM system that used semi-continuous densities.

1. Estimate semi-continuous HMMs models for the reference speaker.
2. Given the adaptation speech, estimate new means for the Gaussian components within the forward-backward algorithm, while keeping the discrete densities (mixture weights) fixed.
3. Use the new means as a new VQ codebook in a discrete density system.

The result was a factor of three reduction in error relative to cross-speaker recognition. Rtischev also performed an experiment in which he trained the system on 10 speakers and found that the SI recognition performance was surprisingly good. When he adapted the codebook means using the technique described above, there was a further small gain.

Unsupervised Codebook Adaptation

Furui [12] proposed an unsupervised codebook adaptation algorithm. This is much more difficult, because only the overall statistics of the speakers are known, rather than having a frame-to-frame correspondence. The basic method is to create a hierarchical mapping from the spectral space of the reference speaker to the space of the target speaker.

1. Shift the overall reference codebook mean to the target mean
2. Divide each data cluster in half
3. Assign each resulting codebook mean to the nearest cluster from the other speaker.
4. Shift each reference codebook mean by the corresponding mean difference
5. Apply continuity constraints at cluster boundaries
6. Go to 2

In mapping the space hierarchically, he is basically assuming that the principal components of the two spaces correspond to the same phonetic directions. This turned out to be a good assumption for the first few bits of the mapping, since these mostly correspond to the first two cepstral coefficients. The result was reduced error rate when only steady-state cepstra were used. However, when differential cepstra were also used, there was no longer any significant improvement over the cross-speaker model.

Probabilistic Codebook Mapping

Next we describe the technique used within the BYBLOS system, which is a probabilistic VQ mapping from the reference speaker to the target speaker [33, 9]. We assume that a deterministic mapping of the codebook indices or means is not sufficient to describe the differences between two speakers. Instead, we assume that, given a particular VQ index, k_i, produced by the reference speaker there is a probability distribution for each of the possible target VQ indices, k'_j, for the exact same phonetic sound.

We can view the discrete pdf of each state "s" as a probability row vector,

$$\mathbf{p}(s) = [p(k_1|s), p(k_2|s), \cdots, p(k_N|s)] \tag{4}$$

where $p(k_i|s)$ is the probability of spectral index k_i given state s of the HMM model. We then denote the probability that the new target speaker will produce quantized input spectrum k'_j, given that the prototype speaker produced spectrum k_i as $p(k'_j|k_i)$ for all i, j. The set of conditional probabilities $P(k'|k)$, which form an $N \times N$ matrix, $\mathbf{T}$, constitutes the state-independent mapping required to perform the probabilistic transformation from one speaker's spectral space to another's. The critical step in this approach is to estimate the matrix T that maximizes recognition performance. We can then compute each discrete pdf $\mathbf{p}'(s)$ for the new speaker as the product of the row vector $\mathbf{p}(s)$ and the matrix $\mathbf{T}$.

$$\mathbf{p}'(s) = \mathbf{p}(s) \times \mathbf{T}; \quad \mathbf{T}_{ij} = p(k'_j|k_i) \tag{5}$$

There are many ways to estimate $\mathbf{T}_{ij}$. In [33] we showed that the mapping can be thought of as part of an HMM in which the original parameters are fixed. Therefore, the mapping could be estimated by aligning the input quantized spectra against the prototype HMMs of whatever was said using the forward-backward algorithm. Thus the algorithm was supervised, but could be text-independent. The algorithm worked quite well on a low perplexity task. However, the accuracy degraded significantly with loosely constrained grammars.

A second, much simpler algorithm turned out to result in higher performance [9]. Figure 8 illustrates the algorithm that we use to estimate the speaker transformation matrix.

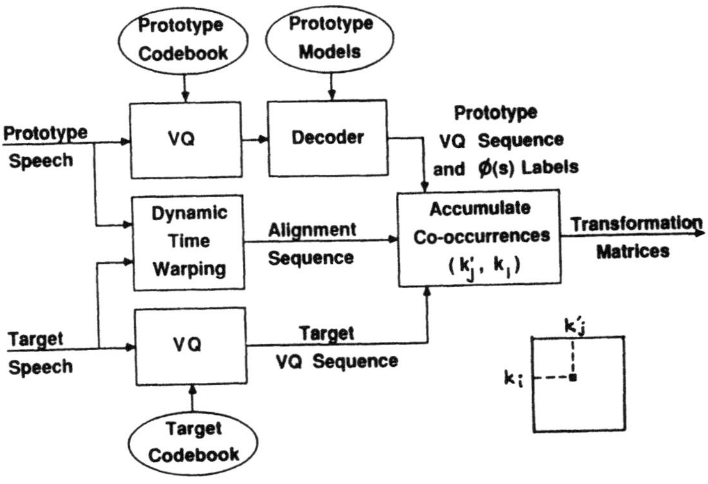

Figure 8. Text-Dependent estimation of transformation matrix using dynamic time warping

The algorithm requires that we record the same sentences from the target and reference speakers. We use DTW to align the phonetically similar frames of the two utterances (in the same manner as Shikano). Then, we simply count the cooccurrences of spectra between the two speakers, and convert these directly to VQ transformation probabilities. (We also keep a separate transformation matrix for each phoneme. However, the resulting phoneme-dependent matrices are not well-trained, and must be interpolated with the phoneme-independent matrix.)

The recognition accuracy was significantly higher than with the previous estimation method. In particular, with 30 minutes of speech from a single reference speaker and 2 minutes of adaptation speech from the new speaker the word error rate on the DARPA RM task was 6.5%, which was essentially equivalent to the error rate of a speaker-independent system with 109 speakers and a total of 3-1/2 hours of speech. The clear advantage of this adaptive scenario is that we do not need to collect speech from over 100 speakers in order to bring up a new system. It also results in much lower error rate than the SI model for nonnative speakers or when using microphones and environments that are different from those in the training.

Probabilistic Fuzzy VQ Mapping

The method described above requires a substantial delay between when the adaptation speech is collected and when the adapted models can be used. This is the time necessary to multiply all of the HMM pdfs by the probabilistic mapping matrix. Nakamura [24] combined the fuzzy VQ method of representing speech [41] with the probabilistic spectral mapping method for adaptation. The input fuzzy VQ, which can be thought of as a row vector of probabilities, can be multiplied by the transposed probabilistic mapping matrix. The result is another fuzzy input vector which can then be used directly with the original reference HMM models. The result is mathematically equivalent to multiplying the pdf vector by the matrix, as long as we use a single mapping matrix. Thus the long computation delay can be avoided, at the expense of slightly more computation during recognition.

4 Speaker Adaptation from a Speaker-Independent Training Corpus

Given that pooling data from several speakers reduces the error rate of a cross-speaker model, it is natural to try to produce an adapted model using reference speech data from several speakers. For this scenario the goal is to achieve performance that is better than adaptation from a single reference speaker and also better than the initial speaker-independent performance. There have been previous attempts at speaker adaptation from a speaker-independent corpus [16]. However, the gain from using more than one reference speaker has usually been small.

Our natural inclination would be to try to use similar methods as used in the single-reference adaptation system. In this case, rather than using a small amount of speech from a large number of speakers, it would be more natural to use a large amount of speech from a smaller number of speakers, in a manner more similar to the single-reference adaptation scenario. In particular, we have 1/2 hour of speech from each of 12 different speakers. Therefore, before embarking on this effort we needed to establish a baseline SI control experiment using these speakers.

4.1 Speaker-Independent Control Experiments

Training with Few Speakers

Our initial assumption was that it was desirable to have training speech from as many speakers as possible in order to estimate the best speaker-independent model. In particular, if we could, we thought we would rather have one sentence from each of several thousand speakers, than more speech from fewer speakers. Therefore, we expected that speaker-independent recognition using 1/2 hour of speech from 12 speakers would not perform very well. When we ran the initial experiment in which we simply pooled the speech from all 12 speakers and ran the forward-backward algorithm, together with the triphone cooccurrence smoothing procedure, we found that the error rate was 8.5%, which was better than expected.

Independent Training

Then we reasoned that both the training algorithm and the cooccurrence smoothing algorithm operate under the assumption that all of the data comes from a single speaker distribution. When we train the system on speech from all of the speakers at once the forward-backward algorithm is constrained to find alignments of the data to the models that are consistent with each other. In addition differences in the spectra due to differences in speaker - rather than random differences are used to smooth the distributions. We conjectured that both the forward-backward training algorithm and the smoothing algorithm were really most appropriately *speaker-dependent* operations. We found that the best results were obtained by performing both of these operations on the speech of only one speaker at a time, as shown in figure 9.

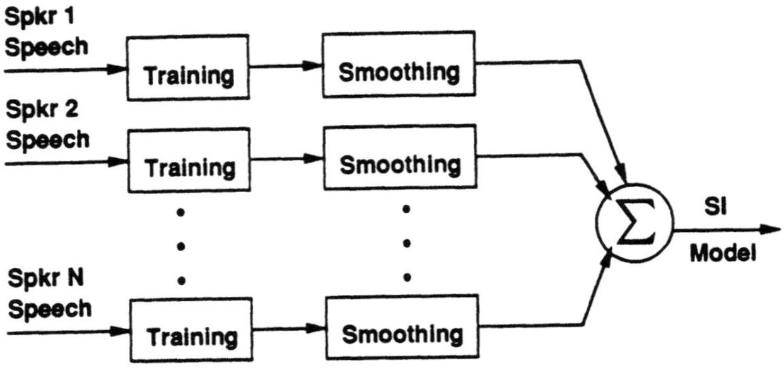

Figure 9. New Speaker-independent training paradigm. Models for each training speaker are independently trained and smoothed before averaging

Thus, the forward-backward algorithm is free to find a sharp discriminating alignment and model for each speaker. We postpone combination of the data for as long as possible, and at the end, we average the discrete observation densities for the different speakers. While the model produced is no longer the maximum likelihood model, it is somehow more appropriate for recognition. The result was that the word error decreased to 7.4%, which is quite similar to that for the SI-109 speaker training set.

There are several major advantages for having a speaker-independent system trained in this way on a few speakers. First, it is much easier to collect a large amount of speech from a few speakers than to arrange for, set up, and collect a few sentences from a large number of speakers. Second, it is no longer necessary to run the training algorithm on all of the data at once. This becomes an important issue because the computer time needed to train the models is often several days. In addition, when we get training from another speaker after we have an initial model, it is simple to average it in, without having to retrain using all of the speech again. Third, in contrast to the usual scenario, there is a very large incentive for those speakers who were the training speakers. These speakers have the option of using their speaker-dependent models with three times lower error rate than with speaker-independent models made from other speakers. Alternatively, our most recent experiments show that the error rate for training speakers who use the averaged models (multi-speaker mode), is close to the speaker-dependent error. Fourth, separate models allow further speaker-specific operations such as adaptation via a speaker transformation.

4.2 Multi-Reference Speaker-Adaptation

We have tried two different methods for using the speech of many speakers as reference speech for speaker adaptation. We must remember that it would be incorrect to try to adapt the entire speaker-independent model to a new speaker. The speaker-independent model has been broadened to cover the space of all likely speakers. Thus, the distributions are necessarily smeared. Shrinking this entire model down to the space of the new speaker could not remove this smearing. (Some limited experiments have confirmed this expectation.) Instead, we must somehow transform the speech of each of the reference speakers independently.

Normalized Features Model

The first method that we tried was based on parameter normalization using the 109 speaker corpus. We record a large amount of training speech from a single speaker who is defined as the prototypical reference speaker. Then, we transform the speech parameters of each of the reference speakers independently into the common prototypical space, using a similar supervised DTW-based technique to that of Shikano. The normalization for each reference speaker is deterministic, and is estimated using sentences that are in common between the prototype speaker and each reference speaker. Then, we pool all of the normalized speech as if it came from the prototype speaker, and use this model in the same way that we normally perform single-reference adaptation. This process is illustrated in Figure 10.

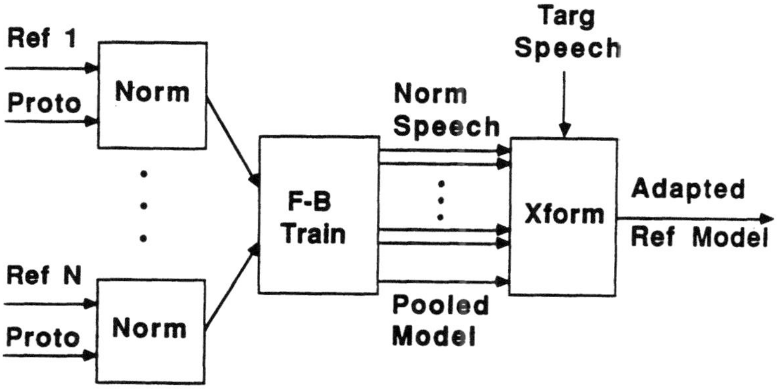

Figure 10. Normalized Pooled Parameters. The parameters from each reference speaker are first normalized into the space of the prototype speaker. All of the pooled normalized data is then used to estimate a model which is used for adaptation

Unfortunately, the result using either the 109 speakers or the 12 speakers was only marginally better than the speaker-independent model.

Averaged Adapted Models

In the second technique, which is illustrated in Figure 11, we transform each of the speaker-dependent models to the new target speaker independently using our single-reference adaptation, and then average the adapted models. As can be seen, this process is quite similar to the process we have suggested for speaker-independent recognition which is illustrated in Figure 9.

We performed several experiments with different amounts of training, always keeping the amount of adaptation speech fixed at 2 minutes. Table 2 shows the error rate for speaker adaptation with 1, 4, and 11 reference speakers compared with speaker-independent recognition using 12 speakers. (Unfortunately, two different test sets were used in these experiments, due to the constraints on the speakers available. However, several previous experiments have verified that the two test sets are equally difficult.) The results show that the speaker adaptation improves somewhat when the reference speech is longer. They also show that using more reference speakers reduces the error rate (25% for 4 speakers, 33% for 11 speakers). Finally the multi-reference adaptation condition has 4.1% word error which is 45% lower than the (12-SI) speaker-independent result.

If we examine the results more carefully, we notice that the number of speakers who

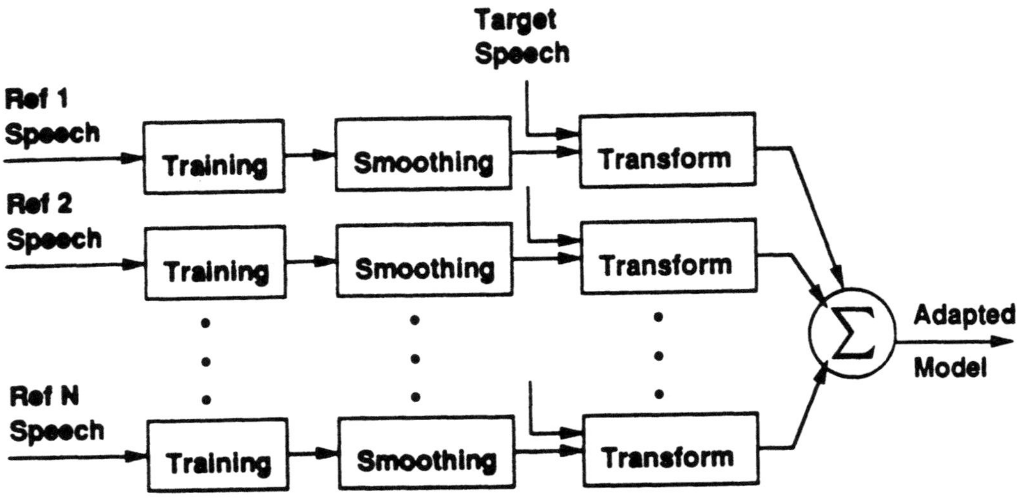

Figure 11. Average Adapted Model. The models of each reference speaker are independently adapted to the target speaker, and then averaged

# Refs	Training	Test Set	% Word Error	% Sent Error
1	30 min	June 88	6.2	31
1	2 hr	Feb 89 SD	5.1	26
4	8 hr	Feb 89 SD	4.2	23
11	5.5 hr	June 88	4.1	23
12 SI	6 hr	Feb 89 SD	7.4	37

Table 2. Speaker adaptation results under 4 conditions, compared with SI recognition

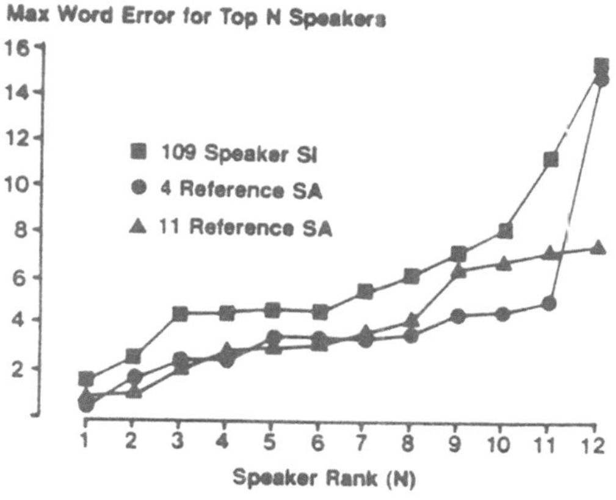

Figure 12. Cumulative results for three conditions. The SA models have fewer outliers than the SI models

	Same Sex	Opposite Sex	Average
No Adaptation	10	77	46
With Adaptation	5.5	6.1	5.8

Table 3. Effect of Reference Gender. Speaker Adaptation is insensitive to gender

are still outliers varies with the method. Figure 12 shows, for each of three conditions, the cumulative number of speakers with error rate below a certain value.

The worst condition is the SI-109 speaker-independent condition with 109 training speakers. Not only is the performance worse, but there are two speakers that are not well covered by the 109-speaker model. Thus, 109 carefully chosen speakers is not sufficient coverage of the possible speakers, even though all of the test speakers were native speakers of American English. (In our experience, any speakers with mild accents experience severe degradation with such a model.) The 4-reference adapted condition is generally the best condition, even though only 4 speakers are used. However, there is still one speaker who cannot be modeled well by 4 adapted references. Finally, the 11-reference adapted condition seems to have no outlier speakers. (We have also found that adaptation to speakers with foreign accents works about as well as on native speakers.) Thus, one of the powers of speaker adaptation, in general, and this method in particular, is that by adapting to the target speaker, we can often normalize out differences that are not covered even by a large number of speakers.

Finally, in the table below we compare cross-speaker recognition and single-reference adaptation when the reference speaker is the same, or opposite sex. As can be seen, there is essentially no sex bias remaining in the adapted model.

Summary of Results

In summary, we present a table of several different training conditions and algorithms, sorted by error rate. The data can be organized into four groups: cross-speaker, speaker-independent, speaker-adaptive, and multi-speaker/speaker-dependent. First, we observe that

Training Scenario	% Word Error
Cross-Speaker	46
Sex-Dependent Cross-Speaker	10
2 Speaker SI	10
12 Speaker SI	7
109 Speaker SI	6
1 Reference Adaptation – 30 min	6
1 Reference Adaptation – 2 hr	5
11 Reference Adaptation – 30 min	4
4 Reference Adaptation – 2 hr	4
12 Reference MultiSpeaker	2.5
Speaker-Dependent – 30 min	1.7

Table 4. Summary of results for different training scenarios. In all cases the models and parameters used were the same type

speaker-independent recognition has not progressed very far beyond sex-dependent cross-speaker recognition. (Note that we can build a 2-speaker-independent system simply by having 1 male and 1 female model in parallel.) The single-reference adaptation schemes are very economical alternatives to SI recognition, with the same performance. Adaptation from multiple references improves over the SI condition. Finally, there is still a large difference between the best speaker-independent or speaker-adaptive systems and speaker-dependent systems. Furthermore, it can easily be argued that the speaker-dependent performance is not nearly good enough for real systems. Thus, substantial work is still needed.

Conclusions

We have reviewed the typical uses of Hidden Markov Models for Continuous Speech Recognition. Good performance clearly requires that we incorporate basic speech knowledge within the model. We showed that the N-Best Paradigm provides a simple and efficient way of integrating speech recognition with natural language processing. The Word-Dependent N-Best Algorithm was shown to be as accurate as the exact algorithm, while requiring much less computation. We introduced the Forward-Backward Search Algorithm, which reduces the computation needed for the N-Best search by an additional factor of 40. These new search algorithms have made it possible to implement an integrated speech and natural language system that runs in real time on a commercially available workstation.

We characterized several different speaker adaptation techniques according to the training scenario required and the methods used. We showed that, contrary to our expectations, we can build a speaker-independent model using only two speakers, and that with a large amount of speech from 12 speakers the accuracy is essentially equivalent to that for a small amount of speech from 109 speakers. We also showed that separate training and smoothing of models from different speakers results in higher accuracy. This result will make speaker-independent recognition much more practical in terms of data collection effort, flexibility, and the accuracy for the frequent users. Finally, we described a method for adapting a speaker-independent model to a new speaker, using the new speaker-independent training paradigm described above. The resulting error rate is almost half that of the speaker-independent system.

Acknowledgement

This work was supported by the Defense Advanced Research Projects Agency and monitored by the Office of Naval Research under Contract No. N00014-89-C-0008.

References

[1] Aubert, X.L. (1990) "A Fast Lexical Selection Strategy for Large Vocabulary Continuous Speech Recognition", *NATO Advanced Study Institute on Speech Recognition and Understanding* Cetraro, Italy, July 1990.

[2] Bahl, L.R., Brown, P.F., de Souza, P.V., and R.L. Mercer (1986) 'Maximum Mutual Information Estimation of Hidden Markov Model Parameters for Speech Recognition.' *IEEE Int. Conf. Acoust., Speech, Signal Processing*, Tokyo, Japan, 1986, Vol 1 pp. 49-52, Paper No. 2.3.1.

[3] Bahl, L.R., Brown, P.F., de Souza, P.V., and R.L. Mercer (1988) "A New Algorithm for the Estimation of Hidden Markov Model Parameters", *IEEE Int. Conf. Acoust., Speech, Signal Processing*, New York, New York, April, 1988, pp. 493-496, Paper 11.2.

[4] Bahl, L.R., de Souza, P., Gopalakrishnan, P.S., Kanevsky, D., and D. Nahamoo (1990) "Constructing Groups of Acoustically Confusable Words". *Proceedings of the ICASSP 90*, April, 1990.

[5] Baum, L. (1972) *An inequality and association Maximization technique in Statistical Estimation for Probabilistic Function of Markov Processes*, Inequality, Vol III, 1972, pp 1-8.

[6] Bellegarda, J. and D. Nahamoo (1989) "Tied mixture continuous parameter models for large vocabulary isolated speech recognition" *IEEE ICASSP89*

[7] Brown, P. (1987) "The Acoustic-Modeling Problem in Automatic Speech Recognition" *PhD Thesis, CMU, 1987*

[8] Chow, Y-L. and R.M. Schwartz (1990) "The N-Best Algorithm: An Efficient Procedure for Finding Top N Sentence Hypotheses". *Proceedings of the DARPA Speech and Natural Language Workshop* Cape Cod, October 1989.

[9] Feng, M., Kubala, F., Schwartz, R., and J. Makhoul (1988) "Improved Speaker Adaptation Using Text Dependent Spectral Mappings", *IEEE ICASSP-88*, Apr. 1989, paper S3.9.

[10] Fissore, L., Micca, G., and R. Pieraccini (1988) "Very Large Vocabulary Isolated Utterance Recognition: A Comparison Between One Pass and Two Pass Strategies". *Proceedings of the ICASSP 88*, pp. 267-270, April, 1988.

[11] Furui, S. (1986) "Speaker-Independent Isolated Word Recognition Based on Emphasized Spectral Dynamics," *IEEE ICASSP-86, pp. 1991-1994*

[12] Furui, S. (1989) "Unsupervised Speaker Adaptation Method Based on Hierarchical Spectral Clustering", *IEEE ICASSP-89*, May 1989, paper S6.9.

[13] Gerstman, L. (1967) Classification of Self-Normalized Vowels", Conference on Speech Communication and Processing, Massachusetts Institute of Technology, 1967.

[14] Gillick, L. and R. Roth (1990) "A Rapid Match Algorithm for Continuous Speech Recognition", *Proceedings of the DARPA Speech and Natural Language Workshop* Hidden Valley, June 1990.

[15] Gupta, V.N., Lennig, M., and P. Mermelstein (1987) "Integration of Acoustic Information in a Large Vocabulary Word Recognizer", *IEEE ICASSP-87*, Apr. 1987, pages 697-700.

[16] Hattori, H., Nakamura, S., and K. Shikano (1990) "Supplementation of HMM for Articulatory Variation in Speaker Adaptation", *IEEE ICASSP-90*, Apr. 1990, paper S3.6.

[17] Huang, X.D. and M.A. Jack (1989) "Semi-continuous hidden Markov models for speech recognition" *Computer Speech and Language*. Vol 3, 1989

[18] Jelinek, F., and R.L. Mercer (1980) "Interpolated Estimation of Markov Source Parameters from Sparse Data" in E.S. Gelsema and L.N. Kanal (editor), Pattern Recognition in Practice, pages 381-397, North-Holland Publishing Company, Amsterdam, 1980

[19] Lee, K.F. (1988) "Large-Vocabulary Speaker-Independent Continuous Speech Recognition: The SPHINX System", PHD dissertation, Carnegie-Mellon University, Apr. 1988, CMU-CS-88-148

[20] Lee, K.F., Hon, H., and M. Hwang (1989) "Recent Progress in the Sphinx Speech Recognition System", *Proceedings of the DARPA Speech and Natural Language Workshop*, Morgan Kaufmann Publishers, Inc., Feb. 1989, pp. 125–130.

[21] Lowerre, B. (1977) "The Harpy Speech Recognition System", *Doctoral Thesis* CMU 1977.

[22] Makhoul, J., Roucos, S., and H. Gish (1985) "Vector Quantization in Speech Coding" *Proc. of IEEE, Vol. 73, No.11, pp.1551-1588*

[23] Mariño, J. and E. Monte (1989) "Generation of Multiple Hypothesis in Connected Phonetic-Unit Recognition by a Modified One-Stage Dynamic Programming Algorithm", *Proc. of the European Conf. on Speech Communciation and Technology, Paris, Sept. 1989, Vol. 2, pp. 408-411*

[24] Nakamura, S. and K. Shikano (1989) "Speaker Adaptation Applied to HMM and Neural Networks", *IEEE ICASSP-89*, May, 1989, paper S3.3.

[25] Peterson, G. and H. Barney (1952) "Control Methods Used in a Study of the Vowels", JASA - Vo. 24, No. 2 - March 1952, pp. 175-184.

[26] Price, P., Fisher, W.M., Bernstein, J., and D.S. Pallett (1988) "The DARPA 1000-Word Resource Management Database for Continuous Speech Recognition," *IEEE Int. Conf. Acoust., Speech, Signal Processing*, New York, NY, April 1988, pp. 651-654.

[27] Rigoll, G. (1989) "Speaker Adaptation for Large Vocabulary Speech Recognition Systems Using Speaker Markov Models", *IEEE ICASSP89* May, 1989, paper S1.2, pp. 5-8

[28] Rtischev, D. (1989) "Speaker Adaptation in a Large-Vocabulary Speech Recognition System", Masters thesis, Massachusetts Institute of Technology, Jan. 1989.

[29] Schwartz, R. (1971) "Automatic Normalization for Recognition of Vowels of All Speakers", BS thesis, Massachusetts Institute of Technology, June, 1971.

[30] Schwartz, R.M., Chow, O., Roucos, S., Krasner, M., and J. Makhoul (1984) "Improved Hidden Markov Modeling of Phonemes for Continuous Speech Recognition", *Proceedings of the ICASSP 84*, paper 35.6, March, 1984.

[31] Schwartz, R.M., Chow, Y., Kimball, O., Roucos, S., Krasner, M., and J. Makhoul (1985) "Context-Dependent Modeling for Acoustic-Phonetic Recognition of Continuous Speech", *Proceedings of the ICASSP 85*, pp. 1205-1208, March, 1985.

[32] Chow, Y.L., Schwartz, R., Roucos, S., Kimball, O., Price, P.J., Kubala, G.F., Dunham, M.O., Krasner, M.A., and J. Makhoul (1986) "The Role of Word-Dependent Coarticulatory Effects in a Phoneme-Based Speech Recognition System", *IEEE Int. Conf. Acoust., Speech, Signal Processing*, Tokyo, Japan, April 1986, pp. 1593-1596, Paper No. 30.9.

[33] Schwartz, R., Chow, Y.L. and F. Kubala (1987) "Rapid Speaker Adaptation using a Probabilistic Spectral Mapping", *IEEE ICASSP-87*, Apr. 1987, paper 15.3.1.

[34] Schwartz, R., Kimball, O., Kubala, F., Feng, M. Chow, Y., Barry, C., and J. Makhoul (1989) 'Robust Smoothing Methods for Discrete Hidden Markov Models", *IEEE ICASSP-89*, May 1989, paper S10b.9.

[35] Schwartz, R. and Y.L. Chow (1990) "The N-Best Algorithm: An Efficient and Exact Procedure for Finding the N Most Likely Sentence Hypotheses", ICASSP-90, April 1990, Albuquerque S2.12, pp. 81-84.

[36] Schwartz, R.M., and S.A. Austin (1990) "Efficient, High-Performance Algorithms for N-Best Search", *Proceedings of the DARPA Speech and Natural Language Workshop* Hidden Valley, June 1990.

[37] Schwartz, R.M., and S.A. Austin (1991) "A Comparison of Algorithms for Finding Multiple Sentence Hypotheses", to be presented at *IEEE ICASSP-91*, May 1991.

[38] Schwartz, R.M., and Austin, S.A., Placeway, P. (1991) "The Forward-Backward Search Strategy for Real-Time Speech Recognition", to be presented at *IEEE ICASSP-91*, May 1991.

[39] Shikano, K., Lee, K.F., and R. Reddy (1986) "Speaker Adaptation Through Vector Quantization", IEEE Int. Conf. Acoust., Speech, Signal Processing, Tokyo, Japan, April 1986, pp. 2643-2646, Paper No. 49.5.

[40] Steinbiss, V. (1989) "Sentence-Hypotheses Generation in a Continuous-Speech Recognition System," *Proc. of the European Conf. on Speech Communciation and Technology, Paris, Sept. 1989, Vol. 2, pp. 51-54*

[41] Tseng, P., Sabin, M., and E. Lee (1987) "Fuzzy Vector Quantization Applied to Hidden Markov Modeling", *IEEE ICASSP-87*, April, 1987, paper S15.5.

[42] Young, S. (1984) "Generating Multiple Solutions from Connected Word DP Recognition Algorithms". *Proc. of the Institute of Acoustics, 1984, Vol. 6 Part 4, pp. 351-354*

A 20,000 word automatic speech recognizer
Adaptation to French of the US TANGORA system

H. Cerf-Danon, P. de La Noue, L. Diringer, M. El-Beze, J.C Marcadet

IBM France Scientific Center
5 Place Vendôme, 75001 Paris, France

Introduction

In this article we describe the adaptation to the French language of the Tangora system, developed by the IBM Yorktown Speech Group for American English. This work is a continuation of the activity of our group which developed the PARSYFAL system [1], a 200,000 word speaker-dependent syllable based dictation system which word recognition accuracy is about 90%. The French Tangora system uses exactly the same hardware as its US counterpart. It is a single box autonomous system running on an IBM PC AT machine with 4 signal processing cards (Albert) and an INTEL 80386 processor [2]. It handles a vocabulary of 20,000 words and is designed to take down dictated press agency dispatches. The dictation and training are done in isolated word mode.

System Architecture

Signal processing

The signal, sampled at 20,000Hz, is pre-emphasized and 12 bit quantified. An FFT is performed every 10ms on a window of 512 samples (25.6 ms) and outputs a 20 dimensional vector representing the energy in 20 frequency bands between 200 and 8000 Hz. This first part, fully described elsewhere [3,4], is speaker independent. Each of the acoustic vector thus produced is then clustered in one of the 200 speaker-dependent labels computed from the speaker's training corpus. This sequence of labels (100 per second) is the input to the decoder.

The acoustic processing, running on one of the Albert cards is microphone independent. Hence the microphone can be changed during training and recognition without perceptible loss for the user.

Decoding Strategies

As described in [5], the decoding is carried out in 3 steps. In a first step, an acoustic Fast Match (FM) builds a list of candidate words for each sequence of labels. This step is phonetic: the input sequence of labels sent by the signal processor is evaluated in regards of all the different phonetics present in the dictionary. Each phoneme, as described in [6], is modeled by a hidden markov machine, and a complete word is modeled by a concatenation of such machines. Even though a faster sub-optimal strategy which looks for the most likely sequence of phonemes [7] has given very good results, it still seems more reliable to compare the phonetics of all words with the label sequence.

NATO ASI Series, Vol. F 75
Speech Recognition and Understanding. Recent Advances
Edited by P. Laface and R. De Mori

Secondly, the list of potential words ranked by likelihood which has been created by the Fast Match is refined with a Language Model (LM) component which uses the hypothesized word history. The words present in the FM-LM list are then sent to the acoustic Detailed Match (DM) step. It is not possible to recognize a word that is missing from this list.

This third step, the DM, is where all the words of the FM-LM list are compared one by one with the arriving sequence of labels and where the log probability of the most probable path of the stream of labels through each of those words states sequence is computed. For each word it uses its fenonic baseform [8], which is a sequence of 2 state hidden markov machines taken from an alphabet of 200 machines.

For each word w fenonically modelled by u we compute $p\,(f_1^n\,/u)$ where $f_1^n = f_1 f_2 ... f_n$ is the sequence of incoming labels. A Viterbi algorithm is used:

$$p\,(f_1^n\,/u) = \underset{a_1 a_2 ... a_n}{\text{Max}}\, p\,(a_1 a_2 ... a_n/u) \prod_{i=1}^{n} p\,(f_i \mid a_i)$$

where $a_1 a_2 ... a_n$ is a string of arcs, i.e. a path through the markov model of u. The equation is maximized on all possible paths through u.

The above probability is then combined with the FM and LM scores. A stack decoding algorithm [9] extends the most likely not yet extended branch of the tree made up of the succession of potential words. The FM, LM, DM are recursively applied on the most likely node.

Training

Tangora is a speaker dependent system: any new user has to go through an enrollment procedure which takes less than 1 hour during which the user records 110 sentences.

Training is carried out in 3 steps: a K-means clustering algorithm is performed on the acoustic vectors of the training to create 200 personalized clusters (prototypes). Then, the output and transition probabilities are estimated for the fenonic model used in the DM [10]. Finally, the output and transition probabilities for the Phonetic Model used in the FM are computed. The phonetic machines [11] are 7 states HMM.

Vocabulary and Language Model

A vocabulary of 20,000 words has been collected from a corpus of AFP dispatches over a 13 month period. The corpus contained 35M words, including 215,500 different ones. The static coverage is 96.3%.

Language Model

The LM is a trigram LM which is well adapted to English but less so to French. Nearly 10% of the words have been put in classes and all the words in a given class are regarded as the same entity. A word can be in only one class. We have classes for first names, family names, cities, months, etc. A deleted interpolator model [12] combining trigrams, bigrams, unigrams and uniform probabilities, is used to smooth the probabilities.

$$p(w_3/w_2, w_1) = \lambda_t p_t(w_3/w_2, w_1) + \lambda_b p_b(w_3/w_2) + \lambda_u p_u(w_3) + \lambda_0 p_z$$

$p_t(w_3/w_2, w_1)$ is the trigram probability, $p_b(w_3/w_2)$ the bigram probability, $p_u(w_3)$ the unigram probability and. p_z the zerogram probability.

The smoothing statistics are collected from a held-out corpus. The λ are computed iteratively by reestimation: for instance λ_t is iteratively computed as:

$$\lambda'_t = \sum \lambda_t \frac{p_t(w_3/w_2, w_1)}{p(w_3/w_2, w_1)}$$

where the sum is on all trigrams w_3, w_2, w_1 of the held-out corpus and the probabilities p_t, p_b, p_u are computed from the training corpus.

We collected the trigrams from a corpus of 35M words out of which we obtained 7.2M different trigrams, and used a held-out corpus of 5.3M words containing 1.8M different trigrams. However, for the coverage and all the decoding computations we used the trigram probabilities p_t only if the trigram occurred more than 1 time in the training. There were only 2.5M such trigrams.

Text	Unigram	Bigram	Trigram
Training	96.7%	93.6%	80 %
New Text	96.3%	90.6%	70.3%

Table 1: Uni, Bi and Trigram coverage.

Table 1 presents the coverages on the training text as well as on a new text (another month of news report of AFP).

Vocabulary

Each word of the vocabulary is associated to its spelling, its phonetic transcription(s) and its fenonic baseform(s). The phonetics of the words were taken from a 200,000 word dictionary developed by the group. (Each word can have several phonetic variants.) The words that were missing from that dictionary (surnames, some inflected words, etc) were automatically phonetized. The phonetization had to be reread since there were a lot of proper names from different linguistic origins.

One of the major problems of the French language as far as a trigram LM is concerned, is the huge number of homophones. For example, there is rarely a difference in pronunciation between singular and plural nouns and the same situation sometimes arises for masculine and feminine adjectives and past participles.

This would not be a problem if we could get a very high trigram coverage but trigrams which have not been met in the training text are often pronounced. In these cases the LM uses bigram or unigram probabilities to make a decision; this is likely to lead to grammatical errors especially concerning the agreement of gender and number.

The following example illustrates this situation. If you said: "**Barbara est venue**" ("Barbara has come") and if we have not met this trigram in the training, knowing that in French the trigram "**Barbara est venu**" is impossible (hence never met), the choice between the two above homophones trigrams will be done from the bigram probabilities, if

$p_b($ " **est venu** " $) > p_b($ " **est venue** " $)$
then the decoder will choose "**Barbara est venu**" instead of "**Barbara est venue**" .

In order to reduce these errors the dictionary is built incrementally to make sure that it contains the masculine, feminine, singular and plural forms of the nouns, adjective and past participles, at least when those forms have the same phonetic and have been encountered in the corpus more than a given number of times. The consistency of the phonetic variants of all the words with their grammatical variants is also checked.

Markov Machine Modeling

Since a lot of words are homophones, there are only 13,000 different phonetic forms out of a total of 20,000 words in the dictionary. Each phonetic is associated with its sub-word fenonic baseform used at the Detailed Match step. For each of the 13,000 different phonetic phone sequences (phonetic baseform) a corresponding fenonic phone sequence (fenonic baseform) is built up using a dictionary.

This dictionary gives the sub-word fenonic baseform for all possible phonetic phones according to a classification of their acoustic environment (6,312 classes). The dictionary has been built using 453 phonetically balanced sentences uttered by 8 speakers in isolated syllable mode and a list of 6,000 syllables already used in the PARSYFAL project. From this corpus about 400,000 phones were extracted and for each of the classes, a sub-word fenonic baseform was computed.

Hence it has been possible to build both a phonetic and a fenonic baseform for all the words of the dictionary without having to pronounce them all, i.e. it was not necessary to record the 20,000 word dictionary.

Results and Prospects

Speaker	Subst	Del	Ins	Error-Rate
MEB	62	0	1	6.2%
AMD	48	2	2	5.0%
LDI	43	0	1	4.3%
JCM	41	0	1	4.1%
PLN	59	3	3	6.2%
BYM	55	0	1	5.5%
MSA	55	6	2	6.1%
SAD	54	1	4	5.5%

Table 2. Substitution, Deletions, Insertions and Error Rate

With a 110 sentence long training text (1800 words) tested on 50 sentences (1000 words) the word error rate on 8 speakers is 5.4% (Deletion + Substitution / Number of words in text). There were 7 male speakers and among them 2 swiss one and 1 female speaker. The table beside details these results.

Error analysis

It appears that approximately 20% of the errors are acoustic, 60% are due to the fact that it is a tri**GRAM** LM, 20% are due to the fact that it is a **TRI**gram LM. It seems that tripos [13] , [14] or trilemmas [15] could be used to correct a consistent percentage of grammatical agreement spelling mistakes.

An additional advantage of a triclass model is that it enables the inclusion and deletion of a word from the vocabulary. Some other adaptive method [16] could also be studied.

The phonetic model could be improved by increasing the number of phonetic machines. We currently use 40 phonemes: the use of contextual phones [11] would improve the FM list. Even though the FM is already pretty good, it is important to refine it since a high likelihood given by the Fast Match to a word will decrease the decoding time by simplifying the choices of the stack decoding and enabling a reduction of the size of the FM list.

Acknowledgements

We are very much indebted to all the IBM Yorktown Speech Group and especially to S. De Gennaro for his general help on understanding the architecture of TANGORA.

We received a great deal of help from the IBM Rome Speech Group and especially M. Ferreti and G. Maltese who provided us with the programs to collect the vocabulary, the language model and initialize the construction of the fenonic baseforms.

Bibliography

[1] H. Cerf-Danon, A.M. Derouault, M. El-Beze, B. Merialdo: **Speech recognition in French With a Very Large Dictionary,** *Eurospeech 89 Vol 2, September 1989, pp 150-153.*

[2] A. Averbuch et al.: **An IBM-PC based large vocabulary isolated utterance speech recognizer,** *Proceedings of ICASSP, Tokyo April 1986, pp 53-56.*

[3] J. Cohen: **Application of auditory Model to Speech Recognition,** *Digital Signal Processing Workshop, Chatham MA*

[4] A. Nadas, D. Nahamo, M. Picheny: **Speech recognition using noise-adaptive prototypes,** *Proceedings of ICASSP, New York, April 1988, pp 517-520.*

[5] A. Averbuch et al.: **Experiments with the TANGORA 20,000 word speech recognizer,** *Proceedings of ICASSP, Dallas, April 1987, pp 701-704.*

[6] H. Cerf-Danon, A.M. Derouault, M. El-Beze, B. Merialdo, S. Soudoplatof: **Speech Recognition Experiment with a 10,000 word dictionary,** *NATO Advanced Institute on Pattern Recognition, June 18-20 1986 pp 203-209.*

[7] S.E. Levinson, A. Ljolje, L.G. Miller: **Continuous Speech Recognition from Phonetic Transcription,** *Proceedings of ICASSP, Albuquerque, April 1990, pp 93-97.*

[8] L.R Bahl, P.F. Brown, P.V. de Souza, R.L. Mercer , M. A. Picheny : **Acoustic Markov Models used in the TANGORA speech recognition system,** *Proceedings of ICASSP, New York, April 1988, pp 497-501.*

[9] F. Jelinek: **The development of an experimental discrete dictation recognizer,** *Proc. IEEE, Vol 73, No. 11, March 1985, pp 158-169.*

[10] B. Merialdo: **Speech Recognition with very large size dictionary,** *Proceedings of ICASSP, Dallas, April 1987, pp 364-367.*

[11] A.M. Derouault: **Context-Dependent Phone Markov Models for Speech Recognition,** *NATO Recent Advances in Speech Understanding, 1988 Vol F46 pp 172-175.*

[12] F. Jelinek, R. Mercer: **Interpolated estimation of Markov source parameters from sparse data,** *Workshop on Pattern Recognition in Practice, E.S. Gelsema and L.N. Kanal, Eds, North-Hollland, Amsterdam.*

[13] A.M. Derouault, B. Merialdo: **Language Modeling at the Syntactic Level,** *7th International Conference on Pattern Recognition, August 1984, Montreal.*

[14] A.M. Derouault, B. Merialdo: **Natural Language modeling for Phoneme-to-Text Transcription,** *IEEE Trans on PAMI, PAMI-8 No 6, November 1986.*

[15] M. EL-Beze, A. M. Derouault: **A Morphological Model For Large Vocabulary Speech Recognition,** *Proceedings of ICASSP, Albuquerque, April 1990, pp 577-581.*

[16] A.L. Gorin, S.E Levinson, L.G. Miller, A.N. Gertner, A. Ljolje , E. R. Goldman: **On Adaptive Acquisition of Language,** *Proceedings of ICASSP, Albuquerque, April 1990, pp 601-605.*

Automatic adjustments of the Markov models topology for speech recognition applications over the telephone

Denis Jouvet, Laurent Mauuary, Jean Monné

Centre National d'Etudes des Télécommunications, LAA/TSS/RCP,
Route de Trégastel, 22300 LANNION, FRANCE

Abstract : This paper presents some automatic adjustments of the structure of Markov models with the objective of either reducing model complexity, or improving recognition performance. These modifications are tested on a 36 word vocabulary recorded by more than 500 speakers over the telephone network. The reduction of the model complexity is carried out by merging the similar gaussian functions using an iterative procedure. A 40 % reduction of the number of gaussian functions is obtained on word based models without altering recognition performance.

The improvement of the recognition performance is obtained by dynamically expanding the Markov model. This is achieved mainly by splitting the gaussian functions which make the highest contribution to the observation probability of the training set and by discarding the infrequently used transitions. After some iterations (involving the splitting and discarding operators) a 30% reduction of the word error rate is achieved using pseudo-diphone based models.

Keywords : Speech recognition; Markov modelling; Probability density functions (pdf); Pseudo-diphone units; Modification of the model topology;
Merging gaussian pdf's; Splitting gaussian pdf's; Discarding transitions.

Introduction

Although the structure of Markov models is usually defined a priori, it is one of the factors which influences speech recognition performance. Firstly, this paper will compare two basic structures, and then describe some topology adjustment procedures used either to reduce model complexity, or to improve recognition performance.

When using standard word models it appears that, at least for isolated word recognition, increasing the size of the models (i.e. number of states and pdf's) yields better performance. However, this also increases the computation time, due to the number of observation probabilities that must be computed for each frame. In this context, the reduction of the model complexity is useful, especially for real time applications, and is obtained by merging similar gaussian pdf's.

NATO ASI Series, Vol. F 75
Speech Recognition and Understanding. Recent Advances
Edited by P. Laface and R. De Mori

When using sub-word basic units, it is difficult to increase the a priori size of the acoustical models (they may become too long). So, in order to improve recognition performance, we developed an algorithm that dynamically expands the Markov models, based on the splitting of the gaussian pdf's which make the highest contribution to the observation probability of the training set, and on the discarding of infrequently used transitions.

System overview and data base

At the CNET, we work mainly with small vocabularies, speaker independent speech recognition systems for applications over the telephone network [1]. 6 Mel frequency cepstral coefficients, the energy and its derivative, are computed every 16 ms to obtain the input vectors. The observation probabilities are represented by gaussian functions with diagonal covariance matrices, and are linked to the transitions of the Markov chain. We also use a fully compiled network [2] which includes initial and final silence models, and allows various kinds of modelling : either word units, or sub-word units such as phonemes or diphones.

In this study we use a database consisting of 36 french words recorded over the telephone network (mainly long distance lines). The speech data has been collected from 513 speakers representing different regional accents. This data base is divided into two parts. Half the data base (8354 utterances) is used to train the parameters of the models. The other half consists of 8402 utterances and is used for testing recognition performance.

Words versus pseudo-diphones

The simplest models that can be developed are those which consist in taking each word as a single unit represented by a standard model. When the size of each basic model increases, the word error rate decreases accordingly, as shown on figure 1 (where M13 and M27 design word based models containing respectively 13 and 27 states and pdf's per word; the other models have respectively 15, 17, 19, ..., states and pdf's per word). The largest model (M27) provides the best performance, with approximately a 2.7 % error rate.

The other basic units we use in this experiment are called pseudo-diphones. Each unit corresponds to either a central portion of a phoneme, or a transition between two phonemes, or a sequence of strongly coarticulated phonemes treated as a single unit (cluster). Using these pseudo-diphone units, the digit "0" in French is described as the transition between silence (/#/) and /z/, followed by the central part of /z/, followed by the transition between /z/ and /e/, followed by the central part of /e/, and so on : 0 → #~z . z . z~e . e . e~r . r . r~o . o . o~#

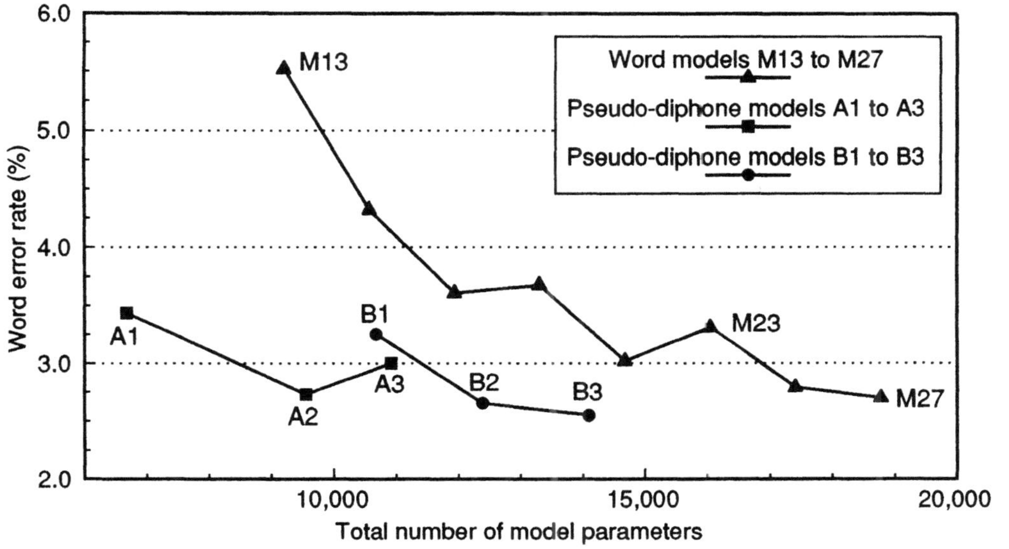

Figure 1 . Word error rate on the test set for 2 different sets of basic units : words and pseudo-diphones. The horizontal axis refers to the total number of parameters of the Markov models, i.e. the transition probabilities and the mean and standard deviation values of the gaussian functions

Various basic structures were defined for the pseudo-diphones units. Figure 2 shows the structures used for the A2 and B2 models. The A1 and A3 models were designed using respectively shorter and longer basic structures than A2. As it appears from the results shown on Figure 1, the length of the basic models cannot be too long (results for A3 are somewhat worse than for A2). The B1, B2 and B3 models were defined with respectively 1, 2 and 3 transitions in parallel (associated to different gaussian pdf's) between the states of the phoneme and cluster units. Thus we have something similar to a mixture of gaussian functions [3] with 1, 2 or 3 components. Figure 1 shows an improvement from B1 to B2, but almost no difference between B2 and B3.

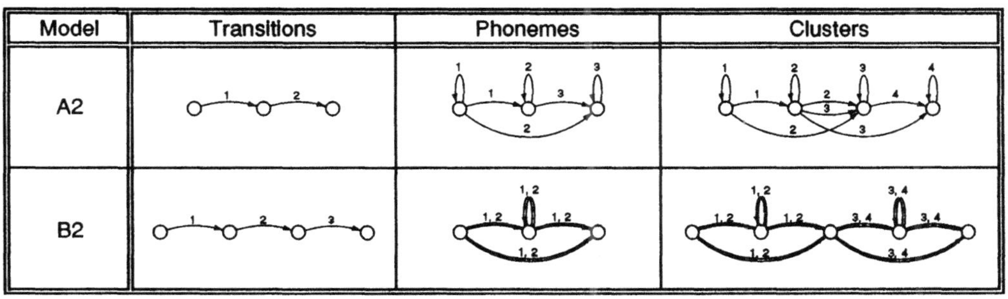

Figure 2 . Representation of the structure of the acoustical models associated to the pseudo-diphones units for the A2 and B2 models. The figures on the transitions describe the tying of the gaussian pdf's inside the basic models

Compression of the word based models

In the case of word basic units, we noticed that increasing the size of models improves the recognition performance. But this also increases the computational load for the recognition process (991 gaussian pdf's for the model M27). However, we can assume that if a given sound is common to several words, the gaussian functions modelling this sound should also be very similar, and that they can be merged without degrading the recognition performance.

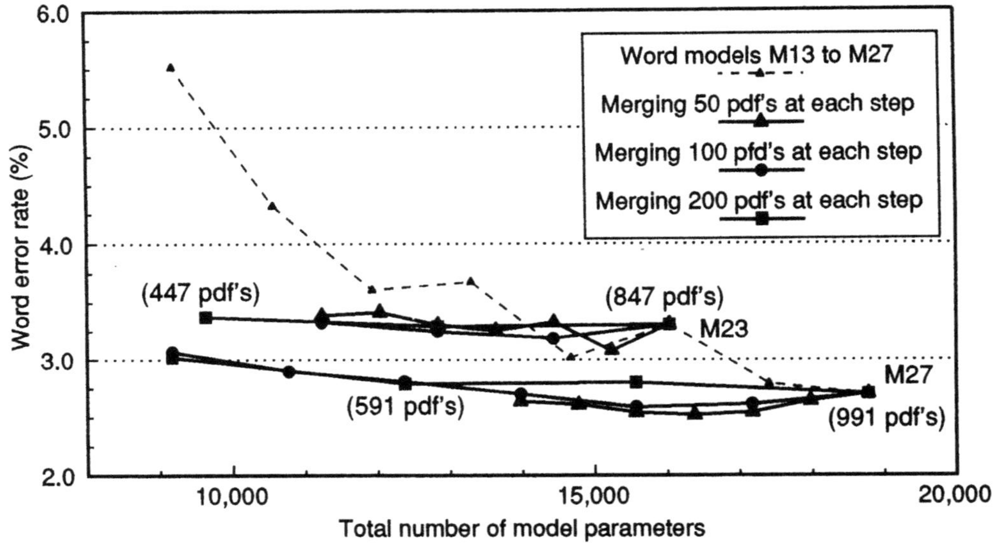

Figure 3. Evolution of the word error rate during the compression of the M23 and M27 word based models. Several compressions, involving the merging of 50, 100 and 200 pdf's at each step, have been iteratively applied

Thus, we have developed an iterative procedure which consists in merging the two most "similar" gaussian pdf's at each step until a criteria is reached. In order to find out the similarity between two gaussian functions, we define the *distance* between two gaussian functions G_i and G_j as the decrease of the logarithm of the observation probability of the training set when they are merged in a single gaussian function G_{ij}^* :

$$\mathrm{D}(G_i, G_j) = \sum_n \mathrm{Log}\,\mathrm{P}(X_n \mid \{G_1 \cdots G_i \cdots G_j \cdots G_N\})$$

$$- \sum_n \mathrm{Log}\,\hat{\mathrm{P}}(X_n \mid \{G_1 \cdots G_{i-1} G_{i+1} \cdots G_{j-1} G_{j+1} \cdots G_N G_{ij}^*\})$$

If we assume that all the frames previously affected (during the Viterbi training) to the gaussians G_i and G_j are now affected to G_{ij}^*, and that all the other alignments are not modified, we can easily compute the parameters of the new gaussian G_{ij}^* and estimate the new probability

$\hat{P}(\cdots)$. This enables the formulation of the distance between two gaussian pdf's as a function of the mean vectors and covariance matrices of the gaussians, and of the number of frames affected to each of them during the training procedure (Viterbi alignment).

The word error rates during the compression of the two word based models M23 and M27, reported on figure 3, show that we can greatly reduce the size of the models by merging up to 40 % of the gaussian functions while at the same time retaining similar recognition performance. Another important point is that we achieved comparable performance when merging either a large number of pdf's (say 200) or iterating twice the merging of 100 pdf's or four times the merging of 50 pdf's.

Expansion of pseudo-diphones based models

In the case of pseudo-diphone basic units, it is difficult to increase the a priori size of the acoustical models (they should not be too long). For this reason, we have developed an algorithm involving the splitting the pdf's which make the highest contribution to the probability of observing the training data, and the discarding of the transitions which are infrequently used.

Splitting some gaussian pdf's

Unfortunately, we have no easy way of estimating the improvement resulting from the splitting of one gaussian pdf. Of course, it would be impractical to split all of them. So we sort out the gaussian pdf's according to their contribution to the total observation probability of the training set, then we split the desired number of pdf's at the top of the list. It is worthwhile mentioning that the gaussians with the largest contributions are generally those which are the most often used.

The splitting of a gaussian G consists in replacing every transition linked to this pdf G by two transitions respectively linked to the new gaussian functions G^+ and G^-. These two functions are created from G by keeping the same covariance matrix (diagonal) and by modifying the mean vector by adding a small random vector. For the 6 pseudo-diphones models defined previously, the splitting of 100 gaussian pdf's leads to a reduction of 10 % to 20 % of the word error rate. However, if we iterate the splitting procedure further, the subsequent steps do not lead to similar improvement for all models.

Discarding transitions

After having applied the splitting procedure, several transitions, linked to different gaussian functions, may occur between two given states of the model. Assuming that they are not all necessary, we have elaborated a procedure to discard the transitions which are infrequently used on the training set (during the Viterbi alignments).

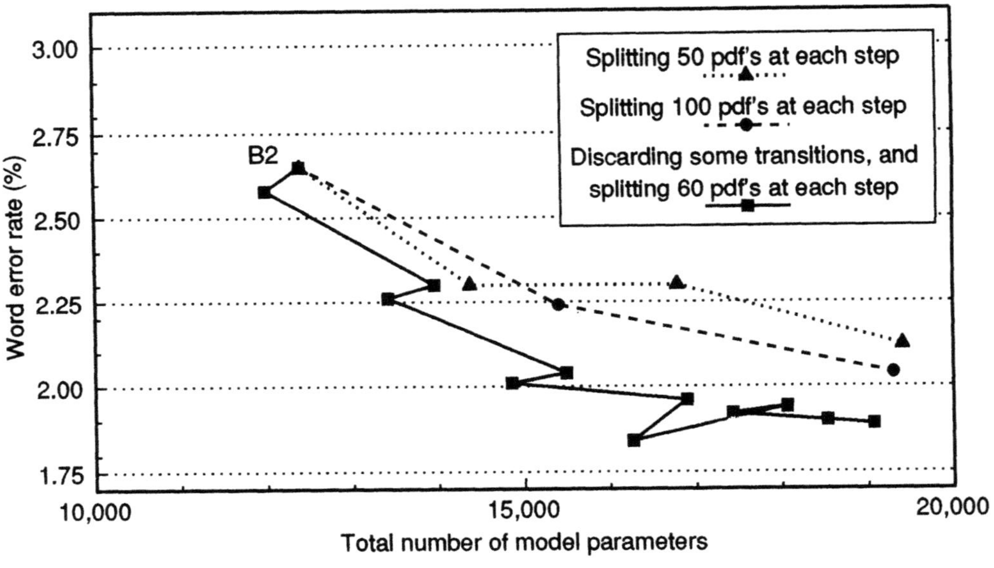

Figure 4. Comparison of several expansion procedures applied to the pseudo-diphone based model B2. The solid line shows the evolution of the error rate after each modification, that is discarding infrequently used transitions and splitting 60 pdf's. The two other expansions (dotted and dashed lines) involve only the splitting operator

The combined use of these two operators results in the largest reduction of the word error rate, as shown on figure 4, which compares several expansions of the pseudo-diphone based model B2. The expansion involving both the splitting and the discarding operators leads to a noticeable reduction of the word error rate (about 30% for the three models B1, B2 and B3).

Conclusion

We have developed three operators (merging, splitting and discarding) in order to automatically modify the topology of Markov models. Although some work still remains to be done in order to define the optimal number of gaussian pdf's that should be merged or split, these operators have already proven effective. In the case of word based models, we can substantially reduce the number of gaussian functions without altering performance. And, in the case of pseudo-diphone based models, the expansion leads to a noticeable improvement in recognition performance.

References

1. Gagnoulet, C. : "*Speech recognition over the telephone : experiments in France*"; Voice Systems Worldwide 1990 conference, London, May 1990, p173-177.
2. Jouvet, D., Monné, J., Dubois, D. : "*A new network-based speaker-independent connected-word speech recognition system*", Proc. IEEE Int. Conf. ASSP 1986, Tokyo, April 1986.
3. Juang, B. H., Rabiner, L. R., Levinson, S. E., Sondhi, M. M. : "*Recent developments in the application of hidden Markov models to speaker-independent isolated word recognition*", Proc. IEEE Int. Conf. ASSP 1985.

Phonetic Structure Inference of Phonemic HMM

Alessandro Falaschi

INFO-COM Dept., La Sapienza Univ., Via Eudossiana 18, 00184 Roma, ITALY

Abstract: An information based, two-stage DP procedure is proposed for phonemic HMM structure inference. It operates on the phonetic description of speech provided by means of an ergodic HMM. The resulting HMM, while comprising a quite large number of states, share the observation densities originated by the EHMM.

Keywords: Phonemic HMM, Structure Inference, Densities Tiing

Introduction

Although the HMMs used for speech recognition basically are finite state automata, their structure generally has only weak relationships with the acoustical structure of the linguistic units being modelled. Intuitive feelings suggest that more realistic models may improve recognition systems performance. Logical separation of the acoustic (phonetic) layer from the linguistic (phonemic) one agrees with the linguistic theories about the distinct roles of phonetic and phonological sciences. So, phonemes should be described not only by their acoustical counterparts, but also by the respective temporal and combinatorial structure pattern. Such dichotomy is providential, because it allows to delimitate the HMM phonetic description problem to the acoustical domain, and not on the amount of linguistic units to be modelled. It must be acknowledged that previous attempts in achieving this goal were based either on semi-continuous HMM [8] or on clustering of HMM densities [5].

Somewhere else in this volume [1] the development of an Ergodic HMM (EHMM) of speech is reported, and its ability in obtaining a segmentation of speech in a finite set of phonetic events documented, each of them being defined by an AR processes and a set of transition probabilities towards the other phonetic events. The present paper uses as phonetic (acoustic) units the segment labels obtained by means of Viterbi alignment of speech against a 64 states EHMM. The label sequences are collected for each occurrence of the same phoneme, and a transition diagram accounting for the main characteristics of such sequences is derived by a two-stage Dynamic Programming procedure. The parameters of the derived HMM structures are re-estimated by a Viterbi Fw-Bw scoring procedure, adopting an intrinsic state

NATO ASI Series, Vol. F 75
Speech Recognition and Understanding. Recent Advances
Edited by P. Laface and R. De Mori

duration model [2], and the resulting models used in a phonemic recognition experiment, achieving about a 70% phonemic recognition score.

The Structure Inference Procedure

This is a quite articulated procedure, each step dealing with a particular aspect of the inference process. The key points of the procedure are the two DP stages. The first provides to align the input sequences to the same length, and accounts for the acoustical (spectral) characteristics of the labels. In between the two a generalization procedure is performed, which collapses together the identical labels which have been aligned to the same instant. The second DP stage performs the selection of the more relevant label sequences for the phoneme, on the basis of an information criterion. Finally, the acoustic symbols occurring along the selected sequences are labelled with state numbers after having checked for the absence of circuits in the graph which is being built.

Data Pre-Processing

Let us indicate as $S_k(n)$, n = 1, ..., L_k, the EHMM state labels for the nth frame of the kth occurrence of a phoneme. All these sequences will be used for the structure inference, but their order of analysis may affect the results. For this reason, sequences are ranked according to a combined criterion which accounts for both the sequences probability and their length deviation from the average. Counting of the labels occurrence allows to evaluate the amount of information carried by label $S_k(n)$ as

$$I_k(n) = - \mathrm{Log}_2\, p(S_k(n)) \quad (1)$$

so that the mean information I_{km} carried by sequence k can be evaluated, and the score on which the sequences ranking is based can be (euristically) defined as

$$\mathrm{Score}(k) = I_{km}*(1+\mathrm{abs}(L_k-L_m/2)/L_m) \quad (2)$$

in which $L_k(n)$ and L_m are the kth sequence length and the mean sequence length over all the k.

First DP Stage

This step iteratively align all the S_k sequences to the same length, which is the length of the best sequence as resulting from (2). Each sequence is aligned against all the preceding ones, by a procedure which can be called a Projected-three-dimensional DP. The adjective Projected is due to the fact that, although one is dealing with a one-to-many DP, only two indexes are involved, being the previous sequences yet warped to the same length. Thus, the alignment of sequence $S_k(n)$ against the equal-length sequences

$S_h(m)$ is done by means of the accumulated distances

$$D(S_k(n),m) = d(S_k(n),m) + \min \{ D(S_k(n-1),m-1), \quad D(S_k(n),m-1) + C_r(S_k(n)), D(S_k(n-1),m) + C_o(S_k(n)) \} \tag{3}$$

where $C_r(S_k(n))$ and $C_o(S_k(n))$ are the repetition and omission costs for the label $S_k(n)$ defined as

$$C_r(S_k(n)) = \alpha \, |n-L_k/2| ; \quad C_o(S_k(n)) = \beta \, 2^{-I(n,k)} \tag{4}$$

in which α and β are two tuning parameters, and $D(S_k(n),m)$ is the minimum Likelihood ratio distortion between label $S_k(n)$ and those labels aligned at time m, as given by

$$d(S_k(n),m) = \min_h \{ d_{LR}(S_k(n),S_h(W_h(m)) \} \tag{5}$$

in which $W_h(m)$ is the warping path that minimizes (3) for sequence S_h. Moreover, the analysis of $W_k(n)$ allows to keep track of how many frames have been associated to the same time position "m", being it a fractional value in the case of label repetitions.

Generalization Step
At this point, the matrix built by the aligned label sequences is collapsed in a Trellis, T(n,m), in which the columns (index m) represent time instants, and all the occurrences of the same label (belonging to different sequences) warped to the same time are collapsed into a single entry, so that each label occurs only once for column.

During this process, two other data structures are kept updated: the trellis nodes Count Trellis C(n,m) which reports about the number of frames associated to each node, and the Incidence Matrix IM(n,m,j) whose elements indicates if the nodes (n,m) and (j,m-1) are connected by any of the training sequences. Finally, a Partial Count array PC(m) stores the sum over n of C(n,m), thus indicating the total number of frames aligned at time m. Fig. 1 a) and b) reports the resulting aspect of T(n,m) and C(n,m) for phoneme /ʃ/.

Second DP Stage
This stage performs the pruning of the derived data structures, in order to infer a model responsible only for the main phenomena occurring among the training data. This feature may be obtained either by adding states and transitions to "tiny" structures [5], or subtracting elements to "huge" ones [6]. The approach here experimented is somewhat different, being it based on an iterating "add-and-delete" philosophy. At each cycle, the "best" path which connects the left to the right side of the Reduced Trellis is searched, and then its "worse" transition marked in the Incidence Matrix, in order to avoid future retrieval of the same path. How

the "best" path and the "worse" transition are defined?

Defining $I(n,m)$ as the information borne by observation of label n at column m of the Trellis, given by

$$I(n,m) = - \text{Log}_2 \ (T(n,m)/PC(m)) \tag{6}$$

the minimum accumulated information in reaching each node is given by the DP recursion

$$D_F(n,m) = I(n,m) + \min_{j:IM(n,m,j)=1}\{ D_F(j,m-1) \} \tag{7}$$

in which the incidence matrix constraints the allowable node sequences as those occurred among the training data and not deleted yet. The path minimizing (7) is also a maximum probability path, as it accounts for the more frequent labels sequence. The method can be repeatedly applied, if one transition of the last path selected is deleted. A method can be deleting the transition for which the information gap is greater [3], being it the less probable along the best path. A better procedure may be outlined after having realized that an entire set of paths become unavailable after deletion of a transition, being it shared among several training sequences. So, for each transition occurring along the best path, the second minimal information path is searched, and its value compared against that obtained for the other transitions. The (second) minimal information path whose overall information is the largest will thus indicate which transition to delete.

It remains to clarify how to compute the second minimal information path scores: this can be done by a Forward-Backward computation mechanism, in which (7) constitutes the Fw equation, being the Bw given by

$$D_B(j,m-1) = \min_{n:IM(n,m,j)=1} \{ I(n,m) + D_B(n,m) \} \tag{8}$$

The sum $D_F(n,m) + I(j,m+1) + D_B(j,m+1)$ gives the minimum accumulated information value among all the paths sharing the node pair (n,m)-(j,m+1). In order to eliminate from comparison the actual best path, eq. (7)-(8) should be evaluated twice, the second time having settled to infinity the information for the starting-ending node of the best path. Combination of the second best ending with the first best start and vice versa will solve the problem. The outlined procedure is iterated until a stop criterion is met, being it based on the amount of original information which is retained by the actually selected paths.

Structure and Model Building

The second DP stage still not resolves the problem of deriving a proper model topology. In fact, usable HMMs require the absence of loops spanning chains of states, and this fact must be checked before proceeding to build the model. This control basically consists in assigning the same state number to different trellis nodes only if they are connected by a transition covered by one of the paths selected at the previous stage, they are labelled with the same acoustical symbol, and no other (selected) path connects the two. Fig. 1 c) reports the state number assignment for phoneme ʃ . Finally, a transition matrix is built on the basis of the state number assignment and the selected paths transitions.

Experimental Set-Up

DB Composition and Signal Analysis

The experiments have been performed on a set of 540 phonetically balanced, phonemically segmented, isolated words uttered by a single speaker. A 64 states EHMM [1] has been trained on such data, and then used for derivation of the finite alphabet acoustical segmentation which feds the inference procedure. The phonetic HMMs inferred were built by a total of 733 states and 1397 transitions, but only 64 (tied) observation densities. Finally, the EHMM used autoregressive densities, while the HMM recognizer was based on MEL-Cepstrum coefficients, so that a spectral representation conversion was necessary.

HMM Re-Training and Recognizer Characteristics

The phonemic HMM parameters were reestimated by a Viterbi forward-backward scoring procedure [4]. In particular, observation density parameters (means and diagonal covariances) were kept tied among all the HMMs sharing the same acoustical label, whereas transition probabilities were state-dependent. The recognizer used also a nice duration model [2], and its parameters (mean and minimum duration) were also estimated as state-dependent. Finally, those transitions which resulted unreestimated were deleted, so that the recognition experiments have been conducted on models containing a total of 1068 transitions.

Results

Table I summarizes the results obtained in these very preliminary experiments. The recognizer was driven by a stochastic phonotax [7] with perplexity 6.23, and the performances scored against either the full phonetic transcription, or a reduced transcription in which some fine phonetic detail were superseded.

Discussion and Conclusions

Although the results given in Tab.I may seem quite poor, it should be remembered that the system is at its very early stage of development, and it uses only 64 observation densities. This number should be augmented in order to improve the spectral resolution for the acoustic space.

N. of Units	%Rec	%Ins	%Del	GER
37	62.9	7.9	9.8	41.6
27	69.6	8.6	10.5	35.9

Tab.I - Preliminary experiments phonemic recognition performances

References

[1] - P.Pierucci, A.Falaschi, M.Giustiniani, "Phonetic Units and Phonological Constraints representation by EHMM", this volume

[2] - A.Falaschi, "Continuosly variable Transition Probability HMM for speech recognition", this volume

[3] - A.Falaschi, P.Pierucci, "Some experiments on HMM Structure Inference", Proc. EUSIPCO 90, Barcellona, SPAIN

[4] - D.Jouvet,J.Monné,P.Dobois, "A new network-based speaker-independent connected word recognition system", Proc. ICASSP-86, Tokyo

[5] - L.R.Rabiner,C.H.lee,B.H.Juang,J.G.Wilpon,"HMM Clustering for Connected Word Recognition", Proc. of ICASSP 1989, Glascow, UK

[6] - F.Casacuberta,E.Vidal,B.Mas,H.Rulot, "Learning the Structure of HMM's through Grammatical Inference Techniques", Proc. of ICASSP 1990, Albuquerque, MN, USA

[7] - A.Falaschi, "A Word-like Phonetic Sequence Statistical Source Model for Large Lexicon Automatic Speech Recognition Systems", 1st ECST, Sept.1987, Edinburgh, UK

[8] - X.D.Huang,M.A.Jack, "Unified techniques for Vector Quantization and Hidden Markov Modeling using Semi-Continuous Models", proc. of ICASSP 1989, Glascow, UK

a) Labels Trellis

54	54	54	54	54	54	54	54	54	44
22	22	62	22	22	22	22	22	22	54
17	62	22	14	14	14	14	14	62	22
62	14	14	62					48	58
33	8	24						24	63
24	33							14	62
40	24								14
39									
14									
2									

b) Labels Occurrence Counts

13.0	8.0	15.0	11.0	9.0	9.0	8.0	9.0	6.0	8.0
9.0	5.0	5.0	4.0	4.0	4.0	5.0	4.0	4.0	4.0
3.0	4.0	3.0	3.0	2.0	2.0	2.0	2.0	3.0	4.0
2.0	2.0	2.0	1.0					1.0	3.0
2.0	2.0	1.0						1.0	1.0
2.0	1.0							1.0	1.0
1.0	1.0								1.0
1.0									
1.0									
1.0									

c) State Numbers Assignment

2	2	2	2	2	2	1	1	1	5
4	7	10	7	7	7	7	7	7	1
	10	7						9	7
12									3
11								6	8
									9

Fig.1 - Resulting aspect of
a) the collapsed labels trellis
b) the trellis node counts
c) the assigned state numbers

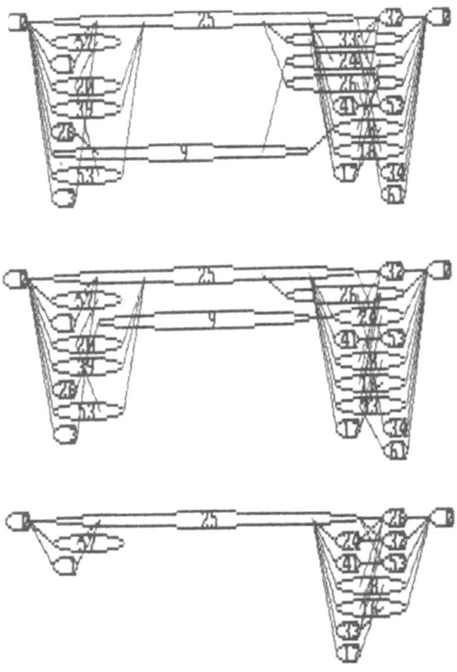

Fig.2 - Phonemic HMM inferred structure for various degrees of pruning for phoneme /d/

Phonetic Units and Phonotactical Structure Inference by Ergodic Hidden Markov Models

Piero Pierucci (*), Alessandro Falaschi (**), Massimo Giustiniani (*)

(*) IBM Italy, Rome Scientific Center, Via Giorgione 159, 00147, Rome, ITALY
(**) Univ.di Roma 'La Sapienza', INFOCOM Dpt., Via Eudossiana 18, 00184, Rome, ITALY

ABSTRACT

A general model for the representation of the spoken language is proposed. It is based on Ergodic Hidden Markov Models and allows the introduction of linguistically consistent sub-phonetic units, accounting for the stationary and transitional behaviour of the spoken message. The main characteristics of the proposed model are analyzed, and some applications to speech recognition and coding are presented.

KEYWORDS Ergodic Hidden Markov Models, Speech Recognition and Coding.

1. INTRODUCTION

Hidden Markov Models techniques are a well established approach to deal with the automatic speech recognition challenges. Since the first HMM introduction as the basis of an isolated word speech recognition machine [1], [2], a number of contributions has been given in order to improve recognition performances and reliability. The most challenging task, that still poses unsolved problems, is continuous speech recognition for large vocabularies. In order to cope with these problems a number of efforts were spent in the direction of modelling the speech production mechanism at the phonetic level through the use of HMM finite state automata. The objective is to find a limited set of units and suitable ways to combine them in order to obtain good recognition performances over a large vocabulary. The common practice generally forces linguistic units to be represented by a fixed number of acoustic events, related to the states of a left-to-right HMM used to represent that unit. Acustic realizations of the phonetic units of the language show, on the other hand, a high variability, mainly due to coarticulation effects among neighboring phones. Recently the use of subword units larger than phonemes, i.e. multiphone units as syllables and demisyllables, has been proposed [3] [4]. The major problems in the use of such subword units are consistency, i.e similar units should have similar acoustic characteristics, and trainability, i.e. the size of the training corpus must be kept as low as possible. Multiphone units did not solve these problems mainly because of their large number and their poor capability in representing contextual effects. The introduction of word dependent phones and context dependent phones [5][6] lead to better results in terms of consistency but the large number of these units still poses trainability problems. In this paper we propose a more general approach to

NATO ASI Series, Vol. F 75
Speech Recognition and Understanding. Recent Advances
Edited by P. Laface and R. De Mori

the representation of the acoustic-phonetic level of the spoken message. The objective is to split the representation of coarticulatory effects and the representation of linguistic-phonetic units, by introducing a general model of speech production wich accounts for the main acoustical events of the spoken language as well as for their relationships. The approach is based on the use of Ergodic Hidden Markov Models, i.e. an HMM that possess a fully connected transition matrix. This model is fairly general in that applications can be found in speech recognition as well as in speech coding. In this paper we will assess the EHMM capability to represent the spoken message in terms of phonemic labels that accounts for phonotactical constraints and acoustic events. The rest of the paper is organized as follows. Section 2 is dedicated to the definition of EHMM and describes the model parameters estimation from speech data. Section 3 reports a study of the main model's features from the point of view of phonotactical constraints representation. Section 4 concerns the state-identification problem, i.e. a study on the phonetic identity of the local models, through a Mutual Information approach. Section 5 reports about the application of EHMM to different speech processing areas. Section 6, finally, describes the open issues and draws the conclusions.

2. MODEL DEFINITION

An EHMM is a stochastic model whose structure consists of a set of K states $Q \equiv \{q_i\}$, a set of K observation densities $F \equiv \{f_i(O)\}$, a transition probability matrix $A \equiv \{a_{i,j} = Prob(q_i^{t-1}, q_i^t)\}$, allowing every state pair sequence, and an initial probability vector $\Pi \equiv \{\pi_i = Prob(q_i^{t=0})\}$, giving the a priori probability of being in each state at the beginning of the encoding process, with $1 < i,j < K$. The experiments here reported refers to a state cardinality of K ranging from 64 to 256. The state observation density function is autoregressive multivariate gaussian; natural candidate to represent the spectrum of an autoregressive source il the LPC representation, used in the form of the autocorrelation method. To each state is hence associated a set of M prediction coefficients and the prediction coefficients autocorrelation function, needed for the re-estimation of the HMM densities parameters. The local model is thus defined by the set of continuous probability density functions defined, for an observation O_t at time t, as :

$$F \equiv \{f_i(O_t) \quad = \quad - \quad 2\pi^{-N/2} \; e^{(-1/2\,\delta(R_{t,i}))}\} \quad , \delta(R_{t,i}) = r^{a_i}(0) r^t(0) \; + 2\sum_{j=1}^{M} r^{a_i}(j) \; r^t(j)$$

where $r^{a_i}(j)$ is the autocorrelation of LPC coefficients of state i, $r^t(j)$ is the autocorrelation of input speech frame [7]. Model parameters are estimated by means of the Baum algorithm [8], on about 700 phonemically compact words [9] pronounced by a reference speaker. When model size approaches values as K=64 or more thus requiring a larger training data set, some modifications to the classical Baum algorithm are required to obtain reliable estimates; we adopted a scaling technique following the approach described in [10]. Speech is sampled at a frequency of 10 KHz; linear prediction analysis has been performed on a frame length of 32 ms., with an 8 ms. frame displacement. The initial EHMM parameters are computed by means of the binary-splitting version of the Lloyd-Max vector quantization algorithm [11], using the Likelihood Ratio distortion measure, in order to be consistent with the observation density functions used in the stochastic model [12]. The transition matrix is initialized by means of a smoothed co-occurrence count statistics of the VQ labels collected for the same training data. Once the model is estimated speech can be labelled using LPC analysis and Viterbi algorithm.

3. TRANSITION MATRIX INTERPRETATION

We start our analysis from the properties of the global model, i.e. the stochastic transition matrix A. During the estimation phase we apply a re-shaping of the transition matrix in order to allow every possible pair of state sequences. At each iteration of the Baum-Welch re-estimation algorithm the non observed couples of state transitions are given a pre-defined low probability value ε, and the rest of A matrix is re-scaled accordingly. This assures that, at the end of the model estimation, state transitions i,j whose matrix entries are $a_{i,j} = \varepsilon$ are not observable from the training data. At the end of the model estimation these values are set to zero, and the rest of matrix A is properly re-scaled. As a first result we have that A is a sparse matrix, thus describing the allowed spectral transitions among the model's states. Finally, since we used a phonetically balanced speech data for training, the transition matrix can adequately represent the phonotactical constraints of the language, embodied in the set of allowed and not-allowed spectral sequences. As the number of states of the model grows (K = 64 --> 256) the transition matrix becomes more sparse, but the model preserves ergodicity, i.e. each state can be reached form all the others. This can be verified multipling the transition matrix by itself $\log_2 K - 1$ times, and checking for the absence of all-zero columns. This assures that at least a path exists between each couple of states, whose maximum length, (i.e. the number of states visited along the path) is K. An insight of the A matrix structure is given in the figure, relative to a model cardinality K = 128. For each state the number of non zero input (t_{in} - x axis) and output (t_{out} - y axis) transitions is plotted. The straight line from (0,0) to (128,128) is tracked as a reference for the $t_{in} = t_{out}$ condition. All states distribute around the reference line, indicating that there are non-absorbing nodes in the model structure. A small number of states possess a small number of interconnections with other states; those states are related to very high self loop transition matrix entries, and can be interpreted as representative of typical vowel-like steady state acoustic phenomena of the speech signal. A limited number of states possess a high number of interconnections with others, representing silence between words, stop consonants, and other flat spectra event. In order to have a more synthetic characterization of these properties we make the following considerations. The adopted statistical framework allows us to define, for the transition matrix, an absolute H^0 and conditional H^1 entropies, represented by

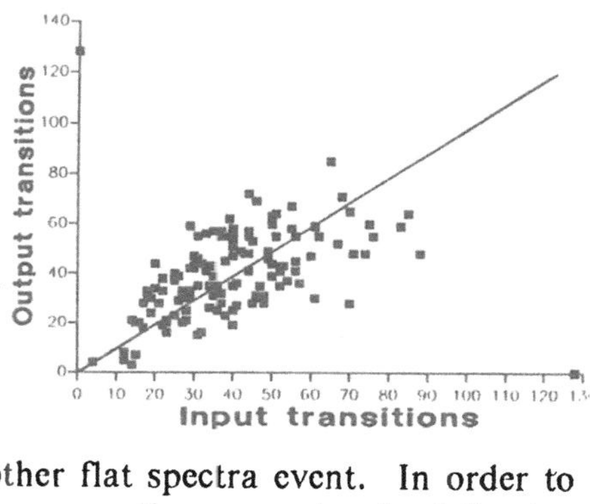

$$H^0 = -\sum_{i=1}^{K} \pi_i \log_2(\pi_i) \quad ; H^1 = -\sum_{i,j=1}^{K} \pi_j a_{i,j} \log_2(a_{i,j})$$

where π_i and $a_{i,j}$ are respectively the absolute and conditional probability of state q^i. The true entropy evaluation theoretically requires knowledges of an infinite numerable order statistics, but yet a simple one-memory Markov model allows an adequate characterization of the source information rate. In table 1 the absolute and conditional entropy values for two different model sizes (K = 64,256) are reported.

Table 1.

EHMM size	H^0	H^1
K = 64	5.43	5.01
K = 256	7.03	1.58

From table 1 we can see that the K = 64 model possess a very high conditional entropy measure, probably indicating that model size is not enough to sufficiently split different acoustic phenomena. On the other hand the K = 256 model shows a very low H^1 value, indicating that a lot of information concerning phonotactical constraints has been captured in the matrix A.

4. STATE INTERPRETATION

In this section we examine the local model. Each local model possess spectral properties described by the autocorrelation coefficients of the gaussian multivariate observation density function. The objective of the state interpretation procedure is to find the relationships among local model spectra and linguistic phonetic symbols, in order to assess the consistency of the phonemic labels obtained from Viterbi alignement of speech through the EHMM. We adopted a Mutual Information approach. We used a 540 isolated words phonetically balanced speech data. The speech data is frame synchronously phonetically aligned (8 msecs). Viterbi algorithm is made run on the speech corpus, providing the labels sequence $S \equiv \{s^t{}_k\}$, $1 < k < K$, $1 < t < T$, and $T =$ total number of labels. Then the probability functions $p(f_j)$, $p(s_k/f_j)$, $1 < j < N_f$, $1 < k < K$ are estimated by frequency count on the speech corpus. The reverse probabilities are computed through the Bayes Rule :

$$p(f_j/s_k) = \frac{p(s_k/f_j)p(f_j)}{p(s_k)} \, .$$

Finally the Average Mutual Information between the set of phonetic units and the phonemic states of EHMM model is calculated :

$$I(f;s) = H(f) - H(f/s) \; ; \; H(f) = \sum_i p(f_i) \; lg_2(1/p(f_i)) \quad ; \; H(f/s) = \sum_{j,k} p(f_j; s_k) \; lg_2(1/p(f_j/s_k))$$

$H(f)$ is the phonetic source entropy, and $H(f/s)$ is the equivocation between the phonetic alphabet and the EHMM phonemic set of symbols. In table 2 are reported the values of the above formulas evaluated for two different EHMM sizes.

Table 2.			
EHMM size	**H (f)**	**H (f / s)**	**I (f ; s)**
K = 64	4.5	2.2	2.3
K = 256	4.5	0.6	3.9

The higher equivocation value of EHMM-64 with respect to EHMM-256 clearly indicates that model size is a crucial issue in order to obtain a reliable phonemic labelling of speech. The EHMM-256 model possess a Mutual Information value that is very close to the intrinsic source entropy. This indicates that the set of symbols produced by an EHMM of sufficient size is appropriate for a reliable labelling of speech data in terms of sub-phonetic labels. Finally the local model interpretation lead us to the definition of a set of acoustic-phonetic labels representing the main acoustical events of the spoken language as well as their relationships, i.e. the transition rules among them.

5. APPLICATIONS

In [13] a completely connected HMM was first proposed for time series modeling; when applied to the speech signal the model showed segmentation properties in terms of broad phonetic classes. In that paper a phisical interpretation of the model as a vocal tract model was proposed, capable of only K (K = 5) different configurations that alternate according to a statistical transition matrix. In [14] a larger EHMM model (K = 64) was proposed for low bit rate speech coding applications. The characteristics of EHMM allowed straightforward coding schemes, and superior speech coding performances were obtained versus classical Vector Quantization schemes. These results were confirmed in [15], where a still larger model (K = 256) has been trained over speech data uttered by 10 speakers; informal listening tests showed the consistency of the approach and the perceptual significance that underlies the EHMM framework. Further studies [16] investigated the use of EHMM as a spectral tracking tool, obtaining smoother spectral estimates of speech signals if compared to classical LPC spectral analysis. Moreover the EHMM framework can constitute an alternative technique to build an acoustic lexicon with respect to Matrix Quantization techniques [17]

The EHMM approach provides an automatic segmentation of natural speech in phonemic units suitable to be used as basic units for applications in the speech recognition filed. An application is described in [18]. The EHMM is used as the acoustic layer of statistical speech unit representations. In this work LTR-HMM phonetic structure models are builded for the italian language. The structure model is represented in terms of states of an EHMM with state cardinality K = 256. Several repetitions of the phone in different phonetic context are tracked by the EHMM, and a Left to Right HMM model is obtained with a technique similar to the one proposed in [19], although the use of the EHMM model allows a reduction of the complexity of the structure inference algorithm and the share of spectral state representations among phone models.

6. CONCLUSIONS

EHMM is an efficient and consistent way to describe the speech signal. The phonemic labelling provided by EHMM is a suitable basis for continuous speech recognition applications. Moreover the general formulation makes the models suitable for low bit rate speech coding applications. Even if encouraging results have already been obtained in the proposed applications, further investigations are required, mainly in the extension of model order, the observation density function, the estimation algorithm and the computational complexity reduction during the estimation phase.

REFERENCES

[1] J.K.Baker, **The Dragon System - An Overview** *IEEE Trans. on ASSP, Vol.ASSP n. 23* February 1975

[2] F.Jelinek **Continuous Speech Recognition by Statistical Methods** *Proc. IEEE, Vol. 64* April 1976

[3] M.J.Hunt, et al. **Experiments in syllable based CSR,** *Proc. ICASSP 88, N.Y., USA* April 1988

[4] A.E.Rosenberg et.al. **Demisyllable based IWR system,** *Transactions on ASSP,vol.ASSP-31,* June 1983

[5] L.R.Bahl et al. **Further results on the recognition of a continuously read natural corpus,** *Proc. of ICASSP 80,* April 1980

[6] Y.L.Chow et.al. **The role of word dependent coarticulatory effects in in a phoneme-based speech recognition system,** *Proc. of ICASSP 86,* April 1986

[7] B.H.Juang, **On HMM and DTW for Speech Recognition-A Unified View,** *AT&T BLTJ, Vol 63, N.7* September 1984

[8] L.E.Baum, **An inequality and associated maximization technique in statistical estimation for probabilistic functions of Markov processes,** *Inequalities, vol III, pp.1-8,* 1972

[9] A.Falaschi, **Automatic selection of phonologically compact phrases,** *Proc. of First SFA Congress, Lyon, France,* April 1990

[10] P.F.Brown, **The Acoustic-Modeling Problem in ASR,** *Phd Thesis, Carnegie Mellon University, USA* April 1987

[11] B.H.Juang, **Design and Performance of Trellis Vector Quantizers for Speech Signals,** *IEEE Transactions on ASSP-36, N.9,* September 1989

[12] A.Buzo, A.H.Gray,Jr., R.M.Gray, J.D.Markel, **Speech coding based upon Vector Quanization,** *IEEE Transactions on ASSP-28, pp.562-574,* October 1980

[13] A.B.Poritz, **Linear Predictive HMM and the Speech Signal,** *IEEE Proceedings of ICASSP, Paris,* 1982

[14] E.P.Farges, M.A.Clements, **HMM applied to very low bit rate speech coding,** *IEEE Proceedings of ICASSP, Tokyo,* 1986

[15] A.Falaschi,M.Giustiniani,P.Pierucci, **A Finite States Markov Quantizer,** *IEEE Transactions of ICASSP, Albuquerque, USA,* April 1990

[16] M.Verola, **Un Modello di Markov per la stima delle proprieta' spettrali del messaggio vocale,** *Tesi di Laurea in Ing. Elettronica, Univ. Roma I,* 1989

[17] C.H.Lee,F.K.Soong,B.H.Juang, **A segment model based approach to speech recognition,** *IEEE Proceedings of ICASSP, New York, USA,* April 1988

[18] A.Falaschi,P.Pierucci **Some experiments on HMM structure inference,** *Proceedings of EUSIPCO, Barcelona,* September 1990

[19] H.Rulot,N.Prieto,E.Vidal, **Learning accurate finite-state structural models of words through the EGCI algorithm,** *IEEE Proceedings of ICASSP, Glasgow, UK,* April 1989

Clustering of Gaussian densities in hidden Markov models

S. Euler

TELENORMA Bosch Telecom, Mainzer Landstr. 128-146, D-6000 Frankfurt a.M.

Abstract: In order to reduce the number of Gaussian densities in a hidden Markov model based speech recognition system, a clustering scheme based on the Kullback divergence and the k-means clustering algorithm is proposed. The approach is tested in speaker independent recognition experiments for a vocabulary of 23 German words. A reduction of 50% in the number of densities can be achieved without degradation of the recognition performance.

Keywords: Hidden Markov models, Kullback distance, k–means clustering

1 Introduction

Hidden Markov models (HMM) with continuous densities have been applied successfully in speech recognition. In particular, in whole-word systems for a small vocabulary they usually perform better than VQ based HMMs. For larger vocabularies, however, due to the increasing number of densities problems arise in computation resources and training data. Both the required computation time and the memory size for the densities increases linearly with the number of words in the vocabulary, assuming that each word model contains the same number of states. Furthermore, the reliable estimation of multidimensional density functions requires a reasonable amount of training data.

Therefore, it is preferable to have a small set of density functions sufficient to cover the whole acoustic space of the words in the vocabulary. Additionally, using a common representation for a given phonetic event in different words can improve the discrimination ability of the recognition system. In this paper we propose a statistically based clustering scheme for density functions in HMM systems. First, density functions are trained for each word individually by means of standard HMM training algorithms. These training algorithms can consider implicitly context constrains. Then, by applying the clustering algorithm sets of similar densities from in general different words are replaced by a common density function.

We use so-called tied density hidden Markov models (TDHMM) as whole-word models together with full covariance Gaussian densities. This type of HMM seems to be particularly well suited for the proposed clustering technique. The technique, however, does not rely on the special properties of TDHMM and could be applied likewise to other continuous density HMMs.

In the next section we will give a brief description of tied density hidden Markov models. The clustering scheme is described in section 3. Results for experiments in speaker-independent, isolated word recognition are discussed in section 4.

NATO ASI Series, Vol. F 75
Speech Recognition and Understanding. Recent Advances
Edited by P. Laface and R. De Mori

2 Tied Density Hidden Markov Models

In continuous density hidden Markov models (CDHMM) a density function $f(\mathbf{y})$ is assigned to each state of the Markov chain used to represent a recognition unit. Using an appropriate training procedure and a sufficient amount of training data the $f(\mathbf{y})$ can yield a very good statistical representation of the acoustic features within each state [5]. With respect to the application in a recognition scheme, however, a more common representation which is joined in all model states, is desirable. Therefore, a generalization was proposed recently, where all density functions are tied together in each model state [2]. For a state q_i the density function p is given by the weighted sum

$$p(\mathbf{y}|q_i) = \sum_{k=1}^{K} b_{ik} \cdot f_k(\mathbf{y}), \tag{1}$$

b_{ik} denoting the probability of finding f_k in the state q_i. Eq. 1 can be interpreted as a generalization of both CDHMM and VQHMM and is in the recognition part identical to semi-continuous or tied mixture HMMs. Results for tied density hidden Markov models in speaker independent, isolated word recognition show significant improvements in the performance as compared to CDHMMs [2]. The number of b_{ik} is proportional to K. Therefore in this approach the computing time and memory required for the word models increase linear with K just as for the density functions.

3 Clustering scheme

In the experiments cited above, the density functions f_k are estimated similarily like in the traditional CDHMMs. The density functions are trained individually for each word in the vocabulary and the number K of densities is equal to the total number of model states over all words. In this section we will introduce the clustering scheme that is used to reduce the initial set of K densities to a subset of L densities.

The clustering algorithm takes into account the statistical distance between two densities. A well-known distance between two density functions f_k and f_l is given by the Kullback divergence $J(k,l)$ [4]

$$J(k,l) = \int (f_k(\mathbf{y}) - f_l(\mathbf{y})) \log \frac{f_k(\mathbf{y})}{f_l(\mathbf{y})} \, d\mathbf{y}. \tag{2}$$

We use Gaussian densities $f_k(\mathbf{y}) = N(\mathbf{y}; \mu_k, \Sigma_k)$, Σ and μ denoting the covariance matrix and mean vector, respectively. In this case a analytic form of $J(k,l)$ is given by [1]

$$J(k,l) = \frac{1}{2}(\mu_l - \mu_k)^T(\Sigma_k^{-1} + \Sigma_l^{-1})(\mu_l - \mu_k) + \frac{1}{2} tr(\Sigma_k^{-1}\Sigma_l + \Sigma_l^{-1}\Sigma_k - 2I). \tag{3}$$

This distance is independent from the training data and the calculation requires only small computing resources. In the case of identical covariance matrices, $\Sigma_k = \Sigma_l$, Eq. 3 reduces to the Mahalanobis distance. It should be noted, however, that for the more general type of density function given by mixtures of Gaussian densities no analytic solution of Eq. 2 is known.

Based on this distance we applied the modified k-means clustering algorithm as described in [7] to the set of f_k. In this algorithm the number of clusters is increased iteratively by splitting the cluster with the largest intra-cluster distortion. The number of iterations can either be specified a priori or adjusted according to the decrease of the total distortion with increasing L.

4 Simulation results

4.1 Experimental setup

The approach was tested in simulation experiments with a 200 speaker data base of 23 German words described in detail in [8]. Three words of the vocabulary, namely NULL, NEUN, and NEIN are phonetically very similar. In general, about half of the observed recognition errors were confusions within these three words. All speech signals had been filtered at telephone bandwidth. The feature vectors consisted of LPC-derived cepstral and Δ-cepstral coefficients. In the experiments described in this paper the order of the LPC-analysis was set to either 8 or 12. We used the segmentation algorithm presented in [3] to obtain initial segment boundaries and corresponding density functions. The word models were left to right models with 8 states without skips and with minimum state duration constraints.

4.2 Clustering

For the setup described above a total of $K = 184$ density functions resulted. These densities were clustered into a given number L of subsets. During the clustering procedure we used the minimax centers as centroids. In this way the mutual distances between all densities had to be computed only once.

Typical results of the clustering are illustrated in Table 1. In this table the members of the nine largest cluster out of a $L = 140$ cluster solution are shown. Furthermore, the phonetic transcriptions according to [6] of the vocabulary words are included. Some of the clusters can be interpreted phonetically. For example, the clusters 4 and 5 correspond to [n] in different positions and the cluster 8 corresponds to [f]. Most other clusters belong to the transitions from silence to speech and vice versa. It is interesting to note that in the case of the short words the densities derived from the silence parts are not clustered together.

4.3 Word modeling

In the following two different modifications of the word models are discussed. In a first recognition experiment for each subset the centroid was used to replace all members of the subset in the TDHMMs. Experiments with either the minimax or the average centroid showed no significant differences in the performance.

Recognition results for different values of L are given in Table 2. The proposed clustering allows a significant reduction of L while maintaining the recognition performance. In all cases, the system with 12-th order LPC give better results than the system with 8-th order LPC. Results for decreasing L, however, show that the more detailed modeling in the 12-th order systems is relatively more effected by the clustering procedure. In the case of $L = 80$ the recognition rates are almost identical. For very small L the system with 12-th order LPC is again significantly better than the system with 8-th order LPC.

In a second experiment the clustered density functions were used in a two-level tree structure. In the first level the likelihood was evaluated for all L centroids. Then in a additional second level the likelihood values for all members belonging to the subset of the best centroid were calculated. For the members of all other subsets the likelihood of their centroids was used. In this way the high resolution is maintained in the region of the acoustic space around the feature vector. Assuming that all subsets have equal number

Table 1. The members of the nine largest clusters (1, ...,9) in a system with $L = 140$ clusters

word	phonetic transcription	model state							
		1	2	3	4	5	6	7	8
NULL	nUl		5						
EINS	aIns						2	3	
ZWEI	tsvaI		2						
ZWO	tsvO								
DREI	draI								
VIER	fi6	9						1	
FUENF	fYnf		8		4		2	3	
SECHS	zEks						2	3	
SIEBEN	"Zi:b@n							1	
ACHT	axt							3	
NEUN	nOYn		5				4		
ENDE	"End@							1	
JA	ja:							1	
NEIN	naIn		5				4		
HILFE	"hilf@					8		1	
PAUSE	"paUz@							1	
WIEDERHOLEN	vi:d6"hOl@n								6
WEITER	"vaIt6	9						1	
STORNIEREN	StOr"ni:r@n	9						4	7
RUECKRUF	"rYkru:f							8	6
WAHLWIEDERHOLUNG	"va:lvi:d6hOlUN								7
RUFUMLEITUNG	"ru:fUmlaItUN								7
KONFERENZ	kOnf@"rEnts							2	6

Table 2. Recognition rates with different number L of density functions. (R23: all 23 words, R10: digits, RC: command words)

L	8 ceps, 8 Δ–ceps			12 ceps, 12 Δ–ceps		
	R23	R10	RC	R23	R10	RC
20	86.6	90.8	97.3	93.0	92.6	97.7
40	94.2	94.7	98.4	95.3	95.5	99.3
60	96.3	96.6	99.1	97.0	97.6	99.4
80	97.2	97.7	99.8	97.3	97.8	99.5
90	97.1	97.7	99.8	97.5	97.8	99.8
100	97.4	98.0	99.8	97.9	98.5	99.8
120	97.4	98.3	99.7	98.0	98.8	99.8
140	97.3	98.2	99.6	98.1	99.1	99.8
184	97.3	98.1	99.7	98.1	98.9	99.9

Table 3. Recognition rates with different number L of density functions using 8 ceps and 8 Δ-ceps. (R23: all 23 words, R10: digits, RC: command words)

L	R23	R10	RC	L_{used}	L_{used}^{theo}
2	63.1	71.0	72.2	40%	51%
4	95.0	96.2	98.9	31%	27%
6	95.7	97.0	99.1	24%	19%
8	94.4	94.7	98.7	18%	16%
10	93.0	95.1	97.8	18%	15%
25	94.7	96.8	98.7	19%	17%
50	96.3	97.3	99.3	30%	29%
75	97.0	97.8	99.6	42%	42%
100	97.6	98.1	99.8	55%	55%
125	97.4	98.2	99.8	68%	68%
150	97.3	98.2	99.7	81%	82%
184	97.3	98.1	99.7	100%	100%

of members, in average only $L_{used}^{theo} = L + K/L - 1$ density functions have to be calculated. The minimum of $L_{used}^{theo} = 2\sqrt{K} - 1$ is found for $L = \sqrt{K}$.

Results of recognition experiments are given in Table 3 together with the number of density functions L_{used} that were used actually. The recognition rates have a local minimum in the range of minimal L_{used}. Configurations with only few branches in the first level give relatively small rates. In particular the binary tree configuration $L = 2$ yields a very poor recognition performance albeit 40% of the density functions are used. Best recognition rates are found with $L = 100$, but the differences as compared to the systems with $L > 100$ are only small.

5 Conclusion

In this paper we proposed a clustering scheme for Gaussian density functions based on the Kullback divergence and the modified K-means clustering algorithm. In the experiments with tied density HMMs the number of density functions was reduced by almost 50% without degradation of the recognition performance.

Comparing the results of the two described modifications of the word models no significant differences were found. In our task a set of about 100 density functions has been sufficient to cover the acoustic space while preserving all information useful for the discrimination between words.

References

[1] M. Basseville. Distance measures for signal processing and pattern recognition. *Signal Processing*, 18:349–369, 1989.

[2] S. A. Euler, B.-H. Juang, C.-H. Lee, and F. K. Soong. Statistical segmentation and word modeling techniques in isolated word recognition. In *ICASSP-90*, pages 745–748, Albuquerque, 1990.

[3] M. Falkhausen, S. A. Euler, and D. Wolf. Improved training and recognition algorithms with VQ–based hidden markov models. In *ICASSP-90*, pages 549–552, Albuquerque, 1990.

[4] S. Kullback and R. A. Leibler. On information and sufficiency. *Annals of Mathematical Statistics*, 22:79–86, 1951.

[5] L.R. Rabiner and J.G. Wilpon. Some performance benchmarks for isolated word recognition systems. *Computer Speech and Language*, 2:343–357, 1987.

[6] J. C. Wells. Computer coded phonetic transcription. *Journal of the International Phonetic Association*, 17(2):94–114, 1987.

[7] J. G. Wilpon and L.R. Rabiner. A modified K–means clustering algorithm for use in isolated word recognition. *IEEE Trans. Acoustics, Speech, and Signal Proc.*, ASSP-33:587–594, 1985.

[8] J. Zinke, S. Euler, A. Buch, and N. Jeck. Realization and optimization of a speaker independent speech recognizer for isolated words on a TMS 320C25. In *EUSIPCO-90*, Barcelona, 1990.

Developments in High-Performance Connected Digit Recognition

Yves Normandin and Régis Cardin

Centre de recherche informatique de Montréal (CRIM)
3744 Jean Brillant, Suite 500
Montréal, Québec
Canada H3T 1P1

Abstract: Recent advances in Hidden Markov Model (HMM) based speaker-independent connected digit recognition have usually tended to make the models more complex. This paper concentrates on improving the training techniques in order to make the most of the available parameters. A new algorithm, *Corrective MMI Training* is introduced. Use of this algorithm resulted in significant improvements in our recognition rates. We now obtain less than 2% string error rate using semi-continuous HMMs with two models per digit.

1. Introduction

Because of its wide-ranging applications, connected digits recognition has always been a problem of great interest among speech researchers. This is especially true because of the availability of the large TI/NIST connected digit database, which allows meaningful comparison of systems from different sites. Recognition rates on the TI/NIST task have been going up steadily. Dynamic (time derivative) speech parameters, which are now almost universally used, have largely contributed to this improvement. Among other techniques that have also contributed, we can mention the use of several models (templates) per word [1, 4], finer time modeling with long HMMs [1], discriminant transforms on the frame speech features [1] and, more recently, the use of Gaussian mixtures with a large number of mixture components.

This paper will focus primarily on improved training techniques. The technique we propose applies the Maximum Mutual Information (MMI) criterion in a corrective-like fashion. We will describe it, along with the refinements which allowed us to obtain both fast convergence and very low error rate.

2. Baseline system

Our baseline system is a standard HMM-based system with discrete output distributions and one model per digit. It uses three codebooks: the first of size 128 for 6 FFT-based, Mel-scaled cepstral coefficients (C_1-C_6), the second, also of size 128 for the corresponding 6 cepstral slopes, and the third, of size 32 for the log energy and its slope. We use the 10 kHz version of the TI/NIST database and

NATO ASI Series, Vol. F 75
Speech Recognition and Understanding. Recent Advances
Edited by P. Laface and R. De Mori

analysis is performed whithout prior endpoint detection, using a frame rate of 10 ms, preemphasis of 0.95 and windows of 256 sample points. The word models are built from a set of sub-word models, using the following lexicon:

1	w-ax + n-tail	7	s + eh + v7 + ax + n-tail
2	t + uw	8	ey + pau + t8
3	th + r-iy	9	n-head + ay + n-tail
4	f + ow-r	oh	ow
5	f + ay + v5	zero	zz + iy-r-ow
6	s + ih-k + pau + k-s		

This unit set is chosen because speech from 13 speakers in the training set had previously been manually segmented according to those units. We use those speakers with the corresponding labels to bootstrap the training procedure, which consists in four iterations of Maximum Likelihood Estimation (MLE) training. For each digit string in the training set, the HMM corresponding to the string is built, with optional silence before and after and optional pauses between each digit, and the Baum-Welch algorithm is applied to this model, using the observation sequence from the string. We found that more more than four iterations usually result in degraded recognition rates on the test set. Unknown string length recognition is performed by applying the Viterbi algorithm on the looped model in Fig. 1. With this system, we obtain a string error rate of about 4%.

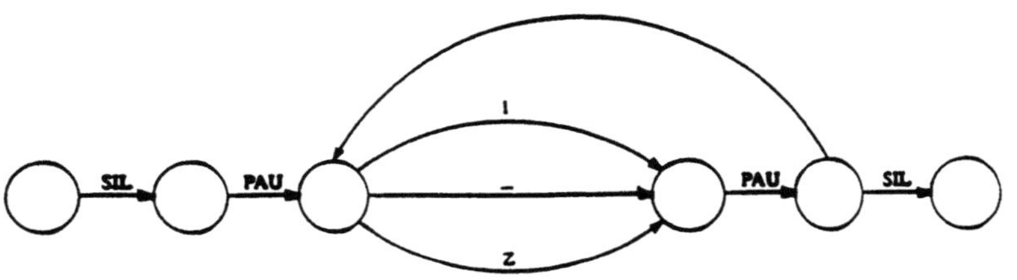

Fig. 1. Looped model

3. Corrective MMI training

Let Y^r be the observation sequence corresponding to the rth digit string in the training set and let M_{Y^r} be the corresponding HMM. Let Θ be the complete HMM parameter set for the task. Then, in MMI training, we are trying to find Θ which maximizes:

$$R(\Theta) = \prod_r P_\Theta(M_{Y^r}|Y^r) = \prod_r \frac{P_\Theta(Y^r|M_{Y^r})P(M_{Y^r})}{\sum_M P_\Theta(Y^r|M)P(M)} = \prod_r \frac{P_\Theta(Y^r|M_{Y^r})P(M_{Y^r})}{P_\Theta(Y^r|M_{loop})} \tag{1}$$

where $P(M_{Y^r})$ is the *a priori* probability of the string, which is fixed and is computed by assuming all digits equally likely. The sum in the denominator is over the HMMs corresponding to all possible digit strings, of which there is an infinite number. Fortunately, this is equivalent to computing the probability of the looped model used for recognition, which is expressed by the last equality in (1) where M_{loop} stands for the looped model.

Traditionally, MMI training has been done using a gradient descent on $-R(\Theta)$, which can be very time-consuming. Recently however, Gopalakrishnan *et al* [2] introduced the following reestimation formula for rational objective function (such as $R(\Theta)$) for discrete HMMs:

$$\hat{\theta} = \frac{\theta\left(\dfrac{\partial \log R(\Theta)}{\partial \theta} + D(\Theta)\right)}{\sum_{\theta'} \theta'\left(\dfrac{\partial \log R(\Theta)}{\partial \theta'} + D(\Theta)\right)} \tag{2}$$

where the sum is taken over all parameters belonging to the same distribution as θ. The formula is guaranteed to converge for any $D(\Theta)$ greater than some unknown constant. However, this constant is very difficult to evaluate so it must generally be approximated. Also it is important to get a good approximation since in the convergence region of $D(\Theta)$, the greater its value, the slower the convergence will be. Gopalakrishnan *et al* proposed the following approximation (in which ϵ is a small positive constant):

$$D(\Theta) = \max_{\theta}\left\{-\frac{\partial \log R(\Theta)}{\partial \theta}, 0\right\} + \epsilon \tag{3}$$

which is what we use in our experiments. However, we found that direct application of (2) resulted in slow convergence and poor results. Following an argument of Merialdo [3], we conjectured that by removing emphasis from the low-valued parameters in the gradient vector, we could move the focus of the optimization from the micro-structure of $R(\Theta)$ to a more global one. From (1), we have:

$$\frac{\partial \log R(\Theta)}{\partial \theta} = \sum_r \left\{\frac{1}{P_\Theta(Y^r|M_{Y^r})}\frac{\partial P_\Theta(Y^r|M_{Y^r})}{\partial \theta} - \frac{1}{P_\Theta(Y^r|M_{loop})}\frac{\partial P_\Theta(Y^r|M_{loop})}{\partial \theta}\right\} \tag{4}$$

which for discrete HMMs reduces to:

$$\frac{\partial \log R(\Theta)}{\partial \theta} = \frac{1}{\theta}(c_\theta - c_\theta^{loop}) \tag{5}$$

where c_θ represents the standard Baum-Welch count for parameter θ and c_θ^{loop} is the corresponding count obtained using the looped model. In his gradient descent based MMI training experiments, Merialdo improved the convergence by replacing (5) by:

$$\frac{\partial \log R(\Theta)}{\partial \theta} \approx \left(\frac{c_\theta}{\sum_{\theta'} c_{\theta'}} - \frac{c_\theta^{loop}}{\sum_{\theta'} c_{\theta'}^{loop}}\right) \tag{6}$$

where the sum is taken over all parameters θ' belonging to the same distribution as θ. We observed similar improvement in our own gradient descent experiments. Using (6) in (2) dramatically improved convergence. We experimented with different variants of (6) based on the same idea and observed similar convergence behavior. However, in all cases, as $R(\Theta)$ gets near its optimum (1.0), divergence is often observed, leading us to believe that (3) is inadequate when $R(\Theta)$ is large.

Our *Corrective MMI training* algorithm, based on the above results, is initialized with the HMMs obtained after four MLE iterations. Subsequently, each iteration is a two-step process. First, recognition is performed on the training set and then reestimation is done using only those digit strings that were incorrectly recognized. The aim here is to correct as many errors as possible from the training set, hoping that it will improve results on the test set. Note that for correctly recognized strings, the two contributions to the counts will tend to cancel out, leaving most of the contribution with the incorrect ones. In practice, the results obtained by training only on errors are similar to the ones obtained by training on the full training set, but at a much lower computational cost. Reestimation is done using (2), (3) and (6), and the HMM parameters obtained are smoothed with the ones from the previous iteration using a weight that is dependent on the number of errors in the training set.

4. Extension to Gaussian densities

It is possible to make a heuristic extension of (2) to the case of diagonal Gaussian densities. This is done through a histogram approximation to each one-dimensional Gaussian density $N(y,\sigma_i,\mu_i)$. Divide the real axis (domain of the densities) into intervals of width Δ and keep only those intervals in which $N(y,\sigma_i,\mu_i)$ is somewhere larger than a certain threshold. Then, each of those intervals corresponds to one possible event of a discrete distribution and has probability $a_i(k) \approx N(\tilde{y}_k,\sigma_i,\mu_i)\Delta$, where $\tilde{y}_k$ is any point in the kth interval. Using (2), we can reestimate the values of the discrete probabilities $\hat{a}_i(k)$, from which we can get an estimate of the new mean and variance as:

$$\hat{\mu}_i \approx \sum_k \hat{a}_i(k)\tilde{y}_k, \quad \hat{\sigma}_i^{\,2} \approx \sum_k \hat{a}_i(k)(\tilde{y}_k - \hat{\mu}_i)^2 \tag{7}$$

First observe that:

$$\sum_k a_i(k) \approx 1, \quad \sum_k a_i(k)\tilde{y}_k \approx \mu_i, \quad \sum_k a_i(k)\tilde{y}_k^{\,2} \approx \sigma_i^{\,2} + \mu_i^{\,2} \tag{8}$$

Since we are not actually dealing with a discrete distribution, the argument for using a biased derivative breaks down so we use (4) which becomes:

$$\frac{\partial \log R(\Theta)}{\partial a_i(k)} = \sum_r \sum_{(l \mid y^r_{l,i} \in k)} \frac{1}{a_i(k)} \left\{ \frac{P_{\Theta,l}(t, Y^r \mid M_{Y^r})}{P_\Theta(Y^r \mid M_{Y^r})} - \frac{P_{\Theta,l}(t, Y^r \mid M_{loop})}{P_\Theta(Y^r \mid M_{loop})} \right\} \tag{9}$$

where $y^r_{l,i}$ is the ith parameter from the lth observation in Y^r and $P_{\Theta,l}(t, Y^r \mid M)$ is the probability of taking the transition t at time l and observing Y^r, given the model M. Combining (7), (8) and (9) gives:

$$\hat{\mu}_i \approx \frac{\sum_r \sum_l \frac{P_l(t, Y^r \mid M_{Y^r})}{P(Y^r \mid M_{Y^r})} y^r_{l,i} - \sum_r \sum_l \frac{P_l(t, Y^r \mid M_{loop})}{P(Y^r \mid M_{loop})} y^r_{l,i} + D\mu_i}{\sum_r \sum_l \frac{P_l(t, Y^r \mid M_{Y^r})}{P(Y^r \mid M_{Y^r})} - \sum_r \sum_l \frac{P_l(t, Y^r \mid M_{loop})}{P(Y^r \mid M_{loop})} + D} \tag{10}$$

and:

$$\hat{\sigma}_i^{\,2} \approx \frac{\sum_r \sum_l \frac{P_l(t, Y^r \mid M_{Y^r})}{P(Y^r \mid M_{Y^r})} y^{r\,2}_{l,i} - \sum_r \sum_l \frac{P_l(t, Y^r \mid M_{loop})}{P(Y^r \mid M_{loop})} y^{r\,2}_{l,i} + D(\sigma_i^{\,2} + \mu_i^{\,2})}{\sum_r \sum_l \frac{P_l(t, Y^r \mid M_{Y^r})}{P(Y^r \mid M_{Y^r})} - \sum_r \sum_l \frac{P_l(t, Y^r \mid M_{loop})}{P(Y^r \mid M_{loop})} + D} - \hat{\mu}_i^{\,2} \tag{11}$$

where, to simplify the notation, we removed the dependence on Θ. Several observations can be made here. First note that the reestimation formulas do not depend on the actual value of Δ used in the approximation. Note also that because of the approximations involved, the variance estimate is not guaranteed to be positive so D must be at least as large as the minimum value that guarantees a positive estimate. We observed fast convergence in all our experiments with (10) and (11).

5. Results

The table below summarizes the results of our experiments. All entries show the error rates obtained on the test set. The baseline column refers to the system described in section 2. The next column shows the effect of applying corrective MMI training to the baseline HMMs. The SCHMM column shows the result after a direct conversion from discrete to semi-continuous HMMs, using the codebook to compute the means and variances. Finally, the last column shows the result of additional training on the models from the SCHMM column.

Models per digit		Baseline	Corr. MMI	SCHMM	SCHMM + train
1 model	word	1.4%	0.86%	0.82%	0.75%
	string	3.9%	2.6%	2.5%	2.3%
2 models	word	-	0.70%	0.63%.	0.63%
	string	-	2.2%	1.9%	1.9%

The results clearly show that, when a large training set is available, our algorithm can result in a substantial improvement over the standard MLE training algorithm alone.

6. Conclusion

Our experiments have shown that it is possible to reestimate efficiently HMM parameters with the MMI criterion without using gradient descent and that this can result in very high recognition rates in a small vocabulary connected word recognition task such as the digits. Further work will try to extend these results to larger vocabularies (such as letters or phonemes), with the ultimate aim of improving training techniques for large vocabulary speech recognition systems.

7. References

[1] G.R. Doddington, *Phonetically Sensitive Discriminants for Improved Speech Recognition*, ICASSP-89, paper S10b.11

[2] P.S. Gopalakrishnan, D. Kanevsky, A. Nadas and D. Nahamoo, *A Generalization of the Baum Algorithm to Rational Objective Functions*, ICASSP-89, paper S12.9

[3] B. Merialdo, *Phonetic Recognition using Hidden Markov Models and Maximum Mutual Information Training*, ICASSP-88, paper S3.4

[4] L.R. Rabiner, C.H. Lee, B.H. Juang, and J.G. Wilpon, *HMM Clustering for Connected Word Recognition*, ICASSP-89, paper S8.5

Robust Speaker-Independent Hidden Markov Model Based Word Spotter

Louis C. Vroomen and Yves Normandin

Centre de recherche informatique de Montréal (CRIM)
3744 Jean Brillant, Suite 500
Montréal, Québec
Canada H3T 1P1

Abstract: Since January 1990, we have endeavoured, in conjunction with the Canadian Department of National Defense, to design a speaker-independent word-spotter. This paper describes the several aspects investigated over the past few months. The open-set nature of the problem requires different techniques than those used in continuous speech recognition. This paper presents the baseline word-spotting system, using hidden Markov models (HMM) to model both keyword and non-keyword speech. In the course of the research, we have found that the features and their transformations greatly affect the performance of the system. Since the task requires speech over a long period of time, various schemes were investigated to produce a robust system with respect to variations in background noise. We present the results of these investigations. Finally, experimental results from the investigations are presented.

1. Introduction

The task of word-spotting is to locate a limited vocabulary within arbitrary speech. Due to the open-set nature of the task, standard feature extraction and recognition techniques used in continuous speech recognition are unsatisfactory. Thus, the main areas of investigation, for word-spotting, become:

- signal processing techniques
- Hidden Markov modeling for keywords
- Hidden Markov modeling for non-keywords
- a method of scoring detected keyword

In the next section, the baseline system, using HMMs, is presented. The techniques of modeling keyword and non-keyword speech is discussed and method of scoring is described. In the following section, we discuss the various signal processing techniques investigated for this task. The current experimental results follow, concluded with an overall discussion of the research.

2. HMM modeling

Several different approaches to word-spotting have been attempted. Early attempts to detect a given set of words in running speech were based on dynamic programming techniques for whole word template matching [1]. More recently, HMM techniques have been used in word-spotting [3].

NATO ASI Series, Vol. F 75
Speech Recognition and Understanding. Recent Advances
Edited by P. Laface and R. De Mori

Using HMMs, word-spotting is achieved by constructing a single HMM network which incorporates the keyword models, a model for the non-keyword speech or alternate model, and a silence model, as shown in Figure 1.

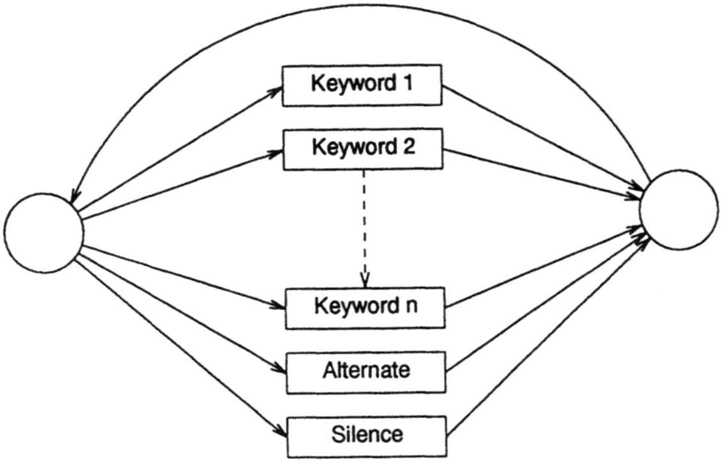

Figure 1. HMM network

First, we describe the training and specifically, the construction of the alternate model. Second, the method of scoring, based on *a posteriori* probabilities is presented.

2.1 Training

The current baseline system is a standard HMM-based system with discrete output distributions. The HMMs are "left-to-right" linear sequences of states and the topology is shown in Figure 2.

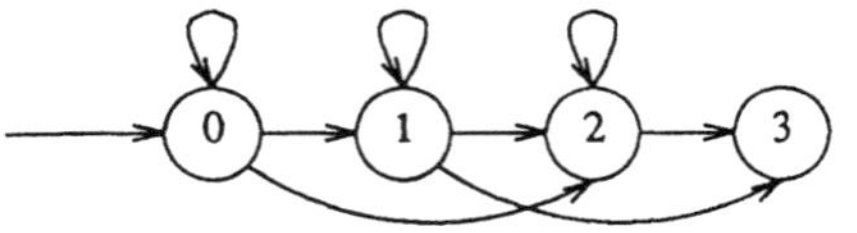

Figure 2. HMM model topology

Currently, whole-word models are used for each keyword. These models are trained using the maximum-likelihood Baum-Welch algorithm. The models have 15 states. In addition to training the word models, a 3 state model for silence is trained.

The training of the alternate model is different. The alternate model is constructed using sub-word units which are derived from the keywords. A lexicon has been developed for each keyword, using sub-word units. For example, boonsboro and conway have been defined as:

$$boonsboro: b + oo + n + s + b + ro$$
$$conway: k + o + n + w + ay$$

All the defined units are trained on keyword speech. The models use the topology of the word models and have 3 states per unit. In addition, "ending" models (*ed*, *ing* and *s*), which also have 3 states per unit, are trained.

The alternate model is then built as a looped model with, in parallel, all the trained sub-word units, the silence model, the "ending" models and a 3 state model called "alt" to model the sounds which have not been modeled by the sub-word units. The constructed looped model is trained using the Baum-Welch algorithm on all the non-keyword speech in the train set.

2.2 Scoring metric

The current word-spotter computes the forward trellis using the looped HMM in Figure 1. The following discussion will use the notation as presented by Rohlicek, et. al. [3].

The updating of this network is achieved time-synchronously using the standard forward algorithm. Thus, given a sequence of observations $x_1, \cdots, x_t$, the score at state i at time t is given by:

$$S_i(t) = p(s_t = i, x_1, \cdots, x_t)$$

An *a posteriori* probability of occupying a given state i can be obtained by:

$$Pr(s_t = i \mid x_1, \cdots, x_t) = \frac{p(s_t = i, x_1, \cdots x_t)}{p(x_1, \cdots, x_t)} = \frac{S_i(t)}{\sum_j S_j(t)}$$

pp It should be noted that $\sum_j S_j(t)$ is over all the states in the trellis.

The determination of the scores is heavily influenced by the alternate model. This is apparent since the denominator sum will be dominated by this model. The scores provide the evaluation of the probability of a keyword ending by:

$$\omega_n(t) = Pr(s_t = e_n \mid x_1, \cdots, x_t)$$

where e_n is the state index of the last state of word n. The local maximum of $\omega_n(t)$ is considered to be the best candidate for the spotted keyword (which includes the alternate model).

The final step is to determine a threshold for each keyword. Currently, the thresholds, T_n are obtained by calculating the average $\omega_n(t)$ for all occurrences of word n in the train set. Thus, a word is spotted when:

$$\omega_n(t) \geq F \times T_n$$

where F is a fractional value, which sets the sensitivity of the word-spotter.

3. Signal processing techniques

Any task which spans a long period of recording time must take into consideration potential variations. Such variation include changes in background noise, room acoustics and the recording hardware. Thus the signal processing techniques used must provide appropriate normalization and adaptation to minimize the effects of the variation.

The next section will present the baseline features used in the system. The following sections will discuss the various schemes used to produce a robust system with respect to long-term effects considered.

3.1 Baseline features

The current baseline system uses three codebooks: the first of size 128 for 6 FFT-based, Mel-scaled cepstral coefficients (C_1-C_6), the second, also of size 128 for the corresponding 6 cepstral slopes, and the third, of size 32 for the log energy and its slope. The data is analyzed using a frame rate of 15 ms, preemphasis of 0.95 and windows of 256 sample points.

The baseline system will be expanded to include the adaptation schemes used in the system. Figure 3 shows the main analysis section of the system.

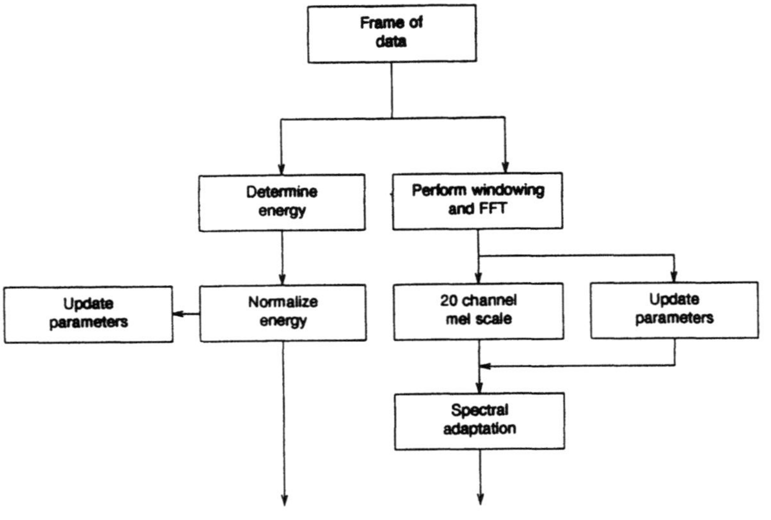

Figure 3. Feature extraction with adaptation

3.2 Speech/noise discrimination

An important aspect to consider when performing adaptation is the use of a robust speech/noise discriminator. It has been observed that histograms of log energy can be modeled by using a bimodal Gaussian distribution [3]. Thus noise can be modeled using a Gaussian, $p_n(e_t)$ and speech+noise as $p_s(e_t)$ where e_t is a given energy observation, at time t.

$p_n(e_t)$ and $p_s(e_t)$ provide a good speech/noise discriminator. A speech/noise threshold ϵ can be derived as the point where the two Gaussians have equal probability, that is $p_n(\epsilon)=p_s(\epsilon)$ Thus, a decision criterion [4] is given as:

$$\delta_t^{no} = \begin{cases} 1 & \text{if } e_t < \epsilon \\ 0 & \text{if } e_t \geq \epsilon \end{cases}$$

Since the background noise can vary over time, the energy histogram must be updated periodically. Also, it should be noted that this discriminator does not model short-duration noise artifacts.

3.3 Energy normalization

The energy normalization attempts to map the variable dynamic range of energy values, due to background noise and the speaker, to a fixed dynamic range. The current transformation is based on the Gaussians derived for the speech/noise discriminator:

$$e_t' = \frac{e_t - \mu_n'}{\mu_s' - \mu_n'}$$

where

μ_s' is the mean of $p_s(\epsilon)$ + an offset
μ_n is the mean of $p_n(\epsilon)$ - an offset

e_t' is clamped such that it cannot go below zero.

3.4 Spectral adaptation

The spectral adaptation developed is based on the work of Van Compernolle [4]. There are two main aspects of the adaptation. The first is noise correction which is used to remove the varying background noise. The second is channel equalization. Equalization performs a compensation for variable microphone and room acoustics.

The noise correction is determined by spectral subtraction. A noise estimate is obtained from all power spectra frames which have have been classified by noise. This estimate is subtracted from the power spectra.

The channel equalization is realized by obtaining the maximum energy which can be expected from each channel, after spectral subtraction. The channel gain becomes the difference between the maximum channel energy and a target value. Thus, spectral adaptation for channel i becomes:

$$A^i = 10 \log_{10} \hat{P}^i + G^i$$

where $\hat{P}^i$ is the noise corrected power spectra, G^i is the channel gain and A^i is the adapted spectrum. The resulting adapted spectra is then transformed into cepstral coefficients.

4. Experimental results

The database currently used in the development of the word-spotter is the Stonehenge or "road rally" database. Marking files are included with the signal files, indicating the location of keywords. The train set, recommended by the designers, is comprised of ten male speakers with each speaker having one conversational file and two read passages files. There is a total of 1265 occurrences of the keywords and 2738.25 seconds of non-keyword speech. The test set has ten male speakers with each speaker having one conversational file. This set has a total of 454 occurrences of keywords and 1405.87 seconds of non-keyword speech.

For the experiments, the markings were augmented to provide the locations of silence in the train set. They were obtained by using the silence/speech discriminator. Thus the new marking files give the locations for silence, non-keyword and keyword speech.

The following table presents our preliminary results from the word-spotter, without HMM smoothing. The results are reported as *recognition rate @ false alarms/per hour (fa/h)*.

System	Train Set	Test Set
Baseline	94.84 @ 5.60	27.53 @ 7.35
Energy normalization	95.20 @ 5.93	33.70 @ 8.80
Spectral adaptation	95.27 @ 6.51	38.77 @ 9.17

5. Conclusions

The results show the word-spotter is still in the beginning stages. Our rudimentary alternate model does not model sounds which are not in the keyword set, thus greatly affecting our results. The choice and use of spectral adaptation, though, has shown that a improvement can be achieved.

Currently, extensive work to improve the baseline system is being carried out. The main area is the improvement and redesign of the alternate model. One possible solution is to use a phonetically labeled database (TIMIT) to bootstrap an alternate model. This would allow the modeling of sounds which do not occur in keyword speech. Also, the use of continuous and semi-continuous HMMs and implementation of a Viterbi back-track is being investigated.

In addition to the improvements to the baseline system, more work is planned for the adaptation procedures and the use of features. One interesting endeavour will be to use an ear model on the signal.

6. Bibliography

[1] Bridle, J. S.: An efficient elastic-template method for detecting given words in running speech. British Acoustic Society Meeting, pp. 1-4. April 1973.

[2] McAulay. R. J. and Malpass M.L.: Speech enhancement using a soft-decision noise suppression filter. IEEE Transactions on Acoustics, Speech and Signal Processing, ASSP-28, pp. 137-145, 1980

[3] Rohlicek, J. R., Russell, W., Roukos, S., Gish, H.: Continuous Hidden Markov Modeling for Speaker-Independant Word Spotting. Proceedings Internation Conference on Acoustics, Speech and Signal Processing, pp. . May 1989.

[4] Van Compernolle, D.: Noise adaptation in a hidden Markov model speech recognition system. Computer Speech and Language, pp. 151-167. Vol. 3, 1989

Robust Speech Recognition in Noisy and Reverberant Environments

H.G. Hirsch

Technical University of Aachen, D-5100 Aachen, FRG

Abstract: A method of digital speech signal processing is introduced to reduce the reverberation of rooms as well as noise. The ability of this method is shown additive to improve speech recognition in such situations.

Keywords: Noise reduction, Dereverberation, Isolated word recognition

1. Introduction

In many real applications of speech recognition systems the recognition rate decreases significantly if background noise is present and/or if the speech is reverberated (e.g. voice dialing in mobile radio or hands-free speech input in office rooms). Several single as well as multi-microphone noise suppression algorithms have been developed to improve the noisy speech in such communication situations (e.g. [1],[2],[3],[4],[5]).

These methods can be applied to speech recognition too. Improvements, using a single microphone spectral subtraction technique, have been reported in e.g. [6],[7]. However, this method is restricted to nearly stationary noise. Furthermore, severe difficulties arise from the required speech pause detection.

This automatic distinction between noise and noisy speech segments is a very difficult task if the noise signal is not stationary. In standard word recognizers one of the major reasons for the increase of error-rate with noisy speech is the failure of the word boundary detection.

NATO ASI Series, Vol. F 75
Speech Recognition and Understanding. Recent Advances
Edited by P. Laface and R. De Mori

2. Reduction of Reverberation

Earlier studies have shown that the recognition rates of word recognition systems deteriorate considerably for a hands free speech input in reverberant environments [8]. The reason for this is the masking of the spectral features of certain phonemes. The dominating vowels mask the following phonemes with less energy and less duration.

A single channel method has been developed to reduce reverberation and to improve speech recognition rates of reverberated speech[8]. The basis is a short-term spectral analysis. It is known that reverberation has a low-pass effect on the temporal subband energy contours [9]. The idea of an inverse high-pass filtering of subband energy signals was introduced to reduce reverberation. The processing steps can be seen in figure 1.

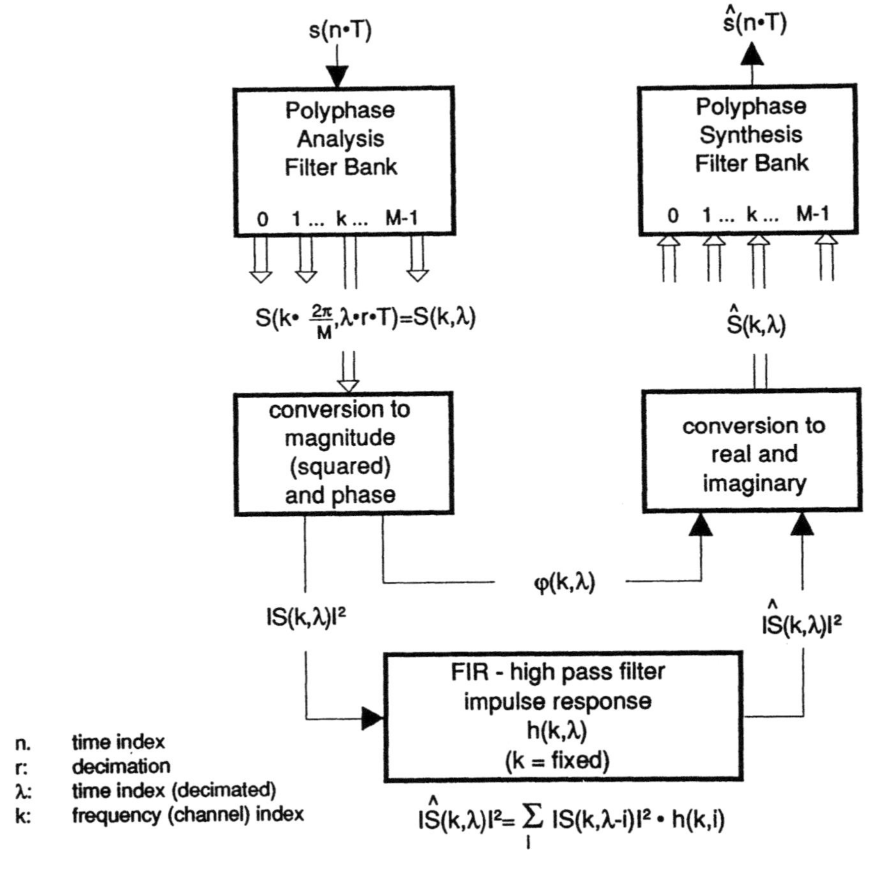

Figure 1. Block diagram of the dereverberation system

A polyphase filter bank is used to analyze the speech signal. The temporal contour of the squared magnitude in each channel is high-pass filtered. The speech signal is resynthesized using the phase of the original (reverberated) signal. A considerable reduction of reverberation can be recognized.

In applying this method as a preprocessing step to a speaker dependent isolated word recognizer, the results shown in figure 2 have been obtained. The recognition system is based on a cepstral speech analysis and on classification with the dynamic programming technique.

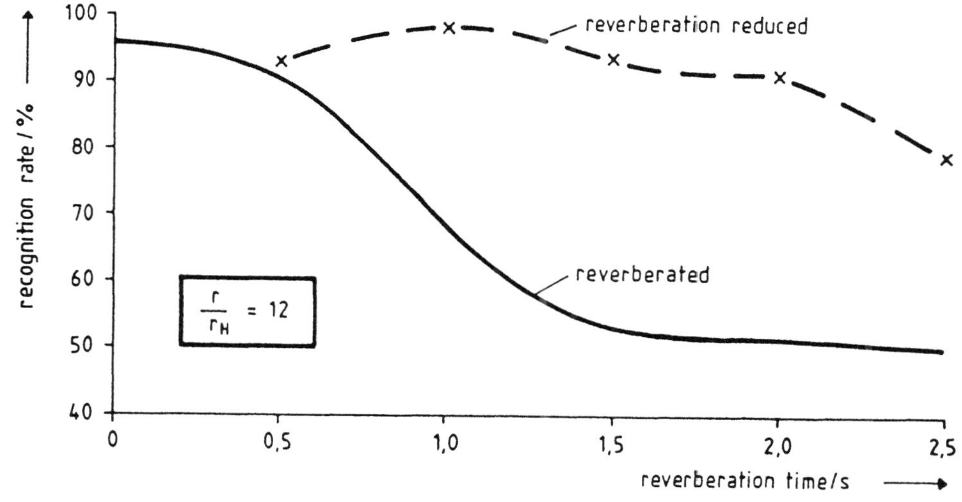

Figure 2. Recognition rates of reverberated speech with and without dereverberation

The reverberated speech has been produced with an artificial reverberation system [10]. A decrease of recognition rate can be seen for increasing reverberation time when using the reverberated speech signals.
A considerable increase of recognition rate can be obtained over a wide range of reverberation time after preprocessing the speech signals by high-pass filtering of the subband energy signals. The high-pass filter was identical for all subbands and all reverberation times. The filter was optimized for a reverberation time of about 1,2 s.

3. Improvement of recognition in noisy situations

This kind of preprocessing can be used for noise reduction too (figure 3). Stationary noise results in nearly constant components for the energy contours of the spectral subbands. Thus a significant amount of noise can be reduced by high pass filtering of subband energy signals. The short-term subband energy is estimated using an analysis filterbank. Then each temporal subband energy signal is filtered with a high-pass FIR filter.The same constant filter is used in each band. No speech pause detection is required.

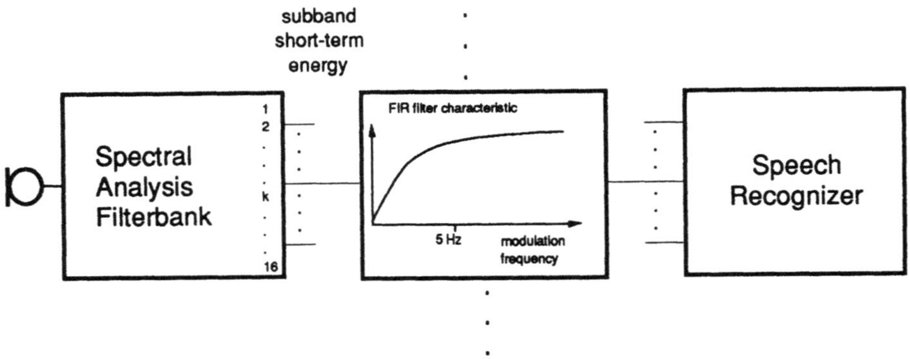

Figure 3. Speech recognition with high-pass filtered subband energy signals

Experimental results are shown in figures 4 and 5 for a speaker independent speech recognizer, based on spectral speech analysis and classification with a dynamic programming technique, using different noise signals at various S/N-ratios.

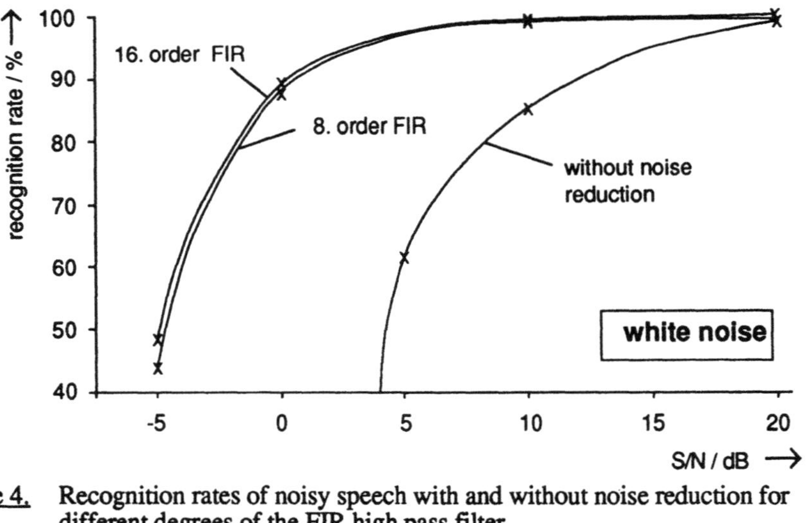

Figure 4. Recognition rates of noisy speech with and without noise reduction for different degrees of the FIR high pass filter

Figure 4 shows a considerable improvement of the recognition rate by using the noise reduction technique. Reference templates have been estimated from undisturbed speech signals in the case without noise suppression. In the other case references were also taken from undisturbed speech but after preprocessing with the high-pass filtering technique.
It can be seen from figure 4 that the results with the filter of 16. order are slightly better in comparsion to the filter of 8. order. A gain of more than 10 dB can be obtained at a recognition rate of 95 %.
The following figure 5 give the results for a car noise.

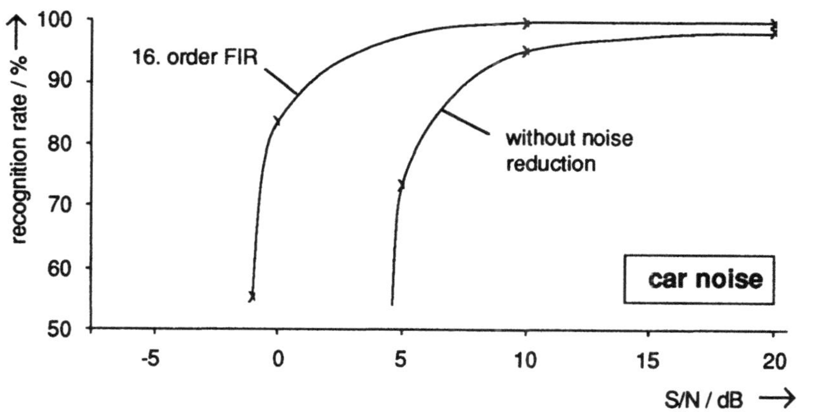

Figure 5. Recognition rates of noisy speech with and without noise reduction

In this application the improvement is not as significant as in the case of using white noise. The reason for this is the fact that the car noise is not stationary due to e.g. changing the gear or using the blink operator.

4. Conclusions

A method to reduce noise and reverberation has been presented which is based on high-pass filtering of subband energy signals. The main advantage of this method is the fact that no speech pause detection is necessary nor any kind of adaption.
A considerable improvement of recognition rate has been obtained with this method for isolated word recognition systems.

References

[1] Lim,J.S.: Speech Enhancement, Prentice-Hall, 1983

[2] Vary,P.: Noise Suppression by Spectral Magnitude Estimation - Mechanism and Theoretical Limits, Signal Processing, 1985, pp. 387-400

[3] Van Compernolle,D. et al.: Speech Recognition in Noisy Environments with the Aid of Microphone Arrays, Proc. European Conference on Speech Communication and Technology, 1989, Paris, pp. 657-660

[4] Zelinski,R.: A Microphone Array with Adaptive Post-Filtering for Noise Reduction in Reverberant Rooms, Proc. ICASSP 88, New York, 1988, pp. 2578-2581

[5] Allen,J.B. et al.: Multi-Microphone Processing Technique to Remove Room Reverberation from Speech Signals, JASA, Vol.62, No.4, 1977, pp. 912-915

[6] Hirsch,H.G. et Rühl,H.W.: Automatic Speech Recognition in a Noisy Environment, Proc. European Conference on Speech Communication and Technology, 1989, Paris, pp. 652-655

[7] Rühl,H.W.et al.: Speech Recognition in the Noisy Car Environment, Proc. European Conference on Speech Communication and Technology, 1989, Paris, pp. 262-265

[8] Hirsch,H.G.: Automatische Spracherkennung in Räumen, Verlag TÜV Rheinland, Köln, 1987

[9] Houtgast,T. et al.: Predicting Speech Intelligibility in Rooms from the Modulation Transfer Function. I. General Room Acoustics, Acustica 46, 1980, pp. 60-72

[10] Schroeder, M.R. and Atal,B.S.: Computer Simulation of Sound Transmission in Rooms, IEEE International Convention Record, Part 7, 1963, pp. 150-155

An ISDN speech server based on speaker independent continuous Hidden Markov Models

Klaus Zünkler

Siemens AG, Corporate Research and Development, Munich, W-Germany

Abstract

In this paper a real time prototype dedicated to single word recognition in ISDN lines is described. This system is speaker independent for a fixed hierarchical command set of totally 61 words. Context dependent continuous density Markov phoneme models are used. To improve recognition rates, a postprocessor based on information measures is proposed, which chooses the best word candidate in respect to transinformation.

In the first part the used speech recognition algorithms are presented. The second part deals with the ISDN speech database, the recording conditions and the achieved recognition rates. In the last part the hardware configuration of the speech server and the implementation of the described algorithms is explained in more detail. An outlook to future work concludes this contribution.

Keywords: Hidden Markov Model, real time speech recognition, information measure, ISDN speech server

1 Model Structure and Recognition Algorithms

The training and recognition process is based on continuous mixture density Hidden Markov Models for phonemes similar as used in the database inquiry project SPICOS (Siemens, Philips, IPO Continuous Speech Recognition) [4, 7]. The main difference is that context dependent phoneme models are used in the ISDN speech server. The Viterbi algorithm is used for both training and recognition [3, 5].

Some assumptions concerning the continuous mixture densities are made to achieve real time processing:

• The contribution of the nearest mixture density is much larger than the contribution of the others. So the probability is computed only for the best fitting mixture density.

• The variances of all mixture densities are considered to be equal. The recognition results will show that this is no limitation, if there are enough mixtures available for modeling the density functions.

• Each mixture density is assumed to be of Laplacian instead of Gaussian type.

With these simplifications the emission probability depends only on the mean of the best fitting mixture density and its mixture weight. After moving to the logarithmic domain, the computation of the probability simplifies to a sort of Euklidean distance. With

NATO ASI Series, Vol. F 75
Speech Recognition and Understanding. Recent Advances
Edited by P. Laface and R. De Mori

the assumption of Laplacian density functions it further simplifies to the computation of the city-block distance.

In an experiment we compared the recognition rate with and without the assumption of equal variance for all mixture densities. In the second case we assumed that the variance depends on the model and on the feature vector component. The variance of each mixture within a specified model and vector component is regarded as equal.

The structure of the used Markov Models is as follows: the Markov models consist of three so-called segments, according to the beginning, middle and end of a phoneme.

Each segment consists initially of two states with tied emission and transition probabilities. Word models are concatenated of phoneme models according to a pronunciation lexicon. During recognition only phoneme sequences contained in the lexicon are used. In a first step we used word-dependent phoneme models. These are derivated from initial context-independent previously trained phoneme models by adapting on the special word context of each word. The recognition results of these word-dependent phoneme models were not very satisfying. The reason for that lies in the missing ability of discriminating confusable words. The pdf's for equal phonemes in different words differed, even in the same context. Thus these phoneme models had to be tied together. This leads to context dependent phoneme models.

We use trigram models on the segment layer, as shown in fig. 1. On the segment

Figure 1. Context dependent word model of word 'drei (three)' (transition probablities not completely displayed)

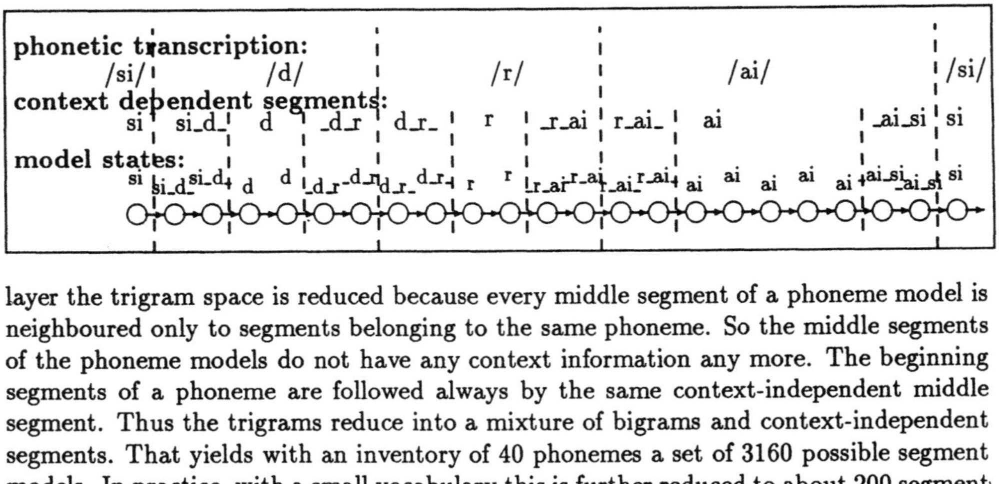

layer the trigram space is reduced because every middle segment of a phoneme model is neighboured only to segments belonging to the same phoneme. So the middle segments of the phoneme models do not have any context information any more. The beginning segments of a phoneme are followed always by the same context-independent middle segment. Thus the trigrams reduce into a mixture of bigrams and context-independent segments. That yields with an inventory of 40 phonemes a set of 3160 possible segment models. In practice, with a small vocabulary this is further reduced to about 200 segment models.

Another improvement is the incorporation of duration into the models. During the training procedure the durations of staying in a segment model are counted by means of histograms. The mean duration will be used then for building the word model. In the example of fig. 1 the mean duration of staying in segment /ai/ is 5, so the segment model gets five states with tied emission probabilities in order to limit the training effort. The resulting probability of segment occupancy can be seen in fig. 2 as dotted line.

The recognition results using this duration modeling are presented in section 3.

Figure 2. Probability of segment occupancy

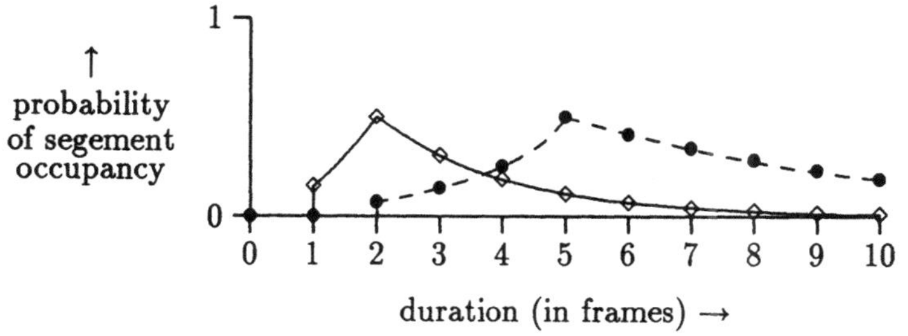

1.1 Recognition using Information Measures

To handle the problem of confusable words, it is necessary to have a technique to discriminate these words better.

This can be done with a postprocessor, looking more exactly at the differences of the best matching word candidates from the HMM classifier. The principle of the postprocessor is as follows: 1) take the ranked output list of word candidates from the HMM decoder, 2) take the top two words, 3) look at the discriminating parts of the two words and compute new probability based on the precomputed transinformation of corresponding parts of the two words, 4) decide for the better candidate and take next candidate from the list of hypotheses, 5) go to step 3) until a specified number of candidates is reached. The recognized word is the remaining candidate after the last comparison.

For discriminating corresponding parts of confusable words the transinformation is used [6]:

$$J_T = \sum_{k=1}^{K} \int_{\vec{x}} p(k) p(\vec{x}|k) \mathrm{ld} \frac{p(\vec{x}|k)}{p(\vec{x})} \, \mathrm{d}\vec{x} \tag{1}$$

In this case in every loop in the postclassifier only two classes are taken into account. This has two advantages: the ability of discrimination between the two phoneme models is maximized and secondly, the n-class classifying problem is reduced to a 2-class problem, which simplifies the computation of the transinformation.

Fig. 3 illustrates this with the example of two highly confusable german digits. These two words are different only in their first part. The indicated paths are the result of the Viterbi search. Now each pair of points at the same time t is compared using the information measure. The weighted scores will be summed up over the two paths and the better word candidate will be taken.

2 Database, Recording Conditions and Recognition Results

The database consists of 61 command words to cover the functionality ISDN offers. It contains 11 german digits (including the two variants of the digit 2: 'zwei' and 'zwo'). Examples for further commands are 'Gebühren (fees)' to ask for the current state of the monthly phone bill, or 'Umleitung (call diversion)' to tell the system that you can be reached under a different number, so that the call is directly diverted to that new number.

Figure 3. Principle of the postclassifier

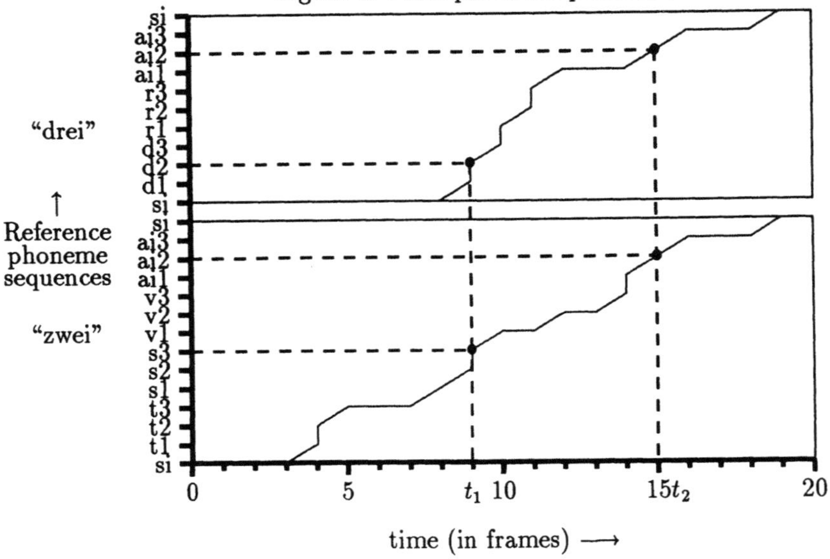

Other commands are dedicated to different information services, as train schedule and so on. We have collected now in average four sessions from 60 speakers. For the recognition tests subsets of the vocabulary are used according to the hierarchical structure of the ISDN application. We used the 11 digits and 13 ISDN commands.

The speech data is recorded with a Codec and the filter circuits of an original ISDN telephone device. The digitized signal is put into the decoder part of the Codec again thus simulating a digital ISDN line. This signal is sampled again at 8 kHz with a 12 bit A/D converter connected to a workstation.

3 Recognition Results

In the recognition tests the ISDN commands are significantly better than the digits, because these words are longer and sound not very confusable. The recognition rates between the test speakers varied strongly due to the small database.

The parameter in the first experiment is the computation with or without variance of the probability density functions in the Hidden Markov Models. The results show no improvement by taking into account the variance. On the opposite, the recognition rate drops from 98.7 % to 95.8 % with the ISDN commands. In the case of digits, the recognition rate drops only by 0.5 %. The reason is that the size of the training data not sufficient for estimating the additional parameters. This becomes obvious when we look at the number of feature vectors contributing to a single mixture mode. Many mixture modes are estimated only by very few feature vectors. So the difficulties to estimate robust variances are clear.

Another parameter is the incorporation of duration modeling. This yields an improvement of the recognition results from 96.9 % to 98.7 % in the case of ISDN command recognition and an increasement from 91.9 % to 94.6 % with the digits.

The last parameter is related to the estimation of mixture densities. In order to decrease the influence of outliers the number of vectors contributing to a mixture density have a lower limit of minimal 1 – 10 vectors. An increasing limit of the minimal number of vectors per mixture decreases the total number of mixtures until the optimal value of 4 vectors is reached and improves the recognition rate for digits from 89.4 to 94.6 %. The additional use of the postclassifier reduces the absolute error rate to 3

4 Hardware Configuration

The prototype system consists of mainly three subsystems. The first one is a PC connected to two ISDN lines using special interface boards. So the system is able to simulate an operator connecting two participants. The incoming speech signal is sent online via ethernet to the next subsystem, a workstation dedicated to the acoustic search and control of the last subsystem. This consists of a set of two special boards developed for signal processing and speech recognition, as described in [1, 2]. The configuration can be seen in fig. 4.

Figure 4. Prototype of ISDN-server

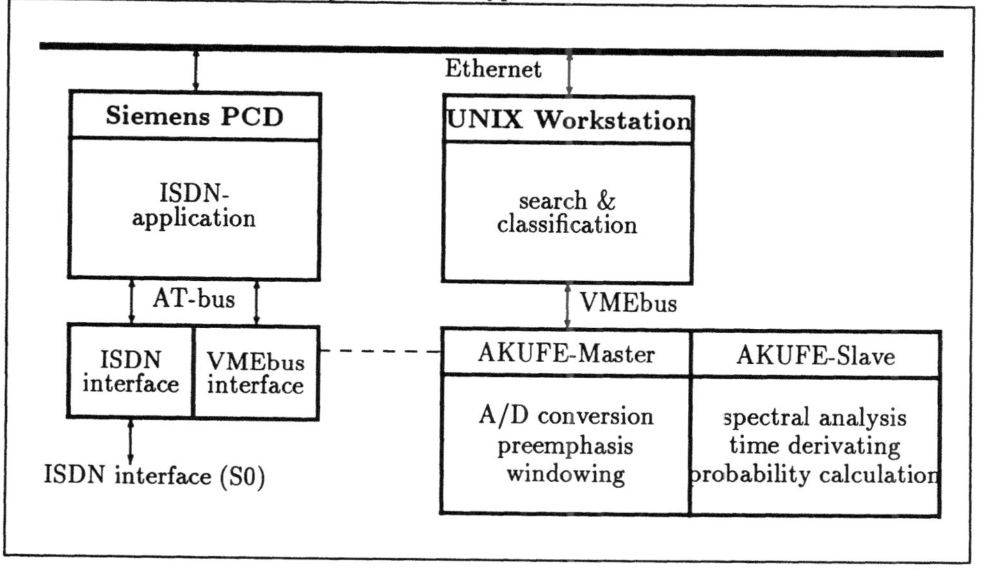

Three signal processors compute the Mel-scaled filter bank energies and their time derivatives and prepare the data for the vector quantizer chip (VQP), which computes the local emission probabilities for the Marcov Models. The chip is designed for the computation of both city block and Euklidean distances corresponding to Laplacian or Gaussian density functions, respectively. So for each used subword model the vector quantizer chip computes the distance to the mean of the best fitting mixture density. The system is able to recognize in real time, since the search on the Workstation is fast enough to evaluate the Hidden Markov Models in parallel.

5 Conclusion and Future Work

There are two main topics of the future work: increasing recognition rate and secondly, integrating the search into the recognition boards thus circumventing the need for a workstation. The work of this paper has shown the usefulness of duration modeling in the task of single word recognition. A refinement of the duration modeling by using not only a mean duration, but additionally variance or more complex distributions would promise even better results. It has been turned out, that the finding of the correct path is crucial for the discriminative postclassifier. So more work has to be done on this point.

To improve the recognition rate it is intended to modify the training algorithms towards discriminative training using information measures of the training material. For this a further method to compute the discrimination weights has to be implemented to reduce the still large training effort.

So in the near future all speech processing including search will be done on the set of recognition boards to get a compact system connected to the ISDN network.

References

[1] A. Aktas and H. Höge. Multi-DSP and VQ-ASIC based acoustic front-end for real-time speech processing tasks. In *Proceedings of Eurospeech*, pages 586 – 589, Paris, September 1989. EUROSPEECH.

[2] A. Aktas and H. Höge. Real-time recognition of subword units on a hybrid multi-DSP/ASIC based acoustic front-end. In *IEEE International Conference on Acoustics, Speech and Signal Processing*, pages 101 – 103, Edinburgh, May 1989. ICASSP.

[3] H. Ney. A script-guided algorithm for the automatic segmentation of continuous speech. In *IEEE International Conference on Acoustics, Speech and Signal Processing*, pages 1209 – 1212, Tampa, 1985. ICASSP.

[4] H. Ney, D. Mergel, A. Noll, and A. Paeseler. *Recent Advances in Speech Understanding and Dialog Systems*, volume F46 of *NATO ASI Series*, chapter Overview of Speech recognition in the Spicos System, pages 305 – 309. Springer, Berlin Heidelberg, 1988.

[5] H. Ney and A. Noll. Phoneme modeling using continuous mixture densities. In *IEEE International Conference on Acoustics, Speech and Signal Processing*, pages 437 – 440, New York, April 1988. ICASSP.

[6] T. R. Vilmansen. Feature evaluation with measures of probabilistic dependence. *IEEE Transactions on Computers*, 22:381 – 388, April 1973.

[7] K. Zünkler. Speech-understanding systems: The communication technology of tomorrow. In H. Schwärtzel and I. Mizin, editors, *Advanced Information Processing, Proceedings of a Joint Symposium Information Processing and Software, Systems Design Automation*, pages 227 – 251. Springer, Berlin Heidelberg New York, June 1990.

RAMSES: A Spanish Demisyllable Based Continuous Speech Recognition System[1]

J.B. Mariño, C. Nadeu, A. Moreno, E. Lleida, E. Monte, A. Bonafonte

E.T.S.I. Telecomunicación, P.O. Box 30002, 08080 Barcelona, Spain

Abstract: A continuous speech recognition system (called RAMSES) has been built based on the demisyllable as phonetic unit and tools from connected speech recognition. Speech is parameterized by band-pass lifted LPC-cepstra and demisyllables are represented by hidden Markov models (HMM). In this paper, the application of this system to recognize integer numbers from zero to one thousand is described. The paper contains a general overview of the system, an outline of the grammar inference, a description of the HMM training procedure and an assessment on the recognition performance in a speaker independent experiment.

Keywords: Speech Recognition, Hidden Markov Model, Demisyllable, Speech Recognition System.

1. INTRODUCTION

During the last two years, a continuous speech recognition system based on demisyllables and discrete hidden Markov models (HMM) has been built in our laboratory. Demisyllables afford a convenient phonetic coding of Spanish utterances, according to the syllabic character of this language. In order to define the demisyllable set, every possible syllable was divided by the strong vowel into an initial demisyllable and a final demisyllable. In our definition the prosodic stress was incorporated to the final demisyllable; accordingly, we distinguished between stressed final demisyllables and unstressed final demisyllables. The main cues of prosodic stress in Spanish are pitch, loudness and syllable length; as pitch and loudness informations are not considered in our system, the main difference between stressed and unstressed final demisyllables is the length of their references. The inventory of Spanish demisyllables is relatively small: less than 750 units. Hidden Markov models have been shown to be a successful tool for describing in a probabilistic way the acoustic features of speech. Our system has been called RAMSES, Spanish acronym for "*automatic recognition by means of semisyllables (demisyllables)*". In this paper we provide a general overview of RAMSES and report its application to recognize the Spanish integer numbers from zero to one thousand, in both multispeaker and speaker independent tasks.

The paper is organized in the following way: in Section 2 the block-diagram of RAMSES is described, Section 3 addresses the task oriented aspects, Section 4 describes the grammar inference, in Section 5 the HMM training procedure is outlined, Section 6 is dedicated to report the recognition experiment results, and finally Section 7 contains the main conclusions.

[1] This work was supported by the PRONTIC grant number 105/88

NATO ASI Series, Vol. F 75
Speech Recognition and Understanding. Recent Advances
Edited by P. Laface and R. De Mori

2. RAMSES' OVERVIEW

Figure 1 shows a general block-diagram of the system architecture. The speech signal is band-pass (100 Hz - 3400 Hz) filtered by an antialiasing filter and sampled at 8 kHz. The utterance is isolated by an end-point detection algorithm and preemphasized. A linear prediction (LP) based parameterization follows: the signal is segmented into frames of 30 milliseconds by a Hamming window at a rate of 15 milliseconds, and every frame is characterized by a LP-filter with 8 coefficients. Afterwards, 12 band-pass lifted cepstrum coefficients are computed [2]; the energy of the frame completes the parameterization. Before entering the recognition algorithm, the system evaluates the spectral difference d(t) corresponding to the frame t by using [1]:

$$d(t) = \sum_{k=-2}^{2} k\ s(t+k)$$

where s(t) is the cepstral vector in frame t. This difference implies a time-average along 90 milliseconds. In a similar way, the energy difference e(t) is calculated. The spectral vector and the spectral and energy differences are vector-quantized separately; in that way, every frame of speech signal is represented by three symbols.

According to the most recent proposals, RAMSES considers energy and time evolution information. However, in our system, the energy is not used directly as a parameter of the signal. This is because the energy depends on the prosody of the sentence and the intensity of the utterance, two very fluctuant features of speech. On the contrary, if the energy is expressed by a logarithmic measure, its difference does not vary with a change in the intensity of the overall sentence, and the variation due to prosodic effects is greatly alleviated.

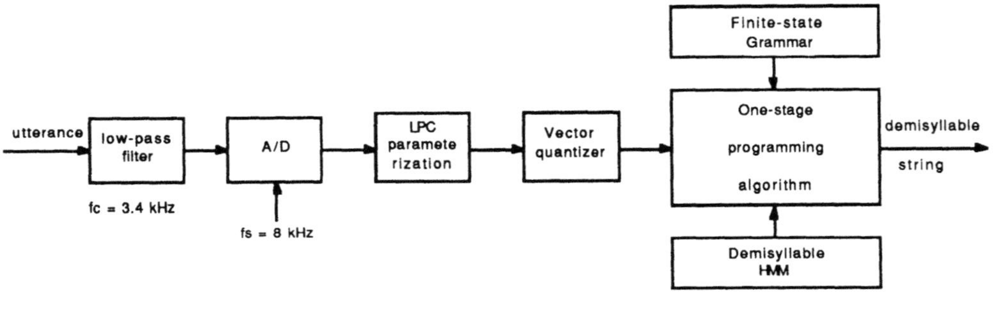

Figure 1. Recognition System architecture

The recognition algorithm performs an one-stage dynamic programming (described for connected word recognition [5]) driven by a finite state grammar. So, the algorithm computes the string of demisyllable models that provides the most likely path of states throughout the utterance and, at the same time, satisfies the grammar constraints. If necessary, a dictionary provides the semantic meaning of the issued sequence of demisyllables.

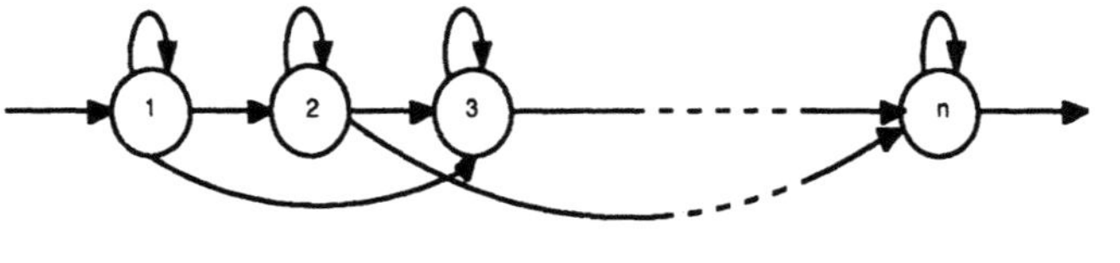

Figure 2. HMM structure

Figure 2 shows the structure used for the hidden Markov models. It is a typical left-to- right structure, that allows to skip one state when the model makes a transition between states. The emission of symbols is associated to the states, that issue three independent symbols (spectrum, spectrum difference and energy difference) when they are visited. The number n of states is a parameter to be determined. During the recognition task, the transition probability between the final state of a model M_i to the first state of the following model M_{i+1} is determined by the duration probability of the demisyllable modeled by M_i. The length probability of a demisyllable is parameterized by the mean and the variance of a gaussian distribution.

This general architecture can be oriented to a specific application by designing the regular grammar and training the demisyllable models.

3. NUMBER RECOGNITION TASK

The recognition of numbers is an interesting experiment in continuous speech recognition research. In addition to its potentially practical application, the recognition of numbers reproduces at reduced scale the general problem of continuous speech recognition: light acoustic differences may correspond to great semantic differences, the numbers exhibit a strong grammatical structure, and there is a relatively important sound variability in the acoustic realization of the numbers.

From ten speakers (6 male and 4 female) a speech data base was acquired in our laboratory. Every speaker uttered one realization of a set of 44 numbers, designed in such a way that: a) every set included at least two samples of each necessary demisyllable in the application, and b) the 44 numbers performed a suitable sampling of the integers from zero to one thousand. The articulation rate of speech spanned from 5 to 7 syllables per second. This data base was segmented by hand into demisyllables and labeled.

D = {uno (1), dos (2), tres (3), kωatro (4), θiŋko (5), sejs (6), sjete (7), otʃo (8), nωeβe (9)}

T = {θero (0), djeθ (10), onθe (11), doθe (12), treθe (13), katorθe (14), kinθe (15), djeθisejs (16), djeθisjete (17), djeθjotʃo (18), djeθinωeβe (19), bejnte (20), trejnta (30), kωarenta (40), θiŋkωenta (50), sesenta (60), setenta (70), otʃenta (80), noβenta (90)}

P = {bejnti (2...), trejnta i (3...), kωarenta i (4...), θiŋkωenta i (5...), sesenta i (6...), setenta i (7...), otʃenta i (8...), noβeŋta i (9...)}

Table 1. Basic vocabularies used to build the Spanish integers from zero to ninety nine.

4. GRAMMAR INFERENCE

Let us consider the vocabularies in Table 1. Vocabulary D corresponds to the digits excluding the zero; vocabulary T includes the zero, integers from ten to nineteen and multiples of ten; finally, vocabulary P is built by the prefixes to form the integers from twenty one to twenty nine (2...), from thirty one to thirty nine (3...) and so on. The vocabulary I of the integers from zero to ninety nine can be obtained by the following combination of the three previous vocabularies:

$$I = (D + T) + (P * D)$$

where the symbol + means the union of vocabularies and the symbol * denotes the cartesian product (every item in P can be followed by every item in D). The grammar G corresponding to I is set up from the grammars of D, T and P by using the tools described in reference [4]. Following a similar procedure, the grammar of Spanish integers from zero to one thousand can be set up. For this application we need 67 demisyllables and a grammar with 118 states. Finally, the grammar of the vocabulary formed by all possible strings of elements of I (the telephone numbers are a subset of such a vocabulary) can be derived by joining the initial and final states in the grammar G into one only state (self-chaining of the grammar G)

In order to describe every item in vocabularies D, T and P as a string of demisyllables a standard phonetic transcription was performed, including the most frequent allophonic variations; thus, one item was allowed to be represented by several different demisyllable sequence. Furthermore, the combination P*D of the vocabularies P and D and the self-chaining of grammar G bring new combinations of sounds, whose coarticulations have to be considered. Our tool for designing regular grammars is well suited to cope with this problematic; the question is reduced either to change a subsequence of demisyllables by another (if the coarticulation effect is considered to happen every time that that combination of sounds appears) or to add a new subchain of demisyllables.

5. DISCRETE HIDDEN MARKOV MODEL TRAINING

Each model was trained independently of the others. Once the samples of every demisyllable were collected from the utterances recorded by the speakers, the Baum-Welch estimation algorithm was applied. At the same time, the mean and the variance of the length of the demisyllable was computed. Finally, the demisyllable models were smoothed according to the co-occurrence probability method introduced in [3].

Previously to apply this procedure, the values for some important parameters of the models had to be determined, i.e., the size of the three codebooks and the number of states. In order to assess the choice, some training and recognition experiments were accomplished. Specifically, the value for those parameters were fixed and the models were trained with the ten speakers; afterwards, the signals in the data base were recognized. Then the parameter values were modified, and the training and recognition procedures were carried out; and so on. As a result of these trials, we drew the following conclusions:

a) as far as the size of the codebooks is concerned, the most suitable choices are: 64 for the two codebooks dedicated to spectral information and 32 for the codebook devoted to energy differences. Although similar performance can be got with other parameter selections, this option requires the minimum codebook sizes.

b) the recognition performance is noticeably dependent on the number of states of the hidden Markow models. Several criteria to determine the most suitable number of states for every model were tested: equal number of states, number of states according to the number of sounds included in the demisyllable, and number of states as a function of the average length of the demisyllable. This third criterion yielded the best performance for almost every experiment carried out, and when it did not lead to the best choice, it afforded a performance near the optimum. As a consequence, we used this criterion in our final design. In Table 2 the definition of the average length criterion is provided.

average length in frames	number of states
≤ 4	2
5,6	3
7,8	4
9,10	5
>10	6

Table 2. Criterion to select the number of states of HMM as a function of the average length of the demisyllables

6. SPEAKER INDEPENDENT EXPERIMENT

Although we acknowledge that our data base is rather reduced, we were interested in carrying out some experiments that allowed to ascertain the ability of RAMSES to cope with speaker independent tasks. To this aim, we made six different training and recognition trials. In each experiment we trained the system with 8 speakers, and then we recognized the speech signals of the other two; in every case, the outside training speakers were taken with different sex.

	before smoothing		after smoothing	
	trained	not trained	trained	not trained
M1*	0.4	0.7	0.0	0.0
M2	2.3	6.8	2.7	6.8
M3	1.4	4.5	3.2	4.5
M4**	0.0	12.1	2.7	6.9
M5	0.0	27.3	6.4	11.4
M6	0.0	2.3	0.5	0.0
F1	0.0	9.0	0.0	3.4
F2	0.0	3.4	0.0	0.0
F3	0.0	4.5	2.3	2.3
F4	0.0	0.0	0.0	0.0
M	0.6	7.0	1.9	3.8
F	0.0	4.9	0.6	1.5
Total	0.4	6.1	1.5	2.8

The total number of utterances of this speaker is: * 136, ** 58

Table 3. Recognition error percentage

Table 3 provides the recognition error percentage achieved for every speaker, when he or she was inside and outside the training set. The results before and after the smoothing of the HMM output probabilities is also shown. We count as one error every number recognized incorrectly, indepently of the number of demisyllables misrecognized. It is worth mentioning that, in most of the cases, the errors affected only one digit in the number (corresponding either the hundreds, or the tenths or the units); for instance, 677 was recognized as 637, or 721 as 621.

From Table 3 we can observe the following facts:

a) The smoothing afforded a remarkable decreasing of recognition errors outside the traininig set; however, the prize to be paid was an increasing of recognition errors inside the training set.

b) For some speakers (for instance, M4, M5 and F1) the RAMSES recognition ability was very different when the speaker was either inside or outside the training set.

c) The recognition performance fluctuated greatly from some speakers to others. This behaviour is much more evident for the speakers outside the training set.

d) The average performance (2.8% of error percentage) was satisfactory.

7. CONCLUSION

Our interpretation of the enunciated observations is twofold. Firstly, RAMSES seems suitable for recognizing number in continous speech, either in a multispeaker task (every speaker inside the training set) or in a speaker independent application (all of speakers outside the training set); and secondly, the data base required for training this latter case must be increased.

Currently, we are recording a new data base, involving 20 new speakers and utterances with strings of integer numbers from zero to one million.

REFERENCES

[1] Furui S., "Speaker-independent isolated word recognition using dynamic features of speech spectrum", IEEE Trans ASSP-34, pp. 52-59: February, 1986

[2] Juang B. H. et al.,"On the use of bandpass liftering in speech recognition", IEEE Trans ASSP-35, pp. 947-954: July, 1987

[3] Lee K. -F. and Hon H. -W., "Speaker-independent phone recognition using hidden Markov models", IEEE Trans ASSP-37, pp. 1641-1648: November, 1989

[4] Mariño J. B. et al., "Finite state grammar inference for connected word recognition", Proc. EUSIPCO'88, pp. 1035-1038: September, 1988

[5] Ney H., "The use of an one-stage dynamic programming algorithm for connected word recognition", IEEE Trans ASSP-32, pp. 263-271: April, 1984

Speaker Independent 1000 Words Speech Recognition in Spanish

J.M.Pardo, H.Hasan, J.Colás

Electronic Engineering Department, E.T.S.I. Telecomunicación, Universidad Politécnica de Madrid, Ciudad Universitaria, 28040 Madrid, Spain

Abstract: In this paper a speaker-independent speech recognition system for 1006 isolated words in Spanish is presented. The approach used is based on discrete Hidden Markov Models. It is a first effort to develop a system with these characteristics in our Department. The initial task has been to recognize isolated words but the final objective is getting a speaker-independent continuous speech recognition system for large vocabularies (more than 1000 words) in our language. The different phases of building our system, its characteristics and used techniques are described. Speech data base characteristics and the process of collecting it are commented. Finally, some relevant experiments which have been achieved are presented and different future improvements will be commented.

Keywords: Speech recognition, speaker-independent, hidden-Markov-models

Introduction

One of the main objectives in automatic speech recognition is to get a system with the ability to decode continuous speech uttered by a speaker using large vocabularies and with an acceptable error rate. Efforts to get a prototype in this sense are being done in our Department for Spanish [1]. In the last years, hidden-Markov-model-based (HMM) algorithms have been the most successful techniques [2,3,4,5,6,7,] used for speech recognition systems. We have started developing a speaker-independent, 1006 isolated words recognition system based on context-independent allophone hidden-Markov-models which are concatenated to build the word HMM for the whole vocabulary (1006 words).

If we wanted to use it as a connected words recognition system (with a pause between words), the perplexity would be 1006 because we don't work with any kind of grammars [4].

Two phases can be considered in the development of the system :

1. Initial phase. A basic approach is created.
2. Final phase. The system optimization is performed.

1. Initial Phase.

We use the Markov techniques to implement our system (forward-backward) algorithms and Baum-Welch re-estimation formulas for training and Viterbi algorithm [8] to recognize). We build, from a graphemic-allophonic transcription dictionary of the 1006 words, the word hidden-Markov-model by concatenating the correspondent context-independent allophone hidden-Markov-models. Also, we collect the speech data base which will be used to train and test the system.

NATO ASI Series, Vol. F 75
Speech Recognition and Understanding. Recent Advances
Edited by P. Laface and R. De Mori

2. Final Phase.

By testing the initial approach, it's possible to detect the symptoms (those causes generate errors) of our system and to test different and appropriate optimization techniques which improve the performance of it. We are now in the second phase, in an advanced point.

System Description

The speech signal is bandlimited to 6.4 kHz and sampled at 16 kHz. Spectral analysis is performed to get 10 Mel Frequency Cepstrum coefficients. The frame energy is calculated and integrated with the MFCC to form a vector of 11 parameters [9]. The frame duration is 16 ms and overlapped Hamming windows are used in this analysis, with 6.25 ms displacement (9.75 ms overlap). We use discrete HMM to model the context-independent allophones (33 allophones) [10] which are in the vocabulary (1006 isolated words). A codebook of 256 prototypes is trained with 60,000 speech frames from different speakers of our speech data base using the Lloyd algorithm [11].

Training Procedure

The system models speech as a time series of vectors generated by a Markov source. The training is made using the forward-backward algorithm and the Baum-Welch re-estimation formulas.

Initial context-independent allophone hidden Markov models are trained with phonemically labeled words by hand (93 words which contain all the allophones). The allophone models are concatenated to build the 1006 word hidden-Markov-models from the dictionary and the forward-backward algorithm is iterated several times. In each iteration, the model parameters are re-estimated using the Baum-Welch formulas. Scaling is used to prevent underflow [12]. It wasn't necessary any speech data segmentation during training although we needed a dictionary with the graphemic-phonemic transcription for the 1006 words of the vocabulary. The re-estimation of the parameters of each phone model involves the sharing of the probabilities among all different words that contain the phone.

The use of units shorter than a word to train is due to the lack of speech data for the process what would produce no robust word Markov models. We use the left to right hidden-Markov-model with three states without skipping arcs.

We have an initial silence model and another final one which we concatenate with the other allophone models to form the word models of the dictionary. We use 25 ms of silence at the beginning and at the end of each word.

Recognition Process

We have used a fast prunning algorithm to select the 200 candidates which are closest to the unknown word from the 1006 word models, so the necessary time to verify the pronounced word is significantly reduced. The vector sequence obtained from the unknown speech input is fed to the recognizer, which applies the Viterbi algorithm, with all the preselected word models. The word that gives the highest accumulated probabilities is recognized.

Speech Data Base

The collection process of the speech data base that we use with our experiments has three main steps:

1. Analog recording of 110 chance chosen words lists from the vocabulary (1006 words) which are pronounced by the different speakers. Two different microphones, three different tape recorders, and five different acoustic quality rooms are used. Digit strings and sentences were recorded too, although they were not used for the experiments reported here.
2. Digitization and storing the speech data.
3. Segmentation, labeling and verification processes.

The vocabulary is formed by the 1006 most frequent words in Spanish extracted from a corpus of 1,500,000 words of spontaneous speech. The speech data base consists of 66 speakers with 110 words each one. There are 45 speakers for training and 21 to test the system.

Optimization Phase

We have added the following elements for improving the performance of the system:
1. Duration model of the allophones.
2. Using more precise models (45 allophone models).
3. Smoothing of the allophone models badly trained.
4. Using differential parameters in addition to static parameters.

Modeling the duration of the allophones

Using an automatic word-segmentation algorithm based in the Viterbi algorithm we segment all the words to train their correspondent allophones and so, some statistics were performed about the duration (in number of frames) of each allophone. With these values, 5 duration histograms were computed and stored in a discrete way in 5 tables which represented all the allophones of our vocabulary. The duration histograms were normalized to have a range between 0.1 and 1.

The resulting formula follows, where the initial log Prob is the Viterbi probability, l_i is the length of the phoneme in frames and W_i is the table value for phoneme i. The weight (25) was calculated experimentally [10].

$$\log \text{Prob} = \text{Initial Log Prob} + (25/N) \log \prod_{i=1}^{N} W_i (l_i)$$

Using more precise models

New prototypes (see tables) were defined to include acoustic events with more detail. The new prototypes of the explosive and affricate allophones represent acoustic units which correspond to parts of sounds that show some stationarity. The hidden-Markov-models of these prototypes are trained with more precision due to the reduction of the variability related to the previous definition of the prototypes.

The HMM of the vocalic prototypes represent their respective units in a more precise way as result of a better definition of the same prototypes including prosody information as fundamental frequency and energy.

Alloph.	Stres.	Nas.	Others
/a/	á	a~	a
/e/	é	e~	e
/i/	í	i~	i
/o/	ó	o~	o
/u/	All repres.	by u	

Allophone	2nd part	1st part
[N/]	TR	N/

Table 1. New prototype set for vocalic allophones and for Ñ

Alloph.	Tension	Friction
[T/]	TAS	T/
[J/]	The same	for both

Alloph.	Tension p.	Explos. p.
[b]	SNR	b
[d]	SNR	d
[g]	SNR	g
[p]	SRD	p
[t]	SRD	t
[k]	SRD	k

Table 2. New prototype set for plosives and affricates

Smoothing of the context independent allophone H.M.M.

The smoothing of the discrete pdf is a technique to compensate the lack of speech data in the training process of the models. These techniques use a confusion matrix between the symbols which contains the correlation between them.

We have tested two techniques to calculate this confusion matrix : the coocurrence method and the Parzen method [15]. We describe the first of them because the best results were obtained with it [10].

$$P_{smoothing}(i) = P_{original}(i) \times T;$$

being : T = confusion matrix, P(i) = discrete pdf associated with state i.

$$P_{final}(i) = w \times P_{original}(i) + (1 - w) \times P_{smoothing}(i)$$

being : w = min [0.99,1/2 log R], R : number of repetitions for that allophone.

So, we take into account the number of repetitions (R) of that allophone. The elements of the matrix T, c_{ij} are computed in the following way :

$$c_{ij} = \sum_m N(m) \times \sum_s OC_{ij}^{sm}$$

being : m = model, N(m) = number of repetitions to train the model m, s = state, OC_{ij}^{sm} = $p(v_i/s) \times p(v_j/s)$ and $p(v_i/s)$ is the probability of vector v_i in state s.

The difference of our approach compared to others [15] is related to the way of using these techniques. Normally, a different matrix T_k is calculated for each context independent allophone model using the information of all their contextual models related with the k model, but we have only used context independent models and only a matrix T for all of them is performed. So, we introduce spectral information only in the poorly trained models (they are few).

Integrating the Transitional Information of the Spectrum

The properties of the transitional information of the spectrum [10] are several:

- The trajectory of the formants is less sensitive to spectral differences and permits to build more robust speaker-independent recognition systems.
- It is more robust to the variations of the transmission channel [14].
- The static and dynamic information of the speech signal brings complementary de tails because they are uncorrelated [10].

So, we have decided to include the delta cepstrum parameters and the delta energy parameter, building a new codebook with the new information, independently of the first codebook and integrating both in the emission probabilities (B matrix of the model) as the product of two probability density functions. In this way, we reduce the distortion error produced by the VQ process which would be greater if all the parameters were integrated in one codebook (22 parameters).

The expression to compute these new coefficients is:

$$\frac{\delta Cm(t)}{\delta t} \approx \blacktriangledown dC(t) = \frac{\sum_{k=-K}^{K} k * C(t+k)}{\sum_{k=-K}^{K} k^2} \qquad K = 2$$

Experiments

In the following table, the obtained results with the system under different conditions are presented. The training was performed with 45 speakers and the test with 21 different speakers. In each experiment, the 110 words which correspond to each speaker are recognized.

Expe.	Param.	N.Models	Duration	Smoothing	Recog.-rate
1	MFCC+Ener.	33	NO	NO	58.4%
2	MFCC+Ener.	33	YES	NO	60.29%
3	MFCC+Ener.	45	NO	NO	63.13%
4	MFCC+Ener.	45	YES	NO	64.35%
5	MFCC+Ener.	45	YES	YES	64.62%
6	MFCC+Ener. Delta Coeff.	45	NO	NO	66.23%

Future Directions

The presented recognition rates in the different experiments we have realized are not very high. The causes are several, one of them is the diversity of conditions of our database. We hope to use different improvements. A first step can be :

1. To build contextual allophone models.
2. To smooth these models conveniently.
3. To increase the number of independent codebooks and the size of the speech data base to train well these codebooks.
4. Using speaker-adaptation techniques.
5. Using continuous HMM.
6. Apply the state duration and word duration in the recognition algorithm.

Conclusions

As we are able to probe, the design and development of a recognition system for large vocabularies, speaker-independent and isolated words is not an easy task. Nevertheless it is a quasi obligated phase in the interesting way towards the development of a continuous speech recognition system.

References

[1] J.M. Pardo and H. Hasan, "Large Vocabulary speaker-independent isolated word speech recognition using hidden Markov models : status report and planned research". EUROSPEECH 89 Conference, Paris September 1989.

[2] Y.L. Chow, M.D. Dunham, O.A. Kimball, M.A. Krasner, G.F. Kubala, J. Makhoul, P.J. Price, S. Roucos, and R.M. Schwartz "BYBLOS : The BBN Continuous Speech Recognition System". Proc. ICASSP-87 pp. 89-92 April 1987.

[3] F. Jelinek "The Development of an Experimental Discrete Dictation Recognizer". IEEE Proc., Vol. 73, No 11, pp. 1616-1624, Nov. 1985.

[4] K. Lee and H. Hsiao-Wuen, "Large Vocabulary Speaker-Independent Continuous-Speech Recognition Using HMM". Proc. ICASSP 88 pp. 123-126 April 1988.

[5] D.B. Paul, "A Speaker-Stress Resistant Isolated-Speech Recognizer". Proc. ICASSP 87 pp. 713-716 April 1987.

[6] L. Rabiner and B. Huang, "An Introduction to Hidden Markov Models". IEEE ASSP Magazine, pp. 4-16, Jan. 1986.

[7] Rabaey J., Brodersen, R., Stoelzle, A., Chen, D., Narayanaswamy, S., Yu, R., Schrupp, P., Murveit, H., and Santos, A., "A Large-Vocabulary Real-Time Continuous-Speech Recognition System". in VLSI Signal Processing, III, edited by R.W. Brodersen and H.S. Moscovitz, IEEE Press, New York, NY, 1988. [8] Levinson, S.E., Rabiner, L.R. and Sondhi, M.M., "An Introduction to the application of the theory of probabilistic functions of a Markov process to automatic speech recognition". B.S.T.J., vol.64, part 1, pp.1211-1234, 1983.

[9] Davis, S.B. and Mermelstein, "Comparison of parametric representations for monosyllabic word recognition in continuously spoken sentences", P. in IEEE Transactions on Acoustics,Speech, and Signal Processing, vol. ASSP-28, no. 4, pp. 357-365, 1980.

[10] Hasan, H., "Speaker-Independent Hidden Markov Model Based Speech Recognition of 1000 Words", PhD thesis, Electronic Engineering Department, Univ. Politécnica de Madrid, July, 1990.

[11] Gray, R.M., "Vector Quantization", IEEE ASSP Magazine 1(2):4-29, April, 1984.

[12] Hasan, H., Ferreiros, J., "Report about scaling in the re-estimation of Hidden Markov Models", Internal Report, Electronic Engineering Department, E.T.S.I. Telecomunicación, Universidad Politécnica Madrid, no.-5-89,Oc. 1989.

[13] Soong, F.K. and Rosenberg, A.E., "On the Use of Instantaneous and transitional Spectral Information in Speaker Recognition", IEEE ICASSP 1986, pp. 877-880.

[14] Weintraub, M., Murveit, H., Cohen, M., Price, P., Bernstein, J., Baldwin, G., and Bell, D., "Linguistic Constraints in Hidden Markov Model Based Speech Recognition". Proc. ICASSP 89 Jersey.

[15]. Schwartz,R., Kimbal, O., Kubala, F., Feng, M.W.,Chow, Y.L., Barry, C.,Makhoul, J., "Robust smoothing method for discrete Hidden Markov Models" IEEE ICASSP-89 pp. 548-551, 1989.

Continuously Variable Transition Probability HMM for Speech Recognition

Alessandro Falaschi

INFO-COM Department, La Sapienza University, Via Eudossiana 18, 00184 Roma, ITALY*

Abstract: A new duration intrinsic model for improved speech recognition by HMM techniques is presented. Assuming an exponentially decaying time dependency of the states loop probability, the duration density can be factorized and a path early pruning theorem demonstrated. As a consequence, computational complexity is greatly reduced with respect to explicit models, whereas recognition performances improve considerably.

Keywords: Duration Models, HMM, Complexity Reduction

Introduction

Improved speech recognition performances can be achieved if speech sounds duration cues are accounted by a detailed statistical model of the HMM states dwell time. In the past few years, two different duration modelling philosophies have been proposed, being them classifiable as explicit and implicit models.

In the first case, a mechanism for Duration Probability Density Function (DPDF) evaluation is given for each state, and its parameters estimated for the best fit with actual data. Approaches of this kind can bring to either a discrete DPDF, as in the Ferguson model [1], or to continuous DPDFs, as by using Poisson [2-3] and Gamma [4] pdf for fitting the observed state duration statistics. The main drawback of these methods is that the computational complexity of the Dynamic Programming algorithm used for recognition increases dramatically, because of the two-fold maximization which has to be performed for dealing with Semi-Markov models. Some proposed shortcuts, as adding duration penalties in a post-processing stage[5], subsequent to acoustical decoding, may lead to sub-optimal procedures if applied in one-stage recognition algorithms, because of the risk of eliminating the correct hypothesis during recognition, being it unrecoverable at the following stage of scoring.

The implicit approach to duration modeling derives a DPDF as the consequence of assumptions on the HMM structure and characteristics. Examples of HMM structure modifications can be found in [6-8]. In particular, it has been shown [7,8] that some particular HMM topologies can well approximate some explicit DPDF. Although one is still dealing with traditional HMMs, the increase in the number of states required for giving a suitable DPDF still complicates the decoding process, thus producing also in this case a remarkable computational complexity increase. Finally, a Semi-Markov model can be defined if the state transition

* This work has been partially founded by ALCATEL-FACE. Only the author is responsible for the ideas and conclusions here reported.

NATO ASI Series, Vol. F 75
Speech Recognition and Understanding. Recent Advances
Edited by P. Laface and R. De Mori

probabilities are allowed to vary in time, as proposed in [9,10]. The aim of the present paper is to illustrate new findings relative to the approach described in [10].

Continuously Variable Transition Probability HMM

In our case, HMM state transition probabilities are supposed to vary according to an exponential decay law, so that the state loop probability (exponentially) decreases as a function of the time spent into the state.

Duration Density Derivation

Definition of non stationary loop and transition probabilities as

$$p(n) = e^{-\mu n} \qquad q(n) = 1 - p(n) = 1 - e^{-\mu n} \tag{1}$$

allows to evaluate the probability of remaining in the same state for k instants as

$$P_D(k) = (1 - e^{-\mu k}) \prod_{n=1}^{k-1} e^{-\mu n} = 2\, e^{-\frac{\mu k^2}{2}} \sinh \frac{\mu k}{2} \tag{2}$$

in which the last expression has been derived by using the equality $1 + 2 + \ldots + N = N(N+1)/2$. Eq. (2) possess a closed form expression for its mean value, given by [12]

$$m_D = \frac{1}{2} e^{\frac{\mu}{8}} \sqrt{\frac{2\pi}{\mu}} \tag{3}$$

which can be effectively utilized for deriving the μ parameter from the observed state mean duration by numerical methods [13]. Fig. 1 reports a comparison among (2) and the geometric DPDF $P_D(k) = (1-p)p^{k-1}$ arising in stationary HMM.

Complexity Simplifications

The basic formulation of the Viterbi algorithm used for recognizing speech by HMM is modified in presence of an explicit duration model, being it now based on the evaluation of the best path probability of entering state S_j at time τ_j and remaining there up to time $t > \tau_j$, indicated as $\alpha_t(j,\tau_j)$, and given by

$$\alpha_t(j,\tau_j) = \prod_{k=\tau_j}^{t} O_j(y(k)) \max_{S_i} \{ a_{ij} \max_{\tau_i<\tau_j} \{ P_{D_i}(\tau_j-\tau_i)\, \alpha_{\tau_j-1}(i,\tau_i) \} \} \tag{4}$$

where O_j and a_{ij} are the observation density and the transition probability respectively. Computation of (4) involves a two-fold maximization over the best previous state and relative entry instant. Moreover, the duration probability is accounted only when the state is exited.

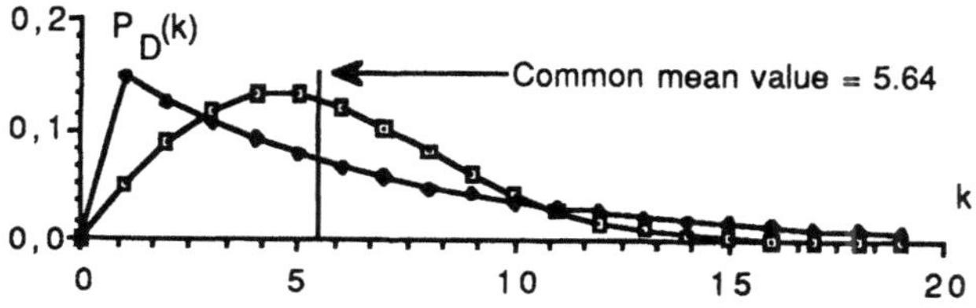

Fig.1 - Comparison among the DPDF shapes for the Stationary (●) and the CVTP (▣) cases

The most relevant feature of the implicit duration model here discussed is its factorization property expressed by the second term of (2), so that (4) may be rewritten as

$$\alpha_t(j,\tau_j) = \prod_{k=\tau_j}^{t} O_j(y(k))\, e^{-\mu_j(k-\tau_j)} \max_{S_i} \{ a_{ij} \max_{\tau_i<\tau_j} \{ (1-e^{-\mu_i(\tau_j-\tau_i)})\, \alpha_{\tau_j-1}(i,\tau_i) \} \} \quad (5)$$

so that the duration model contribution to the score is spread along time. This property allows the comparison of the path scores also for the same state under different entry instant hypothesis. Fig. 2 reports two competing paths for state S_i, being it entered at time h_0 (hypothesis H_0) and h_1 (hypothesis H_1), and which will merge in state S_j at some future (unknown) instant k_1. At that time, the best previous path for S_j will be chosen on the basis of the path probability ratio

$$\mathrm{PPR}\ (j, k_1) = \frac{\alpha_{k_1-1}(i,h_0)}{\alpha_{k_1-1}(i,h_1)} \frac{P(j, k_1 / i, h_0)}{P(j, k_1 / i, h_1)} \underset{H_1}{\overset{H_0}{\gtrless}} 1 \quad (6)$$

which can be written as a function of the PPR evaluated at the present instant k:

$$\mathrm{PPR}\ (j, k_1) = e^{-\mu_i(k_1-k-1)(h_1-h_0)} \frac{1 - e^{-\mu_i(k_1-h_0)}}{1 - e^{-\mu_i(k_1-h_1)}} \frac{\alpha_k(i,h_0)}{\alpha_k(i,h_1)} = \mathrm{DC} \frac{\alpha_k(i,h_0)}{\alpha_k(i,h_1)} \quad (7)$$

In fact, the ratio among present and future PPR values is due only to a duration contribution DC, being the observation density contributions elided. The DC term result to be a monotonically decreasing function of $k_1>k$ for any value of h_0, h_1 and μ, so that its maximum value is taken for $k_1 = k+1$, given by

$$\mathrm{DC}_M = \mathrm{DC}(k_1=k+1) = \frac{1 - e^{-\mu_i(k-h_0+1)}}{1 - e^{-\mu_i(k-h_1+1)}} \quad (8)$$

This quantity represents the maximum duration "gain" for the "older" path H_0, so that it has no chance of resulting the best predecessor of S_j somewhere in the future if

$$\alpha_k(i, h_0) < \frac{\alpha_k(i, h_1)}{\mathrm{DC}_M} \quad (9)$$

This inequality constitutes the basis of the reduced complexity feature of the proposed method: in fact, paths for which (9) holds can be pruned out without affecting the results optimality.

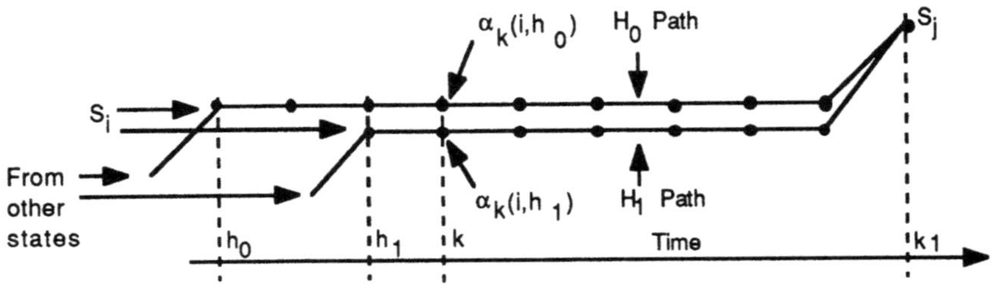

Fig.2 - Illustration of the timing relationships for two competing paths (see text)

Minimum Duration Feature

The DPDF observed in real speech data are not always well fitted by the parametric function proposed. This is the case when the acoustic event possess an intrinsic minimum duration. For this reason, eq. (1) has been shifted, so that

$$p(n) = \begin{cases} 1 & \text{if } n < \tau_m \\ e^{-\mu(n-\tau_m)} & \text{if } n \geq \tau_m \end{cases} \tag{10}$$

where τ_m represents the minimum duration for that state. Obviously, eq. (2) is modified (shifted) accordingly. Moreover, inequality (9) can be evaluated only at time $k \geq h_1+\tau_m$. Finally, it must be observed that new paths can not be generated by paths whose last state is in the minimum duration "region".

Experimental Set-Up

The proposed method has been evaluated in the framework of a speaker independent, telephonic, continuous speech recognition task, by comparing the results against those achievable for stationary HMM, with different search width values.

HMM Characteristics

The recognition vocabulary was composed by 17 words (10 digits plus some command), whose allowable sequences were constrained by a grammar with perplexity 11.2. The whole word HMM structure was built by three skip-less left to right models in parallel, with variable number of states. The observation data are given by the first 8 Mel-Cepstrum coefficients, and each state was associated to a single 8-dimensional gaussian observation density with conditional, diagonal covariance. There were a total of 392 HMM states and 427 transitions. Two different sets of HMMs were considered, either accounting or not for the duration model.

Data Base Composition

Ten speakers (5 male and 5 female) recorded a list of 75 short phrases each, which were subsequently played by an artificial mouth and transmitted trough telephone lines. Six speakers (450 phrases) served as training data, and the remaining four (300 phrases) as test.

DB Partition	HMM Type	%Rec	(Min,Max)	%Ins	%Del	GER
P1	Stationary	60.6	26.4, 95.7	31.9	1.8	58.5
P1	CVTP	70.9	47.1, 95.7	11.6	1.8	37.1
P2	Stationary	73.2	47.1, 88.7	30.8	1.4	49.1
P2	CVTP	81.9	58.6, 95.0	14.4	1.1	29.8

Table I - Performances of the evaluation experiments

Training and Recognition Algorithm
HMM parameters (mean cepstra and variances, transition probabilities, mean and minimum durations) were estimated by means of a forward-backward Viterbi scoring procedure, as proposed in [11]. This avoids typical scaling problems and allows to utilize the same code for both training and recognition. In particular, such code performs a syntax directed, one stage, beam-search Viterbi algorithm, and deals with both stationary and CVTP-HMMs, handling in this last case, during recognition, the early pruning of those paths satisfying (9). The use of the CVTP-HMMs also during training improves the consistency of the estimated quantities.The minimum dwell time parameter was estimated as the minimum duration of each state, along the best paths which passes through it. No word insertion penalties have been used during recognition.

Results

Table I summarizes the average results for two different training/test partitions of the available DB, indicated as P1 and P2, for stationary and CVTP HMM. There is also shown the maximum recognition score deviation for individual speakers, as well as the global error rate figure as defined by [14]. Adoption of the duration model produces about a 30% reduction of the average error rate, as it improves discrimination. The overall relatively low scores may be due to a multitude of factors, as the low spectral analysis order, the absence of spectral derivatives, mixture densities and insertion penalties. Moreover, no check of the proper model initialization has been done.

Fig. 3 reports about the behavior of the recognition score and elapsed time for different beam search values, with partition P1. The two dependent variables are plotted together in Fig. 3-c, in order to evidence that the improved accuracy and the path early pruning feature of CVTP-HMM make it very competitive with respect to stationary HMM. In fact, the best performances of stationary models are achieved in less time, and about a 30% of performances improvement can be obtained in the same elapsed time.

Conclusions

A new intrinsic duration model for HMM has been proposed, and a path early pruning theorem has been demonstrated. This feature makes the approach competitive with respect to stationary HMM, under both recognition performances and computational complexity points of view. Further investigations are planned in order to apply the method to different HMM environments, as phonemic and ergodic models.

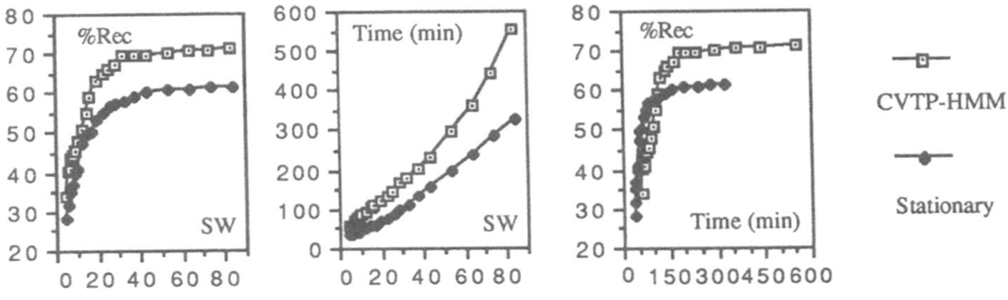

Fig.3 - Recognizer performances comparison as a function of the Search Width

References

[1] - J.D.Ferguson, "Variable Duration Models of Speech", Proc. of Symp. on the application of HMM to Text and Speech, Ferguson jd Ed, Princetown, NJ, 1980

[2] - H.Bourlard, C.J.Welleknes, "Connected Speech Recognition by Phonemic Semi-Markov chains for state occupancy modelling", Signal Processing III, Theories an applications, I.T. Young et al, Elsevier Sc. Pub. B V (North-Holland)

[3] - L.J.Russel, R.K.Moore, "Explicit modelling of state occupancy in Hidden Markov Models for Automatic Speech Recognition", Proc.ICASSP-85

[4] - S.E.Levinson, "Continuously Variable Duration Hidden Markov Models for Speech Analysis", Proc ICASSP-86, p. 1241, Tokyo

[5] - L.R.Rabiner, B.H.Juang, S.E.Levinson, M.M.Sondhi, "Recognition of isolated digits using hidden Markov models with continuous mixture densities", AT&T Tech.J., Vol.64, pp1211-1234, July-Aug. 1985

[6] - M.Codogno, L.Fissore, "Duration modelling in finite state automata for speech recognition and fast speaker adaptation", Proc. ICASSP-87, Dallas

[7] - T.H.Crystal, A.S.House, "Characterization and modeling of speech segment duration", Proc. ICASSP-86, Tokyo

[8] - M.J.Russel, A.E.Cook, "Experimental Evaluation of Durational Modelling Techniques for Automatic Speech Recognition", Proc ICASSP-87, p. 2376, Dallas

[9] - C.H.Lee,"On the use of some robust modelling techniques for speech recognition", Computer Speech and Language, 1989, 3, 35-52

[10] - A.Falaschi, "Phonetic recognition by non stationary HMM", Proc. of FASE Speech 88, Aug. 1988, Edinburgh, UK

[11] - D.Jouvet, J.Monne', P.Dobois, "A new network-based speaker-independent connected- word recognition system", Proc. ICASSP-86, Tokyo

[12] - I.S.Gradshtey, I.M.Ryzhik, "Table of integrals series and products" 4th Ed., Acad. Press 1965 - NY, San Francisco, London

[13] - A.Falaschi, "Decodifica acustico-fonetica del messaggio vocale su basi informatico-strutturali e modelli di Markov nascosti", PhD Thesis, INFO-COM Dpt, La Sapienza Univ. of Rome, Italy

[14] - R.Schwartz, Y.Chow, F.Kubala, "Rapid speaker adaptation using a probabilistic mapping", Proc of ICASSP-87, Dallas

Part 2
Continuous Speech Recognition Systems

Context-Dependent Phonetic Hidden Markov Models for Speaker-Independent Continuous Speech Recognition

Kay-Fu Lee

School of Computer Science,
Carnegie-Mellon University, Pittsburgh, PA 15213

Abstract *

The effectiveness of context-dependent phone modeling for speaker-dependent continuous speech recognition has recently been demonstrated. In this study, we apply context-dependent phone models to speaker-independent continuous speech recognition, and show that they are equally effective in this domain. In addition to evaluating several previously proposed context-dependent models, we also introduce two new context-dependent phonetic units: 1) function-word-dependent phone models, which focus on the most difficult subvocabulary, and 2) generalized triphones, which combine similar triphones together based on an information-theoretic measure. The subword clustering procedure used for generalized triphones can find the optimal number of models given a fixed amount of training data. We demonstrate that context-dependent modeling reduces the error rate by as much as 60%.

NATO ASI Series, Vol. F 75
Speech Recognition and Understanding. Recent Advances
Edited by P. Laface and R. De Mori

Speaker Independent Continuous Speech Recognition Using Continuous Density Hidden Markov Models

Chin-Hui Lee, Lawrence R. Rabiner and Roberto Pieraccini

Speech Research Department, AT&T Bell Laboratories, Murray Hill, NJ 07974, USA

Abstract: The field of large vocabulary continuous speech recognition has advanced to the point where there are several systems capable of providing greater than 95% word accuracy for speaker independent recognition, of a 1000 word vocabulary, spoken fluently for a task with a perplexity of about 60. There are several factors which account for the high performance achieved by these systems, including the use of effective feature analysis, the use of hidden Markov model (HMM) methodology, the use of context-dependent sub-word units to capture intra-word and inter-word phonemic variations, and the use of corrective training techniques to emphasize differences between acoustically similar words in the vocabulary. In this paper we describe a large vocabulary continuous speech recognition system developed at AT&T Bell Laboratories, and discuss the methods used to provide high word recognition accuracy. In particular we focus our discussion on the techniques adopted to select the set of fundamental speech units and to provide the acoustic models of these sub-word units based on a continuous density HMM (CDHMM) framework. Different modeling approaches, such as a discrete HMM and a tied-mixture HMM, will also be discussed and compared to the CDHMM approach.

Keywords: Continuous speech recognition (CSR), large vocabulary recognition (LVR), acoustic modeling, hidden Markov model (HMM), continuous density HMM (CDHMM), segmental k-means, phone-like unit (PLU), context-independent (CI) unit, context-dependent (CD) unit, speaker adaptation

1. Introduction

In the past few years there have been proposed a number of systems for large vocabulary speech recognition which have achieved high word recognition accuracy [14, 16, 24, 27, 38, 50, 54]. Although a couple of the systems have concentrated on either isolated word input [16], or have been trained to individual speakers [16, 54], most current large vocabulary recognition systems have the goal of performing speech recognition on fluent input (continuous speech) by any talker (speaker independent systems). In this paper we describe one such large vocabulary speech recognition system which has been developed at AT&T Bell Laboratories, and discuss the methods used to provide high word recognition accuracy. Details of the recognition system can be found in [24, 26].

The approaches to large vocabulary recognition (LVR), are summarized in the block diagram of Figure 1. There are basically 5 steps in the recognition of speech, namely:

NATO ASI Series, Vol. F 75
Speech Recognition and Understanding. Recent Advances
Edited by P. Laface and R. De Mori

1. Feature analysis in which the speech signal is converted to a sequence of spectral and temporal features representative of the time-varying properties of the speech.

2. Acoustic unit matching system in which the speech features are compared to an inventory of speech units and a "best match" or a series of "best matches" between an arbitrary sequence of the recognition units and the given speech signal, as represented by the feature vectors, is obtained. The choice of speech recognition units is an important issue in the design of any speech recognition system.

3. Lexical decoding of the sequences of speech units into one or more word sequences based on constraints from a lexicon or word dictionary. The lexicon is a description of each word in the recognition vocabulary in terms of the symbols for the relevant speech units. Multiple lexical entries (word pronunciations) are allowed.

4. Syntactic analysis of the word sequences provided by the lexical decoding stage as to appropriateness within the constraints of the recognition grammar which prescribes the allowable sequences of words. The grammar could be a formal grammar (e.g. a covering grammar) or a finite state network (FSN) representation of all word sequences, or a statistical network of word bigram or trigram probabilities.

5. Semantic analysis of all syntactically valid word sequences to determine the sentence which is consistent with the task model. The task model could be a static characterization of the system's knowledge about the recognition task, or a dynamic one which changes with each newly recognized sentence.

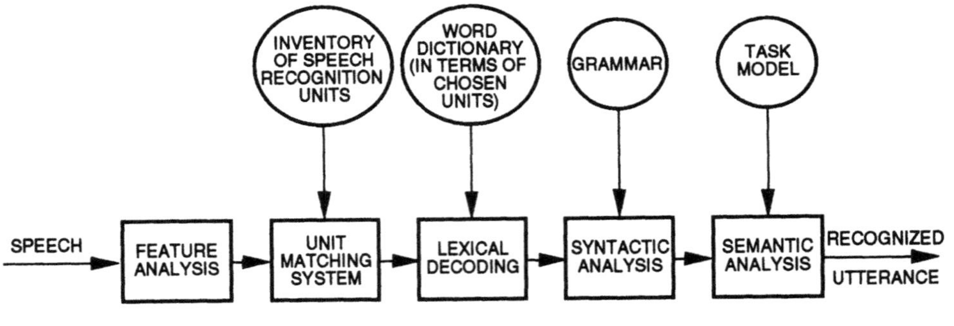

Figure 1. A block diagram of a typical large vocabulary, continuous speech recognizer

The final recognized utterance is the one which has passed through all the stages of recognition and has the highest likelihood with respect to the unit matching system. The approach we adopted to perform continuous speech recognition is a pattern recognition one, in which all acoustic realizations of units, words and sentences are considered as patterns consisting of sequences of feature vectors. Sentence recognition is therefore accomplished by performing pattern matching at unit, word and sentence levels in an integrated manner. Based on the block diagram of Fig. 1, it can be seen that there are several issues to be resolved in order to implement an LVR system. We will address some of these issues in the following sections.

This paper is organized as follows. In Section 2, we discuss the general problem of characterizing acoustic and linguistic spaces and their interactions. We then discuss the important issue of the choice of fundamental sub-word units for recognition and different ways of modeling the set of chosen units. In Section 3, we briefly describe our CDHMM-based recognition system. The focus of Sections 4 through 7 is the presentation of a number of modeling techniques used to achieve high recognition performance. We first describe ways of creating context independent and context-dependent models in Section 4. We then discuss techniques used to model word juncture coarticulation in Section 5. The important issue of improved feature analysis for speech recognition is covered in Section 6, and the use of position-dependent sub-word units to improve unit separation is discussed in Section 7. Finally we summarize the results of our investigations and suggest some further research topics in Section 8.

2. Characterization of Acoustic and Linguistic Spaces of Speech

The characterization of the acoustic space of speech sound is usually achieved by standard signal modeling techniques such as vector quantization (VQ) or matrix quantization. More recently, an acoustic segment model approach [24] has been proposed to find a basic set of segment models that spans the acoustic space. On the other hand, characterization of the linguistic space of speech sounds is usually accomplished by specifying the inherent relations of sounds to form words, words to form grammatically correct (according to syntax), and meaningful (according to semantics) sentences. For applications that do not require interactions between the acoustic and linguistic spaces, e.g. speech coding and spelling checks, it suffices to model the acoustic space and the linguistic space independently.

However, for some applications, such as speech recognition and text-to-speech synthesis, there is a strong interaction between the acoustic and linguistic spaces. A mapping between the two spaces is often required to model the type of interaction (see Figure 2 for a conceptual view of this situation). In some cases, such as isolated word recognition or speaker identification, the acoustic-to-linguistic mapping is one-to-one, i.e. each acoustic signal is labeled uniquely either as a word or a speaker identity. There is no problem in characterizing the mapping process. In other cases the mapping is not straightforwardly defined. For example, in the case of connected digit recognition, an utterance is often associated with a digit string (i.e. block labeled) with no word boundaries specified. The acoustic-to-linguistic mapping can only be approximately defined. However recognition of connected digit strings is still relatively easy [44], because digits are word units and are inherently well-defined and distinct from each other.

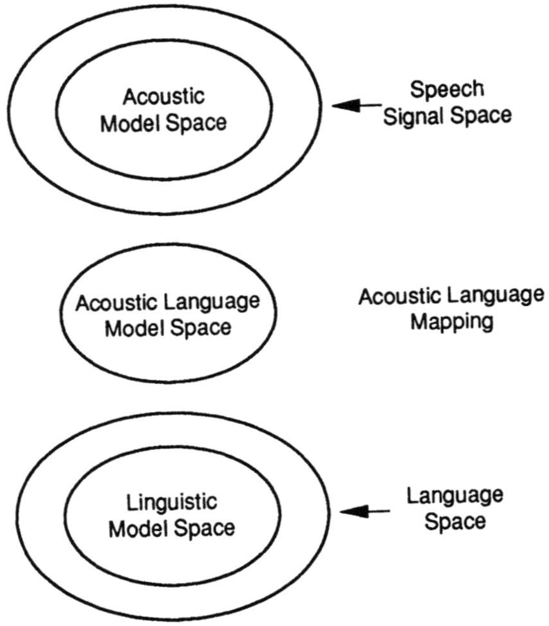

Figure 2. Interaction between acoustic and linguistic spaces

In the case of large vocabulary, continuous speech recognition, the acoustic-to-linguistic mapping is somewhat vaguely defined. On the one hand, the word is commonly accepted as the unit to be recognized. On the other hand, whole-word units are not easily modeled directly because it is not feasible to collect enough speech samples of all the words to be recognized in all possible contexts. Word modeling is often accomplished indirectly using a set of fundamental sub-word units and a lexicon. Therefore we are faced with three modeling issues, namely: (1) the way to model the acoustic space; (2) the choice of the set of fundamental sub-word units; and (3) the interaction between the acoustic and linguistic spaces, which is often achieved through acoustic modeling of the set of fundamental sub-word units using the set of units defined in the acoustic model space. As mentioned above, acoustic space modeling is usually accomplished using VQ or acoustic segment based approaches. We now discuss the issues of the choice of fundamental sub-word units and acoustic modeling of the chosen units.

2.1 Choice of Fundamental Sub-word Units

The choice of sub-word units is made based on issues such as coverage of language, context sensitivity, ease of initialization, and ease of training. There are essentially two choices for the fundamental sub-word units, namely linguistically-based units and acoustically-based units.

Acoustic units are the labels assigned to acoustic segment models [22-23], which are defined on the basis of finding a set of segment models that spans the acoustic space defined by the given, unlabeled training data. It has been shown that about 256 acoustic units are adequate for recognition of a vocabulary of 1109 isolated words in a speaker dependent mode. The acoustic-to-linguistic mapping is defined at the word level through an acoustic lexicon, so that word models can be defined as a concatenation of acoustic unit models according to the definition of the entries in the acoustic lexicon. Word recognition can be carried out similarly once models for all the words in the vocabulary are well-defined. More recently, it has been shown that a set of 32 acoustic units can be effectively used to model speakers in the DARPA resource management training database [46]. Therefore, when performing the acoustic to linguistic mapping on the speaker level, text-independent speaker verification can be accomplished without knowing the linguistic contents of the training and testing data [46]. The acoustic unit approach has been used successfully in isolated word recognition and text-independent speaker verification in which the acoustic-to-linguistic mappings are quite straightforward. The effectiveness on large vocabulary, continuous speech recognition is yet to be evaluated.

The possibilities for sub-word linguistic units include phoneme, diphone, demi-syllable and syllables. For reasons of efficiency of representation, the units chosen for most of the LVR systems are based on the labels of a set of phonemes. To provide some of the advantages of the alternative sub-word unit sets, context sensitive units are also studied. The ways in which these context dependent units are created are discussed later in this paper. Since sub-word segment boundaries are usually ill-defined in continuous speech, it is not feasible to obtain a consistent manual segmentation for those phoneme units over a large database. We therefore obtain a set of phone units which are modeled acoustically based on a lexical description of words in the vocabulary. No assumption is made, *a priori*, about the mapping between acoustic measurements and phoneme labels; such a mapping is entirely learned via a finite training set of utterances. The resulting sub-word units, which we call *phone-like units* (PLU's) [24, 26] are essentially acoustic descriptions of linguistically-based units such as phonemes *as represented in the words occurring in the given training set*. Most current sub-word unit LVR systems are based on units similar to PLU's. Diphone-like Units, demi-syllable-like units and syllable-like units can also be defined similarly.

2.2 Acoustic Modeling of Sub-Word Units

The standard way in which speech units are modeled is as left-to-right hidden Markov models [24, 43]. Within each state of the HMM, the spectral vector is represented by either a discrete density (i.e. a distribution over a spectral codebook) or a continuous density (e.g. mixture Gaussian), or a mixed density (i.e. a continuous density over a codebook of common spectral shapes). We illustrate these differences in Figure 3 which shows an acoustic space covered by vector quantization (VQ) cells (part a), a set of single Gaussian densities (part b) and a set of mixtures of continuous densities (part c).

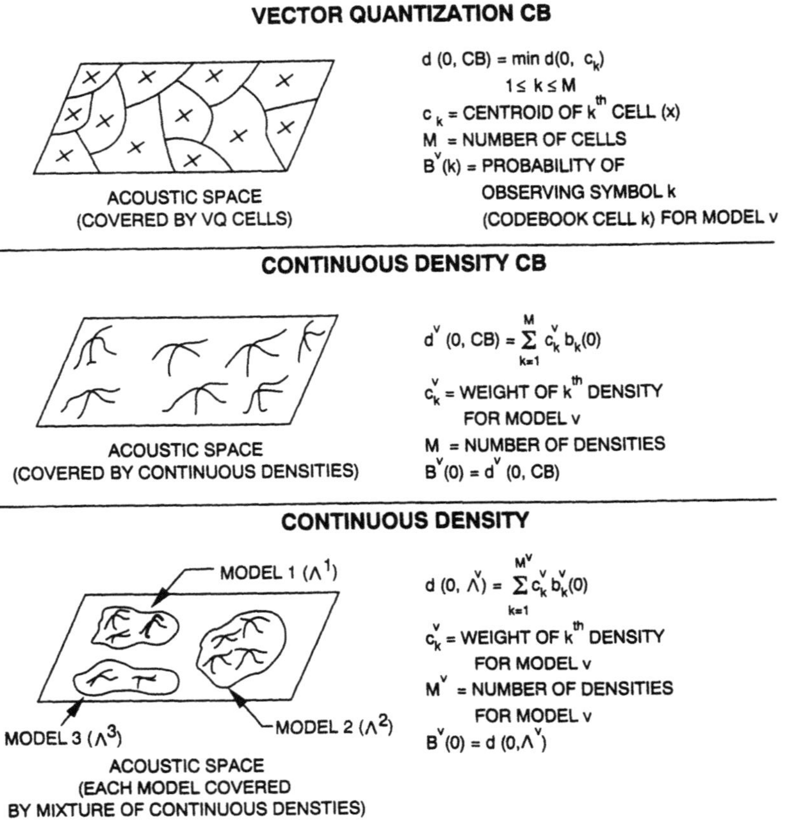

Figure 3. Acoustic Modeling using VQHMM, TMHMM and CDHMM approaches

The VQ codebook (Fig. 3a) represents a partitioning of the entire acoustic space into discrete, non-overlapping cells, usually based on a minimized spectral distortion criterion, with each cell defined by a spectral centroid. The spectral density of a state in a sub-word unit HMM is a discrete density describing the probability that a spectral vector of the unit in that state is closest (in some spectral distance) to the centroid corresponding to each cell. For simplicity we call this approach VQHMM for vector quantization based HMM.

The continuous density method (Fig. 3c) tries to describe the spectral density of each state for each sub-word unit in terms of a mixture of Gaussian densities. Each mixture component has a spectral mean and variance which is highly dependent on the spectral characteristics of the sub-word unit (i.e. highly localized in the total acoustic space). Hence the models for different sub-word units usually are well-separated in the acoustic space, especially if the spectral characteristics of the units are are different.

Finally the mixed density method (called the tied mixture method [3, 38] or the semi-continuous modeling method [9], Fig. 3b) tries to cover the entire acoustic space by a set of independent Gaussian densities. The resulting set of means and covariances is stored in a codebook, and the density in each state of each sub-word model is a mixture of the codebook densities; hence it is a continuous density but one only need store the discrete density of mixture gains to completely characterize the continuous density. Furthermore, since the codebook set of Gaussian densities is common for all models and all states, one can precompute the likelihoods associated with an input spectral vector for each of the codebook vectors, and determine state likelihoods using only a simple dot product with the state mixture gains. This represents a significant computational reduction over the full continuous density case. We will call this approach TMHMM for tied mixture based HMM. It is noted that the tied mixture approach is commonly applied to the entire acoustic space [3, 9, 38]. However, tying can also be accomplished at the sub-word unit level, e.g. phone dependent tying in which multiple sets of common densities are used, with each set covering the acoustic space of each phone [24]. We shall see later in this paper how we can exploit this mixed density concept for representing a set of expanded units, such as context dependent PLU's, based on representations of a set of fundamental units, such as context independent PLU's.

2.3 Representations of Observation Density Within an HMM State

Based on the acoustic analysis, we can represent the spectral density within the jth state of an HMM in one of three ways, depending on whether we used the discrete codebook approach, the continuous mixture density approach, or the mixed density approach. For the discrete density case we get

$$b_j(k) = \text{Prob}(\, O_l \quad \text{and codebook vector } k \mid \text{state } j) \tag{1}$$

whereas for the continuous mixture density approach we have

$$B_j(\, O_l\,) = \text{Prob}(\, O_l \mid \text{state } j) = \sum_{m=1}^{M} c_{jm} \frac{\prod_{d=1}^{D} e^{-\frac{(O_l(d)-\mu_{jm}(d))^2}{2U_{jm}(d)}}}{(2\pi)^{D/2}\left[\prod_{d=1}^{D} U_{jm}(d)\right]^{1/2}} \tag{2}$$

where c_{jm} is the mixture gain for the mth mixture, μ_{jm} is the mixture mean vector, U_{jm} is the mixture covariance vector (we are assuming diagonal covariances), M is the number of mixture components, and D is the number of spectral components in each observation vector. Finally for the mixed density approach we have

$$B_j(\, O_l\,) = \sum_{m=1}^{\hat{M}} c_{jm} \frac{\prod_{d=1}^{D} e^{-\frac{(O_l(d)-\mu_{m}(d))^2}{2U_{m}(d)}}}{(2\pi)^{D/2}\left[\prod_{d=1}^{D} U_{m}(d)\right]^{1/2}} \tag{3}$$

where now the mixture means and covariances are independent of the state and only the mixture gains depend explicitly on the state, and $\hat{M}$ is the number of codebook densities used to represent the entire acoustic space.

A common method of handling the insufficiency of training data for creating statistically reliable context-dependent (CD) PLU's is to smooth individual CD models with individual context-independent (CI) models via a technique like deleted interpolation [15]. In order to use deleted interpolation, both the CD and CI models need to be created based on a common codebook (e.g. discrete observation probabilities) or based on a common set of Gaussian densities (e.g. the mixed density method). If this is the case, then if we denote the spectral density for the CI unit $\$-p-\$$ in state j as B_j^{CI}, and the spectral density for the CD unit p_L-p-p_R in state j as B_j^{CD}, then we create the smoothed spectral density $\overline{B}_j^{\mathrm{CD}}$ as:

$$\overline{B}_j^{\mathrm{CD}} = \lambda B_j^{\mathrm{CD}} + (1-\lambda) B_j^{\mathrm{CI}} \tag{4}$$

where λ is estimated directly from training data which is deleted (withheld) from the training data used to create B_j^{CD} and B_j^{CI}. The forward-backward algorithm can be used directly to estimate λ [15].

The key to the success of Eq. (4) is the commonality of the spectral densities used for the CD and CI units. A slightly different way of exploiting this type of smoothing is to use the mixed density method but localized to each CI PLU. Thus for designing each CD PLU, we assume that within each state the means and covariances of each mixture are the same as those used for the CI PLU model; however we adjust the mixture gains based on the actual occurrences of each CD PLU in the training set. Thus for each state j we do the following:

$$\mu_j(m)^{p_L-p-p_R} = \mu_j(m)^{\$-p-\$}, \qquad 1 \le m \le M \tag{5a}$$

$$\mathbf{U}_j(m)^{p_L-p-p_R} = \mathbf{U}_j(m)^{\$-p-\$}, \qquad 1 \le m \le M \tag{5b}$$

$$c_j(m)^{p_L-p-p_R} \quad \text{estimated from CD training tokens} \tag{5c}$$

We can also apply a form of interpolation which is similar to that of deleted interpolation to the mixture gains (Eq. 9c) by smoothing them with the CI mixture gains, i.e.

$$\overline{c}_j(m)^{p_L-p-p_R} = \lambda c_j(m)^{p_L-p-p_R} + (1-\lambda) c_j(m)^{\$-p-\$} \tag{6}$$

where λ is again estimated from counts of training tokens where the CD model provides a better fit than the CI model. This type of smoothing is especially effective for models created from a small number of training tokens (e.g. less than 30).

2.4 Comparison Between Acoustic Modeling Approaches

VQHMM is the most commonly used modeling approach [16, 27, 38, 50, 54] for large vocabulary speech recognition. As mentioned above, the VQHMM approach is based on a finite size codebook (or multiple codebooks) and a set of discrete HMM's, defined on this common codebook, for each of the sub-word units. The modeling is accomplished in two steps. First, a codebook (or multiple codebooks) is designed without knowing the linguistic content of the training database. Then, assuming a fixed set of codebooks is given, the Baum-Welch reestimation procedure (e.g. [43]) is applied to all the training data to estimate the VQHMM parameters. Due to insufficiency of training data, careful smoothing and interpolation are required in the estimation process.

The size of the codebook is commonly chosen as 256 (or 200 in the IBM systems [16]). When multiple codebooks are used to model the acoustic space, each codebook also is of size 256 [27, 38, 50, 54]. From our experience in speech coding, a 10-bit codebook is often required to quantize the LPC parameters with less than 3dB spectral distortion [17]. It is appropriate to ask the question: why does such a small codebook work so well for speech recognition? Part of the answer is the use of the VQ codebooks. Since all discrete probabilities are defined through the common VQ codebooks, discrimination among different units is inherent in VQHMM. However, due to quantization error and limited acoustic resolution, several approaches have been proposed to improve acoustic modeling. One way to increase acoustic resolution is to increase the codebook size; however this often decreases recognition performance because the discrete HMM parameters are more difficult to estimate and interpolate with a large-size codebook. Another way is to improve HMM parameter estimation by reducing the codebook size; this also results in worse performance due to insufficient acoustic resolution. It seems a codebook size of 256 is a good compromise between the two conflicting goals of improving acoustic resolution and achieving improved parameter estimation.

The TMHMM approach [3, 9, 38] has all the advantages of the VQHMM approach. In addition, the TMHMM approach alleviates some of the problems with quantization. It has the advantage of sharing a small, common set of densities (typically a set of Gaussian densities) among all sub-word units. However, it still suffers from the same problems of limited acoustic resolution and smoothing is still required to estimate the HMM parameters. Some of the recent improvements in performance of the DARPA resource management task are due to the use of TMHMM (e.g. [10]) in place of the original VQHMM.

The CDHMM approach assumes that each sub-word unit can be modeled independently as shown in Fig. 3c. Once the segmentation of the sub-word units is given, HMM parameters are estimated *directly*. There is no need to define a pre-selected set of common densities. Using the mixture Gaussian density assumption, all state observation densities can be approximated to any level of precision. No assumption about the form of the densities is made. One of the advantages of independently modeling each sub-word unit is the capability of modeling different sub-word units with a different degree of acoustic resolution. For instance, a different number of mixture components can be used for different units depending on the availability of training data and the acoustic variability of the training data corresponding to the particular unit to be modeled.

There are three major modeling problems for CDHMM approach. The first is the issue of insufficiency of training data for some of the units because the correlation among similar units is not effectively utilized. The second issue is that discrimination among different units is not enhanced because no common set of densities is used. The above two difficulties can be alleviated somewhat using a phone-dependent tied mixture approach, which we called *CD2* [24]. The idea is to design a set of common densities for each of the context independent units and share the mean and covariance parameters among all context dependent units that are derived from the same CI unit. We will come back to this issue later in the paper. The third issue is the so called *singularity* problem in density estimation or *zero-cell* problem in histogram approximation. Due to insufficient data used to estimate the variance parameter for a particular mixture component, variance underestimation often occurs. To avoid this singularity problem, we use a variance clipping technique [24,45] to set a variance floor such that the value of the estimated variances always exceeds this minimum clipping variance floor. It can be shown that this procedure is the *maximum a posteriori* (MAP) estimate of the variance parameter if the *a priori* information indicates that all the variance estimates are above a minimum value with equal probability [25]. The minimum variance floor values are determined experimentally [24, 45]. We found variance clipping essential when we tried to estimate parameters for a large number of mixture components (say greater than 100,000) when only a limited amount of training data is available. We now describe the design of our large vocabulary recognition system using the CDHMM approach. A block diagram of the recognition system used throughout this study is shown in Figure 4.

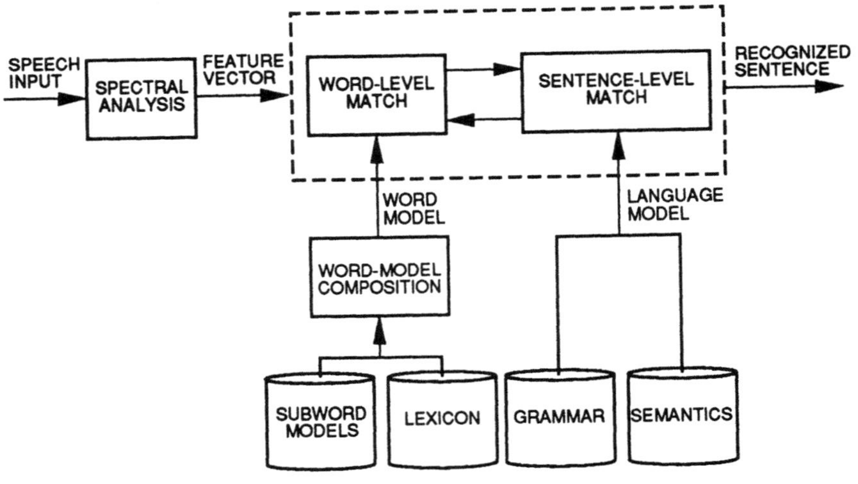

Figure 4. A block diagram of the CDHMM-based continuous speech recognizer

3. CDHMM-Based Large Vocabulary Recognition System

Throughout this study, we use the DARPA resource management task to illustrate the approaches we adopted to large vocabulary recognition. The standard DARPA training material [37] consists of a set of 3990 read sentences from 109 talkers (30-40 sentences/talker). For most of our test evaluations, we used the so-called FEB89 test data which consisted of 300 sentences from 10 new talkers (30 sentences/talker) as distributed by DARPA in February 1989.

As mentioned in Section 1, continuous sentence recognition is usually accomplished by performing a unit match, a word match and a sentence match in an integrated manner (as shown in Figure 4). Details of the recognition system can be found in [24, 26]. We now briefly describe each of system components.

3.1 Acoustic Analysis of Speech

The speech database was originally provided by DARPA at a 16 kHz sampling rate [41]. To make the database compatible with a wide telephone bandwidth, we filtered the speech data between 100 Hz and 3,800 Hz and down-sampled the speech to an 8 kHz rate before analysis. A block diagram of the feature analysis system used in our research is given in Figure 5. The feature vector (or observation vector), O_l, used for recognition and training is the concatenation of the LPC-derived weighted cepstral vector [18, 24], $\hat{c}_l\,(m)$, and the corresponding weighted delta cepstral vector [24, 51], $\Delta\,\hat{c}_l\,(m)$, i.e.

$$O_l \;= \{\;\hat{c}_l\,(1{:}12),\;\Delta\,\hat{c}_l\,(1{:}12)\} \qquad (7)$$

and consists of 24 coefficients per vector. Temporal features such as log energy and its time derivatives features can also be used as part of the observation vector for training and recognition. We will address issues related to feature analysis again in Section 6.

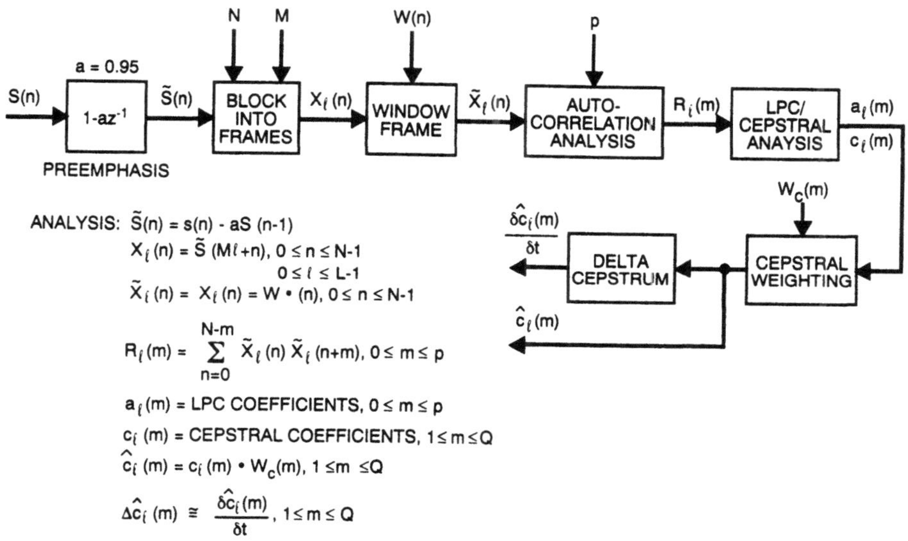

Figure 5. A block diagram of the feature analysis module

3.2 Word Model Composition

We assume that every word in the vocabulary is represented by exactly one entry in a word lexicon, and each lexical entry is characterized by a linear sequence of sub-word units. Each word model is composed as a concatenation of the sequence of sub-word models according to its corresponding lexical representation. A set of 47 context-independent phone labels [24, 26], is extracted from the lexicon and is used as the set of basic speech unit throughout this study to provide a full phone coverage for all the vocabulary words. Context-dependent units are also derived from this basic unit set.

3.3 Language Model Representation

The sentence-level match module uses a language model, based on a set of syntactic and semantic rules of the recognition task, to determine the word sequence in a sentence. In our current implementation, we assume that the language model can be represented by a finite state network (FSN). Two FSN grammar representations of the resource management task [36] are used in this study. In the no grammar (NG) case, any of the 991 words is allowed to follow any other of the 991 words. The perplexity (average branching factor) of this grammar is clearly 991. Since any combination of words can be used to create a sentence, the overcoverage (ratio of sentences generated by the grammar to valid sentences within the task language) of this grammar is extremely high. In the word pair (WP) grammar case, for each word, a finite list of words which can follow it is prescribed. The perplexity of this grammar is about 60 and the overcoverage, while significantly below that of the NG case, is still very high.

3.4 Training of PLU's for LVR

PLU training is accomplished using a variant of the *segmental k-means* training algorithm [24, 26, 42]. Each sentence is represented as a (not necessarily unique) sequence of sub-word units with the option of silence between any pair of words, and at the beginning and/or end of each sentence. First, we express the sentence as a series of optional silences followed by the specified words according to the word transcription of the utterance. Next, each word in the sentence is replaced by entries from the lexicon. Then each lexical entry is replaced by its sequence of sub-word PLU's, as expressed in the lexicon. Finally we allow multiple models for each PLU so we replace each canonic PLU by one or more models in parallel.

The network created by embedding the multiple phone models into each lexical entry, and by embedding the multiple lexical entries into each word, and finally by embedding the word models into each sentence is then used to match the spectral representation of the input via a Viterbi matching procedure. Inter-word context-dependent units can also be incorporated in the phone network and Viterbi decoding can still be applied to segment all training utterances into a sequence of sub-word units. We then collect all the acoustic segments corresponding to a specified sub-word unit, and use k-means clustering on all frames in the same state of the unit to partition the set of frames into a number of clusters which is equal to the number of mixture components for the state. We then use maximum likelihood estimation techniques to obtain estimates of all the parameters of the mixture Gaussian density. The k-means clustering can also be replaced by an EM-algorithm to obtain ML estimates of the parameters of the mixture Gaussian density.

There are at least two issues concerning the training procedure that need further clarification. The first concerns the initial linear segmentation into units and states.

Clearly this is totally inappropriate in general; however in practice it has been shown to be an acceptable way of initialization which does eventually lead to a stable, convergent set of models [24, 38]. The second issue in the training procedure concerns the difference between the segmental *k*-means training (which uses the best segmentation path) and the more traditional Baum-Welch or forward-backward training (which uses the likelihood score over *all* paths). Experience, as well as theory, has shown that the differences in performance between these two training procedures are small and, in general, do not strongly affect estimates of PLU model parameters [24, 34]. As such the segmental *k*-means training is a reasonable method for estimating PLU model parameters.

3.5 Implementation of the Recognizer

In the case when only intra-word units are used in recognition, the FSN representation based on the WP grammar has on the order of 20,000 HMM states and word junction nodes to keep track of at each frame of the input. To reduce computation, a frame-synchronous beam search algorithm [33] is used. The time for computation varied almost linearly with Δ; hence the penalty paid for a large Δ is storage and computation time, but the reward is that the best string is more likely to be included in the beam. Clearly these results show the need for an adaptive beam width algorithm which can reduce its size during regions of good word matches, and increase its size during regions of relatively poor word matches. Such a procedure does not yet exist.

When inter-word units are incorporated in the recognition process, the complexity of the recognizer increases considerably because of multiple fan-out at the end of a word and multiple fan-in at the beginning of the next word which results in a large number of interword connections. To achieve high accuracy, the search algorithm has to keep track of all possible phonetic contexts in which a word can occur and all possible interword connections between every valid pair of words. At this level of algorithm complexity, every single detail of the implementation can have a major impact on the computation requirements. In particular, the data structure used for the network search should be designed in such a way that the memory requirement and computation overhead be kept to a minimum. We have developed an efficient recognizer based on a compiled network approach, and achieved a 10-to-1 reduction in bookkeeping computation. The reader is referred to [39] for a detailed description of the recognizer implementation. In the next four sections we describe the CDHMM-based acoustic modeling techniques used to achieve high recognition performance in our continuous speech recognition system.

4. Intra-word Unit Selection and Modeling

4.1 Modeling of Context-Independent Units

We first discuss issues related to modeling of context-independent units. The use of context independent PLU's has several advantages but also leads to several problems. The advantages include: (1) The models are easily trained using either the segmental *k*-means or the forward-backward algorithm; (2) For any reasonable size training set, the units occur sufficiently often for there to be reasonable confidence in the resulting model parameter estimates – i.e. no parameter smoothing is required; and (3) The units themselves are relatively insensitive to the context from which the training tokens are extracted.

The problems with the context independent units are the following: (1) They do not represent the unit well in all contexts, i.e. for some units the effects of neighboring units are so strong that there is a high degree of acoustic (spectral) variability in the model parameters; and (2) They do not provide high recognition performance for large vocabulary recognition tasks, i.e. no one has achieved over 90% word recognition accuracy for vocabularies of 1000 or more words based solely on using context independent PLU's.

Perhaps the simplest solution to the above problems is to provide more detailed representations of each unit by increasing the number of mixture components in each HMM state. To illustrate the power of this approach, Table 1 shows word accuracy (%) (word recognition rate minus word insertion rate) as a function of the nominal maximum number of mixture components per state, M_{max}, for the FEB89 test set using the word pair grammar. The key point to notice is the sharp increase in word accuracy obtained by going from 1 component/state (Gaussian density) to 18 components/state (typically about 20% absolute error reduction), and then a gradual increase in word accuracy going from 18 components/state to 256 components/state,

M_{max}	1	3	6	9	18	36	128	256
Accuracy	61	72	78	80	81	84	85	86

Table 1. Word accuracies (%) as a function of M_{max} using 47 CI PLU's

This result shows that by increasing acoustic resolution, performance continues to increase so long as there is sufficient training data (as is the case for 47 CI PLU's). It is noted that in typical VQHMM implementations [27, 38, 50, 54], the best performances were achieved with multiple codebooks (up to 4), each with size 256. Therefore, using CDHMM with mixture Gaussian density to model highly variable CI units also requires a similar level of acoustic resolution (say, $M_{max} \geq 128$) to achieve good performance.

4.2 Selection and Modeling of Context Dependent PLU's

Another solution to CI modeling problems is to make use of context information in modeling. The idea behind creating context dependent PLU's is to capture the local acoustic variability associated with a known context and thereby reduce the acoustic variability of the set of PLU's. One of the earliest attempts at exploiting context dependent PLU's was in the BBN BYBLOS system where left and right context PLU's were introduced [48]. The more general case of both left and right context dependent PLU's represents each PLU symbol p as: $p \rightarrow p_L - p - p_R$, where p_L is the preceding PLU symbol (possibly silence) and p_R is the following PLU symbol (possibly silence).

Given the set of 47 PLU's, a total of $47^3 = 103823$ CD PLU's could be obtained in theory. In practice, only a small fraction of them appear in words for a given task. However the number of CD PLU's is still very large (on the order of 2000-7000), making even a reasonable amount of training material insufficient to estimate all the CD PLU models with acceptable accuracy. Several techniques have been devised to overcome these training difficulties, including clustering of models that are similar according to some suitable information-theoretic criterion [27], pooling of contexts having similar acoustic-phonetic properties [29, 47], and smoothing with context-independent units using deleted interpolation [15].

The way in which we create CD PLU models is as follows. We first convert the lexicon from context independent (CI) units to CD units. We then train the set of CD PLU's using the same procedure as used for the CI PLU's, i.e. use the segmental k-means training on the expanded set of PLU's until convergence. The above training procedure leads to one major problem, namely that the number of occurrences of some of the CD units is insufficient to generate a statistically reliable model. There are several ways of dealing with this problem. Perhaps the simplest way is to use a left/right context PLU-to-left or right context PLU-to-no context PLU reduction rule of the form [24, 26]:

Unit Reduction Rule: If $c(p_L - p - p_R) < T$, then

1. $p_L - p - p_R \longrightarrow \$ - p - p_R$ if $c(\$ - p - p_R) > T$
2. $p_L - p - p_R \longrightarrow p_L - p - \$$, if $c(p_L - p - \$) > T$
3. $p_L - p - p_R \longrightarrow \$ - p - \$$

where $c(p_1 - p_2 - p_3)$ is the count in the training set associated with the ordered triplet (p_1, p_2, p_3) ($ is a don't care or wild card phone), and T is the count threshold for applying the reduction rule sequentially through the 3 cases.

To illustrate the sensitivity of the CD PLU set to the threshold on occurrences, T, Table 2 shows the counts of left and right context dependent (LR CD) PLU's, left context (LC) PLU's, right context (RC) PLU's, and context independent PLU's for the 109 talker DARPA training set of 3990 sentences, as a function of T. It can be seen that for a threshold of 50, which is generally adequate for estimating the HMM parameters, there are only 365 intra-word LR CD PLU's (out of a possible 103,823 combinations), and even for a threshold of 1, there are only 1778 intra-word LR CD PLU's; hence only a very small percentage of the possible LR CD PLU's occur in this 3990 sentence set.

Count Threshold (T)	Number of LR CD PLU's	Number of LC PLU's	Number of RC PLU's	Number of CI PLU's	Total Number of PLU's
50	378	158	171	47	754
40	461	172	188	47	868
30	639	199	205	47	1090
20	952	212	234	46	1444
10	1302	243	258	44	1847
5	1608	265	270	32	2175
1	1778	279	280	3	2340

Table 2. Number of intra-word CD units as a function of count threshold (T)

A detailed discussion of the recognition system and the context modeling techniques we used to achieve improved performance is given in [24]. In order to provide a consistent base of performance for comparison with the new results, we used a threshold of 30 in the unit reduction rule for all performance evaluations. When no inter-word units were used, a threshold of 30 resulted in a set of 1090 CD PLU's. With this set of units, we obtained a word accuracy of 91.3% and a sentence accuracy of 58.7% on the FEB89 test set.

A detailed error analysis showed that short function words (e.g. a, the, in, etc.) were involved in about 60-75% of the substitution, insertion and deletion errors. The problems here are related to the fact that the function words are extremely context sensitive. Several possibilities have been proposed for dealing with problems of recognizing function words, including function word dependent PLU's (as used by Lee [27]), inter-word training of context-dependent PLU's, multiple models of function word PLU's, multiple lexical entries for these words and an improved feature analysis. In general, a more detailed unit characterization leads to a significant reduction of function word errors.

5. Word Juncture Coarticulation Modeling

Coarticulation phenomena arising at the boundary between words are a major cause of acoustic variability for the initial and final parts of a word when spoken in continuous speech. If this type of contextual variability is not adequately represented in the recognition system, errors are likely to occur. This is the case, for instance, when words in the dictionary are phonetically transcribed according to their pronunciations in isolation. The solution to this problem is to provide a more precise phonetic representation of the region across word boundaries. Different approaches may be taken to achieve this goal, depending on which type of coarticulation phenomenon is handled.

In [6] we characterized two different types of pronunciation changes at word junctures, namely *soft* and *hard* changes. In soft changes, the alteration of a phone due to neighboring phones is comparatively small and the actual realization is perceived as a variation of the original phone rather than a transformation to a different phone. This alteration is of the same nature as the one observed in intra-word phones. By contrast, in hard changes, a boundary phone may undergo a complete deletion or a substitution by a totally different phone. It was experimentally shown that *phonological rules* were effective in coping with errors generated by hard changes and a 10% error reduction was achieved using only context-independent units [6]. Although the set of phonological rules was applied rarely in both training and in recognition, the application of phonological rules was shown very effective in dealing with the hard word juncture changes which are difficult to model statistically.

However, for soft changes, it has been shown [7, 12, 38, 54] that the use of *context-dependent phone-like units* is most effective. Such units specify context-independent phones according to their context, i.e. the preceding and the following phone. Context-dependent phones are not very effective with hard changes because these changes are comparatively rare and the training material is not sufficient to model the units that would represent them. Soft changes occur more frequently in the training set and hence we can model such changes in a manner similar to the way we model the set of intra-word units.

The methodology we adopted for inter-word unit modeling is described in detail in [7]. Since the initial phone of each word depends on the final phone of the preceding word and similarly for the final phone of each word, there are many more context-dependent, inter-word units than intra-word units. In the 109 speaker training set, there are roughly 5,500 inter-word units compared to only 1,800 intra-word units. The complete set of units, including both intra-word and inter-word units, was chosen according to the same unit reduction rule described in Section 4.

By varying the value of the threshold T different PLU inventories can be obtained and tested. In Figure 6, we plot the number of intra-word units, inter-word units and the sum of intra-word and inter-word units as a function of count threshold $(T-1)$. We can see that the total number of units in the standard DARPA training database is on the order of 7,000, which is too large if we have to model them independently. Based on the unit reduction rules discussed in Section 3, with the same threshold, i.e. T=30, we obtain a set of 1282 units, including 1101 double-context units, 99 left-context units, 35 right-context units and 47 context-independent units.

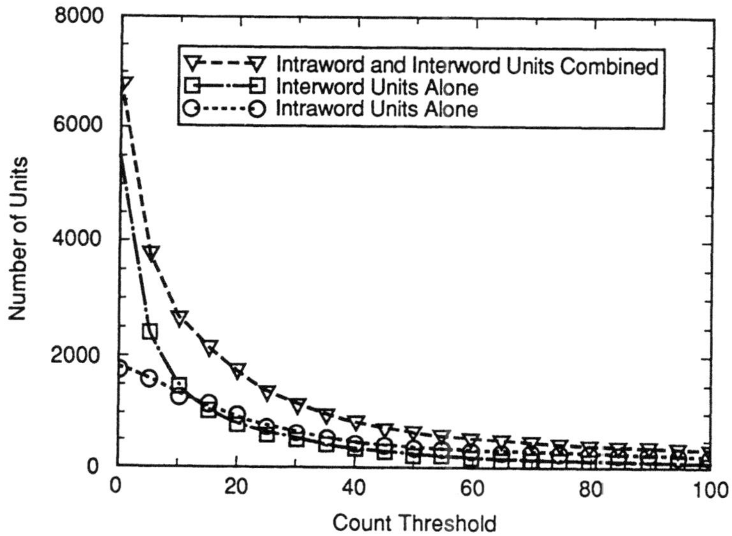

Figure 6. Plot of the number of units as a function of count threshold

A detailed summary of the experimental results using inter-word units can be found in [7, 24]. Based on the set of 1282 PLU's, the recognition performance was 93.0% word accuracy and 63.7% sentence accuracy for the FEB89 test set. A comparison with the results obtained without using inter-word units (i.e. the 1090 PLU set) shows a 20% error rate reduction when inter-word units are applied.

6. Improved Feature Analysis

So far, we have discussed the criteria for the selection of the set of fundamental units, shown how to expand the set to include both intra-word and inter-word, context-dependent PLU's and discussed how to properly model these units. In this section, we focus our discussion on an improved front-end feature analysis. In a VQHMM framework, multiple codebooks are usually designed to circumvent modeling problems due to correlation in the components of the feature vectors. Since we are using a continuous density HMM approach for characterizing each of the sub-word units, it is fairly straightforward to incorporate new features into our feature vectors. Specifically, we found it useful to incorporate higher order time derivatives of short-time cepstral features and log energy features, such as the second cepstral derivatives (*delta-delta cepstrum*), the log energy derivative (*delta energy*), and the second log energy derivative (*delta-delta energy*), into our continuous speech recognition system [26].

6.1 Second Order Cepstral Time Derivatives

The incorporation of first order time derivatives of cepstral coefficients has been shown useful for both speech recognition and speaker verification. Thus we were interested in investigating the effects of incorporating higher order cepstral time derivatives. In particular, we used a window of 3 frames (i.e. an overall window of 70 msecs) to approximate the mth delta-delta cepstral coefficient at frame l, i.e.

$$\Delta_2 \hat{c}_l(m) = K \cdot \left[\Delta \hat{c}_{l+1}(m) - \Delta \hat{c}_{l-1}(m)\right] \tag{7}$$

where $\Delta \hat{c}_l(m)$ is the estimated m^{th} delta cepstral coefficient evaluated at frame l as in Eq. (3), and K is a scaling constant which was fixed to be 0.375 (no optimization was attempted to find a better value of the normalization constant to optimize the k-means clustering part of the training algorithm).

We augment the original 24-dimensional feature vector with 12 additional features derived from Eq.(7) giving a 36-dimensional feature vector. We then tested this new feature analysis procedure on the resource management task using the word-pair grammar. The effect of adding the delta-delta cepstral parameters on recognition performance (using both inter-word and intra-word units) on the set of 1282 PLU's, was that the performance improved from 93.0% to 93.8% on the FEB89 test set. We also observed a big fluctuation in performance from one iteration to the next when delta-delta cepstrum is incorporated. Our conjecture is that the second order cepstral analysis produces very noisy observations based on a 30 msec window and a 10 msec frame shift. Another concern is the effectiveness of each of the additional features. In Ney [35], a pre-selected set of delta and delta-delta cepstral features was used. We believe that an automatic feature selection algorithm, e.g. a principal component analysis, should be used to determine the relative importance of all spectral analysis features.

6.2 Log Energy Time Derivatives

The first order time derivatives of the log energy values, known as delta energy, have been shown useful in a number of recognition systems (e.g. [27]). Most systems use both energy and delta energy parameters as features. In order to use the energy parameter, careful normalization is required. In our system, the energy parameter was normalized syllabically [44]. We did not include the energy parameter in the feature vector [24].

However, we have found that the delta and delta-delta energy parameters are more robust and more effective recognition features, and therefore we augmented the 36-dimensional feature vector with the delta and delta-delta energy features. Similar to the evaluation of the delta cepstrum in Eq.(3), the delta energy at frame l is approximated as a linear combination of the energy parameters in a 5 frame window centered at frame l.

The second order time derivatives of the energy parameters, called the delta-delta energy, are computed similar to the way the delta-delta cepstral features are evaluated in Eq.(7). Again we use a window of 3 frames. Starting with the 24-element feature vector in Eq. (4), by adding delta-delta cepstrum, delta energy and delta-delta energy to the feature set, for every frame l, we have a 38-element feature vector O_l of the form

$$O_l = \{\hat{c}_l(1{:}12),\ \Delta\hat{c}_l(1{:}12),\ \Delta_2\hat{c}_l(1{:}12),\ \Delta e_l,\ \Delta_2 e_l\} \tag{8}$$

where $\Delta_2\hat{c}_l(1{:}12)$, Δe_l and $\Delta_2 e_l$ are the additional 12 delta-delta cepstral coefficients, the delta energy parameter and the delta-delta energy parameter at frame l respectively.

We tested the use of energy time derivatives on the FEB89 test with the word pair grammar. All the tests used the set of 1282 PLU's. We started with a 36-dimensional feature vector. When delta energy is incorporated, we observed a very significant overall improvement to the level of 94.9% word accuracy for the talkers in the FEB89 test. We also note that the improvement due to delta energy varies from speaker to speaker. When delta-delta energy is added to form a 38-dimensional feature vector, we observe the same effect as in the case of delta-delta cepstrum, namely that delta-delta energy is not always beneficial. However, for some talkers the improvement is significant. There is no overall improvement; however all the test speakers achieve over 92% word accuracy. We also found the new feature analysis procedure gave the same level of improvement for the two other standard test sets, namely the OCT89 and JUN90 test sets [26].

7. Position Dependent Unit Selection And Modeling

For all our experiments we have selected the set of basic speech units based on context. However, it is believed that the spectral features of units within words behave in a different manner acoustically from those of units at the word boundaries even when the units appear in the same context. We investigated this conjecture by selecting intra-word and inter-word units independently based on the same unit reduction rule. We call such a selection strategy *position-dependent unit selection.* With a threshold of 30, we obtained a total of 1769 units, including 913 intra-word and 856 inter-word units. For the intra-word units, we have 639 units in double context, 98 left-context and 176 right-context units and 47 context-independent units. For the inter-word units, we end up with 480 units in double context, 310 left-context, 2 right-context units and 46 context-independent units. We use the same modeling strategy described in Section 5, except that when creating the FSN for segmentation and recognition, only inter-word units can appear at the word boundaries and only intra-word units can appear within words.

Using such a position-dependent unit selection strategy, we found that all the sub-word unit models are more focused in the sense that the spectral variability is less than the case in which we combine common intra-word and inter-word unit models. Two interesting observations are worth noting. First, when recognition is performed with the beam search algorithm, the number of alive nodes is much less in the 1769 PLU case than that in the 1282 PLU case. Second, the unit separation (in terms of likelihood) distance is larger for the 1769 PLU set than that for the 1282 PLU set.

For each unit p, the *unit separation distance* is defined as

$$D(p) = \min_{q \neq p} \left\{ L(Y_p \mid \lambda_p) - L(Y_q \mid \lambda_p) \right\}, \tag{9}$$

where p and q have the same middle phone. The minimization is performed over the set of units with the same middle phone as unit p regardless of their context and environment. λ_z is the model for unit z, Y_z represents all the training data segments with unit label z, and $L(Y_z \mid \lambda_x)$ is the average frame likelihood of observing Y_z which has a corresponding HMM source λ_x.

The histograms for the unit separation distances for both the 1769 PLU set and the 1282 PLU set are plotted in Figures 7a and 7b respectively. For the 1769 PLU set, we found a quite remarkable unit separation. Almost all the unit separation distances are larger than 2 (in log likelihood). It is also interesting to note that units appearing in the same context but in a different environment (i.e. intra-word versus inter-word) show the same behavior as units appearing in different context. The average unit separation distance is about 9. For the 1282 PLU set, the histogram plot is skewed to the left and the unit separation characteristics are not as pronounced as those of the 1769 PLU set. Results on the resource management task using the 1769 PLU set showed 6-10% error reduction over the 1282 PLU set using both the WP grammar and the NG grammar [26].

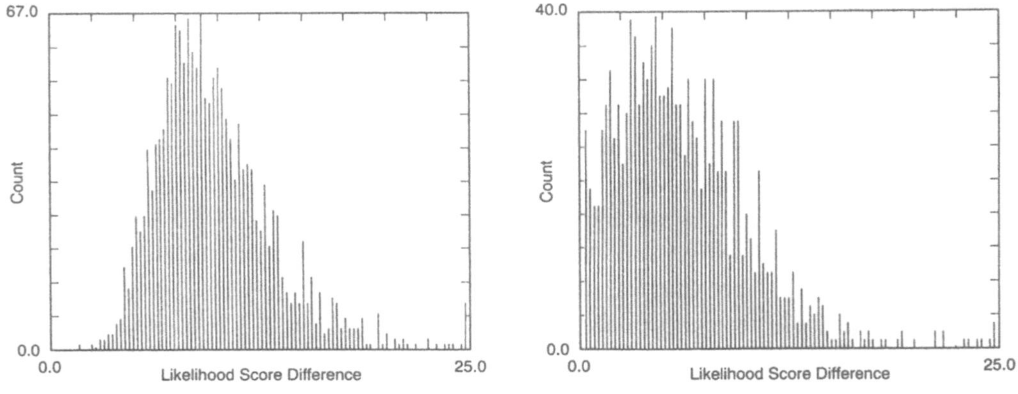

Figure 7. Histograms of the unit separation distances for (a) the 1282 unit set, and (b) the 1769 unit set

8. Summary and Future Research Topics

Based on the CDHMM framework, we have shown that careful selection of context-dependent sub-word units in conjunction with detailed acoustic modeling of the set of sub-word units provided high word accuracies for all three DARPA evaluation sets using the word pair grammar. The results reported in the previous sections are summaried in Table 3 for the FEB89 test set using the word-pair grammar. It is noted that the overall error rate reduction is over 60% going from using only context-independent units to combining all the modeling techniques discussed in the previous sections.

Test Conditions	Error Rate
Context Independent PLU	14%
+ Context Dependent PLU	8.7%
+ Inter-Word CD PLU	7.0%
+ Delta-Delta Cesptrum	6.2%
+ Energy Time Derivatives	5.1%
+ Position Dependent PLU	4.6%

Table 3. Summary of word error rates for the FEB89 test

Given the current level of recognition performance it seems fair to say that when task-specific training data are provided for acoustic modeling of the set of basic speech units using HMM's, high performance can be achieved for a large vocabulary, speaker-independent, continuous speech recognition task with a perplexity of about 60. However, there are still some open issues that need to be addressed. The reader is referred to a recent paper [19] for a discussion of some of those issues related to using HMM's for speech recognition. We list, in the following, a number of modeling issues which we believe to be essential for expanding the capabilities of our current continuous speech recognition system.

8.1 Speech Unit Selection and Modeling

To date, the criterion for selecting the set of fundamental units is based on the rule of observing enough occurrences of each unit in the training data. With the incorporation of position-dependent units we observed that the performance of the system was relatively insensitive to the unit reduction threshold used in deciding the set of units to model. To illustrate this point we show, in Table 4, the word error rates as a function of the threshold for both the word-pair case and the no grammar case for the FEB89 test set. It can be seen from Table 4 that the recognition results are quite stable over a wide range of thresholds. We also observe that the error rates are also stable from iteration to iteration under all test conditions.

Threshold	30	25	20	15	10
Number of Units	1769	2135	2534	2985	3863
WP Word Error (%)	4.6	4.7	4.6	4.7	5.3
NG Word Error (%)	20.8	19.8	19.4	20.6	20.9

Table 4. Word error rates as a function of threshold for the FEB89 test

However, choosing a large set of context-dependent units and modeling the units with enough training data are two conflicting goals that are not easily achievable based on such a simple unit reduction rule. For a general task, the number of context-dependent units required to achieve high performance can be very large, making even a reasonable amount of training material insufficient to estimate all the sub-word models with acceptable accuracy. Several techniques have been devised to overcome these training difficulties, including pooling of contexts having similar acoustic-phonetic properties [47], and model prediction using a linguistic tree structure [29].

Based upon our observations in Section 7, we believe the real benefit of incorporating position-dependent PLU modeling lies in the area of model prediction for units appearing rather infrequently in the training data. We are now experimenting with a *unit expansion rule*, which uses the unit reduction rule first (based on a lower count threshold) to get a set of *auxiliary speech units* which are the units to be expanded into a complete set. The key problem is how to predict the model for units in the auxiliary set. We propose a model prediction strategy such that, for each existing model, we compute the likelihood of observing all the acoustic segments corresponding to each auxiliary unit. Then we assign the model that is the closest to each of the auxiliary speech units. The effectiveness of such an approach is yet to be evaluated.

Another way of accomplishing model prediction, for units rarely appear in the training database, is through a unit dependent tied mixture approach. In [24], we introduced the so called CD2 approach for modeling context-dependent units. The idea is to use the set of context independent units as the basic set and use the set of context dependent units as the expanded set. First, we obtained a very detailed model (e.g. 256 mixture components per state) of all the CI units using all the training data. We then assumed that all context dependent units, which share the same center phone, have the same set of mean and covariance parameters as the CI unit they were derived from. The mixture gains, for each of the mixture components, for each state of all the context dependent units, are then estimated from the training data using a maximum likelihood approach. As an example we plot, in Figure 8, the mixture gains for the resulting models for three CD units, g_ax_t, k_ax_p and l_ax_m. Even without knowing the mean and covariance parameters for $_ax_$, we can see that the mixture gain distributions are very different. This contributes to the discrimination among various CD units. We have shown in [24], that the CD2 method is effective in estimating parameters of rarely seen units when smoothing in eq. (7) is applied. The potential of model prediction based on this idea is yet to be evaluated.

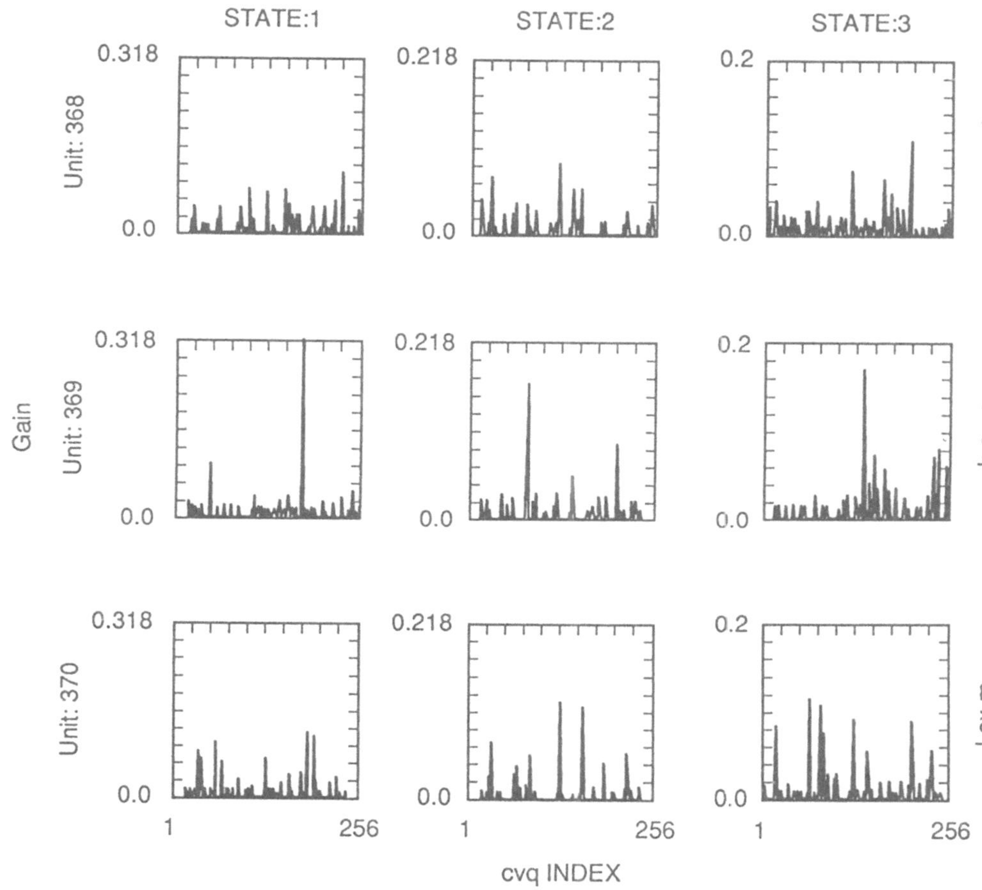

Figure 8. Plot of mixture gains in three contexts in the CD2 approach

8.2 Improved Unit and Word Discrimination

It is well-known that the method of maximum likelihood for estimating the HMM parameters need not be optimal in minimizing error rate for a recognition task. Several alternative estimation and classification criteria have been proposed [1, 4, 32]. The reader is referred to [19] for a comprehensive discussion. Recently, a new training strategy called *corrective training* [2, 28] has been shown useful in practice for HMM-based speech recognition. However, the corrective training procedures was designed primarily for discrete HMM parameters. For CDHMM-based systems, similar corrective training algorithms do not yet exit.

With the goal of improving word discrimination, we have implemented an HMM/LVQ *hybrid* algorithm [20] which uses an HMM for segmentation and a learning vector quantization (LVQ) codebook for classification. Tested on a multi-speaker, E-set recognition task in an isolated word mode, we were able to reduce the error rate from 39%

(using only HMM) to 19% (with the addition of LVQ for classification), i.e. more than a 50% error rate reduction is achieved. We have also developed a theory of adaptive learning for pattern classification [21]. Based on the theory, algorithms can be designed to perform some form of corrective training for continuous density HMM parameters. The effectiveness of such discriminative training algorithms on continuous speech recognition is yet to be evaluated. We also observed, with the use of position-dependent units, that a better unit separation often implies an improved performance. Even such a strategy does not minimize the error rate directly, we believe discriminative training techniques based on maximizing unit separation can still be used to enhance recognition performance [53].

8.3 Lexical Modeling

The lexicon we used in our study was extracted from a dictionary and each word in the vocabulary is represented by a single lexical entry which was defined as a linear sequence of phone units. Based on this simplification, the lexical variability of a word is characterized only indirectly through the set of sub-word models. It has been reported [54] that in continuous speech, the pronunciation of a word can change dramatically especially at word boundaries. Multiple pronunciations or even pronunciation networks [54] can help deal with lexical variabilities directly. However, lexical coverage, defined based only on a linguistic notion, is not enough to cope with the ill-defined mapping between linguistic units and their signal realizations. We believe that a spoken language lexicon, which is derived from real speech data, is preferred to a data independent lexicon.

In a recent study, the use of a *linguistically-defined, acoustically-decoded lexicon* [23] was proposed for isolated word recognition. The same notion has been modified and tested on the DARPA resource management database. By constraining the number of decoded units to be close to the number of phone units in the original lexicon in lexical decoding, we obtained a list of possible lexical entries for each word in the task vocabulary (991 in total). When only intra-word units are used in lexical decoding, we found that close to 40% of the time the top entry agrees with the lexical definition in the original lexicon. However we also found that for short function words, the original lexical definition was often not included in the top five lexical entries. Since more than 60% of the word errors involved function words, we believe that proper lexical modeling of such words can potentially enhance the recognition performance.

8.4 Improved Feature Analysis and Feature Selection

The incorporation of first and second order time derivatives of the cepstral and energy parameters improves recognition performance significantly. However the second order time derivatives are very noisy in the sense that they produces features which are not always beneficial and they are not stable over the segmental k-means training iterations. From our results based on second order cepstral and log energy features, it is clear that better performance can be achieved if we can stablize the features across training iterations. One way to improve the feature analysis is to select features properly so that only those features useful to discrimination are included in the feature vector. Another approach is to combine features effectively through a principal component analysis. We believe that a better feature analysis is one of the techniques we can benefit most from in the area of acoustic modeling for large vocabulary speech recognition.

8.5 Speaker Adaptation

In the case of modeling each sub-word unit with continuous density HMM's, we found, in Section 2, that the most critical issue in estimation was to avoid singularity in the mixture components. We also showed that Bayesian adaptation can be applied to perform variance clipping so that the estimated variances would not fall below a pre-determined floor value. More recently, we found that better performance can be achieved if we use a conjugate prior [25] for the variance component to estimate the variance using Bayesian adaptation techniques.

Another useful feature with the CDHMM approach is that the number of parameters required to characterize the set of fundamental units is still smaller than that required in the VQHMM approach. Therefore, speaker adaptation on CDHMM parameters is rather straightforward. Most of the existing speaker adaptation techniques are limited to the VQHMM approach (e.g. [49]). From a previous study [25], we found that rapid speaker adaptation can be accomplished effectively using Bayesian estimation techniques on adapting CDHMM parameters. We are now applying Bayesian speaker adaptation techniques on the DARPA resource management task. Our preliminary results already show a significant performance improvement over our best speaker independent testing results. The same technique can also be applied to create models for cluster of speakers with similar acoustic characteristics. For example, sex-dependent models are shown effective based on the TMHMM approach [10]. We belive the same level of improvement can be achieved for CDHMM-based approaches using Bayesian adaptation techniques.

8.6 Task-Independent Training and Recognition

To date, the high word recognition accuracy achieved on independent test data is based on task-specific testing using training data collected in the same task environment. However, in practise, it is not feasible to collect a lot of task-specific training data for every new task. The results reported in [8], suggested that the so called "vocabulary-independent" training procedure is effective for a new task when most of the task-specific phone contexts appear in the training corpus. In order to achieve true task-independent recognition, a very large corpus of training material is required and training should be performed only one time. The design of such a task-independent database is still an open problem. Even when such a training set is available, the selection of the set of basic speech units and the acoustic modeling of such a set of units are two important issues yet to be resolved. Another way to make use of the models generated from task-independent training procedures is to perform some form of task adaptation. The ideas discussed above for fast speaker adaptation can also be applied.

8.7 Speech Understanding Versus Speech Recognition

A careful examination of the word error patterns shows that function words still account for about 60% of the word errors. The second dominant category, which accounts for more than 20% of the errors, are confusions involving the same root word (e.g. location versus locations, six versus sixth, Flint versus Flint's, chop versus chopped, etc.) appearing in different forms. This type of errors can easily be corrected with a simple set of syntactic and semantic rules operating in a post-processing mode. For example, for the resource management task, we have developed a *semantic postprocessor* (decoupled from the acoustic decoder) that incorporates a set of simple rules. Our preliminary results [40] indicate that sentence accuracy for the FEB89 test set improved from 70% to close to 90% when the top candidate string decoded using the word-pair grammar is used as input to this language decoder. The word accuracy also improved from 95% to 98%. We assume a recognized sentence is correctly understood if it is semantically unambiguous compared with the reference sentence. Then the *semantic sentence accuracy* is close to 95% when the semantic postprocessor is used to process the output from the speech decoder. When no grammatical constraints are used in speech decoding, the sentence accuracy improved from 24% to 67% and the word accuracy from 77% to over 92%. The semantic sentence accuracy in this case is close to 80%.

Except in cases where some key content words were misrecognized, the simple semantic postprocessor properly decoded the noisy strings provided by the speech decoder without going back to the acoustic domain to request mismatch information. We believe that the language decoder can be more effective if the speech decoder can provide more word and string hypotheses. One way to get more information is to use the *N*-best string search strategies either in a frame-asynchronous mode [13] or in a frame-synchronous mode [51]. More recently, an efficient two-pass tree-trellis search approach, which used a breadth-first search strategy to find the top candidate in the forward pass and a best-first search strategy to obtain the *N*-best candidates in the backward pass, has been proposed for continuous speech recognition. Another way is to construct, in acoustic decoding, a phone lattice and a word lattice [5, 30-31, 55] that contain more word hypotheses, and then generate recognized strings according to the language constraints. The effectiveness of such approaches in real spoken language tasks, such as the DARPA Air Travel Information System (ATIS) task, is yet to be evaluated.

References

1. Bahl, L. R., Brown, P. F., de Souza, P. V. and Mercer, R. L.: Maximum Mutual Information Estimation of Hidden Markov Model Parameters for Speech Recognition. Proc. ICASSP 86, pp. 49-52, Tokyo, Japan, April 1986.

2. Bahl, L. R., Brown, P. F., de Souza, P. V. and Mercer, R. L.: A New Algorithm for the Estimation of Hidden Markov Model Parameters. Proc. ICASSP 88, pp. 493-496, New York, April 1988.

3. Bellegarda, J. R. and Nahamoo, D.: Tied Mixture Continuous Parameter Models for Large Vocabulary Isolated Speech Recognition. Proc. ICASSP 89, pp. 13-16, Glasgow Scotland, May 1989.

4. Ephraim, Y., Dembo, A. and Rabiner, L. R.: A Minimum Discrimination Information Approach for Hidden Markov Modeling. IEEE Trans. on Information Theory, Vol. IT-35, No. 5, pp. 1001-1013, Sept. 1989.

5. Fissore, L., Laface, P., Micca, G. and Pieraccini, R.: Lexical Access to Large Vocabularies for Speech Recognition. IEEE Trans. on Acous., Speech and Signal Proc., pp. 1197-1213, Vol. ASSP-37, No. 8, August 1989.

6. Giachin, E., Rosenberg, A. E. and Lee, C.-H.: Word Juncture Coarticulation Modeling Using Phonological Rules for HMM-based Continuous Speech Recognition. Proc. ICASSP 90, pp. 737-740, Albuquerque, NM, April 1990.

7. Giachin, E., Lee, C.-H., Rabiner, L. R. and Pieraccini, R.: Word Juncture Modeling Using Inter-Word Context-Dependent Phone-Like Units: submitted for publication.

8. Hon, H. W., Lee, K. F. and Weide, R.: Towards Speech Recognition Without Vocabulary Specific Training. Proc. EuroSpeech 89, pp. 481-484, Paris, France, September 1989.

9. Huang, X. D. and Jack, M. A.: Semi-Continuous Hidden Markov Models for Speech Signals. Computer Speech, and Language, Vol. 3, No. 3, pp. 239-251, July 1989.

10. Huang X. D., Alleva, F., Hayamizu, S., Hon, H. W., Hwang, M. Y. and Lee, K. F.: Improved Hidden Markov Modeling for Speaker-Independent Continuous Speech Recognition. Proc. DARPA Speech and Natural Language Workshop, Somerset, PA, June 1990.

11. Huang, E. F. and Soong, F. K.: A Fast Tree-Trellis Search for Finding the N-Best Sentence Hypotheses in Continuous Speech Recognition. Proc. DARPA Speech and Natural Language Workshop, Somerset, PA, June 1990.

12. Hwang, M. Y., Hon, H. W. and Lee, K. F.: Modelling between-Word Coarticulation in Continuous Speech Recognition. Proc. EuroSpeech 89, Paris, September 1989.

13. Jelinek, F.: A Fast Sequential Decoding Algorithm Using A Stack. IBM J. Res. Develop., vol. 13, pp. 675-685, Nov. 1969.

14. Jelinek, F.: Continuous Speech Recognition by Statistical Methods. Proc. IEEE, Vol. 64, No. 4, pp. 532-536, April 1976.

15. Jelinek, F. and Mercer, R. L.: Interpolated Estimation of Markov Source Parameters from Sparse Data. Pattern Recognition in Practice, E. S. Gelsema, and L. N. Kanal, Ed., North-Holland Publishing Co., Amsterdam, pp. 381-397, 1980.

16. Jelinek, F.: The Development of an Experimental Discrete Dictation Recognizer. Proc. IEEE, Vol. 73, No. 11, pp. 1616-1624, November 1985.

17. Juang, B.-H., Wong, D. Y. and Gray, A. H. Jr.: Distortion Performance of Vector Quantization for LPC Voice Coding. IEEE Trans. on Acoustics, Speech, and Signal Proc., Vol. ASSP-30, pp. 294-304, April 1982.

18. Juang, B.-H., Rabiner, L. R. and Wilpon, J. G.: On the Use of Bandpass Liftering in Speech Recognition. IEEE Trans. on Acoustics, Speech, and Signal Proc., Vol. ASSP-35, No. 7, pp. 947-954, July 1987.

19. Juang, B.-H. and Rabiner, L. R.: Issues in Using Hidden Markov Models for Speech Recognition. To appear in Advances in Speech Signal Processing, S. Furui and M. Sondhi editors, Marcel Dekker Inc., New York, 1990.

20. Katagiri, S. and Lee, C.-H.: A New HMM/LVQ Hybrid Algorithm for Speech Recognition. To appear in Proc. GLOBECOM-90, San Diego, CA, December 1990.

21. Katagiri, S., Lee, C.-H. and Juang, B.-H.: A Generalized Probability Descent Method. Proc. Acous. Soc. of Japan, Nagoya, Japan, Sept. 1990.

22. Lee, C.-H., Soong, F. K. and Juang, B.-H.: A Segment Model Based Approach to Speech Recognition. Proc. ICASSP 88, New York, pp. 501-504, April 1988.

23. Lee, C.-H., Juang, B.-H., Soong, F. K. and Rabiner, L. R.: Word Recognition Using Whole Word, and Subword Models. Proc. ICASSP 89, pp. 683-686, Glasgow, Scotland, May 1989.

24. Lee, C.-H., Rabiner, L. R., Pieraccini, R. and Wilpon, J. G.: Acoustic Modeling for Large Vocabulary Speech Recognition. Computer Speech and Language, Vol. 4, pp. 127-165, 1990.

25. Lee, C.-H., Lin, C.-H. and Juang, B.-H.: A Study on Speaker Adaptation of the for Continuous Density HMM Parameters. Proc. ICASSP 90, pp. 145-148, Albuquerque, April 1990.

26. Lee, C.-H., Giachin, E., Rabiner, L. R., Pieraccini, R. and Rosenberg, A. E.: Improved Acoustic Modeling for Continuous Speech Recognition. Proc. DARPA Speech and Natural Language Workshop, Somerset, PA., June 1990.

27. Lee, K. F.: Automatic Speech Recognition – The Development of the SPHINX System, Kluwer Academic Publishers, Boston, 1989.

28. Lee, K. F. and Mahajan, S.: Corrective and Reinforcement Learning for Speaker-Independent Continuous Speech Recognition. Proc. EuroSpeech 89, pp. 485-488, Paris, France, September 1989.

29. Lee, K.-F. et al: Allophone Clustering for Continuous Speech Recognition. Proc. ICASSP 90, pp. 749-752, Albuquerque, NM, April 1990.

30. Levinson, S. E.: Structural Methods in Automatic Speech Recognition. Proc. IEEE, Vol. 73, No. 11, pp. 1625-1650, Nov. 1985.

31. Levision, S. E., Liberman, M. Y., Ljolje, A. and Miller, L. G.: Speaker Independent Phonetic Transcription of Fluent Speech for Large Vocabulary Speech Recognition. Proc. ICASSP 89, pp. 442-444,Glasgow, Scotland, May 1989.

32. Ljolje, A., Ephraim, Y. and Rabiner, L. R.: Estimation of Hidden Markov Parameters by Minimizing Empirical Error Rate. Proc. ICASSP 90, pp. 709-712, Albuquerque, NM, April 1990.

33. Lowerre, B. and Reddy, D. R.: The HARPY Speech Understanding System. Trends in Speech Recognition, W. Lee, Ed., Prentice-Hall Inc., pp. 340-346, 1980.

34. Merhav, N. and Ephraim, Y.: Maximum Likelihood Hidden Markov Modeling Using a Dominant Sequence of States. Submitted for publication.

35. Ney, H.: Acoustic-Phonetic Modeling Using Continuous Mixture Densities for the 991-Word DARPA Speech Recognition Task. Proc. ICASSP 90, pp. 713-716, Albuquerque, NM, April 1990.

36. Pallett, D.: Test Procedures for the March 1987 DARPA Benchmark Tests. DARPA Speech Recognition Workshop, pp. 75-78, March 1987.

37. Pallett, D.: Selected Test Material for the March 1987 DARPA Benchmark Tests. DARPA Speech Recognition Workshop, pp. 79-81, March 1987.

38. Paul, D.: The Lincoln Continuous Speech Recognition System: Recent Development and Results. Proc. DARPA Speech and Natural Language Processing Workshop, Philadelphia, Feb. 1989.

39. Pieraccini, R., Lee, C.-H., Giachin, E. and Rabiner, L. R.: Implementation Aspects of Large Vocabulary Recognition Based on Intra-word and Inter-word Phonetic Units. Proc. DARPA Speech and Natural Language Workshop, Somerset, PA, June 1990.

40. Pieraccini, R., Su, K.-Y. and Lee, C.-H.: unpublished work.

41. Price, P. J., Fisher, W., Bernstein, J. and Pallett, D.: A Database for Continuous Speech Recognition in a 1000-Word Domain. Proc. ICASSP 88, New York, NY, pp. 651-654, April 1989.

42. Rabiner, L. R., Wilpon, J. G. and Juang, B.-H.: A Segmental K-Means Training Procedure for Connected Word Recognition. AT&T Tech. J., Vol. 65, No. 3, pp. 21-31, May-June 1986.

43. Rabiner, L. R.: A Tutorial on Hidden Markov Models, and Selected Applications in Speech Recognition. Proc. IEEE, Vol. 77, No. 2, pp. 257-286, Feb. 1989.

44. Rabiner, L. R., Wilpon, J. G. and Soong, F. K.: High Performance Connected Digit Recognition Using Hidden Markov Models. IEEE Trans. on Acoustics, Speech, and Signal Proc., Vol. 37, No. 8, pp. 1197-1213, Aug. 1989.

45. Rabiner, L. R., Lee, C.-H., Juang, B.-H., Roe, D. B. and Wilpon, J. G.: Improved Training Procedure for Hidden Markov Models. J. Acoust. Soc. Am., suppl. 1, vol. 84, S61, Fall, 1988.

46. Rosenberg, A. E., Lee, C.-H., Soong, F. K. and McGee, M. A.:: Experiments in Automatic Talker Verification Using Sub-Word Hidden Markov Models. Proc. ICSlP90, Kobe Japan, November 1990.

47. Sagayama, S.: Phoneme Environment Clustering for Speech Recognition. Proc. ICASSP 89, Glasgow, Scotland, pp. 397-400, May 1989.

48. Schwartz, R. et al: Context Dependent Modeling for Acoustic-Phonetic Recognition of Continuous Speech. Proc. ICASSP 85, pp. 1205-1208, Tampa, Florida, March 1985.

49. Schwartz, R., Chow, Y. L. and Kubala, F.: Rapid Speaker Adaptation Using a Probabilistic Spectral Mapping. Proc. ICASSP 87, pp. 633-636, Dallas, April 1987.

50. Schwartz, R. et al: The BBN BYBLOS Continuous Speech Recognition System. Proc. Speech and Natural Language Workshop, pp. 94-99, Philadelphia, Feb. 1989.

51. Schwartz, R. and Chow, Y. L.: The N-Best Algorithm: An Efficient and Exact Procedure for Finding the N Most Likely Sentence Hypotheses. Proc. ICASSP 90, pp. 81-84, Albuquerque, NM, April 1990.

52. Soong, F. K. and Rosenberg, A. E.: On the Use of Instantaneous, and Transitional Spectral Information in Speaker Recognition. IEEE Trans. on Acoustics, Speech, and Signal Proc., Vol. 36, No. 6, pp. 871-879, June 1988.

53. Su, K.-Y. and Lee, C.-H.: Robustness and Discrimination Oriented Speech Recognition Using Weighted HMM and Subspace Projection Approaches. Submitted for publication.

54. Weintraub, M. et al: Linguistic Constraints in Hidden Markov Model Based Speech Recognition. Proc. ICASSP 89, pp. 699-702, Glasgow, Scotland, May 1989.

55. Zue, V., Glass, J., Phillips, M. and Seneff, S.: The MIT Summit Speech Recognition System: A Progress Report. Proc. Speech and Natural Language Workshop, Philadelphia, Feb. 1989.

A Fast Lexical Selection Strategy for Large Vocabulary Continuous Speech Recognition

Xavier L. AUBERT

Philips Research Laboratory Belgium, 4 Av. A. Einstein, 1348 Louvain-La-Neuve, Belgium

ABSTRACT

This paper describes an algorithm for performing fast selection of word candidates to speed up the decoding of *continuous speech*. Interacting with word transitions, it fits very well into standard Viterbi decoding. The main features are: a tree-organization of the lexicon, coarse models deduced from the detailed ones, a selection rule compatible with unknown word boundaries, an efficient DP implementation. Various applications show a global reduction of the search cost by a factor 5 to 10 depending on the lexicon size, the error rate being increased by less than 1 %.

Keywords: Continuous Speech Recognition, Large Vocabulary, Fast Match, Lexical Tree, Hidden Markov Models, Dynamic Programming, Beam Search

1 Introduction

Fast lexical selection aims at speeding up the recognition process by performing a detailed search only on a small fraction of the nominal vocabulary which is dynamically pruned according to the coarse acoustic content of the incoming speech signal. As the vocabulary size exceeds several hundreds, the decoding of *continuous* speech involves creating *end of word* hypotheses nearly every time frame. In case of a large perplexity language model, this in turn generates a lot of new word startups, although only a few are worth considering at the current time index. Hence, the interest for selecting the word candidates that appear *acoustically* compatible with the signal *ahead* of the time index, before submitting them to the detailed scoring procedure. However, as opposed to syntactic constraints that improve recognition rate, acoustic predictions are less reliable and may introduce *unrecoverable errors*. Thus, a tradeoff exists between efficiency and accuracy requirements when applying such pruning techniques.

In contrast with *isolated* word utterances, performing a lexical preselection is not straightforward in *continuous* speech since the recognition algorithm has to solve *simultaneously* the problem of word selection together with their location along the time axis. Our strategy developed earlier [1] was based on a presegmentation and labeling of the signal in broad phonetic classes. A subsequent module then performed a fast selection of word candidates compatible with this coarse decoding. This approach had however several drawbacks related to the reliability of the broad phonetic segmentation, since errors at this level are likely to propagate into the word selection step.

This paper presents a new and much more efficient approach inspired from the concept of *Fast Match* recently introduced by the IBM group [2]. The main idea is to

NATO ASI Series, Vol. F 75
Speech Recognition and Understanding. Recent Advances
Edited by P. Laface and R. De Mori

perform the preselection step at the *centi-second* time scale, thus avoiding the need for any kind of presegmentation, and to interact with the word hypotheses. In section 2, we present the general framework and give an overview of the system architecture. Section 3 provides the overall description of the fast selection algorithm. Results are presented in section 4 for two *large vocabulary* continuous speech recognition applications.

2 General Framework and System Architecture

In this study, the system is based on *Hidden Markov Models* (HMM) of phonemes and the Viterbi decoding is performed through a data driven beam search organization [3]. The acoustic, lexical and syntactic knowledge sources are simultaneously taken into account. Each phoneme is modeled by a three state probabilistic machine based on continuous mixture densities [4] and word models are constructed by concatenating the HMMs of their sub-units. Constraints at the syntactic-semantic level are introduced by a *stochastic* language model based on *bigrams* of word categories [5].

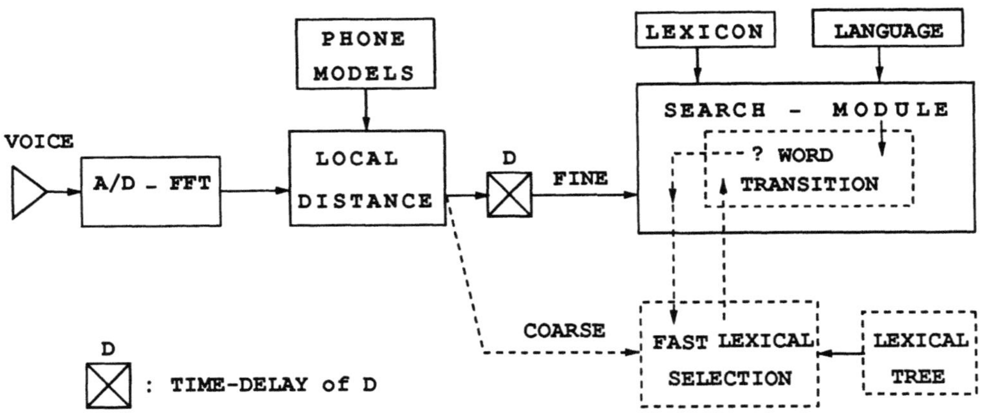

Figure 1. Architecture of recognition system with fast lexical selection

In the course of the left to right decoding process, only the most promising hypotheses are kept active and further developed according to the beam width. Whenever an active path reaches the end of a word model, the language module provides the possible successors: when the linguistic constraints are of stochastic nature, *every* lexicon entry is an admissible candidate, the transition being weighted by its probability. Therefore, the most natural opportunity for interacting with the present search algorithm is given at the *transitions* between words, that correspond to potential segmentation points.

Along this line, the selection task consists in constructing a list of word candidates that are likely to *start at current time index t*, based on the acoustic content of the incoming speech signal in the sub-interval ranging from t to t+D where D is a fixed length. Figure 1 above shows the general system architecture and explains the coupling of the lexical preselection with the main search procedure. As the fast selection is performed on a sub-interval extending *ahead* of the time index, the local scores for this "coarse" stage have to be available with some advance versus the detailed ones. Hence, the reason for introducing the *time-delay D* in the data-flow of the "fine" local distances, this delay being equal to the anticipatory context size.

3 Description of the Fast Selection Algorithm

The efficiency of the lexical preselection -a key point for the success of the task- is achieved by combining various techniques. Four main components may be distinguished:

1. a *tree organization* of the whole lexicon in terms of phone-like units
2. *coarse models* automatically derived from the trained detailed HMMs
3. a robust *decision rule* for selecting word candidates in *continuous* speech
4. a highly optimized *DP alignment* with *subsampling* on the selection interval

3.1 Tree-organization of the Lexicon

Within the detailed search, the lexicon is handled as a simple *table*, each entry being represented separately by its phoneme sequence. However, for the purpose of fast selection, a much more efficient structure follows from treating identical word *stems* in common. This is particularly true for inflected European languages that exhibit frequent word ending variations. The tree-organization of the whole lexicon is obtained by performing *left recombinations* on the original transcriptions made of *context-independent* subword units (see figure 2 hereunder).

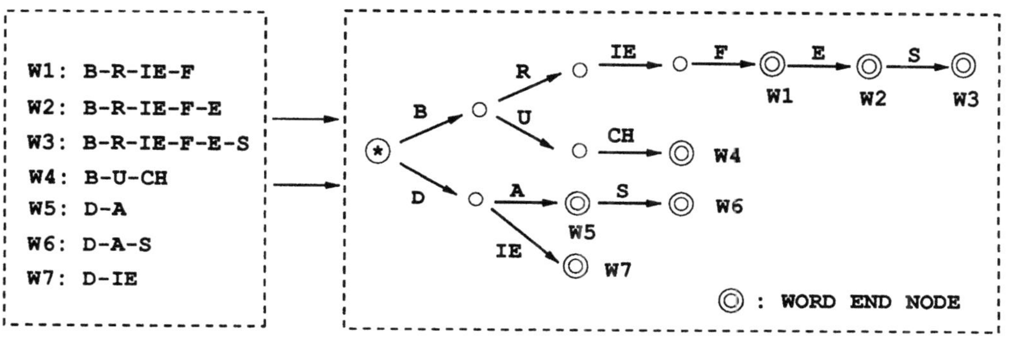

Figure 2. Illustration of the tree organization for a 7 word sub-lexicon

To retrieve the corresponding lexical entries, a flag is attached to each node of the tree, indicating whether it corresponds to a leave and, if so, providing the word identity. The obvious advantage of the tree lies in its intrinsic property of organizing the lexicon in *subsets of words starting with the same phoneme sequence.* To benefit from this saving, it is understood that a full spread through the entire tree will be prevented by rapidly pruning the branches with a poor score during the selection stage. For efficacy, each node is given its precompiled "children" list, i.e. all lexical entries that are produced when expanding the arcs originating from this node up to the terminals. Looking at the number of phoneme arcs in the early node generations of the tree reveals the interest of such structure: for our 10,000 word German lexicon, there are 25 arcs in the first generation and (resp.) 316 and 1464 in the second and third one.

3.2 Coarse Phone Models Aiming at Fast Matching

Starting from the robustly trained *context-independent* phoneme models, a set of *coarse* phone machines is deduced. Each coarse model is a single state machine obtained by pooling the *Probability Density Functions* (PDF) of the three states for the corresponding phoneme. More precisely, the probability of being on a particular "coarse" state given an acoustic vector is taken as the *maximum* over the three PDFs for the same phoneme class. There is thus no need for any kind of additional training. Besides, a *minimal* duration is required for the admissible realization of each coarse phone, the value of which is chosen *speaker-independently* and given very conservative values (30 msec for plosives, 50 msec for most vowels or fricatives and 70 msec for diphtongues). With regard to the detailed three-state HMMs, the observation probability of the associated coarse model is, by construction, always larger due to the maximum operation.

3.3 Decision Rule for Selecting the Word Candidates

Owing to the *fixed* length of the anticipatory interval, the selection rule has to cope with candidates whose realization might be shorter or longer than D. In the course of the fast matching process, a *beam search* technique is adopted which prunes the unpromising phone nodes, eliminating at the same time the acoustically *incompatible* candidates. Hence, the selection is obtained time synchronously: whenever a word end node is activated in the tree, this lexical entry is put on the candidate list stack. When the end of the interval is reached, all words corresponding to the *extension of still active stems* are included using the precompiled "children" lists associated to each node. It must be emphasized that the design of this decision rule gives full flexibility concerning the choice of D: any positive value of the selection window can be used, the depth of the alignment along the tree arcs adapting itself automatically.

The reliability of the lexical selection depends on the *beam threshold choice* that controls the tradeoff between efficacy and accuracy. Careful experiments have shown the usefulness of a variable beam width that is increased along the time axis following a linear law with saturation. The validation of such heuristics rests upon the results obtained under various conditions (see section 4).

3.4 Optimized DP Algorithm in Tree with Subsampling

The preselection algorithm itself is nothing but a particular data driven organization of the *Dynamic Programming* (DP) beam search [3], adapted to the lexical tree structure and designed for the greatest efficiency. The DP recurrence implementation is reduced to the extreme minimum of operations, the trace-keeping being unnecessary for the current purpose. Moreover, it appears that the fast matching may be based on *one* acoustic vector over *two* and this halves the computations of the selection stage without affecting the results.

4 Application in Continuous Speech Recognition

This new fast selection technique has been tested on real *continuous* speech signals recorded in office environment. Two databases have been considered: first, 200 *German* sentences (1391 uttered words) representing database queries spoken by 4 speakers and second, 200 *English* sentences (2886 uttered words) extracted from business letters, pronounced by 2 speakers. In each case, the phoneme HMMs have been trained on a *speaker dependent* mode using a phonetically balanced database, the training vocabulary having a small overlap with the test lexicon. For the German task the training material lasts about 7 minutes while for the English task it comprises 15' of speech.

4.1 Choice of the Anticipatory Interval Length (D)

In continuous speech, it is not clear from the outset which *interval length* will do the best job for the lexical preselection. On the other hand, the implemented algorithm has been designed to cope with any positive D value. Therefore, a systematic investigation has been carried out within a 1,000 word German lexicon for one speaker and one whole test session. The results are summarized in table 1 hereunder.

Selection Interval Length	0.0	0.1"	0.2"	0.3"	0.4"	0.6"	0.8"
Av. Candidate List Size	100%	32%	11%	3.5%	2.2%	1.8%	1.7%

Table 1. Relationship of candidate list size versus selection interval length

A short value of D (0.1 second) leads to a limited preselection effect, yet quite significant. This corresponds roughly to the performances of our previous *Fast Look-ahead* scheme where only the next phoneme was taken into account [1]. When the anticipatory interval is made longer, more words can be eliminated and the average candidate number reduces asymptotically to a few percentage of the lexicon size. However, the selection cost increases also significantly with D. The best compromise is achieved with a selection window of about 0.4 second. Looking at the speaking rate of the six test speakers reveals that this corresponds roughly to the average word duration.

4.2 Definition of the Performance Indexes

The ultimate goal of a fast selection technique is to speed up drastically the decoding without increasing significantly the recognition error rate. Hence, the impact of the present work is made clear by comparing the main performance indexes (search cost & recognition rate) for both the original search procedure and the new one including the fast lexical selection. Two different measurements of the search cost have been considered: first the average number of hypotheses processed at the csec. scale by the detailed match and second, the overall CPU-time used together by the preselection step *and* the main decoding, but not including all computations implied by the acoustic front-end. Thus, the distance calculations are not taken into account in the CPU figures.

On the other hand, additional errors might be introduced: a selection error (though unfrequent) can even lead to more than one incorrect word in the decoded sentence, especially when a long word is falsely rejected. However this worst case happens only in very few "bad" sentences. Word error rates includes deletions, insertions and substitutions. All experiments have been performed on a 3 MIPS workstation.

4.3 Results on Continuous German Task

The recognition of the German sentences has been performed under two different sets of conditions, comprising either a lexicon of 1,000 words or an enlarged one with 10,000 entries. In both cases, syntactical constraints are integrated by means of a stochastic language model, leading to a test set perplexity of respectively, 124 and 388 [5].

For the 1,000 word task, the average number of elementary paths is reduced by a factor of 10 while the search CPU-time is divided by more than 6. The word error rate averaged over the speakers is increased by 0.5 %. For the 10,000 word task, the average number of paths is divided by 20, the search CPU is reduced by about one order of magnitude and the error increase is still around 1 %. Depending on the task, the average candidate list represents (resp.) 2 % and 1% of the vocabulary.

4.4 Results on Continuous English Task

For the recognition of the English sentences, no language model was available, hence the test set perplexity is equal to the lexicon size of 2,437 words which explains the relatively high error rates. Table 2 summarizes the essential figures for two speakers. The selected candidate list comprises 38 words on average, i.e. 1.5 % of the lexicon.

SPEAKER	Male British			Male North American		
	Av.PTS	W.ERR.%	S-CPU	Av.PTS	W.ERR.%	S-CPU
NO LEX.SEL.	15,116	25.2%	14H 50'	16,006	29.3%	18H 08'
+ LEX.SEL.	754	25.9%	1H 42'	944	29.7%	2H 08'
IMPACT	/ 20.1	+0.7%	/ 8.7	/ 17.0	+0.4%	/8.5

Table 2. Performance indexes and impact of lexical selection for English task (P=2,437, W.ERR. stands for Word Error Rate & S-CPU for Search CPU time)

The present algorithm thus offers an appropriate answer to both requirements of reliability and efficiency, for very large vocabulary continuous speech recognition.

References

[1] Aubert X.L., "Fast Look-Ahead Pruning Strategies in Continuous Speech Recognition", Proc. ICASSP, Glasgow, pp. 659-662, 1989.

[2] Bahl L., De Gennaro S., Gopalakrishnan P. & Mercer R., "A Fast Approximate Acoustic Match for Large Vocabulary Speech Recognition", Proc. Eurospeech, Paris, pp.156-158, 1989.

[3] Ney H., Mergel D., Noll A. & Paeseler A., "A Data-Driven Organization of the Dynamic Programming Beam Search for Continuous Speech Recognition", Proc. ICASSP, Dallas, 1987.

[4] Ney H., Noll A., "Phoneme Modelling using Continuous Mixture Densities", Proc. ICASSP, New-York, pp. 437-440, 1988.

[5] Steinbiss V., Noll A., Paeseler A., Ney H. et al., "A 10,000-word Continuous Speech Recognition System", Proc. ICASSP, Albuquerque, 1990.

Performance of a Speaker-Independent Continuous Speech Recognizer

L. Fissore ◇ P. Laface ⋆ G. Micca ◇ R. Pieraccini ∘

◇ CSELT - Centro Studi e Laboratori Telecomunicazioni
Via G. Reiss Romoli 274 - 10148 Torino, Italy
⋆ Dipartimento di Automatica e Informatica - Politecnico di Torino
Corso Duca degli Abruzzi 24 - 10129 Torino Italy
∘ AT&T Bell Laboratories, Mountain Avenue 600, Murray Hill, NJ 07974 USA
*

Abstract

The speaker-independent word recognition component of a Speech Understanding System is described. It is based on Hidden Markov Models (HMM) of phonetic units and produces a lattice of word hypotheses suitable to be parsed by a linguistic analyzer that uses the acoustic scores of words for generating the best interpretation of the spoken utterance given a set of syntactic and semantic rules.

This work presents the results of a set of experiments carried out with the aim of selecting a suitable set of sub-word unit models for an E-mail inquire application. The inquiries have been recorded over a dialed-up telephone line connected to the local PABX. The performance of the system is presented and comparative results are given for Discrete and Continuous Density HMMs.

1 Introduction

The continuous speech recognizer presented in this paper is the acoustic Front-End of a Speech Understanding System being developed by Cselt within a research program on man-machine interfaces to information providing systems through the telephone network.

A voice-activated inquire system to be used over the telephone imposes severe constraints upon the speech recognition component: it must accept input from different, untrained speakers, must be robust to extra-linguistic acts, and has to deal with the problems raised by the reduced bandwidth and by the line noise and distortions. Of course, real time performance is required, coupled with acoustic decoding accuracy. The recognizer has been designed having in mind these goals, in particular the real-time constraint. Its main features, in the present implementation, are the use of differential parameters, the possibility of multi-speaker training, and accurate modeling of context-dependent units and of function-words.

2 Recognizer Overview

The functional architecture of the recognizer is given in Fig. 1.

* *This research has been partially supported by the ECC within the ESPRIT Project n. 2218 "SUNDIAL"*

NATO ASI Series, Vol. F 75
Speech Recognition and Understanding. Recent Advances
Edited by P. Laface and R. De Mori

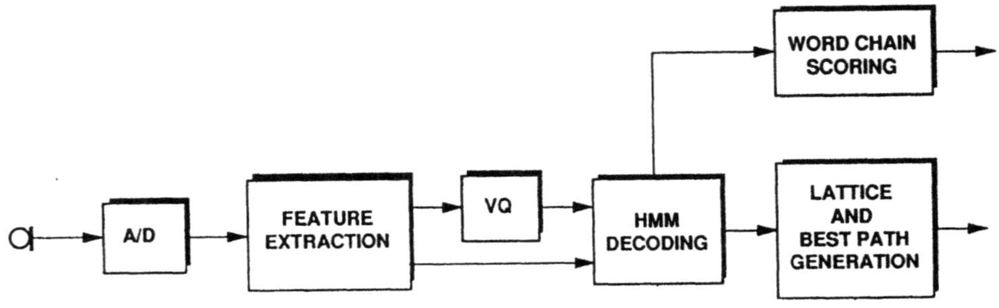

Figure 1. Front-End General Architecture

Both Discrete and Continuous Density HMMs (DDHMMs - CDHMMs) have been used in the simulations, while DDHMMs only are currently implemented in the hardware [2]. HMMs are trained through the *Forward-Backward* or the *Segmental K-means* training procedures [1,11].

The output of the acoustic decoder is the **Word Lattice** and the **Best Scored Word Sequence**. The Word Lattice entries are lexical identifiers with their associated acoustic likelihood score, and estimated time boundaries. The Word Lattice is processed by the Linguistic Modules of the SUS that generates an interpretation of the sentence by means of a score-driven parser[3]. The Best Scored Word Sequence is the sequence of words spanning the utterance along the optimal path of the acoustic decoding. It is so far used only for computing the recognizer accuracy.

2.1 Feature Extraction

The speech signals are collected through a linear electret-type telephone connected to a local PABX, low-pass filtered at 5 KHz, and sampled at 16 KHz by means of a 12 bit A/D converter. No preemphasis is adopted, and 12 cepstral coefficients C_i are computed by applying a Discrete Cosine Transform to the parameters extracted by means of a Mel-based spectral analysis. Differential cepstral parameters [7] are included in the feature vector to better characterize spectral changes over time. An utterance is, therefore, represented by a sequence of vectors $\mathbf{S}(l) = \{\mathbf{C}(l), \mathbf{\Delta C}(l), E(l)\Delta E(l)\}$, where $E(l)$ and $\Delta E(l)$ are the energy of the l-th frame and the differential energy parameter respectively.

2.2 Vector Quantization

Three separate codebooks have been generated for DDHMMs-based experiments: the first one for the cepstral parameters, the second for the Delta cepstral ones, and the last one for the energy and the delta energy parameters taken together. A training database (Section 3) has been used for codebook generation through the full-search LBG algorithm [10]. Codebooks of 8 or 9 bit have been obtained for cepstral and Delta cepstral parameters, while 5 or 6 bit codebooks have been used for the energies. The (9, 9, 6) bit multi-codebook has been extensively tested for simulation experiments, while a (8, 8, 5) bit multi-codebook vector quantization can be easily supported by the current real time implementation of the recognizer.

2.3 HMM Modeling

The current hardware architecture supports DDHMMs only, but as a hardware implementation of the CDHMMs is planned to be shortly available, performance evaluations have been carried out for the latter type of HMMs too.

2.3.1 Phonetic transcription

The sentences are transcribed according to the basic phoneme set for the Italian language. They are represented by strings of phonemes which are then processed by a rule compiler for taking into account coarticulation phenomena, regional pronunciation and alternative pronunciations [6]. The set of rules currently used is composed of 10 coarticulation rules and a branching rule related to the optional presence of silence between adjacent words.

2.3.2 Continuous Density HMMs

First order, left to right Hidden Markov Models with 3 states are used to characterize the units chosen for representing the lexicon. Each model is defined by a state transition matrix $\mathbf{A} = a_{ij}$ where $a_{ij} = 0$ for $j < i$ and $j > i+1$, and a continuous mixture density $b_j(\mathbf{x})$ for each state j of the form: $b_j = \sum_{m=1}^{M} c_{mj}\mathcal{N}(\mathbf{x}, \mu_{mj}, U_{mj})$ where $\mathbf{x}$ is the input frame vector , c_{mj} is the mixture weight for the mth component in state j, μ_{mj} is the mean vector for mixture component m in state j, U_{mj} is the covariance matrix for mixture component m in state j and $\mathcal{N}()$ is the Normal density function. Diagonal covariance matrices have been used in the experiments described in this work.

The training algorithm The parameters a_{ij}, c_{mj}, μ_{mj} and U_{mj} for every model are estimated through the *Segmental K-means* [11] training procedure.

For training context-independent speech units every sentence of the training database is linearly segmented into a number of sections that is equal to the number of phone units representing the spoken sentence (assuming no silence between words). In turn, each segment is linearly apportioned among the states of its corresponding model.

A *K-means* clustering procedure is performed on the speech segments corresponding to every state of the models. M clusters per state are generally obtained. This number is reduced when the number of frames within a state is less than a fixed threshold.

The mean vector and diagonal covariance matrix of each cluster are estimated assuming that the cluster is characterized by a Gaussian probability density. So each cluster represents a single mixture components and its mean and covariance constitute the new μ_{mj} and U_{mj} for that state of that model. The weights c_{mj} are estimated as the relative number of frames associated with mth cluster.

The estimated models are used to segment each training sentence. The segmentation is carried out by decoding the utterance using the Viterbi algorithm. The clustering and reestimation procedures are then iterated for a given number of iterations (generally 5).

Context-dependent units are bootstrapped by using the parameters of their corresponding context-independent models. For instance, triphone /pos/ and diphone /og/ are initialized by the parameters previously obtained by training the context-independent model /o/.

2.3.3 Discrete Density HMMs Training

Discrete Density HMMs are trained by means of the Forward-Backward algorithm.
Context-independent units are initialized by uniformly distributed probabilities, and five iterations are usually required to obtain the convergence of the parameters of the models.
Context-dependent units are bootstrapped,even in this case, by using the parameters of their corresponding context-independent models. Two iterations of the Forward-Backward procedure are generally required to reach model convergence.

The recognition algorithm The recognition module uses the Forward decoding algorithm which has been proven to yield better word accuracy and more accurate word boundary detection with respect to the Viterbi decoding algorithm [4]. Scores are normalized according to the procedure described in [5].

The lexicon of the application is represented by a tree whose nodes represent phonetic units. At an inner level, nodes are substituted by their corresponding Hidden Markov states (Fig. 2). The root of the tree is represented by the *silence* that is modeled by three tied states.

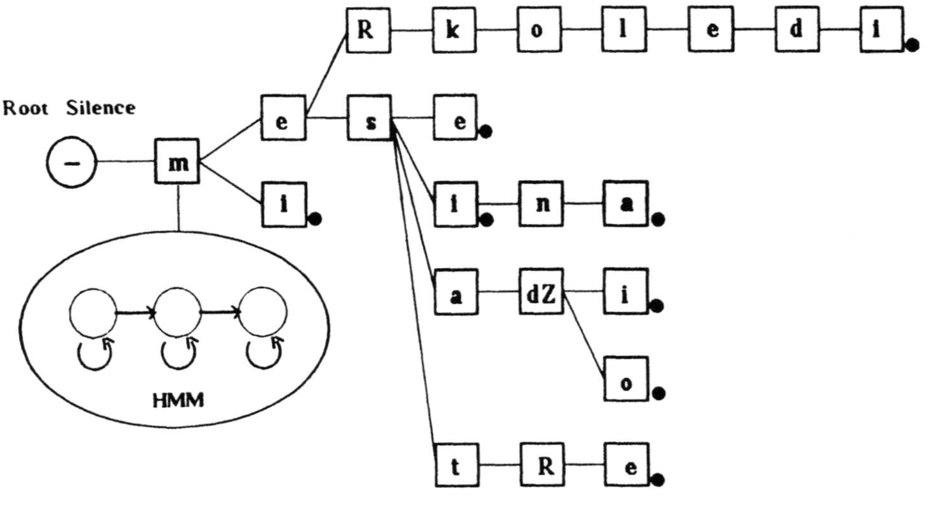

Figure 2. Lexical tree and HMM states.

The decoding process develops the lexicon tree into a trellis structure. The trellis is updated at each frame on the basis of a list of active nodes (ANL). Active nodes are expanded according to the scoring process of the Forward algorithm. Nodes whose score falls below the current best score by more then a Beam Search threshold **B** are eliminated from the ANL list. When a terminal node is taken into account and its score is greater than a Decoding threshold **D**, the associated lexical item is generated into a Word Lattice. However, instantiations of a given lexical item that have the same starting time are not allowed in the word lattice, only the best one is selected. A terminal node is linked back to the tree root, but its score is multiplied by a word penalty factor **WP**, ranging from 0 to 1.0. This factor has been introduced for reducing the generation rate of short words, that are likely to appear as sub-strings of longer words.

3 Speech databases

The speaker independent databases recorded for training and testing the system consist of sentences belonging to a voice operated E-mail appplication, plus a set of phonetically balanced, task independent sentences. Examples of the E-mail sentences are:

- *Leggimi l'argomento di questi messaggi* (Read the subject of these messages).
- *C'e' posta per me da Berlino il diciassette febbraio ?* (Is there mail for me from Berlin on February seventeenth?).

8800 sentences represent the training database, 4400 of which are E-mail sentences, while the others are phonetically balanced ones. 110 male and female speakers of age ranging from 13 to 73 contributed to the database, each speaker uttered 80 sentences with an average length of 7.22 words. The test database consists of 1200 E-mail sentences uttered by 20 speakers that did not contribute to the training set. The average length of test sentences is 5.18 words, speech rate varies from 1.3 to 3 words/sec, and the overall 10000 sentence database corresponds to about 9 hours of speech. The 4400 phonetically balanced utterances were obtained from a set of 128 sentences used for performing speaker dependent recognition experiments [5].

4 Recognition experiments

Since the test database (20 speakers * 60 sentences) was collected a few months later than the (110 speakers * 80 sentences) training database, the first experiments were carried out by using as a test set, a subset of (30 speakers * 10 sentences) from the training utterances. The models were trained with 3200 E-mail sentences from the remaining 80 speakers. The test vocabulary included 969 words. This experimental condition will be referred to, in the following, as *condition A*.

As soon as the test database became available, the complete set of 8800 sentences was used in the training procedure, and the new 1200 test E-mail utterances were used to test the recognizer; in this second phase (*condition B*) the lexicon included 786 words.

In each experiment the number of insertions i, deletions d, and substitutions s has been evaluated by comparing the recognized string of words with the correct one by means of a Dynamic Programming procedure. The Word Accuracy rate **WA** was then computed as

$$\mathbf{WA} = 100 * \left(1 - \frac{substitutions + insertions + deletions}{correct\ number\ of\ words}\right)$$

4.1 Recognition results - Discrete Density HMMs

4.1.1 *Condition A*

A baseline system with *condition A* was set up by generating the models of the basic phoneme set.

Several preliminary experiments were required for tuning some system parameters such as the Parzen smoothing standard deviations $\overline{\sigma} = [\sigma_C, \sigma_{delta-C}, \sigma_{E,delta-E}]$, the Beam Search threshold, the Decoding threshold **D**, and the word penalty **WP**. The latter parameter has influence over the number of deletion and insertion errors: by increasing its value the Forward decoding process produces more insertion than deletion errors, and viceversa.

A Word Accuracy of 42.3 was obtained with the following parameters:

$$\overline{\sigma} = (0.3, 0.4, 0.3) \qquad \mathbf{WP} = 10^{-9}$$

This result gives an improvement of about 10 percentual points in **WA** with respect to the single codebook vector quantization result (**WA** $= 32.66$, $\sigma_C = 1.0$, **WP** $= 0.1$).

To cope with the appearance of a high percentage of errors produced by short monosyllabic words, which are highly affected by between-word coarticulation phenomena, function-word dependent phones have been defined. Function-words occurring more than 50 times in the training set were represented by specialized phone units: word DA (*from*), for instance, was translated as d(DA) + a(DA), where d(DA) and a(DA) represents phones /d/ and /a/ in the context of word DA respectively.

The overall number of phonetic units becomes 84, that are trained bootstrapping their models through the parameters of the 27 phoneme models. Two Forward-Backward iterations were sufficient to reach model convergence. A performance of **WA** $= 51.9$ was measured by using the following set of parameters:

$$\overline{\sigma} = (0.8, 1.2, 0.5) \qquad \mathbf{WP} = 10^{-7}$$

The next step in the refinement of the models was the introduction of units depending on their left and right context. A triphone is represented by a unit of the kind abc that has to be

units	number	selected
triphones	1264	114
biphones	341	77
function word phones	57	57
phones	27	27
total	-	275

Table 1. Context dependent units (*condition A*).

interpreted as phone b appearing with phone *a* on its left and with phone *c* on its right. At the beginning and ending of each word the context is represented by the "-" symbol; therefore, the word POSTA (/posta/), for instance, is translated into triphones as:

$$- \textbf{po}\ pos\ ost\ sta\ \textbf{ta}- \quad (1)$$

Triphones were selected according to their frequency of occurrence in the training corpus.

The procedure for translating the sentences is as follows: first the sentences are represented as context independent and function-word phones, then all the triphonic sequences which occur more than 350 times are substituted by their corresponding triphone model. A set of 251 speech units was so obtained and a **WA** = 65.8 was observed when the following set of parameters was used:

$$\overline{\sigma} = (0.3, 0.5, 0.1) \qquad \mathbf{WP} = 10^{-12}$$

The last set of experiments in *condition A* was performed by modeling also right-context biphones. To this purpose, a new set of 275 units consisting of phonemes, function word phones, 114 triphones and 77 biphones was built. Tab. 1 shows the number of units appearing in the training set and the number of units selected for the different sets.

It must be noticed that, due to the incremental way in which the unit set was built, some biphones may have, as their right-context, a contextual unit: the unit o[*ost*], for instance, corresponds to phone /o/ having triphone *ost* as its right context.

The recognition result using these units was **WA** = 68.5, with

$$\overline{\sigma} = (0.2, 0.3, 0.2) \qquad \mathbf{WP} = 10^{-10}$$

Some considerations can be drawn analyzing these results: specialized units outperform more general units, provided that enough training material is available. Both function-word and context dependent phones yield significant improvement in word accuracy. A second consideration arises by observing the breakdown of the results obtained for each speaker by using specialized models: good improvements are obtained for "bad" speakers, while speakers getting good recognition results by using less specialized units, do not significantly increase their performance. The global effect is a balance of speaker recognition rates, with a noticeable improvement of the average rates.

4.1.2 *Condition B*

In the successive set of experiments (*condition B*), the overall training database was used to train the models, and 1200 test utterances were used for performance evaluation. Recognition tests with a set of 305 and 1157 context dependent speech units have been performed.

The set of speech units was incrementally built by means of the procedure previously described, where the minimum number of occurrences for both triphones and biphones was set to 500. The resulting inventory consists of 305 units, of which 107 are triphones and 113 are biphones. The

	condition A	*condition B*
27 ph, single cod.	32.7	-
+ multiple cod.	42.3	-
+ fw	51.9	-
+ triph.	65.8	-
+ biph.	68.5	67.1
(8, 8, 5) VQ	-	66.1

Table 2. Discrete Density HMMs: summary of the results

percentual coverage of the whole training database in terms of context dependent units is 82.6%, distributed as follows: 11.32% function word phones, 39.16% triphones, 32.12% biphones.

A **WA** = 67.1 was obtained with

$$\overline{\sigma} = (0.2, 0.2, 0.2) \qquad \mathbf{WP} = 10^{-10}$$

This decrease of more than one percentual point in terms of **WA** with respect to the 275 unit set of *condition A* is not surprising, even though these results cannot be strictly compared due to the different experimental conditions. In fact, in *condition A*, *application dependent* models were trained, while in *condition B* one half of the training samples includes phonetically balanced sentences, that are, therefore, *application independent.*

An enlarged set of 1157 units was finally obtained by imposing a minimal occurrence threshold of 50 to context-dependent units; the resulting number of triphones and biphones was 929 and 144 respectively. A Word Accuracy of 66.4 was obtained by using this enlarged set of units. This figure indicates that so large a set of speech units cannot be trained robustly due to the huge number of parameters to be estimated. The statistical weakness of these models, therefore, overwhelms their acoustic-phonetic specialization. This set, however, is useful for estimating reasonable tradeoff between specialization and statistical robustness of models, and for defining phonetic units with a different approach: clustering of specialized units on the basis of their entropy [9].

All experiments so far described are characterized by a (9, 9, 6) bit multiple codebook. Since the hardware architecture of the Front-End currently supports 8 bit vector quantization only for Cepstral and Delta-cepstral coefficients, due to memory and real time constraints, the 305 HMMs were trained using a (8, 8, 5) bit vector quantization of the features. By using 256 codewords and setting the same values for $\overline{\sigma}$ and **WP**, the result is **WA** = 66.1. This corresponds to a slight decrease of performance that can be explained by the higher statistical robustness of the trained models that trades off the larger quantization error. A summary of experimental results using Discrete Density HMMs is reported in Tab. 2.

4.2 Recognition results - Continuous Density HMMs

The experiments using Continuous Density HMMs were performed according to an incremental approach similar to that employed with Discrete Density HMMs. Again, preliminary investigations on the effects of different setting for the Beam Search and the word penalty thresholds were necessary for tuning the recognizer.

Other experiments performed by varying the number of Gaussians per mixture indicated that, even if a limited number of training samples is available for each unit, increasing the number of Gaussians in mixtures still has a beneficial effect at the **WA** level. All experiments were performed in *condition B* described in Section 4.

Mixtures of 15 Gaussians over the 24-dimensional space defined by 12 Cepstra and 12 Delta-cepstral coefficients have been used in these experiments. The 27 phoneme models, bootstrapped from segmentations obtained through previously trained Discrete Density HMMs, gave a baseline performance of 60 of **WA**.

Units	ceps. only	28	84	305	305
DDHMMs	32.66	42.3	51.9	67.1	67.1
CDHMMs	n. a.	60.0	67.8	73.8 (15 G)	75.3 (30 G)

Table 3. Comparison of Discrete and Continuous Density HMMs performance.

The segmentation of these models initialized the 84 function word unit models. Two iterations of the Segmental K-means were run to obtain the final models (**WA** = 67.8).

Then a context-dependent unit experiment was carried out by using the 305 unit set previously described. Again, the 305 unit models were bootstrapped from the segmentation obtained with trained context-independent models; after the first K-means iteration, a performance of **WA** = 73.8 was measured.

The number of mixtures was then increased to 30 and a noticeable improvement of performance - **WA** = 75.3 - has been obtained which indicates that CDHHMs have a greater robustness to statistical insufficiency than DDHMMs. Similar experimental evidence has been reported in [8] where up to 256 Gaussians mixtures per state have been adopted.

A final comparison of Discrete and Continuous HMMs performance is given in Tab. 3.

5 Conclusions

The performance improvement yielded by Continuous HMMs is relevant, although it becomes smaller for more specialized context dependent units. The complexity of the training and recognition algorithms for CDHMMs, however, is about one order of magnitude greater than the complexity of the corresponding algorithms for DDHMMs. A real time hardware implementation must take into account, therefore, computational constraints related to state of the art DSP technology. Furthermore, the Speech Understanding System performance is not related to the word accuracy measure in a straightforward way. The "quality" of word lattices is more important than the best scored sequence of word for understanding a sentence. Semantically irrelevant words can be badly scored in the lattice, because this affects only the search space of the linguistic parser for obtaining the meaning of the sentence. Further experimentation is planned on joined acoustic-linguistic decoding.

References

[1] L.R. Bahl, F. Jelinek, and R. Mercer. **A maximum likelihood approach to continuous speech recognition.** *IEEE Trans. on Pattern Analysis and Machine Intelligence*, PAMI-5(2):179–190, 1983.

[2] A. Ciaramella, D. Clementino, and R. Pacifici. **A Personal Computer Based Continuous, Speech Recognizer for Large Vocabulary Applications.** In *Proc. EUSIPCO-90*, pages 1503–1506, Barcelona, Spain, 1990.

[3] M. De Mattia and E. Giachin. **Experimental Results on Large Vocabularies Continuous Speech Understanding.** In *Proc. of the ICASSP*, pages 691–694, Glasgow, Scotland, 1989.

[4] L. Fissore, E. Giachin, P. Laface, G. Micca, R. Pieraccini, and C. Rullent. **Experimental results on large vocabulary continuous speech recognition and understanding.** In *Proc. of the ICASSP*, pages 414–417, 1988. sect. S10.2.

[5] L. Fissore, P. Laface, G. Micca, and R. Pieraccini. **Lexical Access to Very Large Vocabularies.** *IEEE Trans. Acoust., Speech and Signal Processing*, 37(8):1197–1213, 1989.

[6] L. Fissore, G. Micca, R. Pieraccini, and P. Laface. **ESPRIT PROJECT P26, Advanced Algorithms and Architectures for Signal Processing, Final Report**. Technical Report, Cselt Technical Report n. 88.284, 1988.

[7] S. Furui. **Speaker-Independent Isolated Word Recognition Using Dynamic Features of Speech Spectrum.** *IEEE Trans. Acoust., Speech and Signal Processing*, ASSP-34 (1):52–59, 1986.

[8] C.H. Lee, L.R. Rabiner, R. Pieraccini, and J.G. Wilpon. **Acoustic Modeling of Subword Units for Speech Recognition.** In *Proc. of the ICASSP*, pages 721–724, 1990.

[9] K.F. Lee and H.W. Hon. **Large-Vocabulary Speaker-Independent Continuous Speech Recognition Using HMM** . In *Proc. of the ICASSP*, pages 123–126, 1988. sect. S3.7.

[10] Y. Linde, A. Buzo, and R.M. Gray. **An Algorithm for Vector Quantizer Design.** *IEEE Trans. on Communications*, 28:88–95, 1980.

[11] L.R. Rabiner, J.G. Wilpon, and B.H. Juang. **A Segmental K-means Training Procedure for Connected Word Recognition Based on Whole Word Reference Patterns.** *AT&T Technical Journal*, 65(3):21–31, May/June 1986.

Automatic Transformation of Speech Databases for Continuous Speech Recognition

S. Rieck, E.G. Schukat-Talamazzini, T. Kuhn, S. Kunzmann, E. Nöth

Lehrstuhl für Informatik 5 (Mustererkennung),
Friedrich-Alexander-Universität Erlangen-Nürnberg,
Martensstraße 3,
8520 Erlangen, F.R. of Germany

Abstract: In this paper a method is described to generate automatically the labels for a new speech database from an existing manually labeled speech database. This becomes necessary when new standards are introduced and the speech signals have to be resampled. A dynamic time warping algorithm is used to match the original and the resampled speech signals. The comparison is carried out on mel based features. To improve computation time the search space for the DTW algorithm is restricted. Several experiments were carried out with a normal density Bayes classifier to check the quality of the new labelings. The results showed only a slight decrease in performance when using the new labelings.

1 Introduction

Speech recognition is often done with statistical methods. For this purpose an initial mapping between speech signal and symbolic description is necessary in order to get a maximum likelihood estimation of distributions or emission probabilities. If the symbolic description is on the phone level the speech sample has to be manually labeled by an expert, which is a time consuming process. Fortunately we have a hand labeled speech database at the University of Erlangen. The speech signals are bandpass filtered with a band-width from 0.1 to 3.4 kHz and sampled at a rate of 10 kHz. The sentences are also available as tape recordings. The speech data with this cutoff frequency was used in the EVAR system [6]. For our work in the ESPRIT SUNDIAL project we want to use speech data with different standards: The band-width should be increased from 3.4 to 6.4 kHz and the frame rate is to be changed from 12.8 msec to 10.0 msec. The goal is to transform the labels of the speech database automatically so that they can be used for the training of a Bayes classifier or for the initialization of a bootstrapping procedure for new unlabeled data [3].

2 Automatic Transformation of Labeled Speech

2.1 Overview

The generation of the new labelings is divided into three steps: First the recorded utterances have to be resampled, then two utterances reflecting the same analog recording have to be compared and after that new transcriptions are generated.

NATO ASI Series, Vol. F 75
Speech Recognition and Understanding. Recent Advances
Edited by P. Laface and R. De Mori

The resampling of the whole speech data is necessary, as an increased band-width of the speech signal is not available from mere interpolation. The resampling can be facilitated by doing it in blocks of several minutes and automatically cutting them into single utterances.

In order to compare two utterances the speech signal has to be transformed into feature space. The used features have to be chosen carefully . They must be suitable to ignore irrelevant properties of the speech signals like background noise or different recording levels. Features in the time domain, like zero crossings or peak to peak values, are not suited for comparison, as they depend on the frequency resolution and on the loudness of the speech signal. Therefore short time spectral analysis is applied.

By the use of a dynamic time alignment procedure the positioning and the alignment is carried out simultaneously. The result of the comparison is a mapping between similar time frames represented by feature vectors.

In the last step the precomputed mapping is used in order to generate new transcriptions. Every frame based transcription can be transformed this way.

2.2 Used Features

The used features must be suitable for comparing two speech signals with different bandwidth and loudness. A possible way is to use the Mel Cepstrum Coefficients (MCC). These coefficients are derived from the Mel Frequency Coefficients (MFC) as follows [2, 7]:

$$MFC_n(l) = \log\left(\sum_{k=0}^{255} D(l,k) \cdot \mid FC_n(k) \mid^2\right) \quad l = 0,\cdots,31 \;\; n = 0,\cdots,N-1 \tag{1}$$

$$MCC_n(k) = \sum_{l=0}^{31} MFC_n(l) \cdot \cos\left(\frac{k \cdot (2l+1)\pi}{64}\right) \quad k = 0,\cdots,31 \;\; n = 0,\cdots,N-1 \tag{2}$$

$FC(k)$ denotes a spectral coefficient of the fourier power spectrum, $D(l,k)$ is a coefficient of the lth triangular window.

A pair of utterances has to be compared under the same conditions. Consequently, the analysis window size and the used frequency ranges have to be the same. An equal analysis window size is achieved by taking a corresponding number of sampling values. The frequency range is adjusted by taking an appropriate number of fourier coefficients FC. Table 1 shows an example for 10 kHz and 16 kHz speech samples. In both cases a 256 point DFT is applied.

In order to ignore different loudness $MCC_n(0)$ is not used. $MCC_n(0)$ is a rough measure for signal energy, since it is the sum over all Mel Frequency Coefficients (Eq. 2).

sampling frequency	10 kHz	16 kHz	
analysis window size (12.8 msec)	128	205	samples
used frequency range (5.0 kHz)	128	80	Fourier coefficients

Table 1. Conditions for comparisons

2.3 Mapping

The starting and ending frame of the utterance in the new speech signal are unknown, since the signals are resampled. To combine comparison and detection of starting and end points in the new signal dynamic time warping is used. The mapping is computed using dynamic programming with the following recursion [5]:

$$D^*(j,k) = d(\underline{c}_j, \underline{d}_k) + \min\{\ D^*(j,k-1),\ D^*(j-1,k-1),\ D^*(j-1,k)\} \tag{3}$$

In this equation $\underline{c}_j, \underline{d}_k$ denote feature vectors, $d(\underline{c}_j, \underline{d}_k)$ is the Euclidean distance measure. The optimal path gives the mapping with minimum distance of the two signals. For retrieving the optimal path, only the backpointers have to be stored.

2.4 Restriction of Search Space

The search space can be restricted to an area below and above a line, because the optimal path for the mapping should be a straight line between the starting and ending points of the speech. To do this, the position of the line has to be estimated. This can be done by specifying the inclination and one point. The inclination is approximately the ratio of the frame rates. An obvious point is the starting point, which is the transition from silence to speech in both signals. The transition in the original signal is easy to find as it is marked by the hand labeling. The transition in the new signal is found by doing a threshold operation. For this purpose $MCC_n(0)$ is used now. This coefficient is a measure for signal energy as already mentioned. Before doing this the $MCC_n(0)$ has to be smoothed to remove artefacts from the signal e.g. noises caused by switching on the microphone or the tape recorder. This is done by median filtering.

2.5 Determination of Threshold

Speech and silence values described by $MCC_n(0)$ can be modeled by two one dimensional Gaussian densities $N(\mu_0, \sigma_0)$ and $N(\mu_1, \sigma_1)$ respectively [1]. Fig. 1 shows an example. The intersection of both densities determines the threshold. The two densities and the threshold are simultaneously estimated by the following algorithm, which is a modification of the standard LBG-Algorithm [4]:

```
Initialization: given
                N training values MCC_n(0)
                (N = number of frames in the utterance);
                ε > 0 stop criterium
                θ^(0) initial threshold (mean value of all training values)
                i = 0
Step 1:         Calculate μ_0^(i), μ_1^(i), σ_0^(i), σ_1^(i) using the partition defined by θ^(i).
                Calculate new threshold θ^(i+1) by identifying both densities.
                i = i + 1
Step 2:         If θ^(i) − θ^(i−1) < ε,
                then STOP,
                else repeat step 1.
```

For the determination of the threshold only a few iterations are needed (usually less than 10).

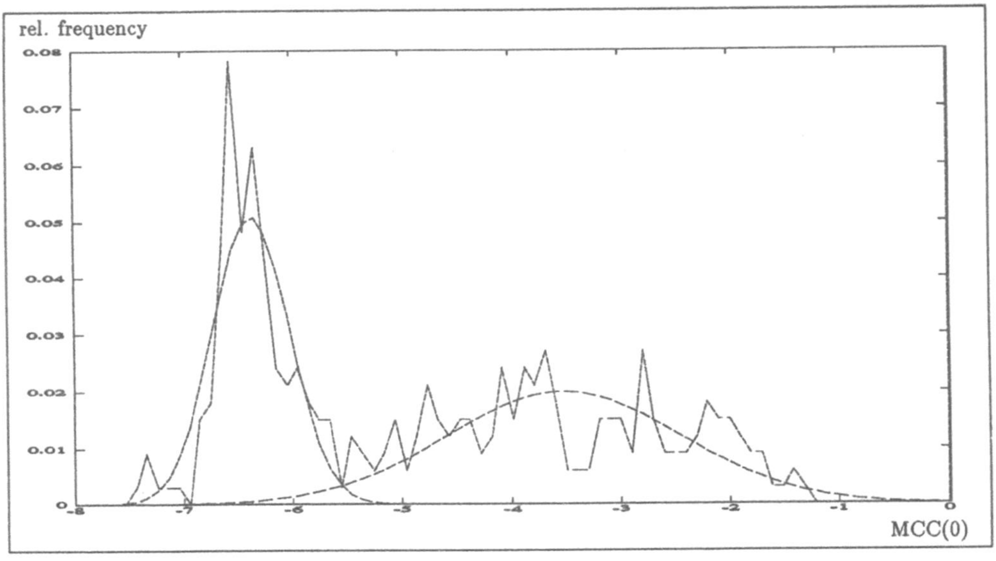

Figure 1. Original distribution and estimated distributions for silence and speech values

3 Experimental Results

The speech database which had to be transformed consisted of 250 utterances from 6 male and 6 female speakers. The speech data are available as a tape recording of approx. 13 minutes. Table 2 shows the parameter values which were changed.

parameter		old	new
sampling rate		10 kHz	16 kHz
band-width	f_{min}	0.1 kHz	0.1 kHz
	f_{max}	3.4 kHz	6.4 kHz
frame rate		12.8 ms	10.0 ms

Table 2. Parameter values to be changed

The following transcriptions for 10 kHz speech data are available: a phone component labeling, a phone labeling and a orthographic labeling. These labelings were based on 12.8 msec time frames. The phone component labeling showed one phone component per frame. The phone and orthographic labeling contained starting and ending frame of each phone or word respectively.

The mapping of two utterances can be seen in Figure 2. As expected the optimal path resulted in an almost straight line. The small deviations originated mostly from different mappings on the same label. The restriction of the search space resulted in a saving of up to 75 % of computation time, depending on the length of the original utterance.

A total amount of 52 MB was processed (20 MB / 10kHz and 32 MB / 16 kHz). The computation was carried out on a DECstation 3100. The execution time was about 3 times real time for the computation of the mapping and the generation of the transcriptions.

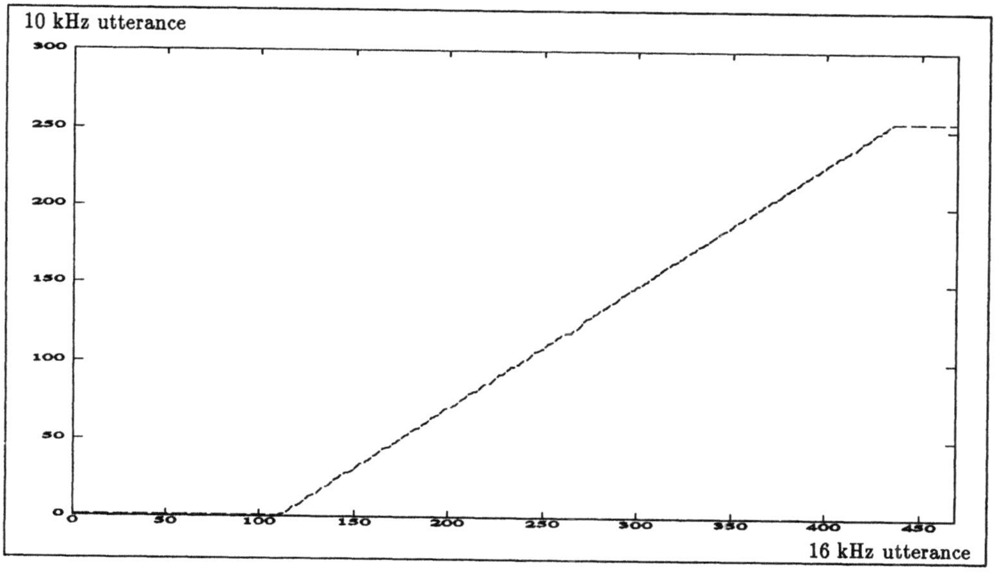

Figure 2. Optimal path

	recognition rate									
# features	15	16	17	18	19	20	21	22	23	24
10 kHz	44.96	45.79	46.13	45.96	45.88	45.89	45.93	45.89	45.87	45.83
16 kHz	43.36	43.25	43.30	43.67	43.82	43.75	44.10	44.25	44.22	44.26

Table 3. Evaluation results of a normal density Bayes classifier trained with 11 speakers and tested with the remaining speaker.

The evaluation was carried out with a normal density Bayes classifier to check the quality of the new labels. As a performance criterion the recognition rate for phone components was chosen. In a first step mel-based delta cepstral features were extracted from the speech signal. The number of components in the feature vector was varied between 15 and 24. As can be seen in Table 3, the recognition rate of the 16 kHz sample is slightly worse. This is probably due to an incorrect mapping in small local areas e.g. of transient phonemes. However, we expect that a more global evaluation criterion like the word recognition rate leads to better results. The corresponding experiments are still to be carried out.

4 Conclusions

The method presented in this paper shows a way how to generate automatically new labeled speech data out of an existing hand labeled speech database. A prerequisite for this is to resample the original speech data with the desired environment parameters. Using a dynamic time warping algorithm it is possible to compute a mapping between old and new speech signals without explicit calculation of starting and ending points of the new speech signal. By restricting the search space the computation time can be drastically reduced. The results of classification experiments show that the new automatic labeling is almost as good as the original one.

Acknowledgement: The work reported in this paper was supported in part by the ESPRIT project P2218 SUNDIAL and by a joint research project of the German " Bundesministerium für Forschung und Technologie (BMFT)". Only the authors are responsible for the contents of this paper.

References

[1] D. Van Compernolle. Noise adaptation in a hidden Markov model speech recognition system. *Computer Speech and Language*, 3(2):151 - 167, April 1989.

[2] S.D. Davis and P. Mermelstein. Comparison of parametric representations for monosyllabic word recognition in continuously spoken sentences. *IEEE Trans. on Acoustics, Speech, and Signal Processing*, 28(4):357–366, August 1980.

[3] T. Kuhn, S. Kunzmann, E. Nöth, S. Rieck, and E.G. Schukat-Talamazzini. Iterative Optimization of the Data Driven Analysis in Continuous Speech. In *this volume*.

[4] Y. Linde, A. Buzo, and R.M. Gray. An algorithm for vector quantizer design. *IEEE Trans. on Communications*, 28(1):84–95, January 1980.

[5] H. Niemann. *Klassifikation von Mustern*. Springer-Verlag, Berlin, Heidelberg, New York, Tokyo, 1983.

[6] H. Niemann, A. Brietzmann, R. Mühlfeld, P. Regel, and E.G. Schukat. The Speech Understanding and Dialog System EVAR. In R. De Mori and S.Y. Sun, editors, *New Systems and Architectures for Automatic Speech Recognition and Synthesis. NATO Series F 16*, pages 271 - 302, Springer-Verlag, Berlin, Heidelberg, New York, Tokyo, 1985.

[7] P. Regel. *Akustisch-phonetische Transkription für die automatische Spracherkennung*. Volume 83 of *Fortschritt-Berichte*, VDI Verlag, Düsseldorf, 1988.

Iterative Optimization of the Data Driven Analysis in Continuous Speech

T. Kuhn, S. Kunzmann, E. Nöth, S.Rieck, E. Schukat-Talamazzini

Lehrstuhl für Informatik 5 (Mustererkennung),
Friedrich-Alexander-Universität Erlangen-Nürnberg,
Martensstraße 3,
8520 Erlangen, F.R. of Germany

Abstract: We present an iterative method to optimize the word recognition rate for a data driven analysis in continuous speech by using a large set of speech samples. After a short description of our system environment a bootstrapping method for an iterative parameter estimation will be discussed. The initialization of the bootstrapping procedure is done by using a limited amount of hand labeled training data to estimate the statistical parameters roughly. In the second step the statistical parameters are estimated more exactly on the basis of unlabeled training data. Some experimental results for the bootstrapping method performed on unlabeled training data in comparison with results achieved by parameter estimation on labeled training data will be given.

1 Introduction

The acoustic phonetic decoding in speech recognition systems is often done by statistical methods like a *Bayes Classifier* or *Hidden Markov Models* (HMM) [6] which requires a large amount of labeled training data at the various linguistic levels. This means that the training data have to be hand labeled by an expert. This is a very time consuming process. Therefore the enlargement of a speech database is expensive. On the other side a large amount of training data is necessary for a robust parameter estimation. But there are further reasons why hand labeled speech samples might be necessary:

- A hand labeling at the various linguistic levels is needed for the evaluation of the recognition algorithms, e.g in our system environment the word recognition rate is computed on the basis of a word hand labeling. Such an evaluation criterion is useful if one wants to locally optimize the word recognition module and in the case of spontaneous speech with hesitations and interruptions. In the latter case the mostly used criterion 'word accuracy' is not appropriate.

- In an investigation that is currently being done at our institute we want to find out how well the position of the focal accent can be predicted with intensity. However it is not clear over which part of the speech signal (syllable nucleus, syllable, word, or phrase in focus) the intensity should be computed. Thus, a hand labeling of the utterances is necessary for this kind of basic research [5].

NATO ASI Series, Vol. F 75
Speech Recognition and Understanding. Recent Advances
Edited by P. Laface and R. De Mori

2 System Environment

Our data driven recognition of a continuously spoken utterance is part of the speech understanding and dialogue system EVAR [4] and can be divided into the following steps:

- Feature extraction and phone component labeling.
- Phone segmentation and classification.
- Generation of word hypotheses.
- Search for constituents with a context free grammar.

In the feature extraction module the speech signal is partitioned into consecutive frames of 12.8/10 msec., sampled with 12/16 Khz, and quantized with 12/16 bit. For each frame a feature vector with 9 cepstral components and 2 temporal derivatives is computed. Each frame is classified into phone components with a *Bayes Classifier* and labeled with the best n choices ($n \leq 5$).

The phone segmentation and classification is divided into four steps which are described in the following [1, 7]:

1. An initial partition is generated by searching for homogeneous ranges in the speech signal using the first alternative of the phone component stream. All frames labeled with the same phone component are grouped together to segments which are normally shorter than a phone because of the coarticulatory phenomena. The segment stream is carefully smoothed in order to reduce the number of potential segments. For instance, frames whose left and right neighbors have the same first alternative are eliminated.

2. In order to recognize plosives and diphthongs which consist of different phone components and in order to take into account classification errors a segment graph is generated by considering alternative endpoints for each starting point. The set of alternative endpoints is given by the initial partition. The size of the segment graph is restricted by limiting the maximum length of a phone. Initial segments with a length greater than the maximum phone length are not partitioned (e.g. periods of silence).

3. Each edge of the segment graph is compared with all phones of an inventory using discrete density HMM's. The result is a scored segment graph. Currently we distinguish between 44 phones which are modeled by 59 phone component HMM's.

4. Using a heuristic search procedure (A^*-Algorithm) the global optimal path is found in the scored segment graph. The output is a linear sequence of segments where each segment is labeled with the five best scoring phone classes.

The stream of phone hypotheses is taken as input data for the generation of word hypotheses with discrete density HMM's [1, 2]. A subword unit approach is used to reduce the amount of training data. All phones are modeled by the same edge oriented elementary HMM. In figure 1 alternative elementary HMM's are shown. The model *SID* was chosen for the experiments. The word model is generated automatically by replacing each phone in the word by the corresponding HMM. In the recognition process a set of word hypotheses is generated via a word spotting technique (*vertical summation* [8]). The prefix equivalence of two consecutive word models is used to reduce the computation time.

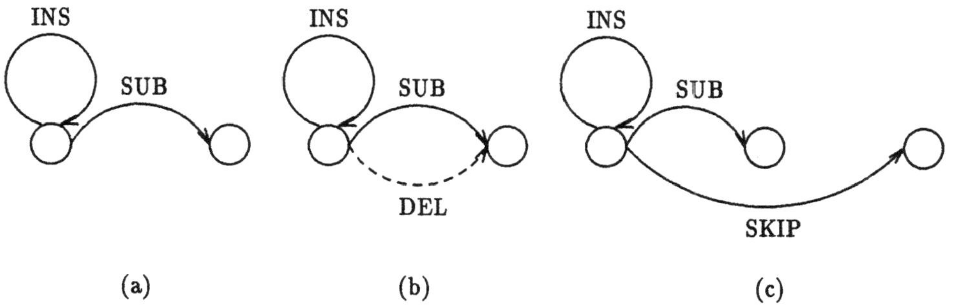

Figure 1. alternative HMM's for subword modeling: *SI* (a), *SID* (b), *SIS* (c)

Finally, a heuristic search procedure (A^*-Algorithm) is used to find word chains (constituents or sentences) in the lattice of word hypotheses. The search space is restricted by a context free grammar. Two word hypotheses are chained if they are adjacent on the time axis and syntactically correct. For the computation of a new score the word chains are modeled by HMM's on the phone level and compared with the overlapped part of the speech signal.

3 Parameter Estimation without Handlabeling

One of the main efforts in speech recognition systems is the collection of large speech databases for statistical parameter estimation because the system performance increases with more available training data [9]. Even more training data is needed for context-dependent phone models (e.g diphones, triphones) or syllable models in the recognition process [3]. Therefore we have made a large effort in expanding our speech database. For each utterance we need a hand labeling

- into phone components for the training of the *Bayes Classifier*,
- into phones for the training of the phone component HMM's and
- into words which we use for the training of the phone HMM's as well as for the computation of the word recognition rate.

Because the process of hand labeling speech is a very time consuming process a procedure for deriving a automatically a symbolic description (hand labelling) between the speech samples and the recognition units as well as a parameter training without any kind of hand labeling was developed.

3.1 Step by Step Coarsening of the Handlabeling

First we investigated the influence of a hand labeling into phones, words, or utterances on our HMM parameter training. In all of the following experiments the frame classification was done with the same *Bayes Classifier* trained with a hand labeling into phone components. The experiments were carried out with the EVAR speech sample for which a hand labeling with respect to phone components, phones and words is available. The EVAR speech sample consists of 243 utterances (about 11 min) of 6 speakers (3 female and 3 male) which was divided into a training set of 128 utterances (about 5.5 min.) and a test set of 115 utterances (about 5.5 min.). We have carried out experiments of parameter training using

- the phone labeled training data (*phone*),
- the word labeled training data (*word*) and
- the word transcription without reference to the speech signal (*sentence*).

The evaluation of the experiments was made on the word level with a lexicon of 549 words. As evaluation criterion we used the rank of the correct word hypotheses normalized by the number of segments. The results show that for the HMM parameter estimation the transcription *sentence* is sufficient because the decrease of the recognition rate can be neglected (see table 1).

	word recogniton rate in percent generating								
	1.0	*2.0*	*3.0*	*4.0*	*5.0*	*7.0*	*10.0*	*20.0*	*50.0*
	hypotheses per segment								
phone	40.7	54.4	62.5	67.3	71.6	76.9	82.3	89.5	92.7
word	41.3	52.3	59.8	65.7	70.3	74.6	81.5	89.5	92.7
sentence	42.0	52.2	58.1	63.2	68.4	74.5	79.3	88.7	92.5

Table 1. word recognition rate for step by step coarsening of the handlabeling

3.2 The Bootstrapping Procedure

Our further work was based on the training of the *Bayes Classifier* without hand labeling into phone components. Therefore we have developed a bootstrapping method which can be described as follows (see figure 2): In the initialization stage, the *Bayes Classifier* is trained with a limited amount of hand labeled speech data. In the second stage, a hand labeling into phone components is generated automatically using a HMM-based alignment procedure. This hand labeling can be used to reestimate the statistical parameters. If there is an improvement of the word recognition rate, the bootstrapping procedure stops, otherwise a new iteration begins.

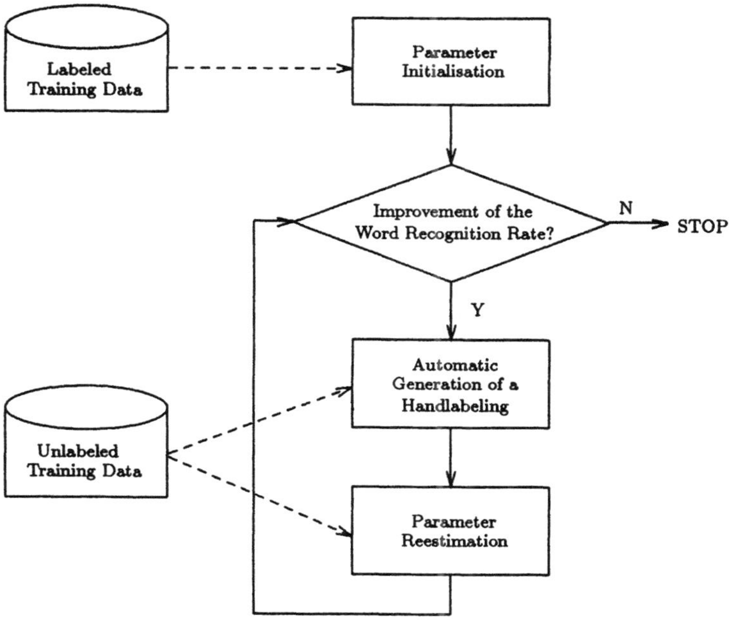

Figure 2. bootstrapping procedure

The bootstrapping procedure was tested with the SIEMENS speech sample for which only a hand labeling into phone components is available. The SIEMENS sample consists of 700 utterances spoken by 4 speakers. Each speaker spoke the SOTSCHEK corpus (100 phonetically balanced sentences) twice. The first recording session was used as training set (400 sentences, about 22 min.) The test set consisted of the remaining 300 sentences (about 16 min.) of the second recording session. For the evaluation a word hand labeling was derived automatically using the same HMM based alignment procedure as in the bootstrapping procedure.

The experiments based on a lexicon with 1300 words were carried out as follows: The parameters of the *Bayes Classifier* were initialized with 100 utterances of the EVAR speech sample (*boot_0*). The bootstrapping procedure was finished after two iterations (*boot_1*, *boot_2*), because no further improvement of the word recognition rate was observed. For comparison, we have computed the word recognition rate that can be achieved if the *Bayes Classifier* is trained with a phone component hand labeling (*hand labeled*). The results of the bootstrapping experiment (see table 2) show that the decrease of the word recognition rate is negligible especially if the high costs for hand labeling speech are taken into account. Here the same evaluation criterion was used as in section 3.1.

4 Conclusions

We have presented a method of estimating the statistical parameters without a hand labeling. At first it was shown that for the HMM parameter training the word transcription

	word recogniton rate in percent generating								
	1.0	*2.0*	*3.0*	*4.0*	*5.0*	*7.0*	*10.0*	*20.0*	*50.0*
	hypotheses per segment								
boot_0	32.7	43.6	49.9	55.1	59.7	64.8	70.0	79.1	85.8
boot_1	46.1	58.5	66.7	70.8	73.7	77.4	81.1	85.4	89.2
boot_2	48.6	60.9	67.6	72.1	75.3	78.2	80.9	86.0	89.4
handlabeled	52.3	65.0	70.2	73.0	75.8	79.1	81.3	84.6	87.5

Table 2. word recognition rate for the bootstrapping procedure

without reference to the speech signal is sufficient. Secondly a bootstrapping procedure for an automatic adaptation of the statistical parameters was presented using unlabeled training data. Only a small amount of a phone component labeling is required for the training of the Bayes Classifier in the initialization stage. The increase of the word recognition that can be achieved by using hand labeled training data is negligible.

References

[1] S. Kunzmann. *Die Worterkennung in einem Dialogsystem für kontinuierlich gesprochene Sprache.* PhD thesis, Technische Fakultät der Universität Erlangen-Nürnberg, 1990.

[2] S. Kunzmann, T. Kuhn, and H. Niemann. An Experimental Environment for Generating Word Hypotheses in Continuous Speech. In H. Niemann, M. Lang, and G. Sagerer, editors, *Recent Advances in Speech Understanding and Dialog Systems*, pages 311–316, Springer Verlag, Berlin, Heidelberg, New York, 1988.

[3] K. Lee, H. Hon, M. Hwang, and S. Majahan. Recent Progress and Future Outlook of the SPHINX Speech Recognition System. *Computer Speech & Language*, 4(1):57–69, 1990.

[4] H. Niemann, A. Brietzmann, U. Ehrlich, S. Posch, P. Regel, G. Sagerer, R. Salzbrunn, and E.G. Schukat-Talamazzini. A Knowledge Based Speech Understanding System. *Int. J. Pattern Recognition and Artificial Intelligence*, 2(2):321–350, 1988.

[5] E. Nöth, A. Batliner, and T. Kuhn. Intensity as a Predictor of Focal Accent. In *XIIème Congrès International des Science Phonétiques*, will be published in 1991.

[6] L. R. Rabiner. Mathematical Foundations of Hidden Markov Models. In H. Niemann, M. Lang, and G. Sagerer, editors, *Recent Advances in Speech Understanding and Dialog Systems*, pages 183–205, Springer Verlag, Berlin, Heidelberg, New York, 1988.

[7] A. Reißer. *Ein zeitsynchrones Segmentierungsverfahren für die Lautklassifikation mit Markov Modellen.* Technical Report, IMMD5 (Mustererkennung), Universität Erlangen-Nürnberg, 1990.

[8] E. G. Schukat-Talamazzini. *Generierung von Worthypothesen in kontinuierlicher Sprache.* Volume 141 of *Informatik Fachberichte*, Springer Verlag, Berlin, Heidelberg, New York, Tokyo, 1987.

[9] R. Schwartz and F. Kubala. Hidden Markov Models and Speaker Adaptation. In this volume.

Syllable-based stochastic models for continuous speech recognition

G. Ruske and W. Weigel

Lehrstuhl für Datenverarbeitung, Technische Universität München, Franz-Joseph-Str. 38, 8000 München 40, Federal Republic of Germany

Abstract: The paper describes an automatic speech recognition system which is based on syllabic segmentation of the speech signal. Stochastic models (HMMs) are used for representing demisyllable segments. The advantages of syllabic processing within the different stages of the system (i.e. segmentation, phonetic classification, word and sentence recognition) are demonstrated and discussed on the basis of experimental results. Word and sentence recognition with a perplexity of 27 reached 74% and 96%, respectively.

Keywords: loudness, syllabic processing, demisyllable HMMs, A*-search.

1. Introduction

One of the main problems in speech processing arises from coarticulation, meaning the overlap of gestures for succesive sounds and their alteration during speaking. As a consequence, speech sounds can not be evaluated independently on the acoustic-phonetic level and phoneme models therefore have to be treated in a context-dependent way. Our solution to this problem is to use processing segments which contain more than one phoneme. In general, the segments should contain basic phonological structures or articulatory gestures. For this reason we apply *demisyllables* as smallest processing units.

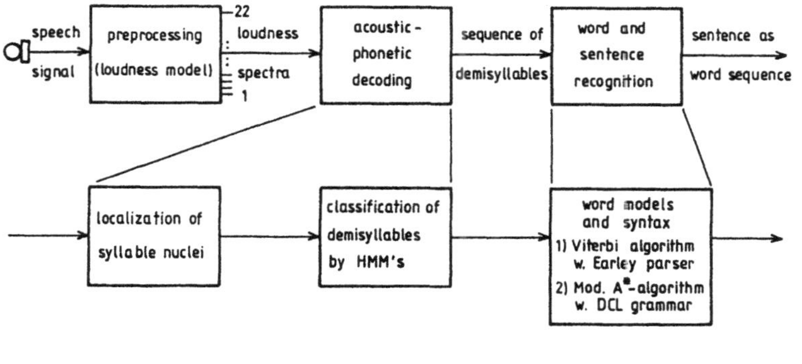

Fig. 1. System overview

NATO ASI Series, Vol. F 75
Speech Recognition and Understanding. Recent Advances
Edited by P. Laface and R. De Mori

An overview of the complete system is given in Fig. 1. The continuous speech recognition system consists of modules which perform the tasks of segmentation, the classification of demisyllable segments, the lexical search, and the recognition of the sentence. After localization of the syllable nuclei the region between two consecutive syllable nuclei contains two *demisyllables*. Classification of demisyllables is based on *Hidden Markov Models* (HMM's). Thereby also the optimal position for the syllable boundary is determined. After that the word and sentence recognition follows.

The words are described by syllable-based *pronunciation models* operating on the level of phonetic symbols. The syntax-controlled sentence recognition is carried out either by means of 1) a Viterbi search (one-stage DP-algorithm) or 2) a modified A*-algorithm.

2. Preprocessing and segmentation

Preprocessing of the incoming speech signal is carried out by a loudness model simulating human loudness perception. This model yields 22 loudness components $N_\nu(t)$ giving a so-called loudness spectrum with loudness components N_ν for ν=1...22 Bark.

In a first step the *syllable nuclei* are localized. For this purpose the modified loudness function $N_m(t)$ is calculated as a weighted sum from the loudness components:

$$N_m(t) = \sum_{\nu=3}^{15} N_\nu(t) - \sum_{\nu=20}^{22} N_\nu(t)$$

This contour is smoothed by a low pass filter and results in a curve as shown by a dashed line in Fig. 2. The local maxima of this curve as well as additional spectral evaluation enable the localization of syllabic nuclei quite well [1,2].

In principal, a crude *vowel classifier* is applied by matching spectral vowel templates at each time instant of the sentence and calculating the minimum distance (from 6 spectra). The local minima of the resulting distance curve are regarded as a second hypothesis for a syllable nucleus.

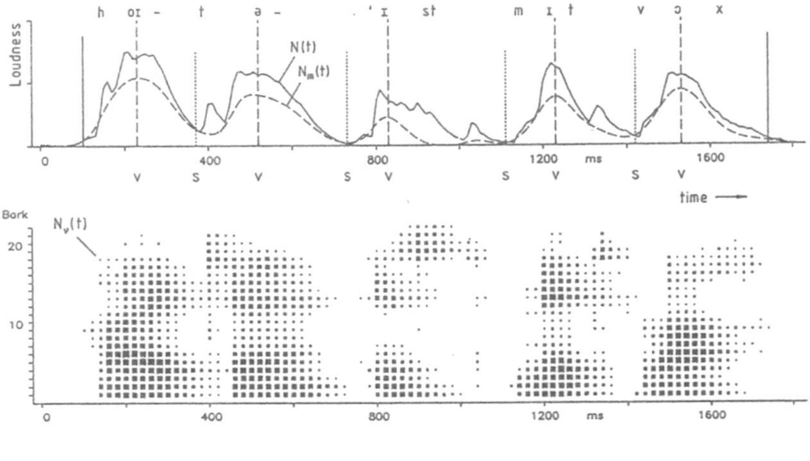

Fig. 2. Loudness spectra of the sentence 'Heute ist Mittwoch'

As an alternative, a neural net (TRACE model) has been used for the localization of syllable nuclei. In order to reduce the number of input units, so-called loudness cepstra have been calculated from the loudness spectra which are very similar to the well-known mel-frequency cepstral coefficients. A gliding window of 5 cepstra with 13 coefficients was fed to the net. The experimental results are displayed in Tab. 1. The neural net performed the localization task (detection rate) with an accuracy of 97.8 % and the classification of vowels and diphthongs with 84.6 %. This was in both cases more than 1 % better as compared to our conventional method using the loudness contour and spectral template matching.

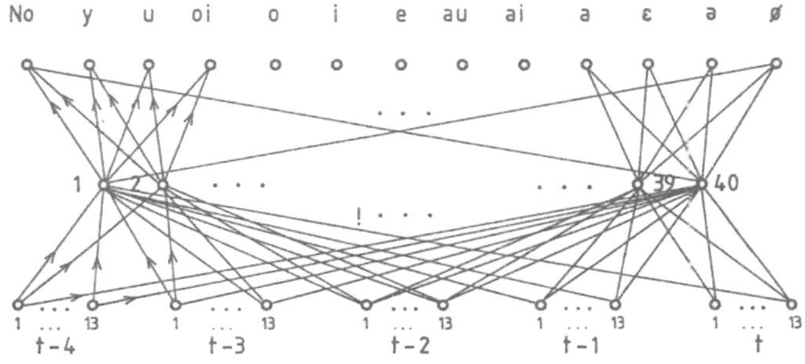

Fig. 3. Localization of syllable nuclei by application of a Neural Net (5 loudness cepstra)

Method:	Neural Net	Segmentation+Recognition
Recognition rate:	84.6 %	83.2 %
Detection rate:	97.8 %	96.2 %

Tab. 1. Comparison of different segmentation methods (% correct)

3. Modelling of demisyllable segments

The segments between two consecutive syllable nuclei are treated as two demisyllables; the first demisyllable contains a part of the vowel and the final consonant cluster, the second demisyllable contains the initial consonant cluster as well as a part of the following vowel.

3.1 Demisyllables as decision units

In a first series of experiments complete demisyllable segments have been modelled by discrete Hidden Markov Models (HMMs) using vector quantization of the loudness spectra and discrete emission probabilities (left-to-right models with 4-6 states). The reasons for this choice were that no assumptions about the distributions are made, that the vector quantization can be easily smoothed when the amount of training material is small, and that the codebook can be adapted to the speaker.

In order to avoid very sparse density functions, the discrete pdf's have been smoothed during the training procedure, similar to a Parzen estimator; in this equation, the quantity f denotes a factor for the 'smoothing' where the 10 nearest neighbours from the code book were trained according to their weight G:

$$G = \exp(\frac{-f \cdot ln2 \cdot (distance)^2}{4 \cdot (intraset\ distance)^2}) \qquad \text{for} \quad f > 0$$

$$G = 0 \qquad \text{for} \quad f = \infty$$

Thus, the factor f can enhance or reduce the contribution calculated for this neighbour.

The syllable boundary can be determined implicitly by calculation of that time instant which gives the best product between the forward probabilities α^* of final demisyllables and the backward probabilities β^* of initial demisyllables; in this way the maximum joint probability denotes the syllable boundary. Additionally a *time restriction* was applied using a Gaussian model which depends on the actual time length t of the demisyllable, the mean value T and the variance v of their durations in the training set. A major problem, however, is to get enough training material for the large number of demisyllable classes.

We selected speech test material which covers all vowels and consonant clusters appearing in the most frequent 1000 words of German. It was possible to find 23 sentences containing all these units. Recognition was carried out using a subset of these 23 sentences which consists of 8 sentences including 143 different demisyllables. The training was done by 4 versions of these sentences resulting in an average of 6 to 7 training patterns per HMM. The test corpus consisted of 3 other versions of the sentences, see Tab. 2. The first column in Tab. 2 gives the size of the vector codebook, i.e. for spectral shape and for energy. The quantity f denotes the factor for the 'smoothing' during training according

Vector-quantization	Factor f	Model structure	Confusions (%) severe	slight	total
Speech material:	8 sentences				
64 / 8	∞	n6j1	11.6	11.7	23.3
128 / 8	∞	n6j1	10.2	10.7	20.9
128 / 8	10	n6j1	11.6	13.4	25.0
128 / 8	1	n6j1	15.8	16.4	32.2
128 / 8	∞	n5j1	11.1	10.4	21.5
128 / 8	∞	n5j0	9.3	11.6	20.9
128 / 8	∞	n4j0	10.8	10.1	20.9
256 / 8	**∞**	**n6j1**	**9.2**	**10.1**	**19.3**
256 / 8	10	n6j1	9.6	11.4	21.0
256 / 8	1	n6j1	9.6	15.6	25.2
256 / 8	∞	n4j0	9.2	11.7	20.9
Speech material:	23 sentences				
256 / 8	∞	n6j1	8.6	12.3	20.9

Tab. 2. Demisyllable recognition errors (in %)

to the weight G. If the factor f equals infinity this means 'no smoothing'. The model structure is abbreviated by 'n6j1', for example, which describes a model with 6 states where always 1 state can be skipped. The best parameters are printed in bold face and resulted in 19.3% confusions for 8 sentences. The last line gives the results for all 23 sentences (consisting of 524 demisyllables where 269 different demisyllables occur). Here, a confusion rate of 20.9 % was reached, which is only a slight increase. It has to be taken into account that the correct classification of a demisyllable means on the average 2-3 recognized phoneme symbols. Thus, the rate of recognized 'phonemes' was even higher than 80 %. These recognition results proved the applicability of the methods used.

3.2 Consonant clusters as decision units

A more economic solution is to introduce HMMs for consonant clusters which describe a demisyllable segment, but yield as classification result only the consonant cluster; each vowel or diphthong in the syllable nucleus is classified separately. By training each consonant cluster HMM in the context with all vowels it can be expected that the model will become independent from the vowel identity. Inspection of the trained model shows that in fact the consonantal part is used for the consonant information whereas the states of the model located at the vowel side mix all vowels together.

A further problem consists in the coarticulation effects at the syllable boundary. For this purpose the models have to be trained in context, too, or have to be provided with context-dependent input and output states which are linked together accordingly. This approach has the advantage that the number of units is much smaller than in the case of using complete demisyllables; there are about 50 initial consonant cluster models and about 70 final consonant cluster models necessary to represent a German vocabulary consisting of more than 1000 words. Experiments with this approach are reported in [4].

3.3 Syllabic word models

At the word level pronunciation models of words are used which are able to describe the pronunciation observed during training as well as possible segmentation errors [2]. The word models can be regarded as special HMMs operating with phonetic symbols in a syllable-based structure. During training the confusions of symbols are stored in common confusion matrices which are utilized in recognition to calculate the similarity between the classified phonetic symbols and the symbols of the word model. In this way a Bayes' classifier can be realized on the word level. Processing of the word models is carried out economically by use of the Viterbi algorithm which operates within the syllable grid and thus is very fast.

4. Sentence recognition

For sentence recognition two methods have been implemented. Firstly, a Viterbi search (one-stage DP-algorithm) is used which operates on the basis of an Earley algorithm and a context-free grammar. Here the problem arises that a lot of copies of lexicon words have to be considered in order to discriminate words appearing in different grammatical contexts during the search.

Secondly, a very efficient sentence recognition procedure was implemented on the basis of an A*-algorithm. This algorithm has been modified (modified A*-algorithm, MA*) so that the algorithm in a first step performs a heuristic search using a cost function which is calculated as the mean of the syllable cost values. This results in a kind of fast 'depth-first search'. In the next step a special version of the A*-search is performed [3]. Here, the necessary optimistic estimation h(n) of the cost values can be carried out exactly by use of the a-posteriori probabilities from the confusion matrices mentioned in Sect. 3.3. The search operates under syntax control exploiting a Definite Clause Grammar (DCL). The sentence recognition utilizes the syllable structure, too, since the hypothesized words can only start at each syllable position; thus the sentence recognition operates also on the syllable grid which enables very economic and fast processing.

Method	Perplexity	Recognition Sentences (%)	rate Words (%)	Substi-tutions (%)	Omis-sions (%)	Insert-ions (%)
A* with Pruning	27	65	71	27	2	4
MA* with Pruning	27	67	92	7	1	1
MA* with optimistic cost estimation	**27**	**74**	**96**	**4**	**0**	**0**
Viterbi algorithm with Earley parser	96	46	88	12	0	2
Viterbi algorithm without syntax	132	35	86	14	0	4

Tab. 3. Recognition results for sentences and words

The complete system was tested with a speech material consisting of 32 sentences spoken 8 times. As displayed in Tab. 3 a word recognition score of about 96 % and a sentence recognition rate of 74 % was achieved using the DCL syntax and the MA*-algorithm; in this case the perplexity of the test sentences was about 27. The economic operation and the good recognition results demonstrate the potential of syllabic processing in continuous speech recognition.

References

1. Ruske, G., Experiments on the use of demisyllables for automatic speech recognition. Symposium on Speech Recognition, Montreal 1986, Mc Gill University, 49 - 50.
2. Ruske, G. and Weigel, W., Automatic recognition of spoken sentences using a demi-syllable-based Dynamic Programming algorithm. 12th International Congress on Acoustics, Toronto 1986, paper A1-5.
3. Schiel, F., Syntaxgesteuerte Erkennung von gesprochenen Sätzen mit Methoden der Künstlichen Intelligenz. Diplomarbeit, Lehrstuhl f. Datenverarbeitung, Technische Universität München, 1990.
4. Weigel, W., Continuous speech recognition with vowel-context-independent HMMs for demisyllables. Intern. Conf. on Spoken Language Processing, Kobe/Japan, 1990.

Word Hypothesization in Continuous Speech Recognition

Andrea Di Carlo

Fondazione Ugo Bordoni - via B. Castiglione, 59 - 00142 Rome, ITALY - FAX: +39 6 54804400/01

Abstract: In this work, we propose the FUB point of view about the word hypothesization in the context of an Automatic Continuous Speech Recognition System. Our starting idea is the impossibility of the modeling the recognition as a sequential process: it is necessary to design an efficient device to locate the words in the continuous context in which they are embeded. We present a quick review of our current efforts on the procedural definition of the Word Hypothesizer and its iteration with a syntactical parsing system developed in our laboratory. We would put some emphasis on the analogy between the word lattice, the output of the word hypothesizer, and the chart, the working data structure of the parser. They are both defined as directed graphs giving the alternate representations - at lexical and at syntactical levels - of input portions. This analogy is the basis of the interaction: the word lattice can be directly mapped into the chart and this becomes the media for information communications among components.

Keywords: Continuous Speech Recognition, Word Hypothesization, Word Lattice, Chart Parser

Introduction

A very general approach to speech recognition is derived from the assumption that it is possible to recognize through the inverse model of the production. So in Isolated Word Speech Recognition a Word Recognizer matches the signal portion between two silence segments to the signal of every lexical item. The best matching word is proposed as the recognized word. All the recognition process is seen as the iteration of the simple matching operation over the signal portions scanned by the silence boundary markers.

Also for the connected speech, it is possible to see the recognition process corresponding to a sequential production process. (For connected speech, we mean the well-defined sequence of words pronounced without explicit boundary markers). So it is possible to design a regular recognition model for words not necessarily separated by silence. The recognition of a word sequence is possible because a global optimization is applied on the language model [5,7].

For the fluent or spontaneous speech, because it contains noises, bad-structured word sequences, aborted and restarted sentences the situation is different. Moreover, the Italian language is an order free language [8]. In its own nature, the Italian continuous speech is not a word sequence and localizing the signal portion promising for the lexical recognition is very difficult .

NATO ASI Series, Vol. F 75
Speech Recognition and Understanding. Recent Advances
Edited by P. Laface and R. De Mori

The Word Hypothesizer functional definition

The task of the Word Hypothesizer is to locate the spoken words embeded in the context of the continuum of the speech. It uses the lexical knowledges and takes in an utterance representation corresponding - in the abstraction level - to the description of the lexical items; it gives out a set of word hypotheses at a similar degree with some utterance part.

The lexicon is a collection and each element, the lexical item, is constituted of an access part, (the abstract sequence of sounds which composes the word) and an information part (grammatical category, meaning, etc) [6].

So the hypothesizer generates a word lattice: i.e. a hypotheses set consisting an directed graph. Each word hypotheses is defined as the aggregation of the starting time, the ending time - to locate the hypotheses in the input utterance -, the lexical item indicator - the label - and, finally, the score [2].

Word Hypothesization phases

We can distinguish three main steps in the Word Hypothesization: the segmentation - to locate, in the symbolic input sequence, the boundaries of some portion promising for the lexical access - the access into the lexicon - to retrieve the lexical items leaving from the segment - and the hypotheses scoring - to compute the matching degree between the homogeneous representations of the lexical item and the segment.

It is possible to design the segmenter on the base of different approaches for the start-markers and end-markers selection. For start-marking, every discrete time or every phonemic symbol is a possible beginning of the word; alternatively, it is possible to apply some sequentiality criterion like the assumption that the words are in sequence and each new start boundary is close to the end boundary of some yet hypothesized word.

Similarly, it is possible to consider two ways for end-marking: the end-boundary is determined top-down from the knowledge about the duration of the vocabulary words; alternatively, the activation of the lexical items decreases as the length of the accessing segment increases: this is the base for the dynamic determination of the end-boundary.

Traditionally, the access into the lexicon is seen as a typical pattern recognition problem: matching between a segmented pattern and a big number of prototypes with the same nature - from here, the importance of the study about complex and efficient algorithms to data structures search, to distance compute, to string correct.

A new way of seeing is to put the problem in the context of associations between different cognitive level patterns: then the new bottom-up approach to the lexical access as the redundant hashing addressing by Kohonen [4] and the neural association. In the Redundant Hash

Addressing, the main idea is the fractionning the correspondence between the lexical item and its phonemic transcription by means the decomposition of the transcription into components which can be found also in the input and can be use as multiple access keys.

With the segmented portion of the input we access into the lexicon and we obtain a word list as set of possible lexical interpretations of the portion of the signal. Given the hypothesized word list at a particular point in the utterance, the matching algorithm compares the acoustic (or phonemic) representation of the input speech with each of the possible homogeneous representation of the hypothesized words to find that word that is most consistent with the observed speech.

Our Implementation of the Word Hypothesizer

We realized two modules for the word hypothesization: an associative hypothesizer and a top-down hypothesizer.

The former [3] segments the input on the basis of an exhaustive selection of the start-markers and of the prediction of the end-markers from the knowledge about the durations of the lexical items; we obtain as many segments as possible from every input symbol and with every lexically possible length. Then it extracts the hash keys from the segment - all the two symbol substring - and retrieves the words by redundant access into the lexicon. Every hash key is associated to a list of words containing it. So it is possible to increase the activation value of each word in the lexicon when a two symbol substring is found. The lexical hypotheses are created from the localization of the keys in the complete input, from the activated word and from the activation value.

The second one is a constrained top-down word hypothesizer. It directly matches every item of its lexicon against an input segment without bottom-up association. It can accept, from the other modules of processing, the boundaries of the segment on which it must operate. Moreover it uses the subset of the main vocabulary defined by a lexical category as lexicon. Usually, it is used to recover lost hypotheses with less strict automatic constraints and more specific knowledge (shorter input, smaller vocabulary, less strict threshold, specification of the lexical category).

We used two different definitions of the distance to score every activated word. The first one, the Edit Distance, is a very effective method in framework whose matching problem is seen as a variational problem. The patterns are deformed by local scale transformation: in our case the local transformations are the edit operations i.e. the substitution , the deletion and the insertion of single symbols. The matching is then performed by considering only small pieces of patterns at one time. The pieces are shifted to each other to find the best warping with the minimum number of edit operations. The Edit Distance is very significant but is quite time consuming because many hundreds of matching are necessary and the whole word hypothesization phase becomes too long. To solve the time problem, we choice another distance, less significant but much simpler and faster: it counts the number of corresponding symbols in the matched patterns. The patterns are aligned in a very broad way: by matching the symbols one-to-one without shift until the last symbol of the shorter string is processed.

Interaction Architecture

The current effort, in our group activity, is on the interconnection between the different parts of a recognition system and we developed some ideas about the interaction between the Word Hypothesizer and a syntactical Chart Parser. In this context, we realized several parsing modules with different running modalities and we are trying different interaction architectures [8,9].

It is suitable to underline that the output of the word hypothesizer, the word lattice, and the working data structure of the parser, the chart, are both defined as directed graphs giving the alternate representations - at lexical and at syntactical levels - of input portions.

This analogy is the basis of the interaction between the components: the word lattice can be directly mapped into the chart - the chart has a bigger representation power and it can completely collect the lexical information: it becomes the media for information communications among levels. In the further generalization, the chart is a data structure global with respect to the system components.

The symbolic sequence, given as output by the acoustic-phonetic decoder, is sent to a bottom-up word hypothesization module. It detects the input parts which can be interpreted as lexicon item. The word lattice is mapped into the initialized chart: every lexical hypothesis is translated into a chart edge between two nodes corresponding to the word boundaries.

After the chart initialization, the left-to-rigth parser starts; it operates in bottom-up mode, with a best-first strategy of searching in a subspace obtained with all the perfect scored chart edges. If it builds, in a very fast way, a syntactical structure covering all the input and perfectly scored, the success analysis will be obtained and the process will be stopped.

If no structure covering the input is obtained then it will be necessary to continue the process.

The chart is reinitializated at the lexical status and the island-driven parser starts to develop the left-side and rigth-side contexts of every island. The island-driven chart parser builds the correct syntactical constituents concatenating contiguous words or sequence of words to the islands. Some islands cannot be extended because there are no correct words which can be joint to them on the left-side or on the rigth-side. In these cases some active edges are however introduced into the chart.

From these incomplete structures, it is possible to build some prediction on the syntactically possible continuations; when this is possible with only one further step in the syntactical analysis obtaining a structure with only one lexical class missing, a stimulus for a new lexical access is created. The new word hypothesization is constrained over a lexicon subset defined by the syntactical class and over the input portion which as necessary to complete the syntactical structure; moreover this phase uses less strict acceptance threshold for the lexical hypotheses.

With the richer chart, the island driven parsing continues until the end condition is reached. The end condition - fail or success - for this phase is the end condition for all the lexic-syntactical process.

Conclusions

We realized two modules for the word hypothesization: an associative hypothesizer and a top-down hypothesizer. For the former, we obtained an algorithm composed of three procedures: an effective word boundaries' selector, a very fast associative access method and an hypothesis evaluation procedure. The second one, the top-down hypothesizer, is used to recover lost hypotheses with less strict automatic constraints and more specific knowledge. The future efforts are oriented the performance evaluation. It will be articulated on the evaluation of the data structure complexity and on the evaluation of the component performances.

References

1. Di Carlo A., Falcone R., "A Blackboard Architecture for a Word Hypothesizer and a Chart Parser Interaction in an ASR System", Proc. ICSLP 90, Kobe, Japan, nov 1990
2. Di Carlo A., "Ipotizzazione di parole in un sistema di riconoscimento di cifre connesse mediante accesso associativo al lessico" - Rel. FUB 5b3487, 1987
3. Di Carlo A., Paoloni A., "An Experiment in Word Hypothesization performed in the Context of a Continuous Speech Recognition System" - Proc. EUROSPEECH 89, Paris, september 1989
4. Kohonen T., Reuhkaka E., "A Very Fast Associative Method for the Recognition and Correction of Misspelt Words, based on Redundant Hash Addressing", Proc. 4th I.J.C. on Pattern Recognition, 1978
5. Levinson S.E., Ljolje A., Miller L.G., "Large Vocabulary Speech Recognition using a Hidden Markov Model for Acoustic/phonetic Classification", Speech Technology, vol. 4 (4), april/may 1989
6. Parisi D., "Problemi di ricerca sul lessico", Istituto di Psicologia C.N.R. Roma, PCA/2, 1986
7. Rabiner L.R., Juang B.H., "An Introduction to Hidden Markov Models", IEEE ASSP Magazine, january 1986
8. Stock O., "Parsing Flexible Languages with a View to Recognizing Continuous Speech", Int. Workshop on Advances and Applications of Speech Recognition, Rome, Italy, 27-29 may, 1986
9. Ward W.H. & al. - "Parsing Spoken Phrases Despite Missing Words" - Proc. ICASSP 88, april 1988

Phone Recognition Using High Order Phonotactic Constraints

Andrej Ljolje

AT&T Bell Laboratories, 600 Mountain Avenue, Murray Hill, NJ 07974, U.S.A.

Abstract: A new phone recognizer has been implemented which extends the (phonotactic) decoding constraint to sequences of three phones. It is based on a structure similar to a second order ergodic hidden Markov model (HMM). This kind of model assumes direct correspondence between the model states and phones, thus constraints on possible state sequences are equivalent to phonotactic constraints. Very high coverage by both left and right context dependent phone models has been achieved using two methods. The first assumes that some contexts have the same or very similar effect on the phone in question. Thus they are merged into the same contextual class. The outcome is a set of 19 left context classes and 18 right context classes. The second assumes that left context mostly influences the beginning of a phone, whereas the right context influences the end of the phone. Each phone (a state in an ergodic HMM) is represented by a sequence of three probability density functions (pdf's), which is equivalent to a three state left-to-right HMM. We generate acoustic models such that first pdf in the model is conditioned on the left context, the middle pdf is context independent, and the last pdf is conditioned on the right context. A large number of such quasi-triphonic acoustic models can be generated providing a good triphone coverage for a given task efficiently utilizing the available training data set. The current implementation of the recognizer described here has been applied to the DARPA Resource Management Task. Since true phone sequences are not available, they are estimated from text from a phone realization regression tree trained on TIMIT database transcriptions. The estimates of the true phone sequences are used in training the models and generating reference phone sequences for scoring. The best phone recognition match between the most likely output of the regression tree and the phone recognizer for the DARPA February 89 test set was 75.5% correct with 79.5% accuracy.

Keywords: phone recognition, speech recognition, phonotactic models, acoustic models

Introduction

Current continuous speech recognition systems are centered around two different approaches: word level recognition with syntactic constraints [2,3,5] and phone recognition with lexical access and syntactic constraints [4,6]. Both use sub-word acoustic models. So far word based recognizers have exhibited superior performance. There are two main reasons for the performance difference.

The first reason for the difference is that long phone sequences are hypothesized every time a new word is hypothesized (equivalent, on average, to the mean word duration in number of phones) adding significant constraint to the ways an utterance can be decoded.

NATO ASI Series, Vol. F 75
Speech Recognition and Understanding. Recent Advances
Edited by P. Laface and R. De Mori

All the current phone level recognition systems only consider the previous phone which is equivalent to the average word duration of two phones. The second reason is concerned with the frequency of hypothesizing different acoustic models, where context dependent models are used much more frequently than the context independent models. Word based recognizers hypothesize words and use appropriate acoustic models to create a word model. Thus best acoustic models (high contextual constraint) are used most often, whereas poor acoustic models (context independent) are rarely used. In phone recognition all phones are hypothesized, and a large difference between very accurate acoustic models conditioned on both contexts and context independent models creates an imbalance and many errors occur.

In this work we demonstrate that high accuracy phone recognition is indeed possible, and is achieved by addressing the two main problems present in current phone recognition systems: phonotactic constraint and high context dependent acoustic model coverage.

The speech recognition task

The database used in these experiments was the DARPA Naval Resource Management (RM) Task database. The training set consists of speech and text recorded by 109 speakers saying either 40 or 30 sentences, giving a total of 3969 usable sentences. The test set used in the evaluations is the official February 89 test set which consists of recordings by 10 speakers saying 30 sentences each with the corresponding text. It is important to note that transcriptions are *not* available for either the training nor the test set.

The estimates of the true phone sequences were obtained from a phone realization tree. This phone realization tree had previously been trained on 3360 transcribed sentences in the TIMIT database. The estimate of the true phone sequence consists of a phone realization network which provides alternative ways an utterance could have been pronounced and associated probabilities. Only the most likely path through the predicted phone network was used in the experiments.

The phone recognition system

The phone recognizer is based on a structure which consists of three independent models: trigram phonotactics model based on the RM task, the phone duration model and the acoustic phone models. The three models are interrelated through a structure similar to a second order ergodic continuous variable duration HMM (CVDHMM). Each state of the CVDHMM directly corresponds to a phone, while the phonotactic model corresponds to the state transition matrix and the phone duration model corresponds to a state duration distribution. Each state/phone is modeled by a time sequence of three pdf's and can be visualized as a three state left-to-right HMM with all transition probabilities set to 1.0. This way a sufficiently rich structure is achieved to accurately model individual phones, but the simplicity and concise mathematical description of a 47 state ergodic CVDHMM to describe the phonetic structure of the English language is preserved.

The phonotactic model

The phone set chosen for the experiments consists of 46 phones plus silence, and is very often used in other speech recognition systems [2-5]. This phone set thus gives us the maximum number of allowed three phone sequences: 47^3. Fortunately, many of these will never be observed in a given task, and many occur much more frequently than others. In order to get a good model of how often we can expect to observe a particular three phone sequence we have to simulate a large amount of transcribed speech specific to the RM task. We achieve this by using an RM task grammar to generate a large amount of *text* (300000 sentences). A phone realization network was then generated for each of the sentences. The probabilities for all the three phone sequences (trigrams) in all the legal paths through the phone network were then accumulated for all sentences. By normalizing the accumulated sums so that their sum is equal to 1.0, we obtain a good approximation of the true trigram probabilities for the RM task.

The acoustic phone models

Traditionally, left and right context dependent models are generated if enough data is available. The remaining data is used to generate either left *or* right context dependent models, and whatever data that remains at the end is used to make context independent models. In this experiment we exploit two new techniques to increase the coverage of context dependent models, particularly models with both left and right context dependency. The techniques are context clustering and quasi-triphone modeling.

Context clustering

Context clustering for context dependent acoustic models exploits the similarity of contextual effects a number of contexts has on the phone we are trying to model. If the effect of the contexts is the same, or very similar, there is no reason to treat them separately. Merging of several contexts into a contextual class pools data that would have been treated separately and thus provides sufficient data to train many more context dependent models.

Context classes for left contexts, and separately for right contexts, have been defined a priori based on known contextual effects. The result is a set of 19 left context classes and 18 right context classes.

Quasi-triphonic acoustic models

Each acoustic model consists of three pdf's, where the first one corresponds to the beginning of the phone, the middle one to the middle part of the phone and the last one to the end of the phone. In a fully context dependent model (triphonic model) all three distributions are conditioned on both left and right context. However, the contextual influences rarely span across the whole phone. Also, the left contextual effects will be much stronger on the beginning of the phone, whereas the right contextual effects will be more pronounced at the end of the phone. We exploit this characteristic by creating quasi-triphonic models. In these models, the first distribution is conditioned only on the left context, the middle distribution is context independent and the last pdf is conditioned only on the right context.

The duration model

Each acoustic phone model has a duration model associated with it. The model consists of a four parameter gamma distribution. The four parameters are the shortest allowed phone duration (the gamma distribution shift), the mean duration, the variance of duration and the maximum allowed duration for the phone. The shortest allowed duration is equal to the shortest observed duration in the training data, the mean and variance are calculated from the data and the maximum duration is calculated as the 95th percentile of the distribution.

Recognition experiments

The following experiments were carried out in order to investigate the importance of the phonotactic constraint on the phone recognition performance. In particular, we wanted to determine the relative importance of: decoding constraint (legal vs. illegal bigrams/trigrams), weighting of alternative decoding paths (based on probabilities of observing given trigrams) and weighting of the phonotactic model with respect to the rest of the recognition system.

The recognition results show phone accuracy (does not include insertions), % correct (includes insertions) and the error breakdown into insertions, substitutions and deletions. All results have been rounded off to the nearest half percent value.

The baseline performance consists of a simple implementation using the recognizer with true probability estimates from the phonotactic and duration models, and pdf's returned by the training procedure. The results shown are for the February 89 test set and a subset of the training data. The training data subset consists of two randomly chosen sentences by each of 109 speakers in the training set.

Results on Feb 89 test set					
Baseline performance					
data	%corr	%acc	%ins	%del	%sub
train	82.0	88.0	6.0	3.0	9.0
test	70.0	77.5	7.0	4.5	18.0
log trigram prob * 1.5					
train	83.0	88.0	5.0	3.0	9.0
test	72.0	78.0	5.5	5.0	17.0
log trigram prob * 1.5 + pauses					
test	72.5	78.0	5.5	5.5	16.5
log trigram prob * 1.5 + pauses + smoothing pdf's					
test	74.5	79.5	4.5	5.5	15.0
Influence of trigram probabilities					
sc_fac	%corr	%acc	%ins	%del	%sub
0.0	68.5	78.0	9.5	3.5	18.0
1.5	74.5	79.5	4.5	5.5	15.0
2.0	75.5	79.5	4.0	6.5	14.5
Effect of higher constraint					
constraint	%corr	%acc	%ins	%del	%sub
bigram	69.5	74.5	5.0	7.5	18.5
trigram	74.5	79.5	4.5	5.5	15.0

Table 1. Recognition results for the investigated recognition systems

These results can be improved with a number of simple modifications to the recognizer and they are listed in Table 1. We can vary the relative weight of different parts of the recognition system. Thus we can increase the importance of the phonotactic model. We achieve this by multiplying the logarithm of all trigram probabilities by 1.5. The result is an increase in the recognition rate. Another modification attempts to modify the trigrams model by allowing transitions into and out of silence to permit pauses in the middle of a sentence. This also brings a modest improvement in the score. The last improvement comes from smoothing the acoustic model pdf's. The smoothing is achieved by adding a constant to all the eigenvalues in all the covariance matrices in the acoustic model pdf's.

Next we attempt to find out what improvement is possible by having an estimate of trigram probabilities. In the first experiment we have a list of legal trigrams but no knowledge of their frequency of occurrence in the RM task. In the second and third experiment we exploit our estimates of trigram probabilities with the phonotactic model weights of 1.5 and 2.0. The outcome is a significant reduction of errors, mostly insertions, with a modest improvement in the accuracy of the recognizer as we increase the influence of the trigram probabilities.

Last we compare the results by using bigrams instead of trigrams. Bigrams are obtained by averaging one dimension of the trigram model and thus reducing the second order constraint to a first order constraint. However, we only allow triphonic models with matching contexts to follow each other, regardless of the phonotactic constraint, thus somewhat reducing the effect of going from a trigram to a bigram phonotactic model. Trigram constraint clearly improves recognition scores compared to bigram results, demonstrating that extra computation is effectively utilized. These results can also be found listed in Table 1.

Conclusions

High accuracy phone recognition has been demonstrated using a new recognizer structure employing context classes, quasi-triphones and a high order phonotactic model. An increase in decoding constraint from bigrams to trigrams has been shown to significantly improve recognition performance in terms of both accuracy and correctness. The error reduction of 5% has been achieved using trigram probabilities. Availability of a good estimate for the frequency of occurrence of different three phone sequences (trigram probabilities) has also been shown to yield significant reductions in the total number of errors. Additionally, an increase of the phonotactic model's influence in the decoding process has been shown to decrease the error rate (when compared to the true probability estimates).

Recognition rates of 83.0% correct and 88.0% accurate have been achieved on a subset of the training data, while the best result on the speaker independent test set was 75.5% correct and 79.5% accurate.

References

1. Juang, B.H., Rabiner, L.R. and Wilpon, J.G.: On the Use of Bandpass Liftering in Speech Recognition. IEEE Trans. ASSP-35, 7, 957-954 (1987)
2. Lee, C.-H., Rabiner, L.R., Pierracini, R. and Wilpon, J.G.: Acoustic Modeling for Large Vocabulary Speech Recognition. Computer Speech and Language, 4, 127-165 (1990)
3. Lee, K.-F.: Automatic Speech Recognition - The Development of the SPHINX System. Boston: Kluwer Academic Publishers 1989.
4. Levinson, S.E., Ljolje, A. and Miller, L.G.: Continuous Speech Recognition from a Phonetic Transcription, Proc. ICASSP-90, Albuquerque, 93-96 (1990)
5. Paul, D.B.: The Lincoln Robust Continuous Speech Recognizer. Proc. ICASSP-89, Glasgow, 449-452 (1989)
6. Zue, V., Glass, J., Phillips, M. and Seneff, S.: The MIT Summit Speech Recognition System: A Progress Report. Proc. Speech and Natural Language Workshop, Philadelphia, 179-189 (1989)

AN EFFICIENT STRUCTURE FOR CONTINUOUS SPEECH RECOGNITION

R. Pieraccini, C. H. Lee, E. Giachin†, L. R. Rabiner

Speech Research Department
AT&T Bell Laboratories
Murray Hill, NJ 07974

† Now with CSELT, Torino, Italy

Abstract: In this paper, we present an efficient data structure for implementing a continuous, large vocabulary, speech recognizer. The recognition system is based on hidden Markov models of phonetic units for representing both intraword and interword context dependent phones. Due to the large number of connections present in the decoding network, the structure of the recognizer must be carefully designed in order to perform experiments in a reasonable amount of computing time.

Keywords: Continuous speech recognition, beam search, recognizer complexity

1. Introduction

It has been demonstrated that large-vocabulary continuous speech recognizers, based on hidden Markov models of phone-like units, achieve high word accuracy when highly detailed speech units are used for characterizing vocabulary words. In particular, when context dependent units are used to represent both intraword and interword phones, a word accuracy higher than 95% has been achieved [2,4,6,7] on speaker-independent recognition of the standard DARPA Resource Management task when a word-pair grammar of perplexity 60 is used to constrain the search.

When interword units are used, the complexity of the recognizer increases considerably because of multiple fan-out at the end of a word and multiple fan-in at the beginning of the next word which results in a large number of interword connections. At this level of algorithm complexity, every single detail of the implementation can have a major impact on the computation requirements. In particular, the data structure used for the network search should be designed in such a way that the memory requirements and computation overhead be kept to a minimum. When this is the case, a full search implementation of the speech recognition algorithm is totally impractical, if not impossible. An effective solution to this problem consists of performing an intelligent search through the grid and using reasonable but effective heuristics to eliminate unlikely path candidates from consideration. Moreover, for a very efficient implementation, the decoding network should be compiled rather than interpreted, meaning that the memory addresses of all predecessors of each individual node in the decoding network must be stored in advance.

NATO ASI Series, Vol. F 75
Speech Recognition and Understanding. Recent Advances
Edited by P. Laface and R. De Mori

2. Review of the Recognizer Structure

A more detailed description of the large vocabulary continuous speech recognition system implemented by the algorithms described in this paper can be found in Ref [3]. Three state, continuous density, HMM's are used to represent each of the units (except silence which is represented by a single state HMM). There are only two transitions leaving from a state, a self loop transition and a transition to the next state. Transition probabilities are not used during the likelihood computation, since it has been experimentally shown that they don't affect the recognition performance. Within each state of each model, the spectral observation probability density is represented by a weighted mixture of multivariate normal density functions.

2.1 Structure of the Lexicon

When using intraword and interword units, every word in the lexicon is represented by three distinct segments, namely a *head*, a *body*, and a *tail*. The head and the tail take into account all the possible types of coarticulation with adjacent words, according to the unit set chosen for representing the vocabulary. Hence the head and tail of a particular word consist of a collection of all the possible conjuction units at the beginning and at the end of a word. The body of each word is a simple linear sequence of units and is assumed to be independent of the neighboring words. It should be noted that words composed of two phones and words composed of one phone are special cases of this kind of model. Fig.1 shows an example of a word model for three cases; namely for a word consisting of more than 2 phones (a), for a 2 phone word (b), and for a single phone word (c). (the self transition of HMM states has not been drawn in the figure). Silence at the beginning and at the end of a sentence is represented as a single phone word.

The connections between the tails of a word and the heads of the following word are based on a precomputed connection matrix $CONN(ph_i,ph_j)$ whose generic element assumes the logic value true if unit ph_j may follow unit ph_i in the conjunction of two consecutive words. The sequence of words defining legal sentences within the task grammar are expressed through a finite state network (FSN) and a list of permitted word pairs. The matrix $CONN(ph_i,ph_j)$ may be modified according to phonological rules [1].

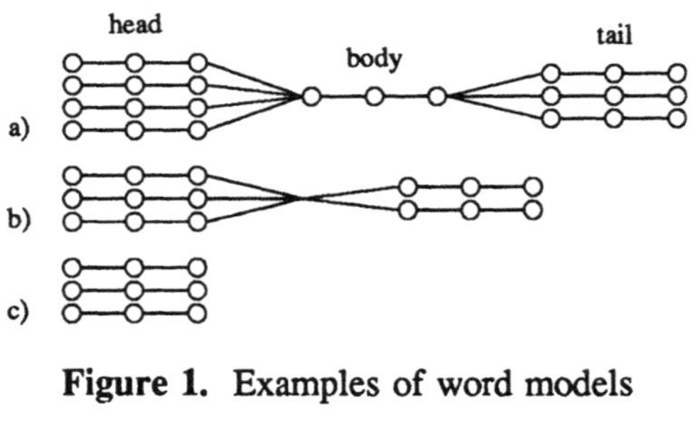

Figure 1. Examples of word models

3. The Decoding Algorithm

A parallel and vectorized computer (an Alliant FX/80) has been used to perform continuous speech recognition experiments. Thus a vectorized compiled representation of the decoding network is necessary for taking advantage of the architecture of the computer. The basic structure into which the decoding network is converted is a *state vector* whose generic element is a particular HMM state of the integrated network. The flow diagram of Fig.2 shows a breakdown of the decoding process for one frame of input speech into functional blocks. In the remainder of this section we analyze the main implementation characteristics of each block.

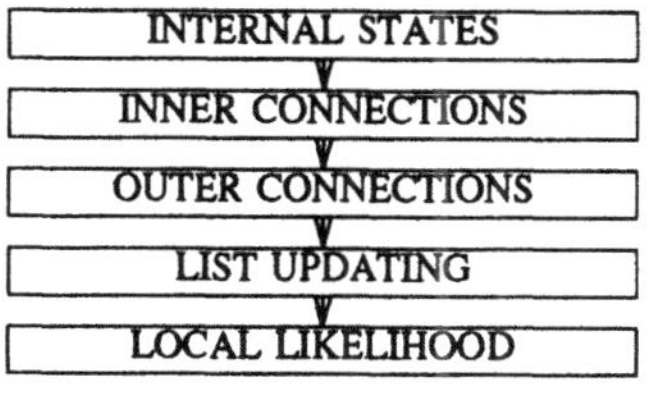

Figure 2. Steps in the decoding of one frame

3.1 Internal states

This block performs the dynamic programming optimization for all the active states in the state vector. The pointers to active states are kept in a list that is updated at the end of the processing of each frame.

3.2 Inner connections

Inner connections are those among the heads, the body and the tails of the same word. Only words with a number of phones greater than or equal to 2 have inner connections. This block performs two basic functions:

1. Finds the best scoring end state among all the possible heads of the word and propagates the corresponding path to the first state of the body if it has a greater score. This is a vectorized and parallelized operation.
2. Propagates the path corresponding to the last state of the body to the first state of each tail of the word, if it has a greater score. This is a parallelized operation.

3.3 Outer connections

Outer connections are those among the tails of a word and the heads of another word. This operation is performed only for active arcs. An arc is considered active when at least one tail end state is active. The pointers to active arcs are kept in a list that is updated at the end of the processing of each frame.

3.4 List updating

The beam search [5] pruning of states is performed in this block. In the standard implementation, at the $i-th$ frame, a pruning threshold Θ_i is set as:

$$\Theta_i = \Lambda^{max_i} - \Delta$$

where Λ^{max_i} is the maximum likelihood among the active states at the $i-th$ frame and Δ is a fixed constant. Hence the likelihood of all states that are active at the $i-th$ frame is compared with Θ_i. A state is then included in the new list if its likelihood is greater than the threshold Θ_i. An additional operation performed by this block is the updating of the list of active arcs.

3.5 Local likelihood

While local likelihood computation in discrete density HMM's is a simple table lookup, with mixture density HMM's it becomes one of the major computational loads of the entire decoding procedure. A particularly optimized version of the state local likelihood computation has been implemented, emphasizing the vectorized structure of the computation. Also, the local likelihood is computed only for active units.

3.6 Guided search

We have also developed a particularly efficient version of the recognizer suited only for experimental assessment of speech recognition accuracy. When assessing performance on a test database, the correct string that has to be recognized is known a priori for every spoken sentence. The forced alignment of the test speech, with the network representing the sentence actually spoken, produces a path whose frame-by-frame score (self score) may be used to further reduce the size of the search space. The concept behind the guided search is that the overall best path will have a final score that cannot be inferior to the final score of the forced alignment procedure. This is not true for the local score along the overall best path and the forced alignment path. It may happen that the overall best path drops below the score of the forced alignment path at a certain point in the search, eventually attaining a better score later in the search. Moreover, if the guided search is performed in a frame synchronous fashion, i.e. the forced alignment is carried out frame synchronously with the recognition, we actually don't know which is the best path in the forced alignment. A non frame synchronous version that first performs the forced alignment, then performs a backtracking along the best path and saves all the local score values, would require too much memory to store the backtracking information during the alignment phase. Thus, only the score of the locally (not globally) best path is available during the frame synchronous alignment. The threshold is computed by decrementing the score of the locally best path by a fixed amount in order to take into account the above mentioned sources of error.

3.7 Timing experiments

Timing experiments have been performed during the development of the algorithm to assess the efficiency of the entire speech recognition system. All the performance scores reported in this section were obtained during the recognition of several sentences using a phone set of 1172 sub-word units. The maximum number of mixture components per state was 16, while the dimension of the observation vector was 24. The guided search strategy was used in all the experiments.

Table 1 shows the average time (in seconds) per sentence (TPS), and average time per decoded frame (TPF), in 3 different versions of the recognizer. In REC1 all the connections are explored at every frame; in REC2 only connections coming from active arcs are explored; and REC3 has the same features as REC2, but uses a compiled version of the connection list.

RECOGNIZER	TPS	TPF
REC1	555	1.8
REC2	326	1.1
REC3	65	0.2

TABLE 1. Average time (in seconds) per sentence (TPS) and per frame (TPF) in three different implementations of the recognizer

Table 2 shows the time breakdown for the five modules of Fig.3 when REC3 is used. The numbers shown are the percentages of time spent in each module during the decoding of one frame.

Operation	Time %
Internal states	14.8
Inner connections	11.8
Outer connections	5.0
List updating	13.2
Local likelihood	55.2

TABLE 2. Percentage of time spent in each module during the decoding of one frame

Table 2 shows that the local likelihood computation accounts for more than 55% of the total decoding time followed in order of decreasing proportion of time by the dynamic programming optimization on the active states, the list updating, and the propagation of scores for inner connections. The propagation of score to outer connections takes only 5% of the whole computation. In fact, even though the number of potential connections is very large (240024 in the experiment), only a small fraction of them are actually used at each frame.

4. Conclusions

This paper provided a detailed presentation of all aspects of the implementation of a large vocabulary speaker independent continuous speech recognizer which is used as a tool for the development of recognition algorithms based on hidden Markov models and Viterbi decoding. The complexity of HMM recognizers is greatly increased by the introduction of detailed context dependent units for representing interword coarticulation. A vectorized representation of the data structures involved in the decoding process, along with compilation of the connection information among temporally consecutive words, has led to a speed up of the algorithm of about one order of magnitude. An average recognition time of about one minute per sentence (on the computer configuration used in the experiments), although far from real time, allows us to perform a series of training experiments and to tune the recognition system parameters in order to obtain high word accuracy on complex recognition tasks such as the DARPA resource management.

5. Acknowledgements

The authors gratefully acknowledge the helpful advice and consultation provided by Fil Alleva of Carnegie Mellon University on the implementation details of the SPHINX system and Douglas B. Paul of Lincoln Laboratories, MIT, for the guided search idea.

6. References

1. Giachin, E. P., Rosenberg, A. E., Lee, C. H.: Word juncture modeling using phonological rules for HMM-based continuous speech recognition. *Proc, ICASSP 90*, pp. 737-740, Albuquerque, NM, April 1990.
2. Huang X., Alleva, F., Hayamizu, S., Hon, H. W., Hwang, M. Y., Lee, K. F.: Improved Hidden Markov Modeling for Speaker-Independent Continuous Speech Recognition. *Proc. DARPA Speech and Natural Language Workshop*, Somerset, PA, June 1990.
3. Lee, C. H., Rabiner, L. R., Pieraccini, R., Wilpon, J. G.: Acoustic modeling for large vocabulary speech recognition. *Computer Speech and Language*, 4, pp. 127-165, 1990
4. Lee, C. H., Giachin, E., Rabiner, L. R., Pieraccini, R., and Rosenberg, A. E.: Improved acoustic modeling for continuous speech recognition. *Proc. DARPA Speech and Natural Language Workshop*, Somerset, PA, June 1990.
5. Lowerre, B. and Reddy, D. R.: The HARPY speech understanding system. In *Trends in Speech Recognition* (Lee, W. ed.), 340-346. Prentice-Hall Inc., New York, 1980.
6. Murveit, H., Weintraub, M., Cohen, M.: Training Set Issues in SRI's DECIPHER Speech Recognition System. *Proc. DARPA Speech and Natural Language Workshop*, Somerset, PA, June 1990.
7. Paul, D. B.: The Lincoln Tied-Mixture HMM Continuous Speech Recognizer. *Proc. DARPA Speech and Natural Language Workshop*, Somerset, PA, June 1990.

Search Organization for Large Vocabulary Continuous Speech Recognition

Fileno A. Alleva

School of Computer Science, Carnegie Mellon University, Pittsburgh Pennsylvania, 15213, USA[1]

Abstract: This paper describes the search organization of the *Sphinx* continuous speech recognition system. The search is organized as a hybrid Viterbi beam search that uses Viterbi beam search [2] to control activation of words and sub-word units at the high level and Viterbi optimal search [3] to evaluate hidden Markov models at the low level. This hybrid Viterbi search allows the Sphinx search running on a Sun 3/60, to evaluate the February 1988 Resource Management test set in 3.0 times real-time.

Keywords: Search, Sphinx, Viterbi Search

Introduction

Improved CSR search organization is important because recent trends (fluent speech, co-articulation modeling, complex language models) in automatic speech recognition, have exacerbated the problems associated with providing sufficient memory bandwidth and computational power for acceptable (real-time) performance of the search component. A variety of computer architectures (message passing, shared memory, distributed shared memory, parallel) provide solutions to the problems of increased memory bandwidth and computational power. Therefore it is up to the search component of CSR systems to be organized in such a way so as to obtain full value from these and future developments in the area of computer architecture. I describe a hybrid Viterbi search organization for the *Sphinx* [1] speech recognition system. The paper is divided into two parts. In the first part I present an analysis of the Viterbi evaluation that leads to the hybrid Viterbi beam search used in *Sphinx*. In the second part I present the hybrid Viterbi search used by the Sphinx search component.

Notation

The following notation is commonly used to describe search in speech recognition systems and will be used through out this paper.

$$v_j(t) = \begin{cases} 0.0 & t=0 \wedge j \neq I \\ 1.0 & t=0 \wedge j=I \\ \text{argmax}_i \; v_i(t-1) a_{ij} b_{ij}(y_t) & t>0 \end{cases}$$

1

This research was supported by the Defense Advanced Research Projects Agency (DOD),, and monitored by the Space and Naval Warfare Systems Command under Contract N00039-85-C-0251, ARPA Order No. 5167.

The views and conclusions contained in this document are those of the authors and should not be interpreted as representing the official policies, either expressed or implied, of the Defense Advanced Research Projects Agency or the U.S. government.

NATO ASI Series, Vol. F 75
Speech Recognition and Understanding. Recent Advances
Edited by P. Laface and R. De Mori

$v_j(t)$	Viterbi probability of being in state j at time t.
a_{ij}	Probability of transition from state i to j where $\sum_{j=1}^{K} a_{ij} = 1.0$.
$b_{ij}(y)$	Probability of emitting symbol y on transition ij where $\sum_y b_{ij}(y) = 1.0$.
y_t	Acoustic symbol observed at time t.
I	Initial state.
F	Final state.
K	State count.

Analysis of the Viterbi recursion

The following algorithm reformulates the Viterbi recursion as an iterative search over the graph A for the observed acoustic sequence $\overline{y}$. Since A is sparse and speech is a left to right process the iterative solution is both more efficient and more natural in this domain.

1. Allocate state space S such that for state i at the tth stage of the search $s.v_i(t)$ holds the Viterbi probability of state i and $s.p_i(t)$ will hold the back pointer for state i.
2. Initialize the search, $t=1$, $s.v_i(0) = \begin{cases} 1.0 & i=I \\ 0.0 & i \neq I \end{cases}$
3. $\overline{s}.v_i(t) = 0.0$.
4. **Foreach** state i
   a. **Foreach** state j such that $a_{ij} \neq 0.0$
      i. **If** $s.v_j(t) < s.v_i(t-1)a_{ij}b_{ij}(y_t)$ **then**
         1. $s.v_j(t) = s.v_i(t-1)a_{ij}b_{ij}(y_t)$
         2. $s.p_j(t) = s.p_i(t-1)$
5. **If** $t=T$ **then** terminate **Else** $t=t+1$, goto step 3.

Figure 1. Viterbi Optimal Search

The Viterbi search algorithm has one main component. Step 4.a.i computes the **argmax** of the current probability of state j and the probability on arc ij for time t. I implement **argmax** as a conditional since there is no need to write the contents of state j if they are not changed.

Figure 2 shows the data flow diagram for performing the viterbi evaluation on one arc of the search graph. In this diagram each state consists of a Viterbi probability, $s.v$, and a path back pointer, $s.p$. An analysis of this data flow diagram considers memory reads, memory writes, ALU operations and branches. Since there is a conditional branch there are two cases to consider. The first case is when $s.v_j(t)$ is **argmax** the other case is when $s.v_i(t-1)a_{ij}b_{ij}(y_t)$ is **argmax**. The two cases are illustrated in table 1.

argmax	
$s.v_j(t)$	$s.v_i(t-1)a_{ij}b_{ij}(y_t)$

`read` $s.v_j(t)$
`read` a_{ij}
`read` $b_{ij}(y_t)$
`read` $s.v_i(t-1)$
`ALUmultiply`
`ALUmultiply`
`ALU compare`
`Branch`
	`read` $s.v_j(t)$
	`read` a_{ij}
	`read` $b_{ij}(y_t)$
	`read` $s.v_i(t-1)$
	`ALUmultiply`
	`ALUmultiply`
	`ALU compare`
	`Branch`
	`write` $s.v_j(t)$
	`read` $s.p_i(t-1)$
	`write` $s.p_j(t)$

Totals	
$s.v_j(t)$	$s.v_i(t-1)a_{ij}b_{ij}(y_t)$

`4 Memory Accesses`
`3 ALU operations`
`1 Branch`
	`7 Memory Accesses`
	`3 ALU operations`
	`1 Branch`

Table 1. Viterbi Update Evaluation cost

When examining table 1 note the *small* number of operations required to evaluate a single arc of the search graph. Any change in the algorithm that changes the number of operations performed in step 4.a.i will have a significant impact on the execution speed of the search. From table 1 we can compute the ratio of memory accesses to ALU operations (including the branch). This ratio gives us an estimate of the required memory bandwidth based on the clock speed of the ALU operations. In the best case illustrated by the first column of the table the ratio is 1:1, in the worst case illustrated by the second column of the table the ration is 7:4. The point here is that the Viterbi evaluation may be memory bandwidth limited, especially on architectures that have very high CPU clock rates. An illustration of this can be seen in table 2. In the next section I describe a pair of optimizations to reduce the instruction count and alleviate memory bandwidth limitations.

Viterbi Initialization Transition

Initially I considered a Viterbi search based on state expansion which consists of evaluating the outgoing arcs from state i to its successor states j such that $a_{ij} \neq 0.0$. We note the probability of $\bar{s}.v(t)$ is initialized to 0.0 at the start of each iteration. As a result the worst case analysis always applies for the first transition to state $s.v_i(t)$. This is cause for some concern since the typical number of incoming arcs to a state i is on the order of two or three. However if we know prior to the evaluation of arc ij that this is the first transition to state j then we can apply an *initialization* transition rather than the standard *update* transition. An *initialization* transition corresponds to the data flow diagram in figure 3.

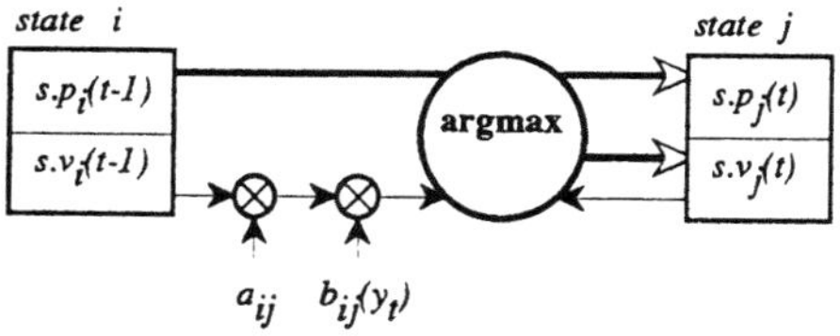

Figure 2. Viterbi Update Data Flow

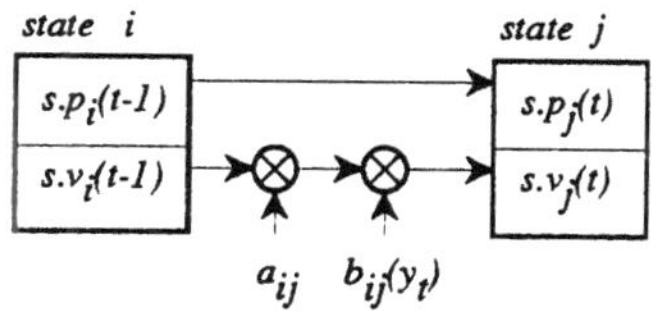

Figure 3. Viterbi Initialization Data Flow

In figure 3 we see a data flow diagram similar to the one in figure 2 accept that the **argmax** has been deleted. The **argmax** is not need by the *initialization* data flow if this is the first time that an arc with state j as its destination has been evaluated because the probability of $s.v_j(t) = 0.0$. Therefore we may immediately replace $s.v_j(t)$ with $s.v_i(t-1)a_{ij}b_{ij}(y_t)$. The savings realized from using the *initialization* data flow is computed with respect to the worst case of the *update* data flow. As a result 1 ALU compare, 1 Branch and 1 memory read are saved. We note with interest that the ratio of memory accesses to ALU operations is 3:1 for the *initialization* data flow.

In order to be able to apply the initialization transition we must know the order of evaluation of the states in the search graph. There are two ways to achieve a known ordering of the state evaluation. Apply a dynamic ordering of the states or arrange for the order of evaluation to remain static. The use of dynamic ordering is impractical because the overhead of any ordering function will overwhelm any savings achieved by applying the *initialization* transition. The algorithm given in figure 1 represents a static ordering of state evaluation over the entire state space making it trivial to apply the initialization transition. However, in most cases it is impractical to use a plain Viterbi search due to the size of search space. Before considering how to make use of the initialization optimization we need to consider the most common optimization to Viterbi search which is Viterbi beam search shown in figure 4.

In Viterbi Beam search the set of states to be evaluated is dynamically controlled by a pruning threshold, ε, which controls the size of the active state list, *ASL*, which is built from the next active state list, *NASL*. This occurs in step *8*. As result of this pruning the static ordering of the original Viterbi search is lost making it difficult to apply the *initialization* transition. Another and perhaps more important draw back of dynamic control of state evaluation is the potential loss of memory locality. If we originally organized the state space such that state i is close to its successor states $j, \forall_j a_{ij} \neq 0.0$, then any benefit that was achieved from from the locality may

1. Allocate state space S such that for state i at stage $t-1$ $s.from.v_i$ holds the Viterbi probability and $s.from.p_i$ holds the back pointer and at stage t $s.to.v_i$ holds the Viterbi probability and $s.to.p_i$ holds the back pointer.
2. Allocate two lists, *ASL* (active state list) and *NASL* (next active state list).
3. Initialize the search, $t=1$, append I to *ASL*, $s.from.v_i = \begin{cases} 1.0 & i=I \\ 0.0 & i \neq I \end{cases}$
4. Initialize stage t, $\bar{s}.to.v_i = 0.0$ and clear *NASL*.
5. **Foreach** state i in *ASL*.
    a. **Foreach** state j such that $a_{ij} \neq 0.0$
        i. **If** $s.to.v_j < s.from.v_i a_{ij} b_{ij}(y_t)$ **then**
            1. **If** $s.to.v_j = 0.0$ **then** append j to *NASL*
            2. $s.to.v_j = s.from.v_i a_{ij} b_{ij}(y_t)$
            3. $s.to.p_j = s.from.p_i$
            4. **if** $s.to.v_j > BestScore$ **then** $BestScore = s.to.v_j$
6. **If** $t=T$ **then** terminate
7. Clear *ASL*
8. **Foreach** i in *NASL*
    a. **If** $s.to.v_i > (BestScore - \varepsilon)$ **then** append i to *ASL*
9. Swap $s.from$ and $s.to$
10. $t=t+1$, goto step 4.

Figure 4. Viterbi Beam Search

now be lost since the successors of state i may be far down on the list of active states with respect to state i. As already pointed out a dynamic reordering of the active state list is not practical and a complete static ordering is ruled out by the nature of Viterbi beam search. The alternative used in the Sphinx search component is based on partial static ordering.

The Sphinx system models words as a sequence of context dependent triphones that are based on the hidden Markov model (HMM) topology depicted in figure 5. The state space for the Sphinx search therefore consists of a concatenation of the HMM's for each word.

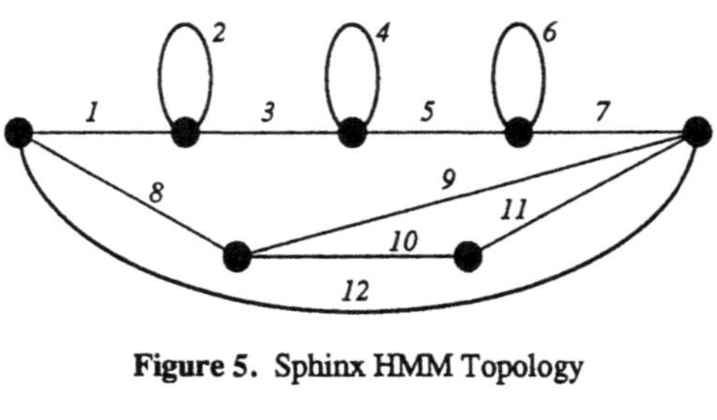

Figure 5. Sphinx HMM Topology

Since the topology of the HMM is the same for all triphones it is reasonable to consider embedding the topology so that it is implicit in some evaluation function. This is completely analogous to the way dynamic time warping implicitly encodes its graph traversal constraints on the warping space. By implicitly encoding the topology of the HMM the expense of pointer following is saved. Pointer following requires one memory read for each arc, a test and a branch to control the loop over the outgoing arcs of each state i. This still doesn't solve the order of evaluation problem since some states in a model may not be active. Our solution therefore is to require that either *all* of the states of a model be classified as active or *none* of the states be classified as active. We now have a hybrid search strategy that uses the optimal Viterbi search for the model state graph and the Viterbi beam search for the word level model graph. A comparison of Viterbi beam search and Viterbi optimal search over a the state graph for a single model follows. In this comparison we assume that all of the states in the model state graph are active.

Viterbi beam search

Best Case					
worst case update trans.	(7 mem, 3 ALU 1 Br)	x 6 =	42 mem,	18 ALU,	6 Br
best case update trans.	(4 mem, 3 ALU, 1 Br)	x 6 =	24 mem,	18 ALU,	6 Br
pointer following	(1 mem, 1 ALU, 1 Br)	x 12 =	12 mem,	12 ALU,	12 Br
Total			78 mem,	48 ALU,	24 Br

Worst Case					
worst case update trans.	(7 mem, 3 ALU, 1 Br)	x 12 =	84 mem,	36 ALU,	12 Br
pointer following	(1 mem, 1 ALU, 1 Br)	x 12 =	12 mem,	12 ALU,	12 Br
Total			96 mem,	48 ALU,	24 Br

Viterbi optimal search

Best Case					
initialization trans.	(6 mem, 2 ALU, 0 Br)	x 6 =	36 mem,	12 ALU,	0 Br
best case update trans.	(4 mem, 3 ALU, 1 Br)	x 6 =	24 mem,	18 ALU,	6 Br
compilation optimization			-15 mem		
Total			45 mem,	30 ALU,	6 Br

Worst Case					
initialization trans.	(6 mem, 2 ALU, 0 Br)	x 6 =	36 mem,	12 ALU,	0 Br
worst case update trans.	(7 mem, 3 ALU, 1 Br)	x 6 =	42 mem,	18 ALU,	6 Br
compilation optimization			-15 mem		
Total			63 mem,	30 ALU,	6 Br

For Viterbi beam search the best case is calculated based on six worst case update transitions since there are six initial transitions to each of the the six states in the HMM. For Viterbi optimal search six *initialization* transitions replace the six worst case update transitions. Finally a *compilation optimization* applies to Viterbi optimal search since all six states of the model are evaluated at the same time. The savings is achieved since $v_j(t)$ is read only once for each of six states rather than once for each of 12 transitions and $b_{ij}(y_t)$ is read only three times rather 12 since there are only three unique functions of b_{ij} mapped across the 12 arcs of the HMM. While this is a peculiarity of the Sphinx HMM it results in a significant savings.

Hybrid Viterbi Beam Search

Figure 6 gives the algorithm for the hybrid Viterbi beam search. This search reformulates Viterbi beam search by replacing the inner loop over the active states with a loop over the active models. Note how step 7.a.ii handles transitions between models. This step determines if the the final state F of model m is with in the pruning threshold. If so it copies the probability and back pointer from state F of model m to state I of all the successor models of m and adds those models to the next active model list. The overall efficiency of the search is improved because, the state expansion is now twice as efficient at the model level, the cost of state management is reduced by a factor of six since states are now managed at the model level and memory locality is preserved during state evaluation since all of the states of a model are evaluated at the same time.

1. Allocate state space S such that $s.from.v_{m_i}$ holds the Viterbi probability and $s.from.p_{m_i}$ holds the back pointer for state i in model m at stage $t-1$ and $s.to.v_{m_i}$ and $s.to.p_{m_i}$ correspond to stage t.
2. Allocate two lists, *AML* (active model list) and *NAML* (next active model list).
3. Initialize the search, $t=1$, append initial model *IM* to *AML*, perform Viterbi initialization on *IM*. (Step 2 from Figure 1).
4. **Foreach** model m in *AML*.
    a. Apply Viterbi Evaluation to m. (Step 4 in figure 1)
    b. **If** $\text{argmax}_i(s.to.v_{m_i}) > BestScore$ **then** $BestScore = \text{argmax}_i(s.to.v_{m_i})$
5. **If** $t=T$ **then** terminate
6. **Foreach** model m in *AML*
    a. **If** $\text{argmax}_i(s.to.v_{m_i} > BestScore - \varepsilon)$ **then**
        i. append m to *NAML*
        ii. **If** $s.to.v_{m_F} > BestScore - \varepsilon$ **then**
            1. **Foreach** n such that $(a_{ij} \neq 0.0) \wedge (i = m_F) \wedge (j = n_I)$
                a. Initialize model n, Append model n to *NAML*.
7. Swap *s.from* and *s.to*, Swap *AML* and *NAML*, Clear *NAML*, t=t+1, goto step 4.

Figure 6. Hybrid Viterbi Beam Search

Summary

Table 2 gives a summary of the performance for several different commercial workstations. Note that in the final column I give the relative dhyrstone[2] performance for each machine. The IBM 530 does not have the best dhyrstone performance but is almost twice as fast as the DEC 5100 on the Hybrid Viterbi search. This is because the IBM 530 has a sophisticated data cache that takes advantage of memory spatial and temporal locality in the HMM while the DEC 5100 only takes advantage of temporal locality.

Machine	Clock	CPU	OS	Time (Secs)	Relative Performance	xReal-Time	Relative Performance Dhyrstone
Sun3/60	20Mhz	MC68020	Mach	1572.72	1.0	3.01	1.0
NeXT	25Mhz	MC68030	Mach	1390.62	1.13	2.67	1.63
DEC 3100	20Mhz	MIPS	Mach	1063.12	1.47	2.04	5.60
DEC 5100	40Mhz	MIPS	Mach	700.9	2.24	1.35	9.82
IBM 530	25Mhz	Power	AIX	392.3	4.01	0.75	9.48

Table 2. Performance of hybrid Viterbi search on the February 1988 RM test set.

Bibliography

1. Lee, K.F. *Large-Vocabulary Speaker-Independent Continuous Speech Recognition: The SPHINX System.* Ph.D. Th., Computer Science Department, Carnegie Mellon University, April 1988.

2. Lowerre, B. T. *The HARPY Speech Recognition System.* Ph.D. Th., Computer Science Department, Carnegie Mellon University, April 1976.

3. Viterbi, A. J. "Error Bounds for Convolutional Codes and an Asymptotically Optimum Decoding Algorithm". *IEEE Transactions on Information Theory IT-13*, 2 (April 1967), 260-269.

[2]The dhrystone is an integer benchmark sometimes used to compare CPU performance. Note that the Hybrid Viterbi search is an integer only algorithm since it uses log probabilities.

Part 3

Connectionist Models of Speech

Neural Networks or Hidden Markov Models for Automatic Speech Recognition: Is there a Choice?

John S. Bridle

Royal Signals and Radar Est., St. Andrews Road, Great Malvern, WR14 3PS, UK

Abstract: Various algorithms based on "neural network" (NN) ideas have been proposed as alternatives to hidden Markov models (HMMs) for automatic speech recognition. We first consider the conceptual differences and relative strengths of NN and HMM approaches, then examine a recurrent computation, motivated by HMMs, that can be regarded as a new kind of neural network specially suitable for dealing with patterns with sequential structure. This "alphanet" exposes interesting relationships between NNs and discriminative training of HMMs, and suggests methods for properly integrating the training of non-linear feed-forward data transformations with the rest of an HMM-style speech recognition system. We conclude that NNs and HMMs are not distinct, so there is no simple choice of one or the other. However, there are many detailed choices to be made, and many experiments to be done.

Keywords: speech recognition, neural networks, alphanet, discriminative training

Introduction

Automatic Speech Recognition (ASR) is a major challenge to pattern recognition and linguistics. Most of the progress in the last 15 years has come from the development of methods based on Stochastic Models (*e.g.* Linear Prediction Analysis, Hidden Markov Models (HMMs) [27] and Stochastic Grammars[11]). There are still large gaps between performance we can achieve and performance we would like to achieve, and also between our knowledge about the structure of speech signals and our ability to use it in speech recognition algorithms. Because of this, any new approach to pattern processing is likely to be seized on and applied to ASR.

There has recently been a flurry of interest in approaches to ASR based on "Neural Networks" (NNs). We need to ask "What is it?", "Is it new", "Is it different?" and "Is it better". We shall then be in a position to decide whether we need to abandon current

NATO ASI Series, Vol. F 75
Speech Recognition and Understanding. Recent Advances
Edited by P. Laface and R. De Mori

stochastic model based methods. (We ignore the cynical question "are there more backers if we call it neural networks?"!)

Neural Network (NN) based techniques have a particular appeal for ASR, because we know that speech is decoded successfully by the brains of people, and brains are made of neural networks. Unfortunately, although we do know the location of some parts of the brain that seem to be specialised for speech and language processing, we know so little about their structure, let alone the algorithms used, that so-called Neural Network approaches to ASR are hardly ever presented as serious models for brain organisation. I conclude that we may get some inspiration from neurobiology, but NN approaches to ASR should be judged by engineering criteria such as "does it do the job?".

The main attractions of NN approaches to ASR seem to be the inherently discriminative nature of the training methods, the possibility of more general non-linear structures than tractable stochastic model based methods offer, and a more intuitively accessible formalism.

Hidden Markov Models and Neural Networks

This section summarises some of the main differences between stochastic model based methods (particularly HMMs) and neural network methods (particularly recurrent networks).

There are many interesting differences between Sm and NN approaches to ASR. Here we consider the form of the assumptions which we implicitly make when we decide that the approach is applicable, methods for learning and recognition, the status of the theory, and methods of handling uncertainty and time.

In the Stochastic Model paradigm we start with assumptions about the form of the *source* of data: that it is a stochastic system of a certain form, such as a probabilistic function of an underlying Markov chain. The form of the *recognition process* is then derived as being appropriate given the source assumptions (using Bayesian inference for instance). Note that we have to restrict our assumptions about the source to those for which we can derive an appropriate recognition process. The parameters of the recognition process are usually derived directly from parameters of the source, and the latter are estimated (learned) from the training data. (Note that it is not necessary to believe that the assumptions about the source are *true* — only that the methods they lead to might work well enough!)

In contrast, in the Neural Network approach we start with assumptions about the appropriate form of the *recognition process* (a cascade of non-linear functions of weighted sums for instance). The learning process adjusts the parameters of the recognition process (the values of the weights for instance).

It is not clear whether, for any particular application, it is easier to produce useful assumptions about the form of the source or the form of the recogniser. Possible sources of ideas for NN recognisers include neurobiology, psychology and approximation theory. In the present paper we suggest using the form of the recognition processes associated with stochastic models.

There is a significant difference in the learning (parameter estimation) methods available for SMs and NNs. Most NN-style methods are based on derivatives of a gobal criterion, typically squared error between outputs and desired outputs for a training set. In contrast, the widespread Baum-Welch re-estimation (E-M) procedure used for HMMs is much more efficient because it can take large steps and guarantee an improvement in the (observation likelihood) criterion being used. Unfortunately the B-W procedure is not applicable to arbitrary criteria such as we may wish to use for encouraging discrimination between classes, and we are driven back to gradient-based [2] and heuristic [4] methods (but see [17].

The recognition algorithms actually used for mainstream ASR solve a search for the best explanation of the data in terms of the model, and are based on Dynamic Programming. They are quite efficient, and in many cases can be shown to produce the optimal (most likely) explanation. They are also compatible with some forms of hardware paralellism.

The form of the recognition algorithm for NNs is already assumed. Typically it is in a form which invites a massively parallel implementation (although this is rarely needed for speech processing).

It is important to understand the distinction between the model, which is a statement about the assumed source of the data, and the recognition calculation, which corresponds to a neural network.

Recurrent "Neural" Networks

Perhaps the ideal form of "neural" network for speech recognition would accept input vectors as they become available, and have some form of internal state which contains all the information about past inputs which is needed to deal with current and future inputs. The internal state would be a function of the current input and the previous internal state. Various "recurrent networks" have been proposed and tried. Usually they have been derived by adding feedback connections to a feedforward network, often a network with one layer of hidden units. Among arrangements that have been tried are

- local connection around each input unit or the hidden units or the output units (making these units into non-linear first-order low-pass filters),
- full recurrent connections around a single layer ,
- feedback from one layer to a previous layer .

In most cases it is not clear what these systems can compute. The general idea is that the network should accumulate information about the input pattern, by combining its internal state with the incoming data at each time. Robinson [28, 14] builds on this notion (of transforming the current state plus input into new state plus output) by using a simple loop of semi-linear logistic units whose inputs are the current speech vector plus the outputs from last frame, and whose outputs are the outputs of the system (one per phoneme in his system) and the 'state' outputs which are fed back to the input of the loop. (We must not confuse the notion of the state of such a network with the notion of the state of an HMM, which is a generator, not an analyser.)

One of the simplest most successful types of recurrent network for ASR has the simplest recurrent structure. The so-called "time-delay neural network" [32, 22] has one local recurrent loop around a linear unit for each class. This loop integrates the evidence for and against the class, which comes from quite elaborate feed-forward structures incorporating shift-registers. At the other extreme, Watrous [33, 34] has constructed recurrent networks with quite complicated structure, based on insight into the kinds of feature detectors which may be appropriate for speech recognition.

Training of recurrent networks can be done using the Backpropagation method for partial derivatives, but the propagation is backwards through time as well as down towards the inputs [29]. This needs the storage of intermediate results on the forward pass, is inconvenient in computer programs for long inputs, and is perhaps rather contrary to the spirit of neural networks. Some people claim to manage by back-propagating only to the previous frame (or few frames).. Kuhn [20, 21] and Zipser have shown that it is possible to compute the correct derivatives without backpropagation through time, at the expense of a large extra set of partial derivatives propagated forwards.

In this paper we concentrate on the backpropagation-through-time method, and explore its relationship with the Backward Pass used in HMM calculations.

Viterbi/Network Hybrids

The main difficulty with applying NNs to ASR is dealing with the sequential nature of speech patterns, and their forté is discrimination between classes based on static patterns. Several workers have grafted together the dynamic programming algorithm for time-alignment ("dynamic time warping" or Viterbi alignment) with MLPs (usually looking at a few frames at a time) [23, 15, 25] [31, 19, 18]. The usual training scheme is to alternate time alignment and network training. Lippmann [24] has pointed out that the scoring for the Viterbi algorithm can be implemented in a special network, but it is not clear how to train such a network.

A Missing Link: Alphanets

The hybrid approach may be adequate, but it is not very satisfying. There are also worries about what is being optimised in the training process. Some neural networkers would be happier with a unified recurrent network of appropriate form for the job, for which we could compute partial derivatives of a plausible training criterion. My solution (see also [26, 36] [7, 8, 6]) is to start with a computation (closely related to dynamic programming) which is used in training HMMs: the full likelihood calculation (forward pass or alpha pass of the forward-backward algorithm).

For a given HMM and a given observation sequence

$$\boldsymbol{y}_1^T \triangleq \boldsymbol{y}_1, \boldsymbol{y}_2, \dots \boldsymbol{y}_T,$$

we compute the likelihood of the model generating all the observations $P(\boldsymbol{Y}_1^T = \boldsymbol{y}_1^T \mid \text{Model})$ via the joint partial sequence likelihoods

$$\alpha_{jt} \triangleq P(\boldsymbol{Y}_1^t = \boldsymbol{y}_1^t \text{ and State}_t = j \mid \text{Model})$$

using the forward recurrence

$$\alpha_{jt} = b_j(\boldsymbol{y}_t) \sum_i a_{ij} \alpha_{i,t-1}. \tag{1}$$

Where $b_j(.)$ is the observation likelihood function for the jth state and a_{ij} is the transition probability from state i to state j.

If there is a distinguished finishing state, N, then the likelihood of the whole sequence is

$$P(\boldsymbol{Y}_1^T = \boldsymbol{y}_1^T \mid \text{Model}) = \alpha_{NT}.$$

If we have a *set* of models, perhaps one per word, we can compute the (posterior) probabilities of the models:

$$Q_w \triangleq P(W = w \mid \boldsymbol{Y}_1^T) = \frac{P(\boldsymbol{Y}_1^T \mid W = w) P(W = w)}{\sum_v P(\boldsymbol{Y}_1^T \mid W = v) P(W = v)}.$$

(Q_w is the probability that the data was generated by the wth model, assuming it was generated by one of them.)

If we assume all words are equally probable *a-priori*, and label the final states of the word models N_w, we can write

$$Q_w = \frac{\alpha_{N_w T}}{\sum_v \alpha_{N_v T}} \triangleq \frac{L_w}{\sum_v L_v}. \tag{2}$$

These Qs have the properties we are looking for in the outputs of a 'neural network' classifier: they are between 0 and 1, and the most probable word has the highest output. They are computed by a a simple recurrent relationship (equation 1) which we can think of as a (single) recurrent network whose internal state at any time t is the vector of real numbers $\alpha_{1t}, \dots, \alpha_{N_W}$ across the possible states of the assumed generator HMMs. There is a feedforward network which computes the b_js from the input at each time. The recurrent loop at the heart of an alphanet is unusual for a 'neural' network, because it contains no squashing non-linearities (except for an optional normalisation[1]) but it does include the product of two signals. At the output of the network we have a normalisation stage, with an optional set of weights for the priors.

What happens if we now forget the origins of this form of this whole network (I call it an "Alphanet") and attempt to optimise its outputs? The partial derivatives of output errors with respect to the parameters can be computed by back-propagation through time

[1]The normalisation can be done at the output, as above, or every timestep. The normalised αs are then probabilities of states given data so far [36].

(via $\frac{\partial E}{\partial \alpha_{jt}}$s). We can compute partial derivatives of our favourite error criterion wrt the parameters of the alphanet, and use any gradient-based optimisation method to improve the criterion. The parameters of the alphanet include the usual parameters of HMMs (*e.g.* state transition probabilities and observation probabilities) but also parameters of any feedforward networks we choose to incorporate into the observation density functions $b_j(\boldsymbol{y}_t)$. We could constrain the parammters of the network so that we retain the luxury of an interpretation in terms of HMMs, but as Niles has pointed out [26] we may prefer to let them fall as they may.

If we choose *my* favourite error criterion (and constrain to compatibility with an HMM interpretation) we obtain some interesting relationships with the Baum-Welch re-estimation method, with discrimative training methods which have previously ben used for sets of HMMs, and with Boltzmann machines. In the rest of this section we present some of those relationships. For a fuller treatment see [9, 12].

Consider a general additive error criterion. Since the error is additive across the training set, so are the derivatives, and we can concentrate on derivatives for a single utterance. training pattern:

$$E^{tot}(\boldsymbol{Q},\boldsymbol{T}) = \sum_w E(Q_w, T_w)$$

where the target vector $\boldsymbol{T}$ is normally a one-from-N indicator of identity ($T_w = \delta_{wc}$ where c is the correct class). Let us start by back-propagating through the error function and the normalisation:

$$\frac{\partial E^{tot}}{\partial L_w} = \sum_v \frac{\partial E(Q_v, T_v)}{\partial Q_v} \frac{\partial Q_v}{\partial L_w}. \tag{3}$$

The form of first term in the sum depends on the particular error function. From equation 2 the second term in equation 3 is

$$\frac{\partial Q_v}{\partial L_w} = \frac{(\delta_{vw} - Q_v)}{L_w}. \tag{4}$$

For the backprop through time we find from equations 1 and 2 that

$$\frac{\partial L_w}{\partial \alpha_{i,t-1}} = \sum_j a_{ij} b_j(\boldsymbol{y}_t) \frac{\partial L_w}{\partial \alpha_{j,t}},$$

which is the same as the expression for the computation of the backward probabilities in the forward-backward algorithm:

$$\beta_{i,t-1} \triangleq P(\boldsymbol{Y}_t^T = \boldsymbol{y}_t^T \mid \text{State}_{t-1} = i) = \sum_j a_{ij} b_j(\boldsymbol{y}_t) \beta_{j,t}.$$

So the main backward-propagation in an alphanet is the same computation as the backwards pass of the Baum-Welch procedure.

The usual error measure for back-prop networks is

$$E_{sq}^{tot}(\boldsymbol{Q},\boldsymbol{T}) = \sum_w (Q_w - T_w)^2,$$

but a more appropriate measure for comparing two probability distributions (such as $\{Q_w\}$ and $\{T_w\}$ purport to be) is the relative entropy [1, 16, 30, 5, 35][2]. This is also equivalent [10] to the so-called Mutual Information measure used in [2, 13] to train sets of HMMs discriminatively. The relative entropy measure reduces to

$$E_{re}^{tot} = \sum_w -T_w \log Q_w,$$

or if we know the true identity, c of each training pattern, so $T_w = \delta_{wc}$

$$E_{re}^{tot}(\boldsymbol{Q}, c) = -\log Q_c = -\log L_c + \log \sum_v L_v. \tag{5}$$

In this case

$$\frac{\partial E_{re}^{tot}}{\partial Q_w} = \frac{-\delta_{wc}}{Q_w} = \frac{-\delta_{wc}}{Q_c}$$

If the word nets are separate, we can express the derivatives with respect to the parameters of underlying hidden Markov models in terms of the Baum-Welch re-estimates [9].

For example, in the case of simple Gaussian output distributions with unit covariance matrix, parameterised only by the mean $\boldsymbol{m}_j$ for each state, and using the entropy score, we find

$$\frac{\partial E_{re}^{tot}}{\partial \boldsymbol{m}_j} \propto (Q_w - T_w)\bar{\gamma}_j^c(\widehat{\boldsymbol{m}}_j - \boldsymbol{m}_j),$$

where w is the word model that state j is part of (we assume separate models here), $\bar{\gamma}_j^c$ is the posterior probability of state j ($\bar{\gamma}_j^c = \sum_t \bar{\gamma}_{jt}^c$, where $\bar{\gamma}_{jt}^c = \frac{\alpha_{jt}\beta_{jt}}{\sum_k \alpha_{kt}\beta_{kt}}$) and $\widehat{\boldsymbol{m}}_j$ is the Baum-Welch re-estimate for $\boldsymbol{m}_j$,

$$\widehat{\boldsymbol{m}}_j = \frac{\sum_t \bar{\gamma}_{jt}^c \boldsymbol{y}_t}{\sum_t \bar{\gamma}_{jt}^c}.$$

The parameters of the true model always move towards the Baum-Welch re-estimates, but whereas in the Baum-Welch Iteration the parameters of the other models are untouched, in the present method the parameters of all the wrong models move *away* from their Baum-Welch re-estimates for that utterance [3]. We can see that this is of the same general form as the Corrective Training method [4].

In the case of training on example data which needs to be explained in terms of sequences of sub-units (for example phrases in terms of words, or words in terms of phonetically-motivated building blocks) the derivatives can be expressed [12] in terms of differences between $\bar{\gamma}$s for two syntaxes, which we call 'clamped' and 'free' (following Boltzmann Machine terminology [1]). The subsripts w, v now each refer to a *sequence* of words.

The free syntax (producing $\sum_w L_w$ and $\bar{\gamma}^f$s) is the one we would use for recognition, allowing any valid sequence of sub-units. The clamped syntax (producing L_c and $\bar{\gamma}^c$s) imposes on the free syntax the extra information we have about the training data.

[2] Also known as discrimination information, cross entropy, asymmetric divergence, directed divergence, I-divergence, and Kullback-Leibler number.

For any parameter θ, the derivatives can be expressed as

$$\frac{\partial E_{re}^{tot}}{\partial \theta} = \frac{\partial \log Q_c}{\partial \theta} = \frac{\partial \log L_c}{\partial \theta} - \frac{\partial \log \sum_w L_w}{\partial \theta}.$$

For example, the error derivatives for the simple Gaussian output distributions are now

$$\frac{\partial E_{re}^{tot}}{\partial \boldsymbol{m}_j} \propto \Delta_j(\widehat{\boldsymbol{m}}_j^{\Delta} - \boldsymbol{m}_j),$$

where we have introduced

$$\Delta_{jt} \triangleq (\bar{\gamma}_{jt}^f - \bar{\gamma}_{jt}^c), \tag{6}$$

$$\Delta_j \triangleq \sum_t \Delta_{jt}, \tag{7}$$

$$\text{and} \quad \widehat{\boldsymbol{m}}_j^{\Delta} \triangleq \frac{\sum_t \boldsymbol{y}_t \Delta_{jt}}{\Delta_j}. \tag{8}$$

Note that $\widehat{\boldsymbol{m}}_j^{\Delta}$ is similar in form to the Baum-Welch re-estimate for $\boldsymbol{m}_j$ but using Δs instead of γs.

TDNNs and Alphanets

So-called Time-Delay Neural Networks [32, 22] are apparently one of the most successful NN approaches to ASR [32] . We can think of TDNNs as at a different extreme from conventional HMM systems: Whereas conventional (continuous single density per state) recognisers have simple ways of computing $b_j(\boldsymbol{y}_t)$, (based on one or a few frames) and a relatively complicated method of integrating evidence across time (making explicit use of the sequential order of states), a TDNN (of the Waibel/Lang type) puts almost all the effort into the transformation from several frames of observations to an indicator of membership of the class (phoneme or word) which is then integrated in a very simple way. (A TDNN looks very like an alphanet with a complicated shared network to produce the b_js and just one state per class!)

Local score functions which look at several frames of observations can be sensitive to *dynamics* of speech patterns. Almost every modern high-performance HMM-based ASR system uses observation likelihood functions which are sensitive to more than one frame. State-dependent temporal integration (as used in HMM-based ASR) deals explicitly with an underlying *sequential* structure.

Dynamics and sequence are both important in the speech code. There is no clear dividing line, and mechanisms for handling one can help with the other, but we should be aware of the differences.

We can look forward to future ASR systems that will have complex non-linear pre-processor stages looking at dynamic aspects of the acoustic pattern over say 50ms, coupled to HMM-DP style processing which accumulates evidence for theories about sequences. One of the open questions is the form of the intermediate representation: the network could produce scores for sound theories (b_{jt}) directly, or it could produce the equivalent of the output of a (soft) vector quantiser, or it could produce a vector to be used much like a conventional mfcc vector into a continuous density output distribution HMM. Back-propagation of partial derivatives can be done through all these systems.

Concluding Remarks

Here are my current opinions on the relationship between NN and HMM approaches to ASR

- The term "neural networks" includes many techniques which might be useful for ASR.
- Stochastic model approaches are not limited to what has been done yet.
- The main difference between conventional (within-class) methods of training HMM-based speech recognition systems and most classifier network methods is in the training criteria, but the relative entropy score (*a la* Boltzmann machine), which is a natural score for networks, is equivalent to MMIE (one of the best-known discriminative training methods for sets of HMMs).
- SM and NN *approaches* may be distinct, but the resulting algorithms need not be: there are speech recognition algorithms that can be understood in terms of recurrent NNs or in terms of HMMs.
- By treating HMM algorithms as networks we obtain several interesting results:
  - another way of looking at HMM recogniser calculations
  - a recurrent network structure that has attractions for dealing with sequential structure
  - A way of understanding TDNNs as one-state-per-class HMMs but with very elaborate output distributions.
  - HMM generalisations which may be useful.
  - discrimination-based training methods, some of which have already been applied to HMMs.
  - another way of looking at Maximum Mutual Iinformation Estimation and Corrective Training for HMMs
  - the opportunity to train non-linear preprocessor stages so that the performance of a complete speech recognition system is improved.

There are many exciting possibilities for speech algorithm research, with many choices to be made, but it is not a matter of choosing between HMMs and NNs, because they are not distinct: we CAN have the best of both!

References

[1] D H Ackley, G E Hinton, and T J Sejnowski. A learning algorithm for Boltzmann machines. *Cognitive Science*, 9:147–168, 1985.

[2] L R Bahl, P F Brown, P V de Souza, and R L Mercer. Maximum mutual information estimation of hidden Markov model parameters. In *Proc. IEEE Int. Conf. Acoustics Speech and Signal Processing*, pages 49–52, 1986.

[3] L R Bahl, P F Brown, P V de Souza, and R L Mercer. Speech recognition with continuous-parameter hidden Markov models. *Computer Speech and Language*, 2(3/4):219–234, December 1987.

[4] L R Bahl, P F Brown, P V de Souza, and R L Mercer. A new algorithm for the estimation of HMM parameters. In *Proc. IEEE Int. Conf. Acoustics Speech and Signal Processing*, pages 493–496, 1988.

[5] E B Baum and F Wilczek. Supervised learning of probability distributions by neural networks. In D Anderson, editor, *Neural Information Processing Systems*, pages 52–61. Am. Inst. of Physics, 1988.

[6] H Bourlard and C J Wellekens. Multilayer perceptrons and automatic speech recognition. In *Proc. IEEE First Intl. Conf. Neural Networks*, 1987.

[7] H Bourlard and C J Wellekens. Links between Markov models and multilayer perceptrons. In D S Touretzky, editor, *Advances in Neural Information Processing Systems 1*, pages 502–510. Morgan Kaufmann, 1989.

[8] H Bourlard and C J Wellekens. Speech pattern discrimination and multilayer perceptrons. *Computer, Speech and Language*, 1989.

[9] J S Bridle. Alphanets: a recurrent 'neural' network architecture with a hidden Markov model interpretation. *Speech Communication*, Special "Neurospeech" issue, February 1990.

[10] J S Bridle. Training stochastic model recognition algorithms as networks can lead to maximum mutual information estimation of parameters. In *Advances in Neural Information Processing Systems 2*. Morgan Kaufmann, 1990.

[11] J S Bridle and L Dodd. Formal grammars and Markov models. Technical Memorandum 4051, Royal Signals and Radar Establishment, October 1987.

[12] J S Bridle and L Dodd. An Alphanet approach to optimising input transformations for HMM-based continuous speech recognition. In *Proc IEEE Int Conf Acoustics Speech and Signal Processing*, Toronto, April 1991. (Submitted).

[13] P F Brown. The acoustic-modeling problem in automatic speech recognition. Technical Report RC 12750, IBM TJW Research Center, 1987.

[14] F Fallside, H Lucke, T P Marsland, et al. Continuous speech recognition for the TIMIT database using neural networks. In *Proc. IEEE Int. Conf. Acoustics Speech and Signal Processing*, Alberquerque, 1990.

[15] M Franzini, K-F Lee, and A Waibel. Connectionist viterbi training: A new hybrid method for continuous speech recognition. In *Proc. IEEE Int. Conf. Acoustics Speech and Signal Processing*, pages S8.4: 425–428, Alberquerque, 1990.

[16] L Gillick. Probability scores for backpropagation networks. Personal communication, July 1987.

[17] P S Gopalakrishnan, D Kanevsky, A Nadas, and D Nahamoo. A generalisation of the Baum algorithm to rational objective functions. In *Proc.ICASSP89*, pages 631–634, 1989.

[18] Y Hirata and S Nakagawa. Comparison among time-delay neural networks, LVQ2, discrete observation HMM and continuous observation HMM. In *Proc. IEEE Int. Conf. Acoustics Speech and Signal Processing*, Alberquerque, 1990.

[19] Ken ichi Iso and Kakao Watanebe. Speaker -independent word recognition using a neural prediction model. In *Proc. IEEE Int. Conf. Acoustics Speech and Signal Processing*, pages S8.8: 441–444, Alberquerque, 1990.

[20] G M Kuhn. A first look at phonetic discrimination using a connectionist network with recurrent links. SCIMP Working Paper 4/87, Institute for Defense Analysis, Communications Research Division, 1987.

[21] G M Kuhn, R L Watrous, and B Ladendorf. Connected recognition with a recurrent network. *Speech Communication*, 1990. Special "Neurospeech" issue.

[22] K Lang and G E Hinton. The development of tdnn architecture for speech recognition. Technical Report CMU-CS-88-152, Carnegie-Mellon University, 1988.

[23] E Levin. Word recognition using hidden control neural architecture. In *Proc. IEEE Int. Conf. Acoustics Speech and Signal Processing*, pages S8.6: 433–436, Alberquerque, April 1990.

[24] R.P. Lippmann and B. Gold. Neural classifiers useful for speech recognition. In *IEEE First International Conference on Neural Networks*, pages IV–417, June 1987.

[25] N Morgan and H Bourlard. Continuous speech recognition using multilayer perceptrons with hidden Markov models. In *Proc. IEEE Int. Conf. Acoustics Speech and Signal Processing*, pages S8.1: 413–416, Alberquerque, 1990.

[26] L Niles and H Silverman. Combining hidden Markov model and neural network classifiers. In *Proc. IEEE Int. Conf. Acoustics Speech and Signal Processing*, pages S8.2: 417–420, Alberquerque, 1990.

[27] A B Poritz. Hidden Markov models: a guided tour. In *Proc. IEEE Int. Conf. Acoustics Speech and Signal Processing*, pages 7–13, 1988.

[28] A J Robinson and F Fallside. A dynamic connectionist model for phoneme recognition. In *Proc. nEuro'88*, Paris, 1988.

[29] D E Rumelhart, G E Hinton, and R J Williams. Learning Internal Representations by Error Propagation. In D E Rumelhart and J L McClelland, editors, *Parallel Distributed Processing: Explorations in the Microstructure of Cognition. Vol. 1: Foundations.* MIT Press, 1986.

[30] S Solla, E Levin, and M Fleisher. Accelerated learning in layered neural networks. *Complex Systems*, January 1989.

[31] J Tebelskis and A Waibel. Large vocabulary recognition using linked predictive neural networks. In *Proc. IEEE Int. Conf. Acoustics Speech and Signal Processing*, pages S8.7: 437–440, Alberquerque, 1990.

[32] A Waibel, T Hanazawa, G Hinton, K Shikano, and K Lang. Phoneme recognition: neural networks vs. hidden Markov models. In *Proc. IEEE ICASSP88*, volume 1, pages 107–110, 1988.

[33] R L Watrous. Connectionist speech recognition using the temporal flow model. In *Proc. IEEE Workshop on Speech Recognition*, June 1988.

[34] R L Watrous, B Ladendorf, and G Kuhn. Complete gradient optimisation of a recurrent network applied to /b/, /d/, /g/ discrimination. *J. Acoust. Soc. Amer.*, 1989. to appear.

[35] E Yair and A Gersho. The Boltzmann Perceptron Network: a soft classifier. In D Touretzky, editor, *Advances in Neural Information Processing Systems 1*. San Mateo, CA: Morgan Kaufmann, 1989.

[36] S J Young. Competitive training in hidden Markov models. In *Proc. IEEE Int. Conf. Acoustics Speech and Signal Processing*, Alberquerque, 1990.

Neural Networks for Continuous Speech Recognition

Frank Fallside

Cambridge University Engineering Department, Cambridge CB2 1PZ, UK.

Abstract: The paper reviews a number of methods for continuous speech recognition, concentrating mostly on work at Cambridge University. The methods reviewed are a 'sound' and phoneme recogniser using duration sensitive nets; the modified Kanerva model for phoneme recognition; a recurrent net for phoneme recognition; a Classification and Regressive Tree (CART) for phoneme recognition; together with methods for lexical access including the NET-gram, the modified Kanerva model, and the 'Compositional Representation' approach

Keywords: speech recognition, neural networks, phoneme recognition, lexical access.

Introduction

Very significant advances have recently been made in continuous speech recognition using Hidden Markov Model (HMM) technology, for example [5],[10]. These have followed a path of developing the basic HMM classifier to its present stage of accurate sub-word, phone, recognition followed by word recognition and linguistic constraints on words. Integral with this development has been the use of increasingly large training and testing databases such as the US DARPA databases TIMIT [7] and the Fleet Management Database [16]. As a result of this error rates of less than 5% are now achievable for multiple speaker, medium vocabulary, limited task domain word recognition in continuous speech.

Artificial neural networks constitute an alternative approach to the same problem, although interestingly it is increasingly being shown that there are close links between the two techniques and they may ultimately converge [2],[21]. Research into the problem of continuous speech recognition using neural networks is much more recent than that using HMMs and so the extent and number of results are much smaller at this time.

However the same path is being followed, of developing and improving network structures, of a two part approach of sub-word unit recognition followed by word recognition and linguistic constraints. Also increasingly large databases are being employed to extend the vocabulary and speaker range. This paper reviews some of the methods being used in this development and demonstrates that neural network technology is beginning to show the same potential as HMMs did earlier and in some cases providing the same performance.

NATO ASI Series, Vol. F 75
Speech Recognition and Understanding. Recent Advances
Edited by P. Laface and R. De Mori

The literature is expanding at the same time. In this paper no attempt is made to give a comprehensive coverage of the new methodology but rather attention is mostly concentrated on work from the author's research group.

The paper is in broadly two parts - firstly sub-word unit recognition and secondly word recognition with a few examples of combining the two.

Phoneme Recognition using a Static MLP & Durational Sensitive Net

This work [8] arose as a development of using a static multilayer perceptron (MLP) for vowel recognition. It has several individual nets but divides broadly into two parts - a 'sound' recogniser followed by a phoneme recogniser - as shown in Fig.1. Broadly the first part uses MLP to recognise 'sound' sub-units such as steady state phonemes or the separate quasi-stationary parts of phonemes such as in a dipthong, and the second part applies durational constraints to the 'sound' string to establish the phonemes.

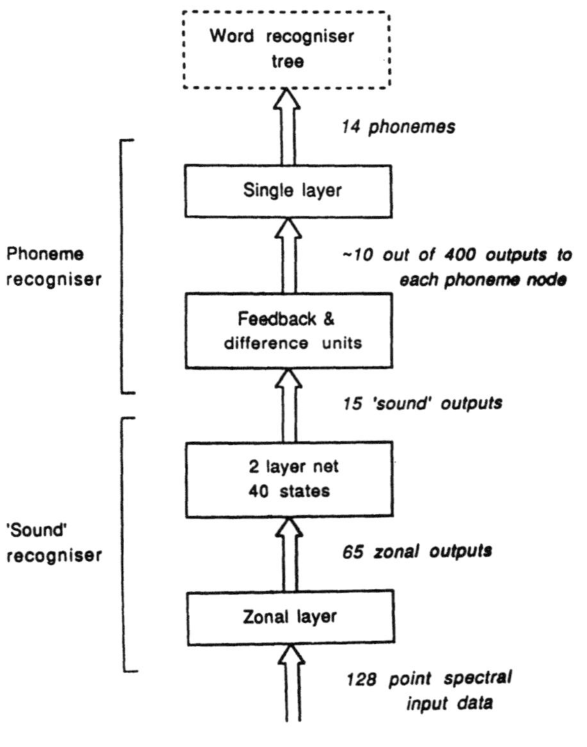

Fig. 1 Phoneme recognition using static MLP & duration sensitive nets

Phoneme	Example
4 stop consonants	
p	Pet
t	Tip
b	Bet
d	Dip
2 fricative consonants	
s	Set
f	Fit
5 pure vowels	
i	bIt
e	bEt
&	bAt
3:	bIRd
@	sIR*
3 diphthongs	
i@	bEER
e@	dARE
ai	bUY

* (unstressed)

Table 1 The reduced phoneme set

Sound	Phoneme(s)	Comments
3 stop consonants (initial)		
p	**p**	syllable
t	**t**	initial
	d	position
b	**b**	only
4 stop consonants (final)		
P	**p**	syllable
T	**t**	final position
B	**b**	only
D	**d**	(pre-burst)
2 fricatives		
s	**s**	
f	**f**	
5 vowels		
i	**i**	whole vowel
	i@	start only
	ai	end only
e	**e**	whole vowel
	e@	start only
&	**&**	whole vowel
@	**3:**	whole vowel (long)
	@	whole vowel (short)
	i@	end only
	e@	end only
a	**ai**	start only
1 silence		
–		speech absent

Table 2 The sound set

The phoneme sub-set investigated, shown in Table 1, consists of 14 phonemes representative of stop consonants, fricative consonants, pure vowels and dipthongs.

The acoustic preprocessing employs sampling at 10kHz, 10ms non-overlapping windows, Hamming windowing, followed by a 128 point FFT. The power spectrum is further processed, using logarithmic compression, normalisation of the energy over a phase or sentence and high frequency emphasis. The preprocessing is given by:

$$b_i = 20(1+0.004i)(\ln(1+a_i/n) - \ln(E/100)) \quad (1)$$

where b_i is the ith output from the preprocessor, a_i is the ith component of the FFT, n is a number related to the noise present and E is the average energy per phrase.

The 'sound' sub-units used are shown in Table 2. These are seen to be closely related to the phoneme set, for example the phoneme /s/ consists of a series of frames all classified as the sound {s}, whereas a dipthong /i@/ begins with a succession of {i} frames followed by a succession of {@} frames with an intermediate transition tapering from one to the other.

The MLP for sound recognition is shown in Fig.2. This was developed from work on vowel recognition which compared performance for an increasing number of layers. It was found that the inclusion of an initial layer of so-called 'zonal' units significantly improved performance. The zonal units are characterised by being connected to a small number of adjacent input nodes rather than to all input nodes, they are somewhat analogous to critical band processing. In the model 65 zonal units were each connected to groups of 8 input nodes, apart from the edges. The number of hidden nodes was found by experiment and the 15 output nodes correspond to the 15 sounds to be recognised.

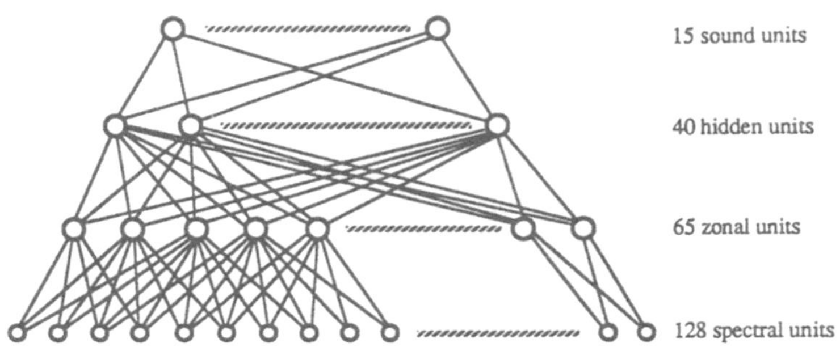

Fig.2 The 'sound' recognition network

The sound classifier was trained using a corpus of hand-labelled continuous speech using words and pseudo-words containing only the phonemes in the reduced set. For single sounds one output target was set to 0.9 and the rest to 0.1. For frames labelled as a combination of sounds the corresponding two units had values summing to 1 and in the case of dipthongs linear transitions were assumed. The zonal layer was pretrained auto-associatively by doubling it up as shown in Fig.3 and then retaining only one layer. The zonal units were trained this way on 25,000 frames with initially randomised weights. It was then embedded in the complete network which was then trained with 500,000 frames.

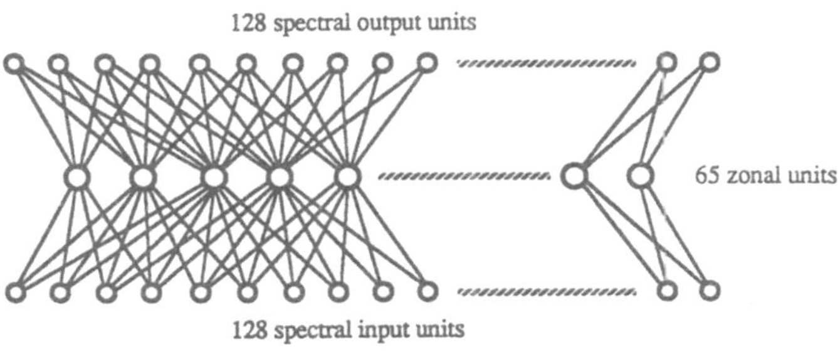

Fig.3 Auto-associative training of the zonal units

Typical test results are shown in Fig.4 for the pseudo-word tib. For clarity only node activations above 15% are shown and as can be seen a 'sound' lattice is being produced together with insertions. The task of the second stage of the phoneme recogniser is to apply durational constraints to prune this lattice.

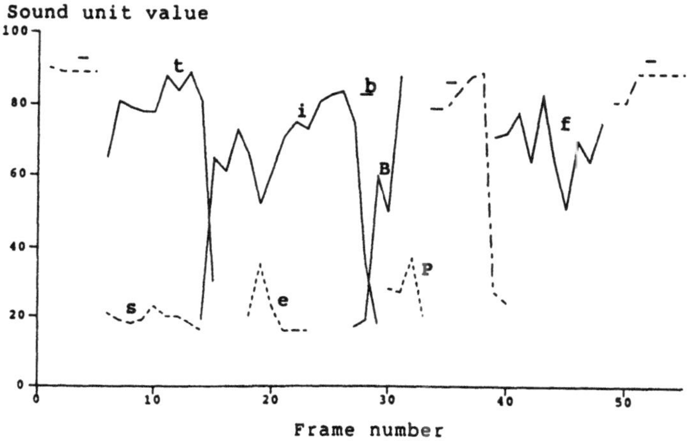

Fig.4 Recognition of the pseudo-word tib

The first component of the phoneme recogniser stage is a combination of 'feedback' and 'difference' units as shown in Fig.5. Each feedback unit has a forward path with a gain $(1-a_i)$ and a unit delay feedback path with a gain a_i. It can be shown that if $a_1=0.25$ and $a_i=a_{i+1}^2$ then the structure shown in the figure has the property that when a unit step at $t=0$ is applied as $f(t)$ the outputs of the network are as shown in Fig.7 with peak values occurring at times 1, 2, 4, 6, 8, 16. This forms the basis of a durational sensitive network with non-linearities added to enhance its performance. A typical result for its behaviour with a noisy pulse of 32 frames frames duration is shown in Fig.7 with a dominant output from the seventh output. Such networks allow the classification of input durations of 1 to 32 frames or 10-320ms which is well suited to the range of phoneme durations.

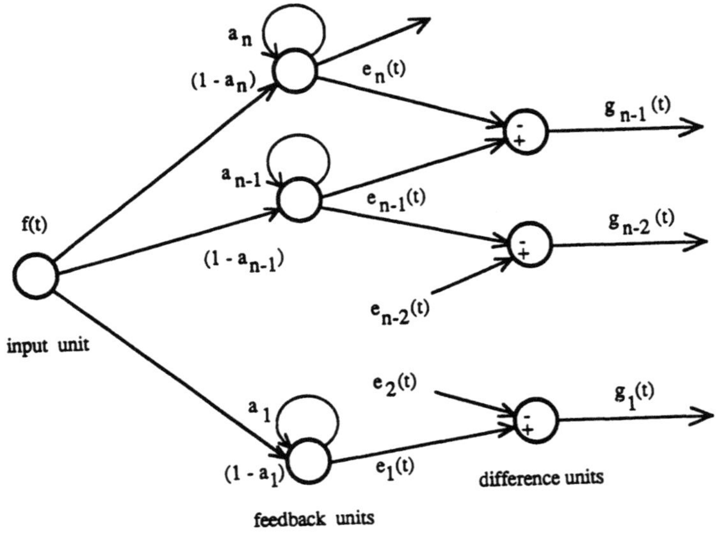

Fig.5 Difference units

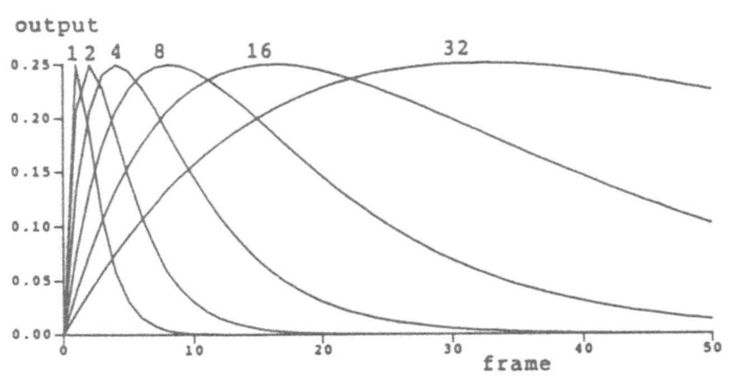

Fig.6 Response of difference units to step function

The final phoneme recogniser system has 15 such units each with its input f(t) connected to one of the sound recogniser outputs. Each unit has 27 outputs (the seven outputs for delays of 1-32 being augmented by other signals, e.g. end of pulse) resulting in a total of 405 outputs for the stage.

Finally the 405 outputs are taken to a single layer perceptron with 14 outputs, one for each phoneme to be recognised. In principle the weights can be trained by back propagation but in practice it is simpler to choose the weights, many of which are zero, by inspection of the sound sequences and their duration corresponding to a particular phoneme.

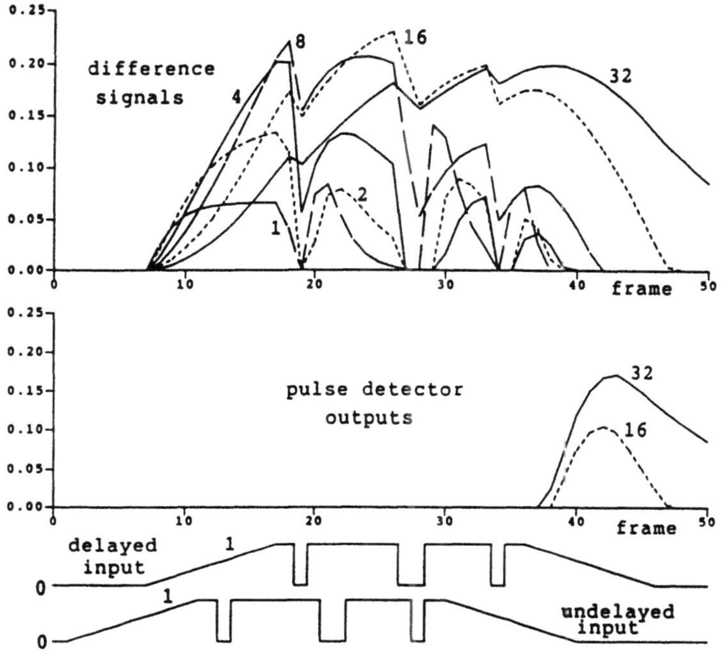

Fig.7 Response for long noisy pulse with skirts

An example of the system output for the pseudo-word tib, showing how the sound lattice is pruned is shown in Fig.8.

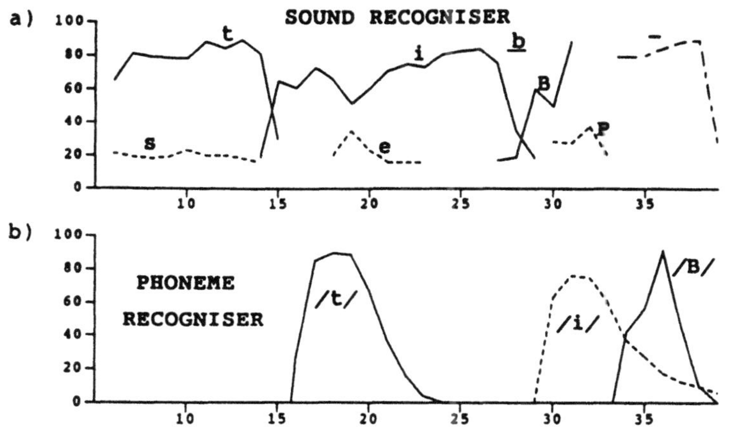

Fig.8 Phoneme detection in the pseudo-word tib

Finally results are shown in Table 3 for the phoneme recognition performance of the system for a single speaker. Test results employed two types of stimuli C_1VC_2 and C_1D, with D a dipthong, which using all possible phonemes makes a total of 90 pseudo-words. A number of versions were used giving a total of 162 words.

	test data		training data	
phoneme	correct(%)	sample	correct(%)	sample
initial consonants C_1				
p	100	15	100	27
t	87	15	70	27
b	47	15	96	27
d	100	15	78	27
s	100	15	100	27
f	87	15	100	27
all	**87**	**90**	**91**	**162**
final consonants C_2				
P	67	18	97	36
T	67	18	81	36
B	94	18	100	36
D	78	18	89	36
all	**76**	**72**	**92**	**144**
vowels V,D				
i	100	24	100	48
e	88	24	98	48
&	96	24	98	48
i@	100	6	100	6
e@	100	6	100	6
ai	100	6	100	6
all	**96**	**90**	**99**	**162**
TOTAL	**87**	**252**	**94**	**468**

Table 3 Phoneme recognition results

The overall recognition accuracy of the test set was 87% and in some phonemes the accuracy was much higher.

This first type of phoneme recogniser is seen to combine classification by a static MLP with durational constraints in a fairly direct and designed manner, next we move to types which avoid such elements of design and are wholly trained from data.

Modified Kanerva Model for Small Word Recognition

The Kanerva model was derived as a model of high dimensional classification using very large memory [9]. It was modified in a way to make it suited to the requirements of speech recognition for phoneme recognition [14], morph recognition [6] and then 'small word' recognition [15]. Here we report the latter most recent formulation which has been implemented for real-time recognition.

The overall system has four basic components as shown in Fig.9. Here a broad input network is used which takes in 1 second of speech. In this parallel real-time implementation 50 adjacent, non-overlapping, 20ms windows of speech are processed by 50 filter banks each with

20 channels. This provides an input vector of dimension 1000 integer values. The overall system was trained to recognise 133 'small words', somewhat related to morphs, associated with the middle 1/10th of a second of speech. Each of the 133 small words was associated with an output node, using binary coding, in the single layer adaptive dot-product output component. Between the 1000 element input layer and the 133 bit output layer a location layer of 9600 units was constructed.

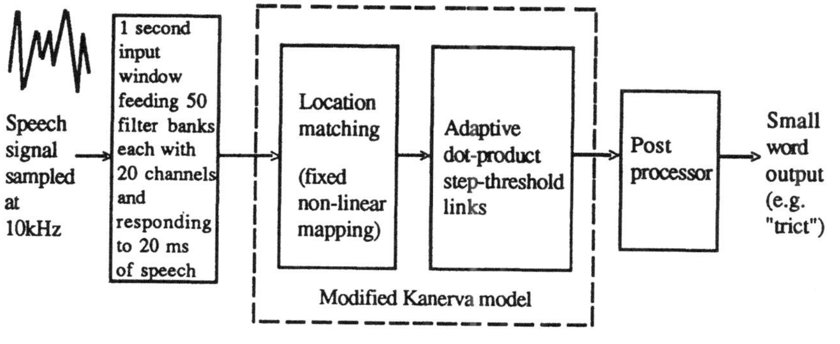

Fig.9 Modified Kanerva model for small word recognition

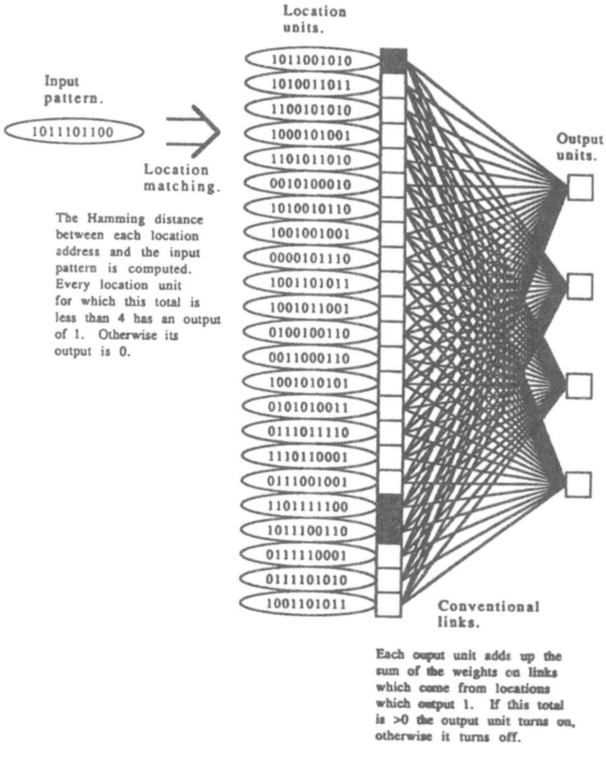

Fig.10 Kanerva algorithm

The basic Kanerva algorithm for a binary input vector is shown in Fig.10. Here the object is to classify an input vector by activating the location values (which are randomly distributed across the location layer) which lie within some metric, for example the Hamming distance, from the input vector; such locations are set and the others reset and then the weights of the adaptive dot-product layer are trained to establish the resulting output pattern, in this case with one bit set to 1 out of the 133 outputs and the remainder reset to zero. In the present implementation the input vector has integer rather than binary elements and a hypercube metric is used with the radius chosen such that about 10% of the locations are active for any given input pattern.

The recogniser thus consists of two fixed mappings, the filter bank and the location matching layer, followed by a simple adaptive mapping, the layer of dot-product step threshold links. Additionally a post-processor element is used consisting of two parts. In the first the 133 binary output is placed in a five slot circular buffer and the total '1' bits for each small word added up over 5 frames. If a clear winner is available with three at least of the five bits set this word was passed to the second stage. In the case of a tie a list of the words previously passed on was used to eliminate all but one of the current tied candidates. The second stage simply removes multiple occurrences of the same small word.

The Cambridge Hotel database was used to train and test the system. This used seven speakers (5M, 2F) uttering two versions of 25 sentences. A total of 347 sentences labelled to a small word level was used to train the system and a total of 350 different unlabelled sentences was used to train it. A list of the sentences with words split up into small words is shown in Table 4.

we wish to see the tall est room
what did you tell me
I need to re pair this car
could you send a taxi to meet the bus
I must re turn to the air port
yes I shall need a credit card
should I wear boot s on the see shore
can you see past the mesh
can you direct me to the shop s
the popular black pool rock can be bad
can you send the picnic par ty to the see shore
this book cost s six ty yen
I need a ride to the red light dis trict
we all need sun shine and a palm tree
my little red car might possib ly be rus(t) ty
see that pre(t) ty red pot under my bed
my solicit (t)or shall be in con tact soon
this room might be a little bit to dark
did you put yes on the sheet
you might un wind and re lax there
where do tour ist s re teat by night
you can ex pect a can teen at the castle
we all in sist on the aquarium
I wish to dis cuss add ition (a)l treat ment
we re mark ed on the ob scure club across the street

Table 4 Cambridge Hotel database

In training, 450 passes were made through the database. In presenting the data successive sentences were offset by 0.8ms up to a maximum offset of 19.2 ms when the offset was reset to zero and the process repeated. When trained, of the 2945 training small words presented to the recogniser 98.8% were correctly recognised from the training data.

In testing, a total of 2968 small words are presented, 424 coming from each speaker and the results are shown in Table 5.

Modified Kanerva model test results: small words recognised in Hotel Corpus.			
Speaker	Correct	Wrong	Missing
rwp	408	18	16
mha	404	16	20
doc	411	17	13
reb	413	8	11
mvs	407	28	17
rjn	414	17	10
jcn	382	44	42
Totals	2839	148	129

Table 5

It can be seen that the recogniser detects 95.6% of the small words present in the test speech and 95% of the output tokens are correct.

This recogniser which involves no explicit linguistic constraints, apart from those implicit in the data, is dealing with a much larger speaker population and vocabulary than the previous recogniser. It is clearly accounting for context using the 'memory' in its broad input and for the temporal behaviour of the speech signal during the small words. We now turn to the recurrent network structure which employs a narrow input window but which incorporates internal memory using delayed state feedback from its hidden nodes.

Phoneme Recogniser using Recurrent Neural Networks

This structure has a narrow input window but employs internal memory to retain left context information. Further it is well suited to the classification of dynamic signal since it is based on the classical, state-variable representation of linear multivariable dynamic systems, used in the control and identification of such systems.

Such a model relates an input vector **u** to an output vector **y** via a system state vector **x** and for discrete-time systems has the equations:

$$\mathbf{x}_{t+1} = \mathbf{A}\mathbf{x}_t + \mathbf{B}\mathbf{u}_t \quad (2a)$$

$$\mathbf{y}_t = \mathbf{C}\mathbf{x}_t + \mathbf{D}\mathbf{u}_t \quad (2b)$$

Taking these over to non-linear systems such as the MLP, the coefficient matrices A,B,C,D now become non-linear operators with

$$x_{t+1} = A[.]x_t + B[.]u_t \tag{3a}$$

$$y_t = C[.]x_t + D[.]u_t \tag{3b}$$

A recurrent net obeying these equations, with no direct connections between input and output, i.e. with D[.]=0, is shown in Fig.11. This is representative of a network with an input layer to the state or hidden units $\mathbf{x}_{t+1}$, together with feedback of the previous state vector $\mathbf{x}_t$, via a unit delay matrix $z^{-1}\mathbf{I}$, and an output layer to the output vector $\mathbf{y}_{t+1}$.

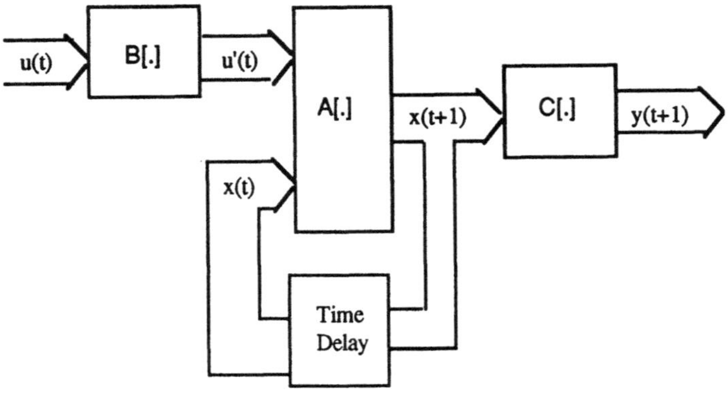

Fig.11 Architecture of recurrent network

Studies of this structure [17] led to the realisation of the simpler form shown in Fig.12, employing only a single layer with inputs $\mathbf{u}_t$ and $\mathbf{x}_t$ and outputs $\mathbf{x}_{t+1}$ and $\mathbf{y}_{t+1}$ together with unit delay feedback z^{-1}I. Each node also has a bias input and employs a logistic function non-linearity.

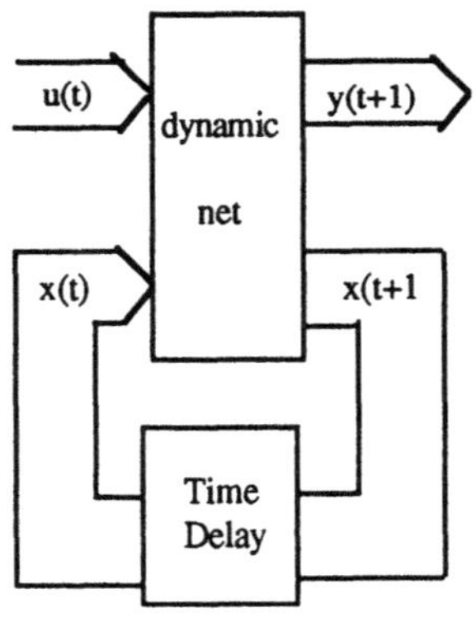

Fig.12 Architecture of reduced recurrent network

Such networks can be trained by 'stretching them out' as shown in Fig.13. The forward path is established by applying a sequence of input frames, for example as shown in the Figure, $\mathbf{u}_{t-2}$, $\mathbf{u}_{t-1}$, $\mathbf{u}_t$ to establish the outputs $\mathbf{y}_{t+1}$, $\mathbf{x}_{t+1}$. Then the errors are backpropagated through the network starting from the network at t using the y_{t+1} targets and assuming the errors associated with $\mathbf{x}_{t+1}$ are zero. This provides the errors associated with $\mathbf{x}_t$ and those from y_t are picked up from its targets. This provides the errors for $\mathbf{x}_{t-1}$ and so on.

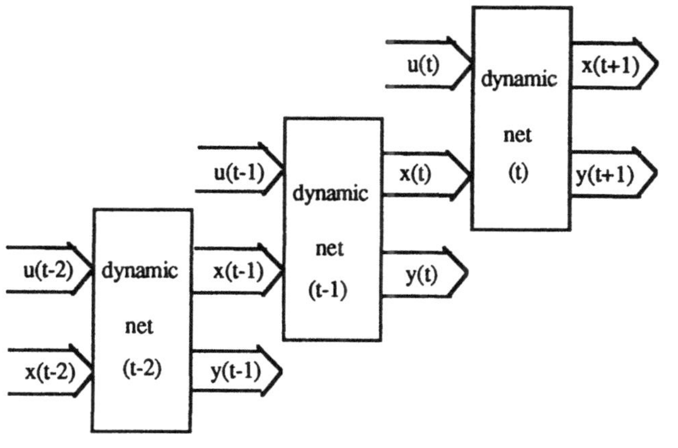

Fig.13 Training of recurrent network

A number of experiments have been carried out with different data. For example comparisons are given with HMM results using the Fleet Management database [20]; for the TIMIT database comparisons were made with HMM results for phoneme recognition [18] and this has been extended using a range of different types of preprocessor including an auditory model [19]. Here we concentrate on the second of these experiments.

These experiments used a 61-output phoneme vector $\mathbf{y}$, 128 states $\mathbf{x}$ and a 31 component input vector $\mathbf{u}$. The 31 components input vector $\mathbf{u}$ consists of: the number of zero crossings per input frame, 4 cube roots of energies in quarter frames, 20 cube roots of energies in the Bark scaled FFT, the position and amplitude of the pitch peak in the cepstrum, 4 positions of peaks in the smoothed spectrum and a bias. This results in a single layer network with 128 + 61 = 189 nodes, each with weights from 128 + 31 = 159 inputs, resulting in a total of 30,051 weights.

The cost function used is the cross-entropy measure

$$\log (p^{(t)}) = \sum_{i=o}^{M-1} d_i^{(t)} \log (y_i^{(t)}) + (1-d_i^{(t)}) \log (1-y_i^{(t)}) \qquad (4)$$

where d_i refers to a target output and y_i to an actual output.

The network was trained on a 64-transputer array using SIMD and weight adjustment was carried out with a variant of an adaptive algorithm [3].

The results were post-processed using dynamic programming matching and bigram and durational constraints and reduced to the 39 phoneme CMU/MIT reduced set.

Results are shown in Table 6 comparing them with the Sphinx results for the same data [11].

Values	Recurrent Net	CMU/SPHINX
correct	75.1%	73.8%
accuracy	68.9%	66.1%

Table 6 Comparison of recurrent net and SPHINX HMM results

It can be seen that the recurrent net results are as good as and slightly better than the SPHINX results. This gives one significant reference comparison between neural net and state-of-the-art HMM phoneme recognition and implies that recurrent nets can be expected to provide the same performance as HMM technology in speech recognition.

Classification & Regression Trees (CART) Neural Networks in Speech Recognition

One problem in the design of neural networks for recognition purposes is the choice of the size of the network. This is usually done by experiment, as was the case in the three networks described so far, but this is necessarily a slow process. This section describes a CART network for phoneme recognition which automatically adjusts its size and allocates a sufficient number of nodes for the classification task given to it [4].

The architecture used is shown in Figs.14 and 15. The overall structure is the binary tree shown in the first Figure, each of whose nodes is of the form shown in the second Figure. When trained the current speech vector **i** is passed down the tree starting with the root node at the top, at each non-terminal node visited the classification function of the node is to determine whether to pass the vector down to the left or to the right, this continues until the terminal node is reached corresponding to the phoneme token for that vector. For a given training set only a tree size sufficient for that data is developed and in use the method is fast and computationally efficient since only an appropriate part of the tree is visited for any input vector; it has been found that only approximately 3MFlops are needed to run a phoneme classifier in real time.

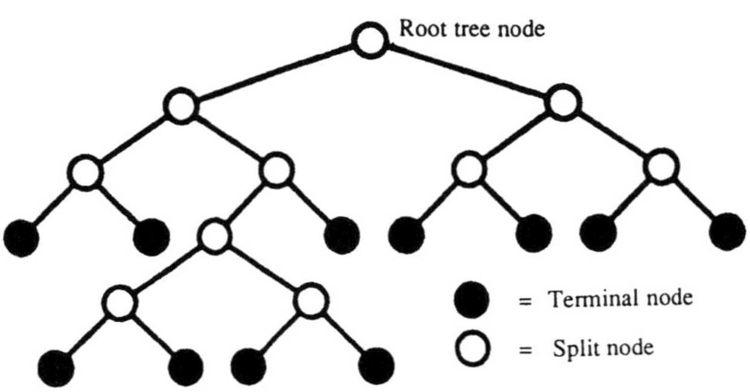

Fig.14 Binary Classification and Regression Tree

Each node has four units as shown in Fig.15. These consist of a sigmoid unit, two radial basis units and a KxK matrix Ω, where K is the dimension of the incoming spectral vector **i**, with K=20 in the present experiments. The purpose of the matrix is to map the input space such that the axes of the two radial basis functions are aligned to the input patterns, which are assumed to be hyper-ellipsoidal in shape. The purpose of the radial basis functions is to determine whether the incoming pattern should be passed down the tree to the left or right, specifying this by the magnitude of the output from the sigmoid function.

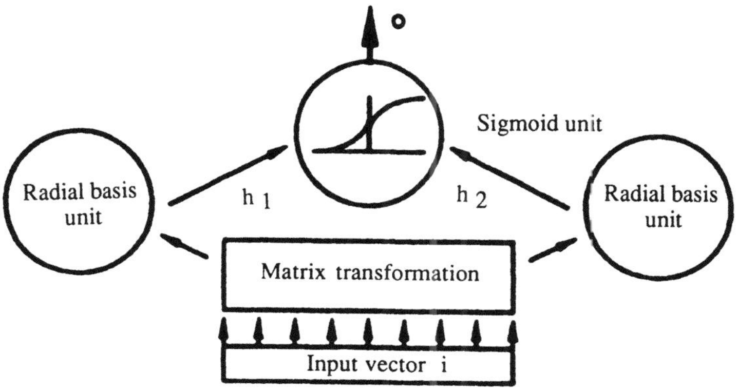

Fig.15 Neural network classifier

For a transformed input

$$\bar{\mathbf{i}} = \Omega \mathbf{i} \tag{5}$$

the radial basis unit is defined as

$$h_m = \exp\left(- \sum_k (\bar{i}_k - \mu_{m.k} / \sigma_{m.k})^2 \right) \tag{6}$$

where

μ_m is the centre vector for the mth radial unit

k is the input dimension index

m is the radial unit index, $m = 1,2$

δ is the radius for the kth axis

$\bar{i}_k$ is the kth element of $\bar{\mathbf{i}}$

A radial basis unit is on, with its output value near 1.0, if the vector $\bar{\mathbf{i}}$ is near its centre μ. The output of the sigmoid unit is

$$0 = \frac{1}{1 + \exp\left[- \left(\sum_{m=1}^{2} w_m h_m + \theta \right) \right]} \tag{7}$$

where w_m are weights and θ is the bias value. The vector **i** is passed to the left or right depending on $0 \geq 0.5$.

The training set consists of J phoneme classes, with N_j example vectors of phoneme j. At each tree node, starting with the root node, training consists of four steps. In the first step the purity P_t of the node, [ ], is established from

$$P_t = \sum_j N_j^2(t) / (\sum_j N_j(t))^2 \tag{8}$$

where $N_j(t)$ is the number of vectors in class j in the training set L_t and j is the phoneme class index. If the node purity is above a threshold (0.9 in the present experiments), the node is made a terminal node and assigned the phonetic class which minimises a particular misclassification cost function.

The second step is to determine which branch, left or right to assign to each of the J phoneme classes. The mean $\mu_{j.t}$ and covariance $\Sigma_{j.t}$ of each class is calculated from the node data L_t, forming a class cluster. A distance matrix between all J class clusters is calculated using a modified Bhattacharyya distance measure. This increases the distance between classes with a large number of training frames, increasing the chances that the two classes will be separated. The two class clusters which are the furthest apart are defined to be the initial left and right branch clusters. Then of the remaining class clusters the one with the smallest distance to either of the branch clusters is merged with that branch cluster and its mean and covariance are updated. All the remaining clusters are merged in turn to either the left or right.

The third step is to train the neural net classifier. The matrix Ω is calculated such that it simultaneously diagonalises the left and right branch cluster covariance matrices such that

$$\Sigma_R \Sigma_L \mathbf{I}_k = \lambda_k \mathbf{I}_k \tag{9}$$

with $\mathbf{I}_k$ and λ_k the orthonormal eigenvector and eigenvalues respectively, k = 1, 2, .. K. Each eigenvector $\mathbf{I}_k$ is then scaled such that

$$\alpha^2 I_k^T \Sigma_l I_k = 1 \tag{10}$$

These scaled column eigenvectors then form the matrix Ω

$$\Omega = [\, I_1, I_2, .. I_K \,] \tag{11}$$

The centre μ of each radial basis function is initially set to the transformed mean of the left and right clusters, the radius of the left radial basis unit is set $\sigma_{1.2} = 5$ and the radius $\sigma_{2.k} = 5\,\lambda_k^{0.25}$, and initially $\omega_1 = 0.7$ (right), $\omega_2 = 0.7$ (left) and $\theta = 0$. Gradient descent using an adaptive algorithm [3] is then used to complete the training of the classifier by maximising an entropy equation.

In the fourth step the training set for the tree node L_t is split according to $0 \geq 0.5$ to form new training sets for the child nodes and the child nodes are trained in the same way. This is continued until all terminal nodes are of sufficient purity.

In practice the purity threshold is deliberately set to a high value to ensure a larger than necessary tree is grown so that good splits are not prevented from being generated. The resulting tree is then pruned using a pruning strategy [1]. This gives a set of trees of decreasing size, each of which is optimally accurate given the number of terminal nodes in that tree. The best tree can be chosen from cross validation estimates or as in the present experiments by testing each tree in turn with unseen data and choosing the best.

Two experiments were carried out using this technique. In the first experiment 15 utterances, 10 train and 5 test, of the digits 0-9 were recorded from a single speaker. The speech was sampled at 10kHz and DFTs taken for a 256 point window with 128 point overlap, and the spectrum converted to a 20 point Bark scale spectrum to form the vector **i**. Each frame was hand labelled with one of 18 phonetic labels. The phoneme recognition accuracy for the optimally pruned trees, versus the number of nodes is shown in Fig. 16. Choosing the tree with 97 total nodes and 49 terminal nodes the phoneme recognition accuracy is as shown in Table 7. Finally an HMM with explicit Poisson durational modelling was trained on the phoneme tokens generated by the CART net and gave the digit recognition accuracy shown in the Table.

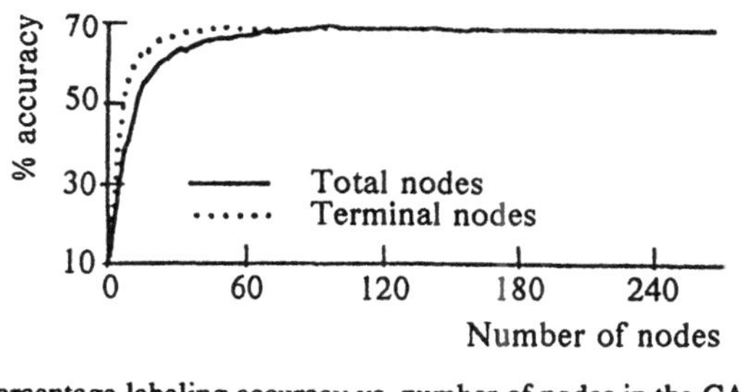

Fig.16 Phoneme percentage labeling accuracy vs. number of nodes in the CART network for Exp.1

In the second experiment 20 utterances of 29 words were recorded from the same speaker. These were processed as before and hand labelled with 37 phoneme classes. A CART tree was trained on data from 15 utterances and tested with data from the remaining 5. Choosing a tree with 371 total nodes and 158 terminal nodes, a phoneme recognition score of 66% was achieved on the test data. Finally an HMM was trained as before and using a beam search and push-down automata grammar, the word and phrase recognition scores shown in the Table were achieved.

	Experiment 1		Experiment 2	
Data	Phonemes	Digits	Words	Phrases
Train	93	100	-	-
Test	68	94	98	96

Table 7 % accuracy scores for isolated digits (experiment 1) and continuous phrases (experiment 2)

The CART networks are slow in the training mode but fast in the recognition mode. For example training the network of experiment 1 took about 2 days on a VAX GPXII and of experiment 2 about 4 days. However a real-time implementation of experiment 2 was achieved using two T800 transputers, the network taking on average 23.6ms to generate a phoneme token from the spectral vector **i**.

Lexical Access & Language Models using Neural Networks

As was mentioned in the Introduction the trend in speech recognition using neural networks has been to shadow HMM technology. This has started with sub-word unit recognition as exampled in the above with the next step in progress on word recognition from these sub units. This stage of lexical access is much less advanced and a few preliminary examples, again without attempting to be comprehensive, are given in this section.

One of the earliest examples is the use of static MLPs in NET-grams to form language models [13]. These are a network analogue of statistical N-grams, which given the current N-1 symbols in a string give the best estimate of the next N^{th} symbol in the string, which are used with HMM technology. The authors used a basic bigram net with 89 input and output nodes corresponding to word categories in the Brown Corpus, and two layers of hidden units, and extended this to N=3,4 using an efficient triangular structure with cross coupling. They showed the performance was as good as a conventional N-gram for the Brown Corpus data and that it performed effectively with for unknown test data which never appeared in the training set.

Several approaches have been studied at Cambridge including the NET-gram. In a preliminary study [6] two approaches were taken to lexical access. In one a modified Kanerva model morph-to-word network was combined with a speech-to-morph modified Kanerva model and in the other a new technique termed 'Compositional Representation', (CR) providing phoneme-word recognition was combined with a speech-phoneme recurrent net recogniser.

In the case of the Kanerva systems the same basic structure was used for sub-unit and word recognisation. The Compositional Representation associates an nxn matrix with each sub-word unit, for example $\Psi(A)$ with sub unit A etc. Then a sequence such as RAAK is represented by the nxn matrix product $\Psi(R)\ \Psi(A)\ \Psi(A)\ \Psi(K)$, which is order sensitive, or post multiplying the product by an n unit vector, by the corresponding n vector. This type of representation can be viewed as a linear neural network with the structure shown in Fig.17, where the elements of the matrices correspond to the weights in the network which can be trained by back propagation.

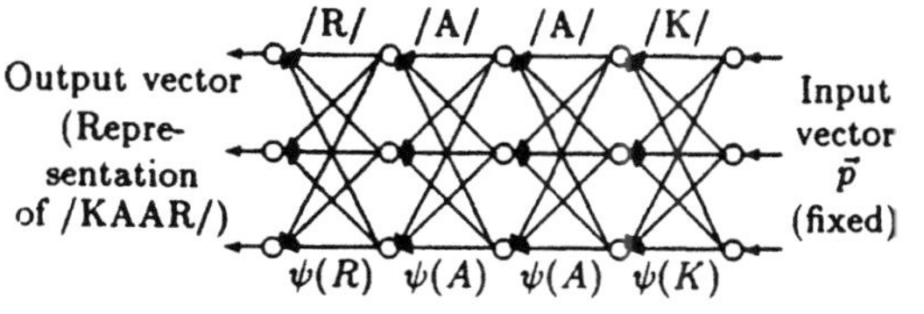

Fig.17 Connectionist interpretation of the Composition Representation

The two approaches were first tested on the small 7 speaker 700 sentence, Cambridge Hotel database. Results are shown in Table 8 for the recognition accuracy of each element of the recognisers - for sub-word recognition and sub-word to word recognition using speech derived data and also text derived data.

Model	Sub-word accuracy	Word accuracy (speech)	Word accuracy (text)
Kanerva	95.6%	90%	98%
Recurrent/CR	70.8%	88.9%	90%

Table 8 Results from HOTEL database

These results are seen to be reasonably accurate. However when tests were carried out on data from the much larger 420 speaker 4200 sentence TIMIT database the accuracies fell off as shown in Table 9.

Model	Sub-word accuracy	Word accuracy (speech)	Word accuracy (text)
Kanerva	66.2%	-	73% (sentence)
Recurrent/CR	62%	9.6%	-

Table 9 Results for data from TIMIT database

One reason for the lack of scaling-up is the reduced accuracy of sub-unit recognition, another results from best-only data from sub-unit recognition being used as the input to the word recognition elements. As a result work has been pursued on modifying the word recognition elements to accept sub-unit lattices instead of only scalar strings. In an early development of this [12], the Computational Network has been augmented to accept phoneme lattices. In these experiments the database consists of 22 words uttered once by 104 speakers over the telephone, providing 2288 utterances. A continuous density HMM phoneme recogniser was used to produce lattices with 46 phoneme labels. Experiments were carried out varying the dimension n of the CR for phoneme string recognition and for phoneme lattice recognition with a branching factor of 3. For n=2 the word recognition accuracy is 58% for string data and 60.8% for lattice data while for n=7 the corresponding results are 86.9% and 90.2%. The last of these results is approaching HMM performance for systems with no language model.

Conclusion

The paper has described some of the work currently being carried out on continuous speech recognition using neural networks. It is clear that this is some way behind state-of-the-art HMM technology and indeed the range of possible architectures is still being explored. In particular the use of neural nets in lexical access and language models is much less developed at present however. However results in a few instances such as acoustic-phonetic decoding using recurrent neural nets where identical performance to HMM results on the same data have been achieved, indicate that neural net technology in speech recognition can be expected to catch up with HMM technology quite soon.

Acknowledgements

The author is indebted to colleagues and former colleagues and in particular, Tim Harrison, Richard Prager, Tony Robinson and Helmut Lucke, without whom this paper could have been written. Acknowledgement is also due to Logica Cambridge Ltd who provided the data and HMM models for the CR phoneme lattice experiments.

References

1. Breiman, L., Friedman, J.H., Olshen, R.A. & Stone, C.J.: *Classification and Regression Trees*, Wadsworth & Brooks/Cole, 1984.
2. Bridle, J.: Alpha-Nets: A recurrent neural network architecture with HMM interpretation, *Speech Communication*, April, 1990.
3. Chan, L.W & Fallside, F..: An adaptive training algorithm for back propagation networks, *Computer Speech & Language*, **3**, 3/4, Sept/Dec 1987, 205-218.
4. Chong, M.W. & Fallside, F.: Classification and regression tree neural networks for automatic speech recognition, Proc. INNC90, Paris, **1**, 187-190.
5. Chow, Y.L., Dunham, M.O., Kimball, O.A., Krasner, M.A., Kubela, G.F., Makhoul, J., Roucos, S., & Schwartz, R.M.: BYBLOS: The BBN continuous speech recognition system, Proc. ICASSP 87, Dallas 1987, 89-92.
6. Fallside, F., Lucke, H., Marsland, T.P., O'Shea, P.J., Owen, M. St.J., Prager, R.W., Robinson, A.J. & Russell, N.H.: Continuous speech recognition for the TIMIT database using neural networks, Proc. ICASSP 90, Albuquerque, 1990, pp.445-448.
7. Fisher, W.M., Zue, V., Bernstein, J., Pallett, D.: An acoustic-phonetic database, Proc. Acoustic Soc. Amer., May 1987.
8. Harrison, T.D. & Fallside, F.: A connectionist model for phoneme recognition in continuous speech, Proc. ICASSP 89, Glasgow, 1989, 417-420.
9. Kanerva, P.: Self-propagating search: a unified theory, Ph.D thesis, Stanford University, Centre for Study of Language and Information, 1984.
10. Lee, K-F.: Automatic Speech Recognition: The Development of the SPHINX System, Kluwer (Boston), 1989.
11. Lee, K-F & Hon, H-W.: Speaker independent phoneme recognition using HMMs, IEE Trans. ASSP, **37**, (11), 1989, 1641-1648.
12. Lucke, H. & Fallside.: Application of the Compositional Representation to lexical access using neural networks, Proc. Intl. Conf. on Spoken Language Processing, Kobe, November 1990, paper 31.13.
13. Nakamura, M. & Shikano, K.: A study of English word category prediction based on neural networks, Proc. ICASSP 89, Glasgow 1989, 731-734.
14. Prager, R.W. & Fallside, F.: The modified Kanerva model for automatic speech recognition, *Computer Speech & Language*, **3**, 1, 1989, 61-81.

15. Prager, R.W., Clarke, T.J.W. & Fallside, F.: The modified Kanerva model: results for real time word recognition, IEE Intl. Conf. on Artificial Neural Networks, London, 1989, 105-109.
16. Price, P.J., Fisher, W., Bernstein, J., Pallett, D.: The DARPA 1000-word resource management database for continuous speech recognition, Proc. ICASSP 88, New York, 1988, 651-654.
17. Robinson, A.J. & Fallside, F.: A dynamic connectionist model for phoneme recognition, Proc. n'Euro '88, ENST, Paris 1988.
18. Robinson, A.J. & Fallside, F.: Phoneme recognition for the TIMIT database using recurrent error propagation networks, Technical Report CUED/F/INFENG/TR42, March 1990, Cambridge University Engineering Department, (submitted for publication).
19. Robinson, A.J., Holdsworth, J., Patterson, R. & Fallside, F.: A comparison of preprocessors for the Cambridge recurrent error propagation network speech recognition system, Proc. Intl. Conf. on Spoken Language Processing, Kobe, November 1990, paper 23.17.
20. Robinson, A.J. & Fallside, F.: Further results from the DARPA resource management database with the Cambridge Recurrent Error Propagation Speech Recognition System, submitted to ICASSP 90.
21. Young, S.J.: Competitive training in Hidden Markov Models, Proc. ICASSP 90, Albuquerque, 1990, 681-684.

Connectionist Large Vocabulary Speech Recognition

Alex Waibel[1]

School of Computer Science, Carnegie-Mellon University, Pittsburgh,PA 15213

Abstract: In this paper, the problem of large vocabulary word recognition is addressed from a connectionist perspective. The problem is not only of practical interest but also of scientific importance, since a workable solution must integrate pattern recognition under consideration of sequential, symbolic constraints. We have developed two large vocabulary word recognition systems based on different speech recognition philosophies. One of the systems exploits the power of neural networks in performing accurate classification, the other the power of producing good non-linear function approximation and signal prediction. We present each system's operation and evaluate its performance. Both achieved respectable recognition scores in excess of 90% correct for vocabularies of up to 5000 words. We suggest further avenues towards improvement of either system and in the process discuss the relative strengths of either approach.

Keywords: neural networkds, connectionism, time-delay neural networks, predictive neural networks

1 Introduction

Recognition of speech by machine has been a fascinating topic of research that has for many years given rise to some of the most innovative and exciting models. It has always been driven by a mix of intuitions relating to system design and engineering on one side and human cognitive modeling on the other. It has always drawn a great deal of ideas, motivation and inspiration from a desire to understand human communication, while imposing the realism of practical engineering constraints and comparative performance measures. Connectionist models or "neural networks" have recently attracted considerable (and renewed) attention in speech recognition as they provide speech scientists with a cognitively plausible model of speech processing while at the same time introducing a novel, yet realistic engineering solution to the problem. A number of initial designs have produced in a short time performance results that compared favorably or exceeded those obtained by traditional speech processing techniques [1, 2, 3, 4]. On the other hand, most of these experiments were limited to small tasks or subproblems of the speech recognition problem such as phoneme classification [1, 2, 5] or small vocabulary word recognition [6, 7, 8].

While these results are encouraging given those limited domains, the question remains to be answered if and how this technology may be used effectively for the design of whole speech understanding

[1]We gratefully acknowledge IEICE for permission to reprint this paper. It has originally appeared in the journal of the IEICE, Vol.J73-D-II, No.8 pp. 1122-1131, (Aug. 1990) by the same author.

NATO ASI Series, Vol. F 75
Speech Recognition and Understanding. Recent Advances
Edited by P. Laface and R. De Mori

systems. Indeed, a common criticism argues that connectionist models are but good classifiers but cannot handle the temporal, sequential nature of speech. As such, connectionist models may be attractive only in limited domains or toy problems, but would scale poorly to large vocabulary speech understanding systems. Although this criticism has been valid for a number of initial simple networks, extensions that overcome these limitations have been proposed and are beginning to produce respectable results on larger problems as well.

In this paper we will describe current research activity that addresses the large vocabulary recognition problem. We present two large vocabulary word recognition systems that illustrate that neural networks 1.) can be used productively for large vocabulary speech recognition by way of classification but also by way of non-linear mapping and system identification 2.) neural networks can be integrated with connectionist as well as non-connectionist strategies to handle temporal, sequential processing to form chains of subword units, words and sentences.

2 The Large Vocabulary Word Recognition Problem

Early on connectionist word recognition experiments were carried out that have exploited the classification capabilities of neural nets by applying an entire word's coefficient matrix to the inputs of static full word networks with output units for each word to be classified. Good results were achieved, but the resulting systems required precise time alignment and a preprocessing stage that determines the endpoints of an input word, both unacceptable requirements in practice in the light of continuous speech, noise and varying speaking rates. Similarly limiting is the fact that only small vocabularies can be handled in this fashion, because network size and training time become prohibitively large and enrollment impractical with increasing vocabulary size.

To overcome the former first set of limitations, networks that model time, temporal distortion (warping) and/or shift-invariance internally have been proposed for small vocabulary recognition. Among them are techniques that integrate neural network based classification with traditional schemes for time alignment and sequence management, such as the Dynamic Neural Net (DNN) [8, 9], word level Time-Delay Neural Networks (TDNNs) [10, 11], hybrid neural net classifiers and Hidden Markov Models [12] and Neural Prediction Models [13]. Most of these models have been tested on small vocabularies (Japanese, French and English digits) and have achieved excellent performance results, but all used dedicated models for each vocabulary word and are in their basic forms not appropriate for large vocabulary recognition.

To extend these models to large vocabulary recognition subword units such as phonemes or syllables must be employed. Since such subword units are limited in number large vocabularies can be constructed as different sequences of these atomic subunits. In large vocabulary word recognition then the task is to identify the most likely sequence of phonetic units that make up a legal word (preferably without requiring segmentation in the process). Several models have been proposed that express sequential constraints in a connectionist framework alone [14, 15, 16, 17, 18]. Alternatively, combinations between the perceived strengths of neural networks at the pattern recognition level with the strengths of traditional methods at modeling sequences such as Hidden Markov Models, Viterbi Decoding, or Dynamic Programming have also been proposed. Such "hybrid approaches" have recently gained in popularity as they appear to offer immediate access to the best of both worlds.

In the following we describe two connectionist large vocabulary recognition systems. They are examples of two different recognition philosophies. We will refer to them as "classification based models" and "prediction based models".

3 Classification Based Models

Neural networks have been shown to implement excellent non-linear classifiers both at the phonetic level as well as at the word level. Large vocabulary systems can therefore be implemented by neural networks that recognize phonemes or parts of phonemes (states) and evaluate how well a sequence of their phonemic output hypotheses match the legal sequence of a word.

3.1 Time-Delay Neural Networks

One of our attempts in doing this is based on the Time-Delay Neural Network (TDNN). This network has been shown to produce excellent phoneme discrimination performance [1]. This network was developed to provide a non-linear non-parametric[2] pattern classifier that can spot features or phonemes independent of precise temporal alignment (shift-invariance property). The network is a multilayer network of units that incorporate current activations from lower layers as well as time-delayed versions of them (context) as input. Fig.1 illustrates a TDNN trained to perform the discrimination task between the voiced stop consonants /b, d, g/ (see [19] for a more detailed description of its operation).

Initial experimentation with this class of networks was performed speaker-dependently on small phoneme sets only (/b,d,g/ discrimination), but extensions to high performance multi-speaker recognition [20] and recognition of all phonemes were soon achieved. Both problems significantly benefitted from modular and incremental learning [20, 21, 2]. By using an integrating supernetwork (Meta-Pi network [20]) to decide on how to gate an appropriate mix of speaker specific network decisions, focus of attention or rapid adaptation to speaker specific classification can be achieved. In *multi*-speaker classification experiments this resulted in speaker-*de*pendent recognition rates - a significant improvement over results from speaker-independent training. Modularity could also be used effectively to overcome problems related to scaling, training time and generalization. By exploiting the featural abstractions in the hidden units of previously trained networks modular training allowed for greater efficiency and flexibility of design while achieving performance greater than or equal to non-modular networks [2].

3.2 Large Vocabulary Recognition by TDNN

Based on a Japanese large vocabulary isolated word database (5240 words) [22, 19, 1] a number of speaker-dependent experiments were carried out to improve the TDNN's performance, particularly in view of large vocabulary recognition [23]. For use in word recognition, speech is to be classified into phoneme output categories over running speech (in this case over entire words spoken in isolation). As the original

---

[2] no assumptions as to the underlying probability distributions need to be made

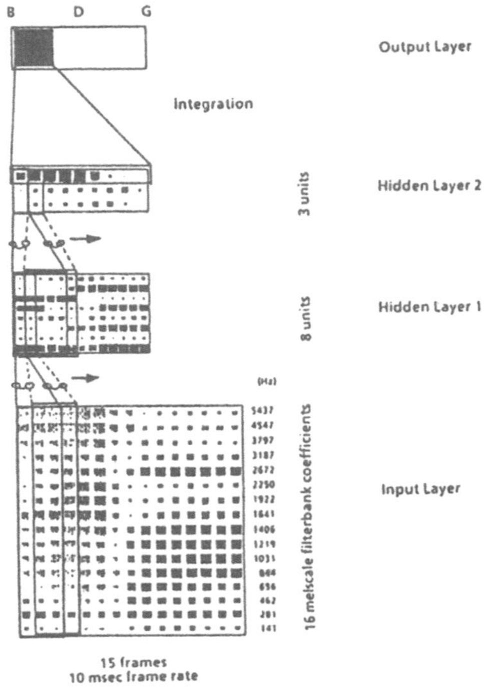

Figure 1. The TDNN architecture (/b,d,g/-task)

TDNNs were trained on excised phoneme tokens only, several modifications were desirable. First, the original excised phoneme training patterns were now artificially misaligned in time by various offsets. It more realistically simulates the absence of precise phoneme labels and segmentation. The resulting introduction of time alignment "noise" turned out not to decrease performance, but lead to noticable improvements instead, particularly for phoneme spotting. Training in this fashion improved generalization and enforced shift-invariant phoneme classification even in transitory regions between phonemes. The resulting phoneme spotting rates of the large scale TDNN's improved from 95.8% to 98.0% and more importantly, the false alarm rates[3] decreased from 62.2% to 23.2%[4]. The performance results of our earlier models and this improved model compared favorably with various other recognition strategies over the same data. For word recognition also a silence category was necessary which was added by modular design to the existing net [23]. Fig.2 shows the resulting large TDNN all-phoneme architecture. Fig.3 shows output activation patterns for the word "wata".

While good phoneme classification performance is indeed encouraging, this will have to be properly integrated and have to translate into good large vocabulary word recognition performance to advance the field. Mature speech recognition technology has already at its disposal a number of elegant techniques for this and similar word-level integration needs to be accomplished in a connectionist frame-work or in

[3]Presumably due to previously undefined transitory regions.

[4]All recognition tests were run on independent test data from the same speaker.

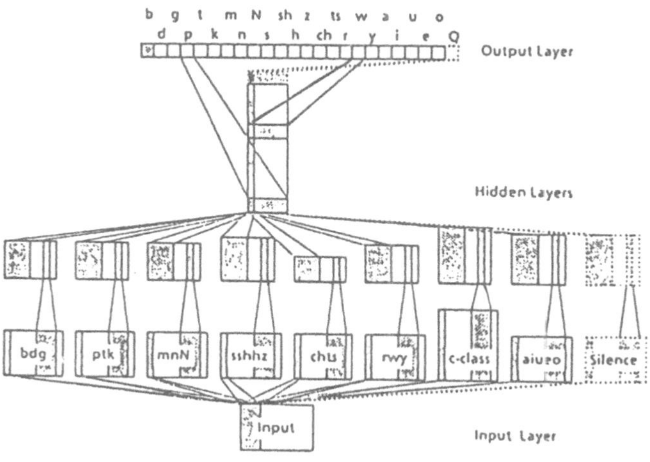

Figure 2. Modular TDNN used to spot all phonemes

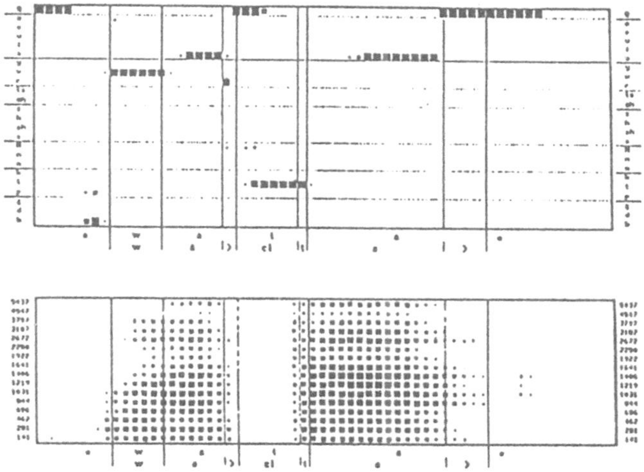

Figure 3. TDNN spotting phonemes in word "wata"

the form of hybrid connectionist/ non-connectionist system design. Neither is necessarily a trivial step to undertake and we shall desribe several successful initial attempts that have been proposed.

Using data from a Japanese isolated word database (as described above) and a TDNN as a front end phoneme level model, a hybrid large vocabulary recognition system was developed [23]. 24 phonemes (5 vowels, 18 consonants and silence) were spotted by shifting TDNNs across time providing the front end for phoneme based word recognition. To recognize a word, the overall likelihood of a word-specific *sequence* of phoneme activations needs to be estimated. To do so, we can approximate the output activations of a TDNN as representing the maximum a posteriori probabilities of a phoneme class given speech at a given time frame [24]. If each phoneme is viewed as a single state with an associated output probability, then a word likelihood can be calculated as the joint probability over all output probabilities over time. Assuming that all states are independent, a word likelihood would be given by the product of framewise outputs. A simple way of implementing this is to evaluate at each time frame the log activation of the output unit that corresponds to a legal phonemic state in the word and summing these log outputs over time. The correspondence between a given time frame and the current active phoneme node is performed by a Dynamic Time Warping (DTW) procedure.

An implementation of this procedure is described by Miyatake, Sawai, Minami and Shikano [23]. Here, a modular TDNN as described above was used, and only one state per phoneme was provided. An LR-parser provided top-down prediction of what set of phoneme transitions are legal to form legal words in the dictionary. For duration control each phoneme state was expanded to the average number of frames of that phoneme before DTW was carried out. Recognition experiments on various vocabulary sizes were undertaken with this system. All experiments were performed vocabulary independently[5] and on independent test data (phonemes not used for training). For a 500-word test vocabulary, first choice accuracy of 98% was achieved. For a large vocabulary of 5000 words, recognition rates as high as 92.6 were obtained. Second and fifth choice rates for the later vocabulary size were 97.6% and 99.1%, respectively, indicating that most confusions occurred among a small group of acoustically similar words (e.g., "itai" -> "ittai").

3.3 Extensions

The performance of the system described does indeed suggest that very high performance can be achieved, independent of training vocabulary and training context. Several problems, however, need to be overcome to further improve large vocabulary speech recognition systems.

Sequencing of Phoneme Internal Events: First, we have already noted that the TDNNs described above were all integrated as single phoneme states. While TDNNs can capture a variety of phoneme specific cues sequential ordering within a phoneme is only imposed within the reach of its fixed duration time-delays. Additional ordering between variably duration subphonemic states must be imposed in the context of word recognition. Variable or adaptive time-delays [25] could be used internally or a sequence of several states [12] per phoneme at its output. This should lead to better performance and duration modeling, particularly in continuously uttered poorly articulated speech.

Stochastic Modeling of Sequences: The most successful and popular approach to stochastic modeling of sequences is given by Hidden Markov Models (HMMs), where a phoneme is given by a stochastic sequence of states that can be linked together into words and from there on into sentences. At each of these levels (lexical, syntactic, etc.) constraints can be applied and probabilities estimated, and their joint probabilities (assuming they are independent) computed. A popular idea therefore is to use the strengths of neural networks at precise pattern classification in combination with the modeling of state sequences and time alignment found in HMMs.

Some of the earlier proposals at this were developed by Bourlard, Wellekens and Nelson [26, 24, 27]. In theoretical and experimental work they had shown that the outputs of a multilayer perceptron

[5]The phonemes used for training were extracted from words of a different vocabulary than the one used for testing.

(feedforward network) trained by backpropagation from a mean square error may be considered to be estimates of the maximum a posteriori probabilities of a given input to belong to its corresponding output class. They have since built on this notion to construct Hidden Markov Model chains where the output activations of a local multilayer perceptrons (MLP) are used as output probabilities for the states in a traditional HMM. Viterbi aligment is performed to assign the framewise MLP firings to corresponding states and to compute an overall word output probability.

Several enhancements were subsequently proposed by several investigators. Morgan and Bourlard [27] achieved significant improvements in recognition performance, by normalizing their network outputs (the a posteriori probabilities) by their respective prior probabilities to eliminate a bias to uneven distributions in the training data. Another technique aimed at optimizing generalization performance is the usage of a cross-validation set. If only limited amounts of training data are available given a net of a given size, this can lead to overfitting to the training data and poor generalization to (poor performance on) new unseen data [27]. Use of an independent pseudo testing set (the cross-validation set) then yields a stopping criterion, that assures that a net is trained with optimal test-set performance in mind. A third enhancement proposed by several researchers is Connectionist Viterbi Training (CVT) [27, 12]. CVT is akin to the segmental k-means training procedure used for Hidden Markov model training [28] and aims at integrated and segmentation free word level training. The idea is to optimize a suitable phoneme (or state) segmentation *together* with the backpropagation network optimization. CVT iteratively finds the best labeling of the input (by way of Viterbi alignment), while the networks attempt to provide better outputs to correspond to these label. These techniques produced good word level recognition performance, that are beginning to compare favorably with other advanced HMMs on continuous sentence [27] and on connected digit [12] tasks.

Research Directions: A host of additional modifications and improvements that are known to work well for HMMs remain to be explored in the context of hybrid connectionist systems. Among them are corrective training (at the word level), choice of best input representation, transition probabilities, choice of optimal HMM topology, optimal neural network architecture, etc. Last not least, work is in progress towards improved training algorithms that generate more meaningful probability estimates at the outputs of local phonetic classification networks to improve *word level* discrimination and overall system robustness.

An alternate exciting research avenue is given by connectionist formalisms that represent sequential constraints altogether internally as connectionist modeling extensions [14, 15, 16, 17, 29, 18]. Such models may relax some of the limiting assumptions made by current recognition strategies and could potentially lead to further improvements in speech recognition system design.

4 Prediction Based Models

The connectionist models that we have discussed so far apply neural nets as classifiers of either word patterns or subpatterns. For classification, the input usually consists of a coefficient matrix and the output approximates a bit pattern representing the classification results. In addition to learning discrete classifications, however, neural networks can implement a variety of other constraint satisfaction tasks. Among them are non-linear function approximation, interpolation and prediction, which generate continuous real-valued output vectors. This can be exploited in speech for various signal mapping and coding applications, including noise suppression [30], speech code mapping [31] and non-linear signal prediction [32]. The use of neural networks as non-linear signal predictors in speech recognition has recently first been shown successfully in the "Neural Prediction Model" proposed by Iso and Watanabe [13] and the "Hidden Control Neural Network" proposed by Levin [33]. Both of these models have so far only been implemented for small vocabulary recognition tasks (i.e., digits), but have yielded high recognition performance speaker-independently. Extensions to large vocabulary recognition are also possible with this approach as we shall see in the following.

4.1 Recognition Using Small Vocabularies

The basic idea is illustrated in Fig.4. A two frame window of input coefficients is input into a multilayer feedforward net trained to produce at its output a frame of coefficients that is as close as possible to the next (future) speech frame. The distance between this predicted frame and the actual next speech frame can be measured as a prediction error or distortion and this distortion is used as error criterion for backpropagation training. Given a set of predictor networks one can imagine training each predictor for a

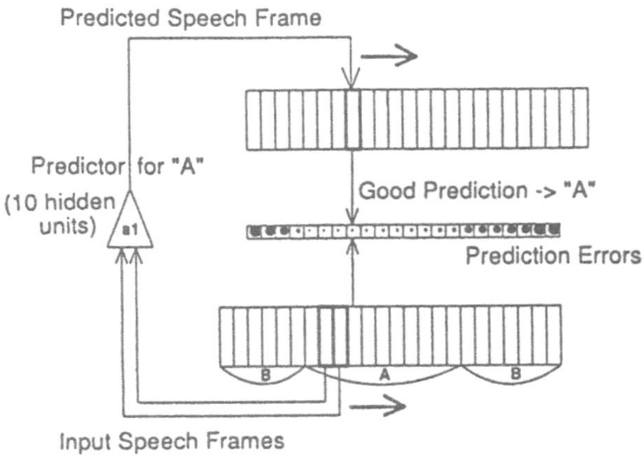

Figure 4. Modeling a phoneme by signal prediction

separate region of an utterance. Each predictor net becomes specialized to best predict this portion of an utterance, such that the prediction error is likely to be lowest in these regions. A word is then represented by the *sequence* of predictor nets that best predicts the actual observed speech. Dynamic Programming is used as a mechanism to optimally apply each predictor sequentially over time to best approximate the actual signal. Fig.5 shows this alignment step based on the matrix of distances between actual speech frames and predicted frames. During training an alignment path is determined by Dynamic Programming. Each predictor is then trained to minimize the error between its output and the speech frames that it was assigned to predict according to the DP-alignment path. During recognition the word whose sequence of predictors minimizes the error between predicted frames and actual signal frames is chosen. Iso and Watanabe [13] used 10 mel scale cepstral coefficients and amplitude change as inputs to their networks. The number of predictors used depended on the utterance and ranged (for Japanese digits) between 9

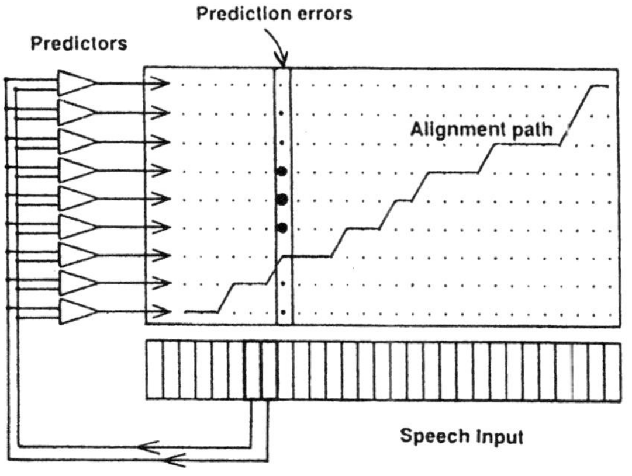

Figure 5. A Neural Prediction Model

and 14. Each predictor net has three layers, an input layer of two 11 coefficient frames, 9 hidden units and 11 predicted output coefficients. Excellent performance (0.2% error) was reported for a Japanese speaker-independent isolated digit recognition task uttered over telephone lines. This result comparèd favorably with other techniques (0.7% for the DNN [34, 8] and 1.1% for DP-matching [35]) tested on the same data.

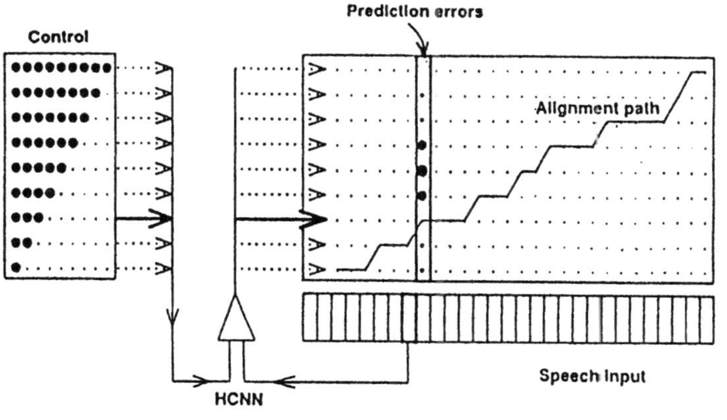

Figure 6. The Hidden Control Neural Network

The model proposed by Levin is similar to the one described above and is illustrated in Fig.6. As before it uses non-linear prediction by neural nets to measure a model's fit to the input data. Unlike the Neural Prediction Model, however, it uses only one single predictor for an entire word and a sequence of varying input flags or "control units" that switch the predictor into alternate modes of operation as time progresses. Similar to "counter nodes"(proposed for spelling correction [36]), these units are used to control the sequential state of the network. The predictor network used 24 speech inputs (12 cepstral and 12 deltacepstal parameters), 30 hidden units, 24 predicted outputs and 8 control input units. The control units turn on sequentially when appropriate and remain on as additional bits are activated ("thermometer" representation). Control transitions (the point at which a new bit is turned on) are determined by Viterbi

Alignment. During training, the Viterbi algorithm determines the state of control unit settings for each speech input frame and applies backpropagation learning to reduce prediction error according to this segmentation. The network was tested on connected digits from the TI-digit database (using male speakers only). Using independent test data but from the same speakers used in training, a word recognition rate of 99.3% was achieved.

4.2 Large Vocabulary Recognition

Large vocabulary word recognition using predictor networks is also possible. For use in large vocabulary recognition, words must here again be decomposed into subword units such as phones or syllables and an optimal model for these units must be trained. Recent work by Tebelskis and Waibel [37] has demonstrated that this can be done without the need for segmentation. In this work, time alignment and connection weights were optimised jointly and the weights of sets of network predictors corresponding to the same phoneme symbols were linked together (as in the TDNN). Experiments with the "Linked Predictive Neural Network" (LPNN) resulted in 94% recognition performance for speaker-dependent isolated word recognition over a database of 234 Japanese words and 90% over a 1000 word vocabulary. The data used in these experiments was given by a confusable subset of the data used for evaluation of the TDNN based system described in the previous section. Performance results on this particular subset were found to be comparable between the two systems.

The operation and training of the LPNN are shown in Fig.7. As before, a set of predictors is assigned to different portions of a word. Here these portions are defined to be phonemes and each occurrence of the same phoneme is modeled by the same set of three predictors. In Fig.7, for example, two words "BAB" and "ABA" may consist of the same phonemes in different order and position. Time alignment of the sequence of predictors is done as before, but all prediction errors assigned to the same phoneme (or portion thereof) train the same predictor net by way of a linkage pattern that defines the legal phoneme sequence of a word. A number of enhancements to this basic scheme have so far been found to be effective. A set of parallel predictors was added to each phoneme model to allow the LPNN to better represent alternate pronunciations and context dependencies. An assignment of each alternate was not predetermined, but the system selects the most suitable alternate based on the prediction errors produced by each alternate. During training the selected alternate is also reinforced by additional training while the others are not. In this fashion, the network automatically generates different models depending on context and pronunciation. A measurable performance improvement was obtained from this technique. Significant improvements were also obtained when phoneme pairs that are only distinguishable on the basis of duration (e.g., in Japanese: "k" vs. "kk") were represented by different sets of predictors. Fig.8 shows an example of processing in the LPNN for an input word "kashikoi". In the top panel, the original spectrogram is shown with 16 spectral coefficients per time frame and time moving from left to right. Underneath, the output predictions of the best predictors (as determined by DTW) at each time frame are displayed. The third panel shows ouput predictions for only one /i/-predictor(s). As can be seen prediction is best in the region corresponding to the final /i/, and degrades in other areas. The final display shows the distance marix obtained for each input frame and for each predictor linked into the word. Alignment is performed based on this matrix and the resulting labeling is shown at the input axis.

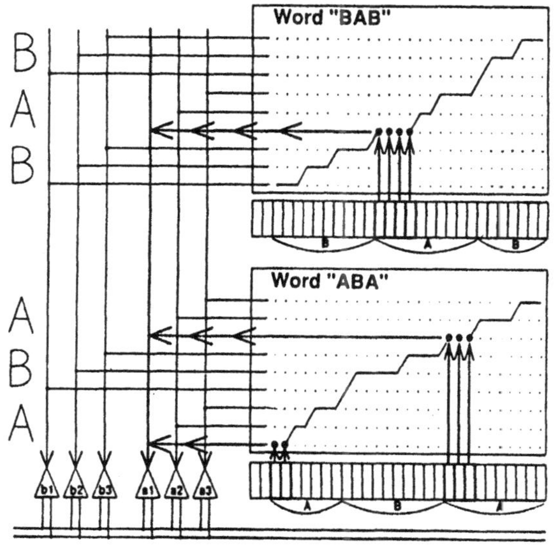

Figure 7. Training a Linked Predictive Neural Net

4.3 Extensions

To further enhance prediction based large vocabulary recognition, several current limitations have to be addressed. The strength of the model described here is that it inherently provides for simple mechanisms for word level integration and optimization. Optimization essentially proceeds top down, in an attempt to suitably represent a word's speech pattern given the phonetic sequence of the word. A possible problem with this approach is the apparent lack of discrimination at the speech pattern level as can be seen in Fig.8 from the relative quality of a single /i/-predictor applied to the entire utterance. This leads to good word level integration, but can result in poor acoustic-phonetic discriminability [38]. The representation is also potentially more sensitive to varying phonetic contexts [38], unless one provides alternate models for alternate contexts or pronunciations. This suggests enhancements similar to those applied to Hidden Markov Models, such as corrective training and context dependent phones. Alternatively, connectionist self-organizing principles could be attempted.

5 Conclusion

In this paper we have reviewed connectionist strategies applied to speech recognition. Reaching beyond mere classification of sound patterns, we have addressed the problem of large vocabulary recognition, where constraints arising from the classification of the underlying speech sounds must be interwoven with the additional constraints of sequentiality and lexical legality. We have on the other hand deliberately limited this discussion to the word level and not addressed sentence level issues that certainly have to be included in complete large vocabulary speech understanding systems (see [39, 40] for further discussion).

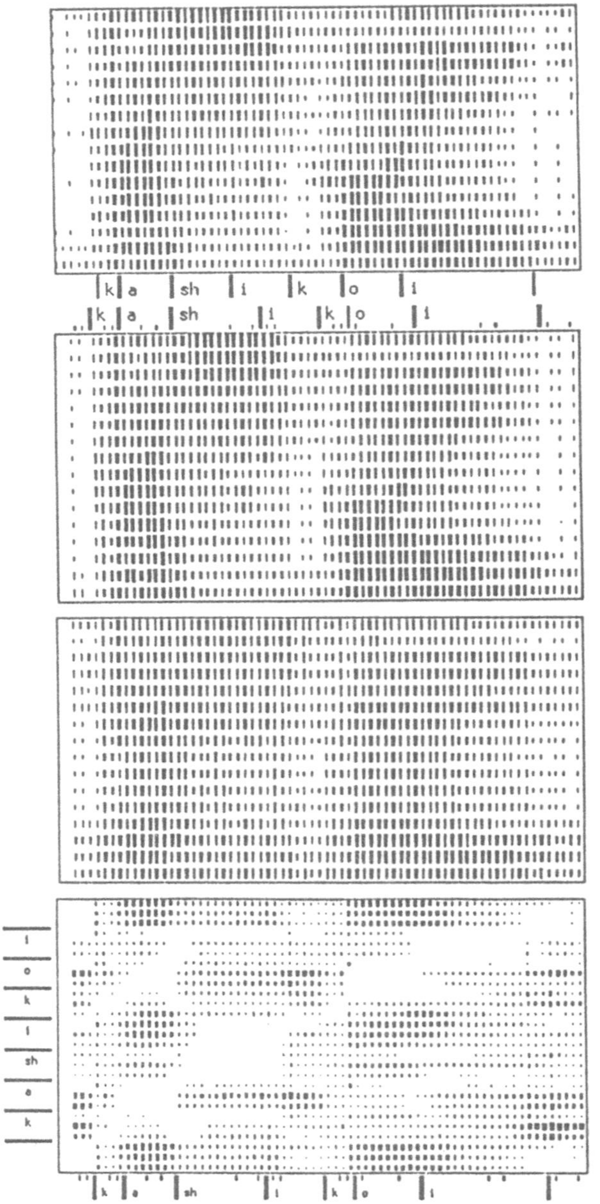

Figure 8. LPNN prediction for the word "kashikoi"
(See text for explanation)

We have developed two different connectionist large vocabulary systems, based on different underlying recognition philosophies. One is based on *classification*, the other on *prediction* of speech. Both strategies achieved excellent recognition performance and performed comparably with respect to each other. Interestingly, either approach displayed different areas of strength and weakness, related to their respective bottom-up or top-down recognition philosophies. While near-term enhancements using either recognition philosophy are being explored, one may wonder what kind of model may ultimately mimick humans' ability to use whatever constraints to recognize speech, be they high level pragmatic or fine-phonetic distinctions. Our search for an understanding of cognitive mechanisms and their realization by machine will undoubtably continue.

References

1. Waibel, A., Hanazawa, T., Hinton, G., Shikano, K. and Lang K., "Phoneme Recognition Using Time-Delay Neural Networks", *IEEE, Transactions on Acoustics, Speech and Signal Processing*, March 1989.
2. Waibel, A., Sawai, H. and Shikano, K., "Modularity and Scaling in Large Phonemic Neural Networks", *IEEE Transactions on Acoustics, Speech, Signal Processing*, December 1989.
3. Moore, R.K. and Peeling, S.M., "Minimally Distinct Word-Pair Discrimination Using a Back-Propagation Network", *Computer, Speech and Language*, Vol. 3, No. 2, 1989, pp. 119-132.
4. Robinson, A.J. and Fallside, F., "A Dynamic Connectionist Model for Phoneme Recognition", *Proceedings of nEuro'88*, IEE, 1988.
5. McDermott, E., Iwamida, H., Katagiri, S. and Tohkura, Y., *Shift-Tolerant LVQ and Hybrid LVQ-HMM for Phoneme Recognition*, Morgan Kaufmann, 1990.
6. Burr, D.J., "A Neural Network Digit Recognizer", *IEEE International Conference on Systems, Man, and Cybernetics*, October 1986.
7. Burr, D.J., "Experiments on Neural Net Recognition of Spoken and Written Text", *IEEE Transactions on Acoustics, Speech; Signal Processing*, July 1988, pp. 1162-1168.
8. Sakoe, H., Isotani, R., Yoshida, K., Iso, K., and Watanabe, T., "Speaker-Independent Word Recognition Using Dynamic Programming Neural Networks", *IEEE International Conference on Acoustics,Speech, and Signal Processing*, May 1989, pp. 29-32.
9. Isotani, R., Yoshida, K., Iso, K., Watanabe, T. and Sakoe, K., "Dynamic Neural Network --- A New Speech Recognition Model Based on Dynamic Programming and Neural Network", *IEICE Technical Report*, September 1988.
10. Bottou, L-Y., "Reconnaissance de la Parole par Reseaux multi-couches", *Proceedings of Neuro-Nimes 88*, November 1988.
11. Bottou, L., Fogelman-Soulie, F., Blanchet, P., Lienard, J.S., "Experiments with Time-Delay Networks and Dynamic Time Warping for Speaker Independent Isolated Digits Recognition", *Proceedings of the Eurospeech*, September 1989.
12. Franzini, M.A., Lee, K.F.,Waibel,A.H., "Connectionist Viterbi Training: A New Hybrid Method for Continuous Speech Recognition", *IEEE International Conference on Acoustics,Speech, and Signal Processing*, April 1990.
13. Iso, K. and Watanabe, T., "Speaker-Independent Word Recognition Using A Neural Prediction Model", *IEEE International Conference on Acoustics,Speech, and Signal Processing*, IEEE, April 1990.

14. Wong, M.K. and Chun, H.W., "Towards a Massively Parallel System for Word Recognition", *IEEE International Conference on Acoustics, Speech, and Signal Processing*, April 1986, pp. 37.4.1-37.4.4.

15. Lippmann R.P. and Gold, B., "Neural-Net Classifiers Useful for Speech Recognition", *IEEE International Conference on Neural Networks*, June 1987.

16. J. L. Elman, "Finding Structure in Time", Tech. report CRL Technical Report 8801, University of California, San Diego, 1988.

17. Bridle, J.S., "Alpha-Nets: A Recurrent Neural Network Architecture with a Hidden Markov Model Interpretation", *Speech Communication*, 1990, (to appear)

18. Young, S.J., "Competitive Training: A Connectionist Approach to the Discriminative Training of Hidden Markov Models", Tech. report CUED/F-INFENG/TR.41, Cambridge University, March 1990.

19. Waibel, A., Hanazawa, T., Hinton, G., Shikano, K. and Lang K., "Phoneme Recognition Using Time-Delay Neural Networks", Tech. report TR-1-0006, ATR Interpreting Telephony Research Laboratories, October 1987.

20. Hampshire, J. and Waibel, A., "The Meta-Pi Network: Connectionist Rapid Adaptation for High-Performance Multi-Speaker Phoneme Recognition", *IEEE International Conference on Acoustics,Speech, and Signal Processing*, IEEE, April 1990.

21. Waibel, A., "Modular Construction of Time-Delay Neural Networks for Speech Recognition", *Neural Computation, MIT-Press*, March 1989.

22. Sagisaka, Y., Takeda, K., Katagiri, S. and Kuwabara, H., "Japanese Speech Database with Fine Acoustic-Phonetic Transcriptions", Tech. report, ATR Interpreting Telephony Research Laboratories, May 1987.

23. Miyatake, M., Sawai, H., Shikano, K., "Integrated Training for Spotting Japanese Phonemes Using Large Phonemic Time-Delay Neural Networks", *IEEE International Conference on Acoustics, Speech, and Signal Processing*, May 1990.

24. Bourlard, H. and Wellekens, C.J., "Speech Pattern Discrimination and Multilayer Perceptrons", *Computer, Speech and Language*, Vol. 3, 1989, pp. 1-19.

25. U. Bodenhausen, "The Tempo Algorithm: Learning in a Neural Network with Adaptive Time-Delays", *Proceedings of the IJCNN*, IJCNN, January 1990, pp. 597-600.

26. Bourlard, H. and Wellekens, C.J., "Links between Markov Models and Multilayer Perceptrons", *Advances in Neural Network Information Processing Systems*, Morgan Kaufmann, 1988.

27. N. Morgan and H. Bourlard, "Continuous Speech Recognition Using Multilayer Perceptrons with Hidden Markov Models", *IEEE International Conference on Acoustics,Speech, and Signal Processing*, IEEE, April 1990, pp. 26.S8.1.

28. Rabiner, L.R, Wilpon, J.G. and Juang, B.H., "A Segmental K-Means Training Procedure for Connected Word Recognition", *AT&T Technical Journal*, May 1986.

29. Niles, L.T. and Silverman, H.F., "Combining Hidden Markov Model and Neural Network Classifiers", *IEEE International Conference on Acoustics,Speech, and Signal Processing*, IEEE, April 1990, pp. 417-420.

30. Tamura, S. and Waibel A., "Noise Reduction Using Connectionist Models", *IEEE International Conference on Acoustics, Speech, and Signal Processing*, April 1988, pp. S12.7.

31. Atal, B., "Non-Linear Mapping between Speech Codes", Personal Communication

32. Lapedes A. and Farber R., "Nonlinear Signal Processing Using Neural Networks; Prediction and System Modeling", Tech. report LA-UR-87-2662, Los Alamos National Laboratory, 1987.

33. Levin, E., "Speech Recognition Using Hidden Control Neural Network Architecture", *Proceedings of the International Conference on Acoustics, Speech and Signal Processing*, IEEE, April 1990.

34. Sakoe, H., "Dynamic Neural Network --- A New Speech Recognition Model Based on Dynamic Programming and Neural Network", *IEICE Technical Report*, December 1987.

35. H.Sakoe, S.Chiba, "Dynamic Programming Optimization for Spoken Word Recognition", *IEEE Transactions on Acoustics, Speech, Signal Processing*, Vol. ASSP-26, No. 1, February 1978, pp. 43-49.

36. Kukich, K., "Back-Propagation Topologies for Sequence Generation", *IEEE International Conference on Neural Networks*, 1988, pp. 301-308.

37. Tebelskis, J. and Waibel, A., "Large Vocabulary Recognition Using Linked Predictive Neural Networks", *IEEE International Conference on Acoustics,Speech, and Signal Processing*, IEEE, April 1990.

38. Endo, T., Tamura, S. and Nakamura, M., "Phoneme Recognition Using Neural Prediction Models", Tech. report TR-I-0107, ATR Interpreting Telephony Research Laboratories, August 1989.

39. Waibel, A. and Lee, K.F., *Readings in Speech Recognition,* Morgan Kaufmann Publishers, San Mateo, CA, 1990.

40. Furui, S. and Sondhi, M.M., *Advances in Acoustics and Speech Processing,* Marcel Dekker, Inc., New York, NY, 1990.

The cortical column as a model for speech recognition: principles and first experiments

Frédéric Guyot, Frédéric Alexandre, Catherine Dingeon, and Jean-Paul Haton

CRIN/INRIA, BP 239, 54506, Vandoeuvre-les-Nancy, France

Abstract: Connectionist models, also known as neural networks, have been widely studied during the past few years. Applications concern various tasks in the fields of pattern recognition and signal processing, especially automatic speech recognition. This chapter presents the basic properties of these models and the different problems in the area of speech recognition to which they have been applied so far. The classical models of neural networks are also briefly recalled. We then concentrate on a particular model grounded on neuro-biological data, the cortical column. The characteristics of the model are given and we then present the architectures of two systems based on the cortical column for solving two different problems of speech recognition, i.e. acoustic-phonetic decoding and isolated word recognition.

Keywords: Connectionism, neural networks, cortical column, acoustic-phonetic decoding, word recognition, learning.

1 Introduction

Automatic speech recognition remains one of the most difficult and challenging problem to pattern recognition and artificial intelligence. Significant progresses have been made in the past two decades with methods essentially based on dynamic programming, elastic matching and stochastic modelling. However performances obtained so far are not yet sufficient for most potential applications.

Recently, so called "neural" network models have become very popular in speech recognition as well as in other fields of automatic perception. These models are ba-

NATO ASI Series, Vol. F 75
Speech Recognition and Understanding. Recent Advances
Edited by P. Laface and R. De Mori

sically networks of elementary, highly interconnected processing elements. Although they have not proved their superiority over classical models, such connectionist models present characteristics that make them attractive: high discriminative power, learning capacities, noise immunity, easily accessible and understandable formalism.

As a matter of fact there exist a large number of techniques and models under the generic term of neural network. We present in section 2 the most common neural models which have been used in speech recognition together with the kind of problems to which they have been applied. A particular approach consists in taking into account neurobiological information about the human cortex. We propose a new model based on this approach, the cortical column. The basic properties of this model are summarized in section 3. We then describe its application to two important fields of speech recognition, i.e. acoustic-phonetic decoding (section 4) and word recognition (section 5). Finally, we present on-going research carried out in order to improve and extend the cortical column model.

2 Neural networks and speech recognition

Although there exists a large variety of neural network models, some basic properties are shared by all these models. They are all made up, according various architectures, of a large number of processing elements. Each element, or "neuron", has a very limited processing power and the connectivity between elements is very high. An important property is that in such models the storage of information as well as all the processing are totally delocalized.

Neural networks present characteristics that make them suitable for use in the fields of signal processing and pattern recognition, especially speech recognition:

- good discriminative power that makes it possible to solve non linear problems through powerful learning algorithms,
- inherent parallelism,
- noise immunity,
- ease of implementation.

However most of these models have problems for taking into account time since this parameter is very difficult to modelize. Several solutions have been proposed within the neural network community for solving this problem. They will be briefly presented later on.

Neural networks have been used so far at different levels of the speech recognition process, especially:

- *parametrization of the speech wave*: experiments were carried out in order to provide networks capable of transforming speech parameters so as to obtain better performances in different situations: adaptation of a speech recognizer to a new unknown speaker, recognition of speech contaminated by a background noise, etc.,
- *feature extraction*: determination of voicing and F0, recognition of place of articulation [12,3],
- *acoustic-phonetic decoding of continuous speech*: an important effort has been devoted to this important problem that will be detailed below when the different connectionist models are presented,
- *isolated word recognition* by whole word template matching. This is the most straightforward use of a neural network that will not be discussed here since it only consists of the recognition of static two-dimensional patterns,
- *lexical access* from phonetic description. An application of the cortical column model to this problem is presented in section 5,
- *linguistic modelling*: neural networks have also been studied for their application to the modelling of a spoken language at different linguistic levels: phonology, morphology, syntactic categories [20,23].

We will now briefly recall the different neural network models that have been used in speech recognition.

Multi-layer perceptrons are the most commonly encountered. The standard form of this model is unable to handle the time parameter. Therefore, several solutions have been proposed in order to extend it:

- time-delay neural networks (TDNN): cf. [24] or the chapter by A. Waibel in this book,

- contextual and recurrent networks: in contextual perceptrons entries include the context in which the input stimulus to be recognized was produced (cf. [5] and the chapter by F. Fallside in this book). Results already obtained show that recurrent nets perform as well as stochastic models for acoustic-phonetic decoding tasks.

Boltzmann machines with sophisticated algorithms based on simulated annealing principles have also been studied in the framework of isolated vowel recognition [21]. Recognition is slightly better than with a perceptron but the computation time is highly prohibitive.

Mixed models have also been recently proposed in order to combine neural networks with classical methods. For instance a model combining a perceptron with non linear time alignment by dynamic programming has been used for word recognition [22]. Similarly, a perceptron has also been coupled with a stochastic time alignment based on Viterbi algorithm [13].

Such hybrid models are interesting but they do not represent the only way of mixing neural networks with stochastic modelling.

Another solution is represented by *alphanets* (cf. the chapter by J. Bridle in this book). It constitutes an attractive solution for combining neural and stochastic models.

Most of the so-called neural networks use neuron-like processing elements that differ greatly from neurons of the cortex. A rather different approach consists in relying more closely to the neuro-biological data which are presently available. One of the first attempts in this direction is the work on unsupervised learning in phonotopic maps which lead recently to an efficient phonetic decoder [16]. A similar solution, using a guided propagation method within a topographic map, has also been proposed for word recognition [4]. The model of cortical column that will now be described also belongs to this category.

3 The Model

The first description of a columnar vertical organization of the cortex has been proposed by Lorente De No at the begining of the century [18]. From a large set of neurobiological data, Burnod [6] outlined a possible modelization of the so called "cortical column", involving the whole cortex and taking into account classical psychological features.

This new automaton for cortex-like networks presents a more sophisticated behavior than the usual formal neuron and thus, can deal with symbolic and large scale problems. We have already described elsewhere the basic mechanisms for computer simulations of the cortical column [1,2,9,10,11]. The reader is invited to refer to these publications for the basic properties of the cortical column. We will now concentrate on the more recent theoretical advances concerning this model.

Architecture

The cortical column automaton distinguishes four kinds of input/output connections (figure 1), each devoted to a specific task, performing various dynamic processing, different memorizations and using particular connectivity patterns. Divisions inside the column enable it to receive and to process information selectively, according to its origin (from the external world, from the same area or from another area of the cortex).

1. The first part of the column (layer I to III of the cortex) carries out all the reciprocal but not symetrical relations with the other columns, in a long range as well as locally. These relations perform temporal memorization and activity propagation throughout the system.

2. The second part (layer IV) receives the information flow which codes the sensory stimuli from the external world and effects output (layers IV and V) toward the external world by various motor actions (speech production,etc.). These connections are organized in receptive fields and encode external non-temporal invariants.

3. The last part is devoted to non-temporal interaction with neighboring columns (interneurons shared by two neighboring columns).

Consequently, the connectivity between columns can be summed up by compact neighboring connections and n long range connections, where 2^n is the number of areas involved in the simulated process. Information propagates from area to area with hypercubic connections, or inside an area from place to place with local connections.

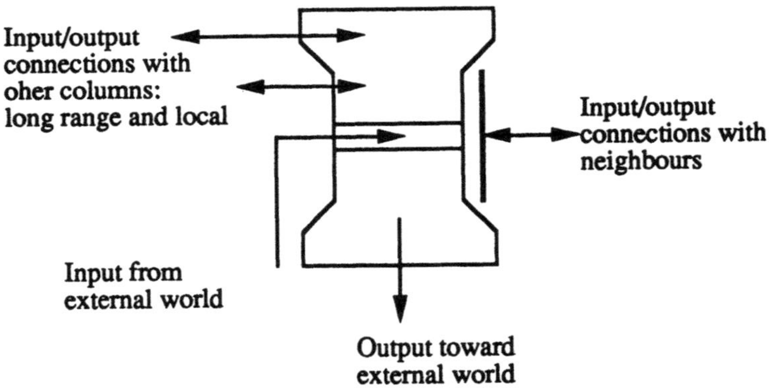

Figure 1. connections

Activation

Whatever the kind of relation may be in use, the states of activity of a column are limited to three values. These states can be interpreted in terms of logic as yes (denoted by E2), no (E0) and perhaps (E1), and, of course, building the dynamic behaviour rules is linked to this interpretation. For each kind of relation, the output of a column is given in a truth table as a function of the inputs, modulated by learning in some cases but more often corresponding to a stereotyped activity propagation.

Learning

Three mechanisms have been developped according to neurobiological data and in order to provide precise behaviors, i.e.:

1. As concerning reciprocal relations (local or long range ones) between two units, a column is able to memorize specific patterns of activation with temporal aspects (before, after). Such spatio-temporal patterns represent causal relation; for example, if a unit B is always active before a unit A, A is viewed as a consequence of B; thus, when A is desired (state E1), it will try to achieve its *sub-goal* B. For long range connections, these rules develop a dynamic logical mechanism, where a problem is solved by recursively calling goals and sub-goals in a sequential way.

On the other hand, it also permits for local connections to memorize dynamic continuous stimuli which are then represented by a succession of neighbouring activations. Another consequence is that, thanks to the propagation of activity from one unit to another, two units can communicate without physical link, preventing us from full connectivity.

2. The division principle is responsible for the auto-organization in the areas and determines interesting invariants of the external world. This principle also provides generalization [6].

   The division phenomenon has been mainly observed in associative areas which are less specified than sensorimotor areas in the child cortex. On the other hand, the parietal associative area, which recombines visual and somatic aspects, will support by learning important modifications in order to represent those invariants that will be useful for somato-visual coordination. The modifications are the following: in associative areas unspecified modules, with a width of around 500 microns and a length of several millimeters, will be recursively differentiated in order to lead to more precise and more intense activity in the area when a stimulus is presented. As long as a full adaptation is not reached, the process continues [6]. We have adapted this auto-organization mechanism to data processing. Before learning, associative areas are not differentiated. When recognition fails, all modules having a non significant relation (middle strength) [1,9] with the solution are recursively split into two half-sized modules. A new module has reduced relations with the external world and can therefore be more representative of a given pattern. On the other hand, a module which has been divided always exists and can trigger a response for another pattern. This mechanism is only used when necessary and yields areas with a maximum contrast state and a minimal division distribution.

3. The last learning process holds on external connections, thanks to lateral inhibition and a local propagation of the receptive fields synaptic weight modifications. These relations act in a self-organizing way and extract local invariants from external data flows. For example, at the time of birth, visual and somatic areas are already specified to detect invariants of the external world. Columns detect orientations

in visual areas [14] and represent parts of the body in somatic areas [25]. During critical developmental periods, deprivation can induce limited plasticity in these maps [19].

The areal structure

As implicitly mentionned above, all cortical columns share the same properties and are only specified by their external inputs. Moreover, the netwok structure is a crucial parameter that determines both qualitatively (depending upon the number of areas) and quantitatively (depending upon the number of units per area) the abilities of the system to solve cognitive problems.

Burnod has also described a model of cortical area organization within an hypercubic structure. Each functionnally distinguished area can be described along 6 binary dimensions: auditive or visual, temporal or parietal, stimuli or time dependent, left or right hemisphere... Thus the whole cortex is assumed to be built on 6 major symetries involving 2^6 functionnal areas.

Behaviour of the network

As for classical formal neuron simulations, each unit evaluates in parallel its inputs and computes its outputs following rules and learning. But a symbolic interpretation of the activities can be easily given, since the unit provides a three valued logic and a precise areal structure.

From the answers, the system has learned some hierarchical goals and sub-goals trees, and is potentially able to trigger them until a stimulus (leave of the tree) is encountered and leads the analysis.

Moreover, inside an area, the data representation goes through a generalization and differentiation process and can be sequentiated through continuous pathways, while the incoming stimuli are progressively gathered into local Gestalt by the receptive fields.

4 Acoustic phonetic decoding of speech

4.1 Position of the problem

A major point lies in the initial encoding which will determine the network processing aspect. Unfortunately, the auditive cortex has not yet been studied as deeply as the visual cortex, because of the difficulties inherent to speech recognition. The first speech recognition experiment we made was more a global recognition of speech spectrograms than a true auditive cortex simulation. We then decided to fit more closely to human features and to extract from different cognitive fields the constraints and the properties necessary to build an efficient distributed speech recognition network.

Biological data.

From speech production to phoneme recognition, the human processing of speech is in three major steps from the internal ear to the cortex through the thalamus:

- The role of the cochlea consists in breaking down the signal into its frequency components and the nervous fibers code the intensity of sounds up to a saturation threshold (around 80 dB), for a characteristic frequency. An adaptation phenomenon makes the fibers react preferentially to a new information brought by the signal [8] (e.g. attack after a silence). Phonetic experiments [8] have shown that the cochlea detects frequency in a non-linear way and has privileged bands of frequency which enhance the relative importance of low frequencies against the high frequencies.

- The thalamic nucleus CGM selectively amplifies the signal according to cortical feedback actions. It is also responsible for intensity contrast enhancement and finally distributes auditive cues onto the cortical surface [15].

- The auditive cortex can be considered as a set of two-dimensionnal maps where the frequency axis is often represented [15]. From these maps, activity propagates in associative areas toward temporal areas for semantic relations with visual areas and toward parietal areas linked to the vocal articulators (tongue, lips...) for speech production. These two parallel information pathways (i.e temporal and parietal) are both involved in speech recognition.

4.2 Application to vowel and fricative recognition

We now describe more precisely the applications we have implemented for the recognition of vowels and fricatives. The pertinent cues vary from one class of phonemes to another; though the recognition principle is the same, initial encoding changes with the class and leads consequently to a specific network for the recognition of each class, instead of a simple global network.

The theory developped in section 3 provides units able to distinguish activity distribution on a bidimentionnal input map. The cues represented in such an area must be carefully chosen. Neurobiological data are very helpful to determine plausible preprocessing and initial encoding. For example, it seems reasonable to have one axis representing a frequency decomposition, close to the type of processing done by the cochlea. Biological observations show that frequency is often represented and recombined in the auditive cortex [15]. That is the reason why we decided to introduce a second frequency axis in our maps, instead of a time axis as in spectrograms. The differences between the two axes are explicited below for the two cases (vowels and fricatives). In order to get a more precise discrimination, we have also used another cortical principle which has been extensively described [14]: in a map, inside a set of columns representing the same indices, local variations make them react to another kind of more precise indices. In our application, such local indices have been determined according to biological considerations (for example, neural structures often evaluate local activity differences) and are also described below for vowel and fricative recognition.

Structure of the network

We now present the structure of the networks for fricative and vowel recognition. This structure is the same for both cases.

- First, the auditory information is extracted and preprocessed outside the network in a way depending upon the class of the phoneme.

- This information is then tonotopically distributed on the auditory area surface. These two points are detailed in the next section.

- The "motor" output area represents a stereotyped distribution of the phonemes: eleven outputs for the french vowels; six outputs for the french fricatives.

- Between the auditory and motor areas, the associative area combines as usual the two representations. In this area, the division principle is used and make it possible to learn a phoneme as a set of paradigms representing its variations.

Two points are important to notice:

(i) All kinds of connections mentioned in section 3 are used in the areas. Feedforward connections carry information from sensory to motor area. The feedback mechanism is used to return back to information when necessary. Intra-areal connections couple or uncouple narrow modules.

(ii) The network is not static but dynamic and receives an asynchronous succession of inputs. Time is not explicitly expressed in the inputs but implicitly managed by the network. It is taken into account with the evolution of activity on the map. This evolution is measured by the time sensitive connections of the column.

Initial encoding

Preprocessing consists in computing a cepstral analysis and in extracting from this analysis primary indices (axes of the area) and secondary indices (local variations for a given primary indice).

For vowels encoding, frequency value of formants has been selected as primary indices. Secondary indices consist of ratios between formant energy maxima. All these data are recombined on the input map. Formants beneath 1300 Hz are reported on the first axis. The second axis supports the other formants.

The same kind of distribution is used for fricatives. The logarithmic scale of the ear is simulated by a strong frequency smoothing above 2500 Hz. Frequency values of maxima of energy and local spectral slopes are extracted as primary indices. The value of these maxima (dB) are placed on the secondary indices.

Experimental results

We have carried two sets of experiments, both for single speaker recognition. For fricative recognition, the network was trained on isolated phonemes but tested on continuous speech (25 words including 50 fricatives). A learning corpus with 7 examples for each fricative leads to 96% of correct recognition. Figure 2 represents the divisions obtained in this case. In the test corpus, the network recognizes 48 fricatives and, moreover, never proposes a response for non fricative phonemes.

An important point is that when the learning corpus consists of only one example per fricative, the associative area is clearly less divided but permits 94% of recognition. This point shows that the first divisions are very important for discrimination. The subsequent divisions are useful for fine adjustments.

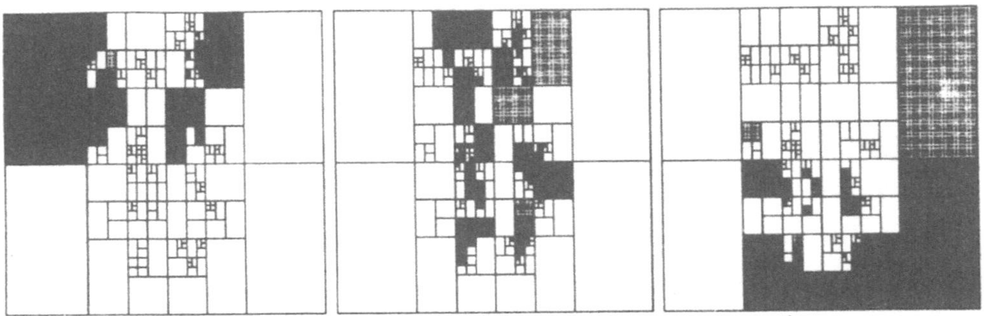

Figure 2. divisions for fricative recognition

Concerning vowels, 20 examples per phoneme were recorded for learning and 10 examples for testing. The recognition rate reaches 99% for the four vowels /a & i o/ and 87% for the seven vowels /a & i o y e eu/. The network required from 3 to 6 samples per vowel class for good discrimination. It takes much less training time than classical supervised classifiers [17]. While concerning information representation and use, the cortical column model is closer to Kohonen's feature maps (cf [16] and section 2) rather than F1-F2 plane multilayer perceptron representation, because of secondary indice discrimination and neighbourhood relations.

5 Lexical representation and word recognition

We now want to get over another stage in speech processing: isolated word recognition. Starting from the acoustic-phonetic decoding net present configuration, we will assume that an output map contains as many modules as phonetic labels previously learned in a supervised way. The activity evolution on the output map, when a speech flow is presented to the input map, forms the succession of the phonetic answers of the net.

In this new problem, data to be processed are no longer relatively temporally stable forms, but temporal successions of recognized shapes.

5.1 Biological data

As recalled in section 3, two cortical modules are able to learn temporal links from cause to effect, one module representing the precondition, and the other the goal. If the goal realization does not depend on external conditions, then it is always the same cause which produces the same goal. This phenomenon allows memorization of inconditionnal temporal chains of several modules, permitting a global goal realization.

For the young child, memorization and recognition of words take place in the temporal auditive areas of the cerebral cortex [6]. Indeed, temporal areas are specialized in memorization of activity sequences organized in a regular rythm. Furthermore, production and recognition of words share the same inconditionnal sequences in these areas. In these chains, the goal consists in activating the meaning circuit of a single word; subgoals are somato-auditive matchings (phonetic production or recognition). Consequently, each word in the lexicon is memorized as an inconditionnal sequence, which is built for production, by successive stages until getting a complete matching with the adult model.

5.2 Application

We complete the previous net with an output map for words, and an associative map between the phonetic output map and the word map (Figure 3). It should be noticed that the word map has been introduced for convenience reasons, whereas in the cortex, a word is associated with a set of images or concepts originating from polysensory recognition algorithms.

Initially, the associative map is divided into as many modules as phonetic answers. This map will be next liable to division mechanism. The learning rules recalled in section 3 allow:

- Memorization of phonemic sequence thanks to temporal links between associative modules.

- Association of a sequence with a word thanks to links between associative modules and words.

Learning of a new word begins with its hearing. The recognition mechanism activates the word output and the corresponding phonemic sequence, the nearest to what has been heard. The new sequence is learned starting from associative phonetic modules of the nearest word. In case of failure, i.e. no complete match with the heard model, then the involved modules divide themselves, and the sequence is learned again; this is repeated until success.

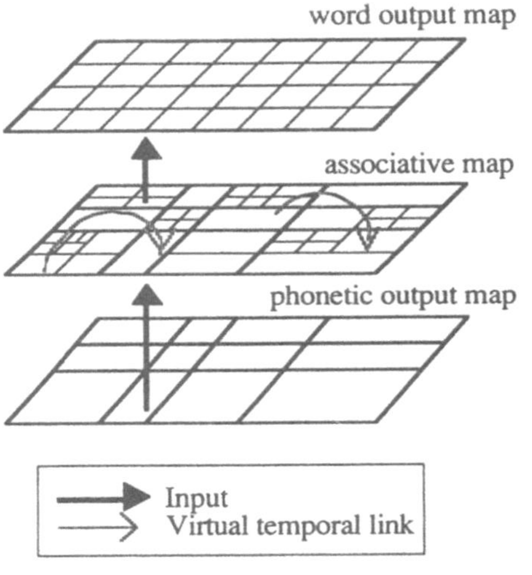

Figure 3. word recognition network

Thus, production and recognition are carried out on the same chains, and a word is learned as a fixed phonemic sequence which amounts to the sequence of articulations necessary to pronounce it. The recognition mechanism overcomes variations of the input sequence compared to the memorized one.

It is important to notice that words are memorized in a way that accounts for their phonetic similarity: similar words will share the same phonetic modules; another consequence of that is to minimize the necessary memory, since modules divide themselves only if needed.

5.3 Experimental results

In order to test our isolated word recognition system, we recorded a corpus of 5 sentences of continuous speech, each pronounced by 9 male speakers. Each sentence contains five words on the average. Word segmentation was made manually.

The system gets as input the results of the APHODEX acoustic-phonetic decoding system, developped in our group [7]. For each speech segment we take into account the best first phonetic answers given by APHODEX. It should be noticed that this phonetic decoding is totally speaker independent.

A first series of tests was carried out on small vocabularies each corresponding to one sentence (5 words on the average). The recognition rate, when taking into account all of the first ex aequo words, is 86% on the average, and 96% when taking the first and second.

A second series was made by gradually gathering the previous vocabularies and their corresponding corpus.

A last series of tests was made by including the all corpus in greatest vocabularies.

The results of these three series of experiments are gathered and shown in figures 4, 5 and 6.

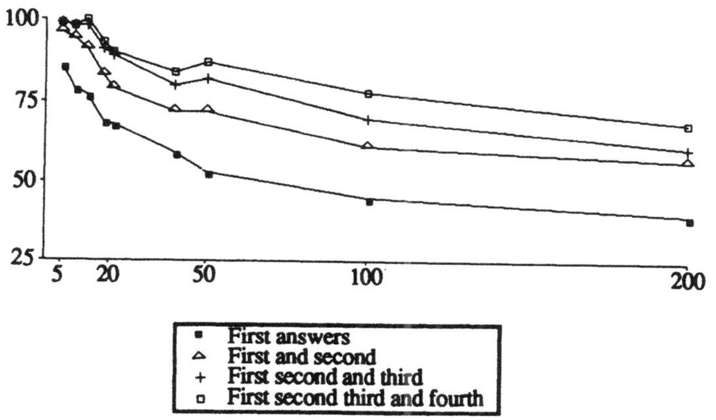

Figure 4. Recognition percentage versus vocabulary size when taking into account the first proposed words, the first and the second ones, etc

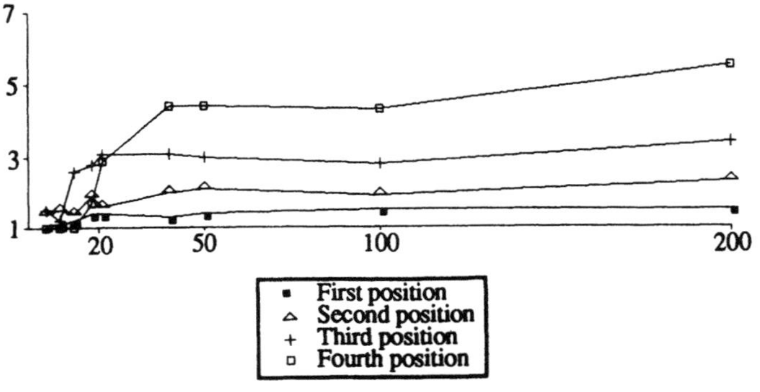

Figure 5. Number of ex-aequo among which the correct answer is found when it is found in first, second etc. position, versus the vocabulary size

Number of	
Words	Answers
5	1
9	2
13	2
18	2
21	3
39	3
51	3
101	5
200	7

Figure 6. Number of answers to be taken into account in order to get exactly 80% of recognition, versus the vocabulary size

It can be noticed that the recognition rate decreases when the vocabulary size increases, since the incertitude on words to be recognized then increases too.

In future works we want to connect the two recognition nets of phonemes and words, in order to make the two nets interact; the word recognition net could in particular take into account the proximities between stimuli on the associative map of the phonetic net.

6 Conclusion

We have presented in this paper a biologically inspired connectionist model for speech recognition. Particular features make this model different from other connectionist approaches. Beyond a good biological plausibility, this model allows for a low connectivity and specific entries to the information flows. The way in which time is managed in the network makes the model well adapted to Acoustic Phonetic Decoding and to Isolated Word Recognition.

It must be noticed that, because of its cortical origin, this connectionist unit is intended to be used for a variety of adaptive and cognitive functions. Up to now, the model has been applied to classical Image Processing problems: character recognition and biomedical image interpretation [1,2,11]. Simulations have shown the unit ability to associative recognition and to more cognitive functions like image interpretation.

We are now investigating new applications for our model. From a speech processing point of view, we can mention two research axes. These points are both mentionned in the biological theorical description of the cortex by Burnod [6]. First, a sequence of phonemes that forms a word can be associated with a meaning (for example, the result of a visual recognition performed by another network). Linking such visual and auditive networks can be very helpful for structured learning in these both fields. Secondly, Burnod proposes a plausible description for two fundamental mechanisms in language acquisition: temporally oriented frontal cortical areas are supposed to support syntactic rules; semantic representation could be localized in specific areas corresponding to the semantic categories (e.g: motor areas for verbs representing actions).

7 References

1. Alexandre, F., Burnod, Y., Guyot, F., Haton, J. P.: La colonne corticale, unité de base pour des réseaux multicouches, C R Acad. Sci. Paris, t. 309, Série III, p. 259-264 (1989)
2. Alexandre, F., Guyot, F., Haton, J.P.: A connectionist network with two complementary visual processing systems for x-ray image interpretation, INNC'90, Paris, 1990
3. Bengio, Y., Demori, R.: Use of Neural Networks for the Recognition of Place of Articulation, Proc. ICASSP-88, New-York, 1988
4. Béroule, D.: Guided Propagation inside a Topographic Memory, Proc. 1st Int. Conf. Neural Networks, San Diego, 1987
5. Bourlard, H., Wellekens, C.: Multi-Layer Perceptrons and Automatic Speech Recognition, Proc. IEEE First Int. Conf. on Neural Networks, San Diego, 1987
6. Burnod, Y.: An adaptive neural network: The cerebral cortex, Masson, Paris, 1988
7. Carbonell, N., Fohr, D., Haton, J. P.: APHODEX, An Acoustic-Phonetic Decoding Expert System, Int. J. Pattern Recognition and Artificial Intelligence, 1, n 2, p. 207-222, 1987

8. Delgutte, B.: Codage de la Parole Dans le Nerf Auditif, Thèse, Université Paris VI, 1984
9. Dingeon, C., Alexandre, F., Guyot, F., Haton, J. P.: Un autre apprentissage cortical: Différentier pour généraliser, Proc. Neuro-Nimes'89, Nimes, 1989
10. Guyot, F., Alexandre, F., Haton, J.P.: Toward a continuous model of the cortical column: application to speech recognition, Proc. ICASSP-89, Glasgow, 1989
11. Guyot, F., Alexandre, F., Haton, J.P.: Principles and applications of the cortical column symbolic neural model, INNS'90, San Diego, 1990
12. Howard, I. S., Huckvale, M. A.: Training Feature Detectors for Use in Automatic Speech Recognition, FASE Speech' 88, Edimburgh, 1988
13. Huang, W., Lippman, R.: A Neural Net Appraoch for Speech Recognition, Proc. IEEE-ICASSP, New-York, 1988
14. Hubel, D. H., Wiesel, T. N.: Receptive fields binocular interaction and functional architecture in the cat's visual cortex, J. Physiol. Lond., 160, p. 106-154, 1962
15. Knudsen, E. I., du Lac, S., Esterly, S. D.: Computational Maps in the Brain, Neurosciences, 10, p. 41-65, 1987
16. Kohonen, T.: The "Neural" Phonetic Typewriter, Computer, p. 11-22, March 1988
17. Lippman, R. P.: Review of neural networks for speech recognition, Neural Computation, 1, 1-38, 1989
18. Lorente de No, R.: The cerebral cortex: architecture, intracortical connections and motor projections, In Physiology of the nervous system, J.Fulton, ed., Lond., Oxford University Press, p. 291-301, 1938
19. Merzenich, M. M., Jenkins, W. M., Middlebrooks, J. C.: Observations and hypothesis on special organisational features of the central auditory nervous system, Dynamic aspects of neocortical functions, G.M.Edelman, W.E.Gall, W.M.Cowan eds, John Wiley New-York, 1984
20. Nakamura, M., Shikano, K.: A Study of English Word Category Prediction Based on Neural Networks, Proc. IEEE-ICASSP, Glasgow, 1989
21. Prager, R. W. et al.: Boltzman Machines for Spech Recognition, Computer Speech and Language, 1, 1, 1986
22. Sakoe, H. et al.: Speaker-Independent Word recognition Using Dynamic Programming Neural Networks, Proc. IEEE-ICASSP, Glasgow, 1989
23. Touretzky, D. S. et al.: Phonological Rule Induction: An Architectural Solution, Proc. 12th Annual Conf. Cognitive Science Soc., Lawrence Erlbaum, 1990
24. Waibel, A. et al.: Phoneme Recognition Using Time-Delay Neural Networks, Proc. ICASSP-88, New-York, 1988
25. Woolsey, C. N.: Organization of somatic sensory and motor areas of the cerebral cortex. In Biological and biochemical bases of behaviour, H. F. Harlow and C. N. Woolsey eds, University Wisconsin Press, Madison, p. 63-81, 1958

Radial Basis Functions for Speech Recognition

Yoshua Bengio

School of Computer Science, McGill University, 3480 University str., H3A2A7, Montreal, Canada

Abstract: the purpose of this paper is to study the application of Radial Basis Functions (RBF) to automatic speech recognition. Results of several experiments with these networks on the recognition of phonemes for the TIMIT database are presented, including an experiment on a recurrent network of RBFs.

1. Introduction

An alternative to networks of sigmoid units trained with backpropagation is a particular type of connectionist model based on a local representation called Radial Basis Functions network (RBF) and introduced in section 2. This type of model has several advantages: solid mathematical foundations [13], neurological [10] and psychological [7] plausibility as well as fast implementation (see section 4.6). Phoneme recognition experiments were performed with this model on the TIMIT database. The results presented in section 3 regard the advantage of spectral *vs* ceptral representation, the advantage of using information about the classes in the preprocessing, the effect of architecture and the representation of context on generalization as well as the combination of a recurrent network of sigmoid units with RBF units.

2. Radial Basis Functions Models

Radial Basis Function (RBF) units produce an output which depends on the distance between the input point and a prototype point in the input space. The output of the network can be written as the weighted sum of the outputs of those RBF units:

$$Fi(\mathbf{X}) = \sum_j W_{ij} \; h(\| \mathbf{X} - \mathbf{P}j \|) \tag{1}$$

$$h(r) = e^{-r^2} \tag{2}$$

where $\mathbf{X}$ is the input pattern vector, $\mathbf{P}j$ is the prototype point associated to RBF unit j and we can choose for h() for example a gaussian, as in (2). The norm in (1) can be weighted by a covariance matrix. Poggio and Girosi [13] show that if our objective is function approximation (approximate a $\mathbb{F}(\mathbf{X})$ with $\mathbf{F}(\mathbf{X},$ some parameters)), seen as hypersurface reconstruction, given a noisy training set of example patterns $\{\mathbb{F}p(\mathbf{X}p),\mathbf{X}p\}$ and a-priori knowledge that the

NATO ASI Series, Vol. F 75
Speech Recognition and Understanding. Recent Advances
Edited by P. Laface and R. De Mori

resulting mapping should be smooth, then generalized RBFs satisfy sufficient conditions to be used for this approximation problem.

2.1 Implementation

The basic implementation advantages of RBFs derive from their representation. Here is how they were trained in the experiments of section 3:

1) Initialize the parameters of the gaussian units (**P**j in (1)) with the results of a cluster analysis (k-means). This step can be interpreted as an unsupervised, competitive learning step which encodes the input pattern in a local representation.

2) Find the output weights of the RBF network (Wij in (1)) with a matrix pseudo-inverse calculation, or with gradient descent. This step is a supervised learning step and can be accomplished rapidly since it is a linear problem, with no local minimum.

3) All the parameters of the system can be tuned to improve performance and perform a global optimization using gradient descent.

3. Experiments of phoneme recognition with RBFs

The results of several experiments are presented here to evaluate the performance of RBFs in a difficult speech recognition problem. The task was phoneme recognition on the TIMIT [6] database, with 39 phoneme classes (as [8]). We used regions 2, 3 and 6 of TIMIT (135 speakers and 292623 frames for training and 28 speakers and 61428 frames for test). The input preprocessing produced 24 mel scaled spectral coefficients plus the energy for each frame of 20 ms, with a step of 10 ms. This resulted in a flow of 25 parameters per frame of 10 ms coming into the network. For the architectures presented below the hidden units look at 4 consecutive frames *i.e.,* a 100 dimensional input. Output units are sigmoid units rather than linear units as in (1). Note that this is equivalent since the sigmoid is invertible but it has the advantage that the outputs are limited to the interval (0,1). In all nets there was a bias unit feeding all sigmoid units in the net a constant 1.0 value.

It is difficult to compare directly the results obtained here with those reported by other researchers for phoneme recognition on TIMIT. The best result reported here was 41.8% error on the test set (section 3.5). This is a frame by frame error obtained by scanning the network on the preprocessed input sentence, and not using any language model, known segmentation, bigrams or duration information. Lee & Hon [8] obtained 41.3% error with HMMs and no language model and 26.2% using a context-dependent bigram model. Leung and Zue [9] obtained 30% error but used the known segmentation and phoneme duration (the network is not placed across phoneme boundaries: it is centered on each phoneme). Robinson and Fallside [14] obtained 24.9% error with a recurrent network plus a dynamic programming postprocessor using a bigram and duration model (the error is about twice that amount without the postprocessor).

3.1 Supervised *vs* unsupervised initialization

Should one use a **completely unsupervised** algorithm to find the cluster centers and spread? What if some points from two classes form two clusters that mostly overlap? K-means will probably choose to represent them with only one cluster. A simple but not optimal solution to that problem is to apply k-means separately for each of the classes. K-means with no class information did not result in as good performance as k-means per class in a simple experiment performed with 78 gaussian hidden units. Output weights were found with pseudo-inverse computation [12]. The result with no class information was 58.1% error on the test set and 52.2% with class information. All other experiments with RBFs described in this paper henceforth used k-means per class.

3.2 Cepstral *vs* Spectral parameters

Previous experiments had shown that spectral representations produced slightly better results than cepstrum coefficients with neural networks of sigmoid units [4]. A similar conclusion was obtained for RBF networks on the recognition of 39 TIMIT phonemes. The networks had 78 hidden units (2 clusters/ class) initialized with k-means and with 3 delays (0, 2 and 4) between the hidden and the output units for the 39 classes. The output weights were obtained with the pseudo-inverse method [5, 12]. The frame by frame error on the test set was 54.5% for the cepstral input and 46.5% for the spectral input. In both cases, we used a mel scale with 24 filters and provided the frame energy to the system, *i.e.*, the network had 25 inputs. With additional direct connections from inputs to outputs, the errors were reduced but the spectrum still produced better generalization: the error on the test set with cepstral input was 48.1% and the error with the spectral input was 45.6%.

3.3 RBFs *vs* Sigmoid Units

Experiments were performed in the same conditions for both a network of RBF units and one of sigmoid units. The networks had 78 hidden units. The network with sigmoid hidden units was initialized with random weights and trained with backpropagation for 22 epochs. The error on the test set was 51.2 %. The network with gaussian hidden units was initialized with k-means, (2 clusters/class, 39 classes). The output weights were found with the pseudo-inverse. The error on the test set was 52.2 %. After doing 10 epochs of gradient descent on the RBF net, *i.e.*, optimizing all the parameters (output weights, cluster centers and cluster spreads), the error on the test set was reduced to 47.8 %.

3.4 Effect of Context and Architecture

Various architectures and ways to represent some short-term context (see table 1) were explored, using multiple delays between the hidden layer and the output layer (as in Waibel's TDNNs [16]).

Table 1. Performance on the TIMIT 39 phonemes recognition task for various sets of delays between the hidden layer and the output layer, using a static network of gaussian hidden units and initialization with k-means per class.

clusters/class	delays	error on test set
2	0	52.2%
2	0,4	48.6%
2	0,4,8	47.7%
2	0,2,4	46.5%
3	0,2,4	45.6%

3.5 Adding a Recurrent Hidden Layer

The improvement shown in table 1 by providing more context with delay links shows the importance of context for the recognition of phonemes. However, it seems natural to consider recurrent networks as a more powerful way to represent context. This motivated the next experiment, in which a layer of recurrent sigmoid units was added to the best net obtained in the previous section.

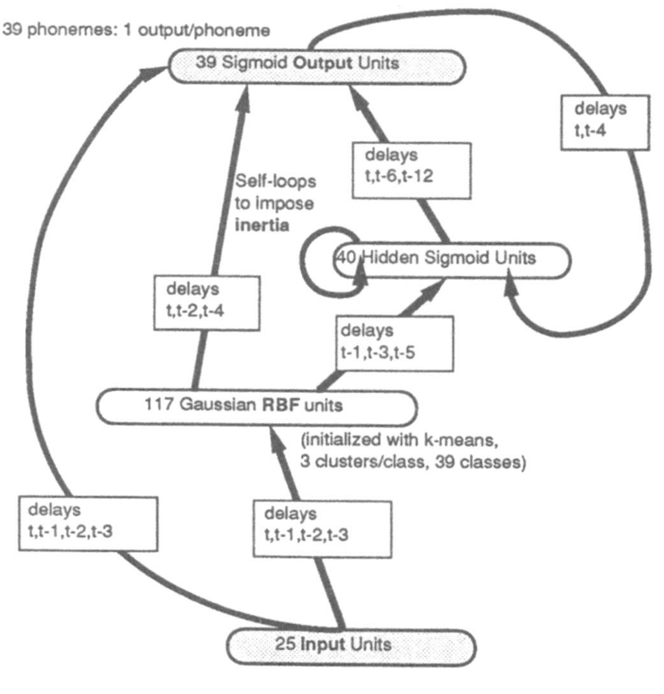

Figure 1. Recurrent neural network with Radial Basis Functions, trained to recognize 39 phonemes of TIMIT, assigning a phoneme category to each frame of 10 ms of the preprocessed signal

The network was trained with the time unfolding algorithm [11, 15]. The initial network was the same as the last one on table 1. The architecture after the addition of a second hidden layer with 40 sigmoid units is shown in figure 1. Cycles are introduced with the outputs feeding the hidden sigmoids with 2 delays (2, 4), and the hidden sigmoids having self-loops with a fixed decay weight of 0.93 which means a time constant (half-life) of 9.5 frames. This represents a very large network with about 54000 weights. The weights of the connections between the hidden sigmoid units and the output units were initialized to low values (in the range [-0.01, 0.01]) so as to not disturb too much the network by the introduction of this layer. However, after a few cycles of training, significant improvement was observed. After 11 epochs on the training set the error obtained with this network on the training set was 40.9% and the error and the test set was **41.8%**.

3.6 Acceleration with a Fast Search

Section 3.3. showed the speed advantage of RBFs w.r.t. sigmoids by initializing the network with relevant values rather than random ones. One can also take advantage of the local nature of the response of the RBF units by searching rapidly for the subset of those units that can respond with an output significantly greater than zero. The method implemented here is based on the use of a grid and table look-ups to find the appropriate subset of clusters. The number of dimensions represented in the grid was reduced by using only the first few principal components of the input patterns (from the training set) to build the grid. This mapping can be expressed as follows:

1) normalization:

$$X'ip = (Xip - AVGXi)/SDXi \tag{3}$$

2) linear mapping through the principal components:

$$X'' = A\ X' \tag{4}$$

where Xip is the ith dimension of a pattern p, AVGX is the average over the training patterns and SDX is the standard deviation of the training patterns (per dimension); A is the matrix formed by concatenating the first few eigenvectors with highest eigenvalues of the matrix B, and B is formed by concatenating all the X'p (*i.e.*, all the normalized input patterns of the training set) as computed in (3). In order to use grid cells more efficiently we used information about the pattern density in each of the reduced dimensions. For each of the reduced dimensions d, a new position is computed from the old position as follows:

$$X'''_d = F_d(X''_d) \tag{5}$$

$$\text{where } F_d(X''_d) = \sum_{i<X''_d} \text{density}_d(i) \tag{6}$$

where $\text{density}_d(i)$ is the fraction of cluster centers falling in the ith interval for dimension d of the principal components space. A very fine quantization of the principal components space is performed in order to compute the density array. The resulting grid is regular in the transformed space (after transformation by (5)) but is not regular in the principal components space. Instead, it attempts to use density information to produce less empty cells and a more uniform distribution of clusters among the cells. With this method, the initialization of the search data structure takes O(N) time and the search is proportional to the number of selected RBF units.

4. Conclusion

In this paper we have studied an alternative to the traditional networks of sigmoid units: RBF networks. The results of several experiments on phoneme recognition yielded several conclusions. It was found that a spectral representation produced better results than a cepstral representation for a RBF network. When initializing the RBF units with k-means, it was preferable to take into account information about the class of each pattern and perform k-means separately for each class. RBF networks performed as well or better than sigmoid networks but required less CPU time. Combining RBF units and sigmoid units in a recurrent network resulted in even better performance.

References

1. Bengio Y., Cardin R., De Mori R. & Merlo E. (1989a), "Programmable Execution of Multi-Layered Networks for Automatic Speech Recognition", in Communications of the Association of Computing Machinery, February 1989, vol. 32, number 2, pp. 195-199.
2. Bengio Y., Gori M. & De Mori R. (1989c),"BPS: a Learning Algorithm for Capturing the Dynamic Nature of Speech", in the Proceedings of the International Joint Conference on Neural Networks 1989, pp. II-417 - II-424.
3. Bengio Y. & De Mori R. (1989),"Use of Multilayer Networks for the Recognition of Phonetic Features and Phonemes", in Computational Intelligence, 5, pp.134-141.
4. Bengio Y., Cardin R., De Mori R. (1990),"Speaker Independent Speech Recognition with Neural Networks and Speech Knowledge", in Advances in Neural Information Processing Systems 2, ed. D.S. Touretzky, Morgan Kaufmann Publishers.
5. Broomhead D.S. & Lowe D. (1988), "Multivariable functional interpolation and adaptive networks", Complex Systems, 2: pp. 321-355.
6. Garofolo J.S. (1988),"Getting started with the DARPA TIMIT CD-ROM: an acoustic phonetic continuous speech database", National Institute of Standards and Technology (NIST), Gaithersburgh, MD.
7. Kruschke J.K. (1990), " ALCOVE: a connectionist model of category learning", Technical Report 19, Cognitive Science Program, Indiana University.
8. Lee K.F. & Hon H.W. (1989), "Speaker-Independent Phone Recognition Using Hidden Markov Models", IEEE Trans. on ASSP, vol. 37, no. 11, pp. 1641-1648.
9. Leung H.C. & Zue V. (1990), "Phonetic classification using multi-layered perceptrons", ICASSP 90, pp.525-528.
10. Mel B.W. & Koch C. (1990), "Sigma-Pi Learning: A Model for Associative Learning in Cerebral Cortex", Advances in Neural Networks Information Processing Systems 2, ed. D.S. Touretzky, Morgan Kauffman.
11. Pearlmutter B.A. (1989), "Learning state space trajectories in recurrent neural networks", Neural Computation, vol. 1, no. 2, pp.263-269.
12. Penrose R. (1955), "A generalized inverse for matrices", Proc. Cambridge Philos. Soc., 51: pp.406-513.
13. Poggio T. & Girosi F. (1989), "A Theory of Networks for Approximation and Learning", MIT A.I. Memo No. 1140.
14. Robinson T. & Fallside F. (1990), "Phoneme recognition from the TIMIT database using recurrent error propagation networks", CUED/F-INFEG/TR.42, Cambridge University Engineering dept.
15. Rumelhart D.E., Hinton G., Williams R.J. (1986), "Learning internal representation by error propagation ", in Parallel Distributed Processing, vol. 1, (eds. Rumelhart & McClelland), Bradford Books / MIT Press.
16. Waibel A., Hanazawa T., Hinton G., Shikano K. & Lang K. (1987),"Phoneme Recognition using Time Delay Neural Networks", Technical Report TR-I-0006, ATR Interpreting Telephony Research Laboratories.

Phonetic features extraction using Time-Delay Neural Networks

Frédéric Bimbot, Gérard Chollet, Jean-Pierre Tubach

Télécom Paris - Dépt SIGNAL, C.N.R.S. - URA 820, 46 rue Barrault, 75634 PARIS cedex 13. FRANCE.

Abstract: A.Waibel introduced Time-Delay Neural Networks (TDNNs) as a specific neural network architecture that is especially well adapted to the "dynamic nature of speech". We propose here to use low-dimensioned TDNNs for discriminating between phonetic features. We give evaluations of the different performances and we comment them. We also compare direct phoneme recognition scores using a sophisticated classical classifier to those obtained with a medium-size TDNN.

Keywords: speech recognition, phonetic features, neural networks, TDNN, classification.

Introduction

Connectionnists networks are alternative tools for performing classification tasks, such as those required for acoustic-phonetic decoding. The non-linear transfer function of elementary neurons makes it possible to render complex shapes for classes, that is to learn different variants of a same pattern, without careful human-time consuming specification of rules, in the training data. Different network structures exist, among which can be opposed fully connected networks and multi-layer ones. The latter are usually faster to train, using back-propagation for instance. The Time-Delay Neural Network is of the second kind, where extra constraints are imposed to the connection structure, and make it well adapted for classifying speech sounds.

The Time-Delay Neural Network (TDNN)

A. Waibel et al. introduced Time-Delay Neural Networks as a specific architecture that can take into account the "dynamic nature of speech". Such a network is able to represent temporal relationships between successive acoustic frames, while providing some invariance under time translation [7]. The basic TDNN proposed by Waibel is composed of an input layer, two hidden layers and an output layer, that are not fully interconnected. Shift invariance imposes extra constraints on connection weights. Therefore, the number of connections in a TDNN is much lower than the one for a similar fully connected perceptron and the number of actual free parameters is even lower.

In recent publications [7], TDNNs proved to be extremely efficient for simple tasks, such as speaker-dependent classification of japanese [b, d, g] utterances within a [b, d, g] set. Waibel et al. also showed that a modular all-phoneme recognition network could be successfully designed on this basis [6]. Equivalent experiments were carried out on a french language data base, not giving exactly as good results, which can be partly explained by the difference in complexity of phonemic systems between both languages [1].

Haffner proposed fast learning methods for TDNNs, that bring down the computing time to very reasonable figures, even on a classical general purpose computer [2]. In particular, Haffner uses a skipping sample strategy which speeds up each training pass (epoch) and makes the whole learning phase reach convergence faster. TDNNs thus provide an appealing starting point for acoustic-phonetic decoding of speech using neural networks, and we investigated them for different kind of tasks and various protocols.

NATO ASI Series, Vol. F 75
Speech Recognition and Understanding. Recent Advances
Edited by P. Laface and R. De Mori

Corpus and protocol

Experiments were all conducted on a set of 200 french phonetically balanced sentences uttered by 1 male speaker. Each phoneme was manually labeled at the center. For all of them, spectra were represented by a 16x16 set of Mel-scaled filter-bank coefficients (8 frames on each side of the label), globally normalized between -1 and +1 with 0 mean on the data base. The corpus of 5270 phones (representing 34 phonemic classes) was halfed between a learning set and a test set. The average number of references per phoneme is 77 in each set.

We have used basic TDNNs, that is those originally proposed by Waibel. All of our TDNN take 16x16-dimension input vectors. The elementary mask in the input layer and in the first hidden layer is made of 4 columns, with a time-shift of 1 frame. The number of output classes then conditions all other parameters, except the number of rows in the first hidden layer. This parameter varies with the experiments and will be given specifically for each of them. Weights of connections coming from a same line in the last but one layer have been ponderated by a gaussian function, so that more importance is given to the center of the phoneme, and less to the borders (which are more influenced by the context). The Mac Clelland Error measure was used [2] instead of the Mean Square Error.

We propose here to use low-dimensioned TDNNs for discriminating between phonetic features. We share the point of view of Howard et al. for instance [3], that phonetics can provide a description of speech that will "appropriately constrain a priori" a neural network. The TDNN structure, on the other hand, warrants the desirable property of shift invariance. Under this approach, there is no hierarchy between the various tasks, whereas Waibel's approach requires a rough pre-classification of the sounds into one macro-class or another before identifying the phoneme within the macro-class itself. We also tested a medium-size TDNN for a direct global recognition of the phonemes. Performance was compared to DTW + k-Nearest Neighbours approaches. The global neural network approach was also tested on vowels only and consonants only sub-corpuses.

K-Nearest Neighbours classifier

Both to check the sufficient consistency of the data and to have some baseline reference scores, experiments using classical classifiers were conducted. We chose the k-Nearest Neighbours strategy for the task of classifying 1 phoneme out of 34.

However, slight time misalignements may cause major errors, and other protocols were used in order to decrease these influences : this problem was adressed by carrying out some Dynamic Time Warping between references and tests and taking the sum of distances along the best path, as a distance measure between patterns. Different DTW strategies were investigated for the 3- and the 9-Nearest neighbours classifiers; description and results (Table 1) are given below :

0 : path along the diagonal only (no DTW), which is a reference value,
1 : path starting at one corner, going to the other, with no other constraint,
2 : path along a restricted corridor following the diagonal,
3 : path inside a diamond centered at the middle of both patterns,
4 : path inside 2 triangles connected by one angle at the middle.

DTW type	9-NN	3-NN
0	**66.2 %**	**64.4 %**
1	**68.4 %**	**65.2 %**
2	**73.6 %**	**73.3 %**
3	**75.8 %**	**74.8 %**
4	**77.5 %**	**77.3 %**

Table 1 . scores for the k-NN classifier after dynamic programming.

These results clearly show the importance of taking into account time distortions and misalignements for better identifying speech patterns, since a well-designed DTW strategy improves the results by more than 10 %. However, if the k-NN approach does not require any learning phase, it takes more than 30 CPU seconds per pattern for the basic k-NN classifier on a DEC station 3100, and DTW improvements are payed by a significant increase of this figure.

Phonetic features

Phonetics describes speech events in terms of features : oppositions are usually binary (voiced vs unvoiced, vocalic vs non-vocalic, grave vs acute,...), but they may also be ternary (front vs central vs back) or even based on 4 levels (opened vs mid-opened vs mid-closed vs closed, for french vowels). Features can be defined according to acoustical properties (voiced vs unvoiced), articulatory properties (front vs back) or even linguistic properties (vowel vs consonant) ; the degree of abstraction of these features increases in this order. Different systems of features exist, that are usually a mixture between acoustical considerations and underlying articulatory characteristics, even though Jakobson proposed strictly acoustical criteria [3].

We have been investigating what kind of oppositions TDNNs are best suited for, by evaluating individual performances for different feature extraction tasks. A set of 13 binary features was built, according to classical phonetic oppositions. This set allows a total description of the french phonetic system (but is not minimal). An other set of 6 discriminative and minimal random features was artificially designed for opposing 34 phonemes with one another. This last set has of course no phonetic background, and is an interesting point of comparison for judging the relevance of the feature-based approach.

A second set of experiments was to investigate non binary oppositions. This lead to a few experiments on ternary features.

A third set of experiments consisted in performing feature distinction on two sub-corpuses of the original one : one composed of vowels and one of consonants.

A special procedure for determining convergence and estimating the generalization capabilities during the learning phase was used and evaluated : the cross-validation strategy.

The cross-validation strategy

The learning set was subdivided into a training set (80 %, i.e. 2108 elements) and a cross-validation set (20 %, i.e. 527 elements). The training set alone was used for training the weights of the TDNN, whereas the cross-validation set was resorted to for deciding when to stop learning.

In reality, a fixed number of iterations was used to optimize the network, and for each iteration, the learning set, the cross-validation set and the test set were passed through the network, hence providing various estimates of the network classification power.

The results given above are thus organized in the same way for all experiments. First figure in bold characters corresponds to scores on the test set, while the self-consistency (i.e the score on the training set) is given in standard characters, between brackets :

- The first score is the score obtained on the test set for the maximum self-consistency of the training data, within all the iterations (cross-validation data excluded),

- The second score is the score obtained on the test set for the maximum score on cross-validation data (and cross validation data only),

- The third score is the best score ever obtained on the test set within all iterations; of course, this score is always greater or equal to all previous ones.

Binary features

For the binary feature extraction task (2 outputs), a small TDNN structure, with 8 rows in the first hidden layer, was chosen. 13 binary phonetic-based oppositions and 6 arbitrary ones were investigated. 500 iterations were carried out. The scores are as follows (Table 2) :

phonetic feature	score for max self-consistency	score for max cross-validation	best score on test-set
I vowel vs non-vowel	**94.2 %** (99.7 %)	**95.8 %** (96.2 %)	**96.7 %** (96.9 %)
II vocalic vs non-vocal.	**95.1 %** (100 %)	**96.5 %** (97.0 %)	**96.9 %** (97.2 %)
III voiced vs unvoiced	**98.7 %** (99.9 %)	**98.9 %** (99.4 %)	**99.1 %** (99.7 %)
IV sonant vs non-son.	**96.3 %** (100 %)	**96.9 %** (98.4 %)	**97.2 %** (98.4 %)
V nasal vs non-nasal	**97.0 %** (100 %)	**97.7 %** (99.5 %)	**98.1 %** (98.3 %)
VI grave vs acute	**88.8 %** (98.5 %)	**90.6 %** (96.5 %)	**91.4 %** (96.9 %)
VII extreme vs central	**84.9 %** (95.1 %)	**84.4 %** (92.1 %)	**87.1 %** (89.3 %)
VIII compact vs diffuse	**89.8 %** (99.1 %)	**91.7 %** (96.0 %)	**92.2 %** (96.0 %)
IX rounded vs unround.	**94.9 %** (99.2 %)	**94.8 %** (97.5 %)	**95.3 %** (97.3 %)
X bemol vs non-bemol.	**90.7 %** (98.9 %)	**90.1 %** (98.8 %)	**91.8 %** (97.6 %)
XI delayed vs non-del.	**94.8 %** (99.9 %)	**97.1 %** (98.7 %)	**97.9 %** (98.2 %)
XII discont. vs cont.	**94.7 %** (100 %)	**97.9 %** (99.1 %)	**97.9 %** (98.8 %)
XIII fricative vs non-fric.	**97.0 %** (100 %)	**97.3 %** (99.9 %)	**97.7 %** (98.9 %)
6 random features (average)	**84.2 %** (96.1 %)	**85.3 %** (92.1 %)	**86.8 %** (90.9 %)

Table 2 . scores for binary phonetic features, and average score for random features extraction using TDNNs.
scores on test set (and self-consistency).

These percentages first prove that the cross-validation strategy is a satisfying test to avoid over-learning, even though it does of course not lead to the best possible score on the test-set. Altogether, scores for maximum cross-validation are 0.74 % under highest possible scores, while those for maximum self-consistency are 1.72 % lower. This 1.0 % is significant, all the more since the absolute scores are close to 100 %.

Scores were significantly improved by the gaussian windowing of the last but one layer.

While modes of articulation are very efficiently detected (binary features n° III, V, XI, XII and XII), those related to place of articulation provide less satisfying results. Actually, place of articulation is a more abstract notion, whereas acoustic correlates for modes of articulation are much more directly rendered on the spectrum.

However, only one of the binary features gives as poor results as the average for random features.

Ternary features

A second step was to test a slightly larger TDNN, with 16 rows in the first hidden layer, for extracting ternary features (3 output classes). 3 ternary phonetic-based oppositions were investigated. 1000 iterations were carried out. The scores are as follows (Table 3) :

ternary phonetic feature	score for max self-consistency	score for max cross-validation	best score on test-set
XIV high vs med vs low	**89.5 %** (97.8 %)	**88.4 %** (96.4 %)	**87.2 %** (94.9 %)
XV front vs central vs back	**85.4 %** (96.8 %)	**83.3 %** (94.6 %)	**81.3 %** (91.5 %)
XVI [p, u] vs [t, a] vs [k, i]	**86.3 %** (97.1 %)	**85.9 %** (92.7 %)	**82.5 %** (92.8 %)

Table 3 . recognition scores of a few ternary phonetic features using TDNNs.
scores on test set (and self-consistency).

All three features are based on articulatory concepts (the third one being slightly more arbitrary than the 2 first ones). The lukewarm results confirm the difficulty for a TDNN to perform very well on abstract oppositions.

Vowel / consonant a priori distinction

The next step was to carry out the same tasks after having split the corpus into a set of vowels (including semi-vowels) and a set of consonants. Note that a TDNN can perform the vocalic vs consonantic opposition, with an accuracy of approximately 96.5 % (binary feature n° II).

Nevertheless, the drawback of such a protocol is the considerable decrease in the volume of data for each experiment : 2952 vowels and 2318 consonants (for training, cross-validation and testing) make the results statistically less valid. Both subsets were however split equally into training and testing, with still 20 % of the training data used for cross-validation (that is only 295 tokens for vowels, and 232 for consonants...).

Not all previous binary features were tested on each corpus, but only those that permit a full and non-redundant discrimination within each subsets. A few ternary features were also tried, as well as a 6-level combination of a binary and a ternary feature.

Results are given synthetically below (Tables 4 and 5), with the indications of type of feature (binary, ternary,...), the number of rows in the first hidden layer (8 or 16) and the number of iterations (500, 1000 or more) :

phonetic feature on consonants	score for max self-consistency	score for max cross-validation	best score on test-set
III (voiced) (bin, 8, 500)	**98.8 %** (100 %)	**98.8 %** (100 %)	**99.1 %** (99.3 %)
V (nasal) (bin, 8, 500)	**97.8 %** (100 %)	**98.2 %** (99.7 %)	**98.8 %** (99.2 %)
XI (delayed) (bin, 8, 500)	**95.6 %** (100 %)	**89.4 %** (99.1 %)	**97.2 %** (99.0 %)
XII (discont) (bin, 8, 500)	**88.1 %** (100 %)	**93.9 %** (97.6 %)	**96.8 %** (98.7 %)
XIII (fricat) (bin, 8, 500)	**95.2 %** (100 %)	**95.1 %** (100 %)	**96.1 %** (95.8 %)
XV (front) (tern, 8, 1000)	**86.7 %** (99.6 %)	**88.1 %** (94.0 %)	**89.9 %** (95.1 %)
I-XV combination (voiced-front) (6 classes, 16, 1800)	**82.0 %** (100 %)	**82.7 %** (99.9 %)	**88.9 %** (99.1 %)

vowels	---	---	---
V (nasal) (bin, 8, 500)	**97.7 %** (99.9 %)	**97.7 %** (99.6 %)	**98.2 %** (99.0 %)
VI (grave) (bin, 8, 500)	**96.0 %** (100 %)	**95.7 %** (99.8 %)	**96.7 %** (98.7 %)
VII (extreme) (bin, 8, 500)	**88.1 %** (97.8 %)	**87.5 %** (96.6 %)	**89.4 %** (92.0 %)
VIII (compact) (bin, 8, 500)	**92.5 %** (100 %)	**94.0 %** (97.8 %)	**94.4 %** (97.4 %)
IX (round) (bin, 8, 500)	**93.8 %** (100 %)	**93.2 %** (94.5 %)	**95.3 %** (97.6 %)
XV (front) (tern, 8, 1000)	**92.3 %** (100 %)	**91.3 %** (99.5 %)	**93.6 %** (98.2 %)

Table 4 and 5 . scores for consonants-only and vowels-only feature extraction using TDNNs. **scores on test set** (and self-consistency).

It must first be noted that self-consistency generally reaches 100 %, probably because the training set is not highly over-dimensioned compared to the classification capabilities of the TDNN. These results also show that the reliability of the cross-validation strategy is considerably decreased, the cross-validation set being to small for both sets of data. Scores for maximum cross-validation and maximum self-consistency only differ by an unsignificant 0.01 % and both are by 1.85 % under best possible scores. The cross-validation strategy may thus not be useful for too small data sets.

It must also be observed that performances do not increase very much for those features that were already well extracted by a TDNN operating on all data. On the opposite, scores are generally improved on difficult features (since a complex problem is divided into 2 simpler ones). But these results are to be considered with caution, in particular because of the non-comparable confidence intervals for the experiments.

All phonemes recognition

We finally performed direct global recognition using a TDNN with a larger first hidden layer (32 rows). This task was tested on the whole corpus as well as on the consonant and the vowel corpuses separately. 1000 iterations were carried out. The results are given below (Table 6) :

corpus	score for max self-consistency	score for max cross-validation	best score on test-set
all phonemes (34 classes)	**78.2 %** (99.7 %)	**78.0 %** (99.7 %)	**81.9 %** (99.0 %)
consonants only (18 classes)	**88.0 %** (100 %)	**88.3 %** (97.6 %)	**90.7 %** (95.8 %)
vowels only (13 classes)	**77.7 %** (100 %)	**80.0 %** (98.9 %)	**81.2 %** (96.7 %)

Table 6 . scores for all phonemes recognition, consonant recognition, and vowel recognition using a TDNN. **scores on test set** (and self-consistency).

Here the cross-validation strategy reveals quite disappointing, probably because the average number of cross-validation elements per class is not high enough. Nevertheless, these scores prove that a TDNN can at least perform as well as the most complex of our k-Nearest Neighbours classifier, on the all phonemes recognition task. And it must be mentioned that the recognition process is also incomparably faster with a TDNN than with the Nearest Neighbours approach (approximately 1000 times), which is in itself of major importance. It is also interesting to note that vowels are significantly more difficult to distinguish than consonants.

Better results on the all phonemes recognition experiment (83.6 %) were obtained on the same corpus in an other laboratory [5], using a mixed TDNN + LVQ architecture, where the last but one layer of a TDNN-derived network is used as an input for the LVQ classifier. This tends to show interesting properties of internal representations in hidden layers of a TDNN.

Conclusions

The aim of this work was to give evaluations of TDNNs, for different kind of tasks, with various level of complexity. Phonetic feature extraction tasks provide very interesting results individually. However, the integration of these different modules into a global network remains to be carried out, as a definite assesment experiment. Direct global identification of phonemes by a single TDNN gives as good results as the most sophisticated combination of DTW and k-Nearest Neighbours algorithm we investigated, with a undoubtful advantage for the neural approach in terms of recognition speed. Incidentally, an a priori distinction between vowels and consonants appears as a good compromise between hierarchical approaches relying on macro-classes and parallel approaches using features.

Acknowledgements and references

This work was carried out in the framework of SPRINT, Esprit project - BRA 3228 (Speech Processing and Recognition using Integrated Neurocomputing Techniques). In the framework of these experiments, some pieces of software were exchanged between RSRE-Malvern, United Kingdom and Télécom-Paris, France. We also wish to thank here Patrick Haffner from CNET-Lannion, France, for the software that he provided us with, for conducting these experiments.

1. Devilliers L. : Reconnaissance monolocuteur des phonèmes du français au moyen de réseaux à masques temporels. XVIIIèmes JEP, pp. 227-231. May 1990.
2. Haffner P. : Fast back-propagation learning methods for neural networks in speech. A.T.R. Report. 1989.
3. Howard I., Huckvale M. : Training feature detectors for use in automatic speech recognition. proc of SPEECH '88. book 4, pp. 1365-1372.
4. Jakobson R. : Essais de Linguistique Générale. Paris. Ed. de Minuit. 1963.
5. SPRINT. Esprit Project BRA 3228. Deliverable of Task 3. 1990.
6. Waibel A., Hanazawa T., Hinton G., Shikano K., Lang K. : Modularity and scaling in large phonemic neural networks. Technical Report TR-1-0006, ATR Interpreting Telephony Research Laboratories. October 1987.
7. Waibel A., Hanazawa T., Hinton G., Shikano K., Lang K. : Phoneme recognition using Time-Delay Neural Networks. IEEE Trans. on Acoustics, Speech and Signal Processing. vol. 37, n° 3, pp. 328-339. March 1989.

Improved Broad Phonetic Classification and Segmentation with an Auditory Model

L. Depuydt, J. P. Martens [1] and L. Van Immerseel

Laboratory for Electronics and Metrology, University of Ghent, St. Pietersnieuwstraat 41, B 9000 Gent, Belgium; Depuydtl @ Imec.be

Abstract: We describe a broad phonetic classification and segmentation algorithm based on neural networks and dynamic programming. The basics of our algorithm are outlined in another paper [7], so here we will focus on the introduction of auditory model features replacing the mel-scale parameters. Our auditory model incorporates critical band filtering, short time adaptation and temporal analysis of the auditory nerve responses. Unlike previously proposed synchrony models, it emphasizes the envelope rather than the instantaneous frequency as the carrier of perceptually relevant information.

Keywords: Auditory model features, neural networks, dynamic programming, speaker independent speech recognition.

Introduction

In the Hidden Markov Model (HMM) approach to continuous speech recognition, the phonetic classification and segmentation are obtained simultaneously, and coarticulations are modelled by means of context dependent phone models [4]. The HMM strategy is very powerful, but it needs a large number of models and parameters, and it has a tremendous amount of segmentations to examine during the recognition process.

We are currently investigating a segment based recognition strategy in which the speech utterance is segmented prior to the phonetic classification. Experiments by Roucos and his colleagues [8] have demonstrated the feasibility of such an approach. The segmentation will be used to generate context-dependent segment descriptions which will be supplied to a neural network classifier. By adopting a multi-level segmentation strategy [3], we hope to avoid making too many deletion errors in an early stage of the recognition process.

In this paper we describe the broad phonetic classification and segmentation component of our recognizer. A neural network generates broad phonetic class (BPC) evidences for each frame. These evidences are used in a dynamic programming (DP) stage to produce an initial segmentation of the speech utterance. The BPC evidences inside the initial segments are then used to compute segment boundary evidences, and to construct a dendrogram of broadly classified segments. For a detailed description of the algorithm we refer to [7]. Here we focuss on the introduction of auditory model features [6].

[1] Research Associate of the National Fund for Scientific Research

NATO ASI Series, Vol. F 75
Speech Recognition and Understanding. Recent Advances
Edited by P. Laface and R. De Mori

Speech Material

We have used a hand-labelled multi-speaker database of continuously spoken numerical strings, uttered by 30 different speakers (15 males and 15 females). The vocabulary counts 24 different morpheme like entries.

The database consists of 300 different utterances (10 from each speaker): 192 utterances from 24 speakers (8 from each speaker) are used for training, 108 are used for testing. The test set is divided into a multi-speaker test set (the remaining 48 utterances from the training speakers) and a speaker independent test set (60 utterances from 6 new speakers).

All utterances were hand-labelled at the laboratory. The labels were similar to those used in the TIMIT database. They correspond to phone components rather than to phones.

Feature Extraction

The speech signals were bandlimited to 4 kHz and sampled at a rate of 10 kHz. A 20-dimensional feature vector was extracted every 10 ms.

For the first experiments a Hamming window of 25.6 ms wide was applied every 10 ms. For each frame a Fourier spectrum was computed and converted into a set of mel-scale log spectrum coefficients [2].

Our auditory model [6] consists of an outer and middle ear lowpass filter (with a 10-dB boost at 4kHz) followed by a decimation unit and a set of 20 parallel analysis channels. Each channel incorporates a bandpass filter (BPF), an Automatic Gain Controller (AGC), an envelope extraction filter (LPF) and an Envelope Component Extractor (ECE) (fig. 1). The input sampling frequency f_s (time index n) is assumed to be 20 kHz. Therefore, the even samples of our speech signals are duplicated to fit in with this assumption.

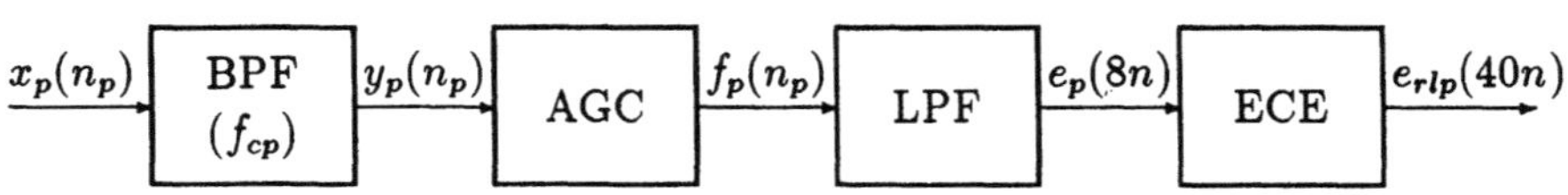

Figure 1. Channel p of the auditory model

The decimation unit produces 4 signals which serve as inputs to the analysis channels operating at sampling rates $f_s/2^k$ with $k = 0..3$. The sampling rate f_{sp} of channel p is determined by the best frequency f_{cp} of its BPF. The corresponding time index is denoted as n_p.

The bandpass filters are 4-th order filters whose best frequencies f_{cp} are spaced equidistantly on a critical band scale, and whose 3-dB bandwidths are equal to one critical band width. In our present model, the best frequencies cover a range from 0.14 to 4.5 kHz. The envelope patterns extracted by the lowpass filters in the different channels are sampled at a fixed rate of 2.5 kHz (which is the smallest f_{sp} used in the analyzer).

The ECE normally extracts 3 different envelope components, but for the present study it is reduced to a first order lowpass filter generating only a loudness/roughness component e_{rlp} which is sampled at a rate of 0.5 kHz. A moving average filter is used to ensure that resampling at 0.1 kHz (which is the intended frame rate) doesn't introduce aliasing products in the feature vector.

Phonetic Classification

The broad phonetic classification of individual speech frames is accomplished by a multi-layer perceptron (MLP) with one hidden layer. For each frame n it computes 5 outputs $o_i(n)$ $(i = 1..5)$ representing evidences for the broad phonetic classes *vowel*, *sonorant*, *fricative*, *burst* and *silence/closure*. These classes describe the manner of articulation of the speech sounds.

It has been demonstrated [4] that differential information is important. Therefore we have constructed different MLP's according to a NETtalk like architecture [9]. To classify frame n, we consider all frames in some observation interval $(n - M_1, n + M_2)$. However, to reduce the number of weights, only some of these frames are actually connected to the hidden layer. The training is performed as described in [7].

We have evaluated the broad phonetic classification performance for different sizes of the hidden layer. Each frame receives a unique label corresponding to the maximum MLP-output, and this label is compared to the one derived from the hand-marked data. Results for the broad phonetic classification task are listed in table I.

		mel-scale parameters			auditory model		
hidden units	weights	train	multi speaker	speaker indep.	train	multi speaker	speaker indep.
8	533	14.5	16.0	16.4	10.9	12.8	13.0
12	797	13.1	14.2	15.5	10.2	12.1	12.6
20	1325	12.6	14.1	14.3	9.5	11.0	11.8
30	1985	12.6	14.3	14.0	8.9	10.8	11.7

Table I. frame error rates (in %)

The input frames were $n - 4$, $n - 2$ and n (see [7] for a motivation). The figures indicate that:

- By introducing auditory model parameters, the speaker independent error rates drop from 14.0 to 11.7% (30 hidden nodes)
- There is some discrepancy between the training and the test database results, but not between the multi-speaker and the speaker-independent test results.
- The frame error rates seem to stabilize as soon as the MLP has more than 20 units on the hidden layer.

From the second observation we conclude that 24 speakers suffice to train a speaker independent broad phonetic classifier.

Phonetic Segmentation

The aim of our phonetic segmentation algorithm is to generate a limited number of segmentations, including one that is better than the unique segmentation produced by the best HMM speech recognizer available. According to [5], deletion and insertion error rates of 6.6 and 7.7 % can be obtained with context-dependent phone models and a bigram phone language model.

Our segmentation algorithm starts by searching for extrema in the individual MLP output functions $o_i(n)$. By constructing segments around the maxima which exceed a threshold o_T [7] an independent set of segment candidates is created for each broad phonetic class.

Clearly, some of the segments from one set can overlap with others from a different set, and some frames may not belong to any segment at all. A dynamic programming algorithm is used to find the sequence of segments (going through the endpoints of the speech utterance) which minimizes a gap/overlap cost

$$E = \sum_{ij} |t_{ei} - t_{bj}|^2$$

For a particular sequence, t_{ei} and t_{bj} represent the endpoint of segment i and the starting point of its successor j. Once the optimum sequence has been identified, the segment boundaries are adjusted. The result is an initial segmentation of the utterance.

In order to construct a dendrogram, an evidence between 0 and 1 is assigned to each boundary [7] of the initial segmentation. The segment merging stage of our algorithm searches for the boundary with the smallest evidence, and it merges the two surrounding segments to one new larger segment. The above procedure is repeated until all boundaries whose evidences are smaller than a threshold e_M have been processed. Once the dendrogram has been constructed, every segment receives a unique BPC label which is derived from the MLP outputs inside its boundaries.

Results

By means of dynamic time warping the segments produced by the initial segmentation stage are aligned with the hand-marked phonetic units. The cost function that was used for this alignment is described in [7]. The best segmentation comprised in the dendrogram is identified by a second dynamic programming stage. This best segmentation is defined as the one with the smallest number of deletion, insertion and substitution errors. Chigier and Brennan [1] defined the depth of the dendrogram as the total number of segments in the dendrogram, divided by the total number of labels in the manual segmentation. This depth is a measure for the average number of alternative segmentations comprised in the dendrogram.

By changing the free parameter o_T it is easy to control the deletion and insertion error rates of the initial segmentation. Starting from different initial segmentations, we evaluate the error rates of the corresponding best segmentations. The depth depends on the number of segments in the initial segmentation and on the value of the threshold e_M. Our best results are depicted on figs. 2a and 2b. They show the segmentation error rates (deletions + insertions) and the total error rates (deletions + insertions + substitutions) versus the depth of the dendrogram.

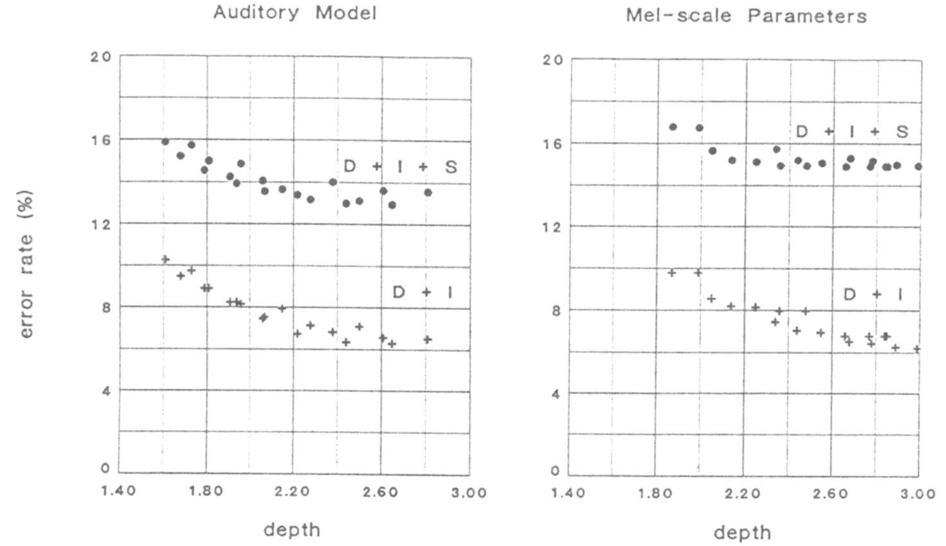

Figure 2. Best segmentation error rates

- If a large depth is accepted, the segmentation error rate can be as low as 6% for both speech representations.
- The minimum total error rate is about 13% if auditory model parameters are used, and 15% if mel-scale spectra are used.
- At lower depths, the segmentation error rate is smaller if auditory model features are employed.
- The difference in total error rate is about 2%, independently of the depth.

As expected, the auditory model parameters yield better segmentations and better classifications of the segments.

Conclusion

We have proposed a broad phonetic classification and segmentation algorithm which is capable of constructing a dendrogram comprising a *limited* number of segmentations, including one that is better than the segmentation generated by the best HMM recognizer available. The dendrogram not only represents a severe reduction of the search space, but it also offers a solid basis for the generation of context-dependent descriptions of the phonetic segments. These descriptions can then be supplied to a statistical phoneme recognizer, e.g., a neural network.

Clearly, the auditory model features constitute a richer representation of the speech utterance than the mel-scale parameters. Both the segmentation and the classification performances improve significantly.

Acknowledgements

We wish to acknowledge the National Fund for Scientific Research for granting CPU-time on an IBM 3090 machine, and IMEC for making it possible to connect to that computer. We also acknowledge Frank Elewaut of IBM, and the supercomputing group at the K. U. Leuven, for their contribution to the vectorization of our software.

Parts of this text present results of the Belgian National incentive-program for fundamental research in Artificial Intelligence, initiated by the Belgian State - Prime Minister's Office - Science Policy Programming. The scientific responsability is assumed by its authors.

References

[1] B. Chigier and R. Brennan (1988), "Broad Class Network Generation Using a Combination of Rules and Statistics for Speaker Independent Continuous Speech", Proceedings ICASSP-88, 449-452.

[2] S. Davis and P. Mermelstein (1980), "Comparison of Parametric Representations for Monosyllabic Word Recognition in Continuously Spoken Sentences",*IEEE trans. Acoustics, Speech and Signal Processing 28, pp 357-366.*

[3] J. Glass and V. Zue (1988), "Multi-level Acoustic Segmentation of continuous speech", Proceedings ICASSP-88, 429-432.

[4] K.F. Lee (1989), "Hidden Markov Models: past, present and future", Proceedings EUROSPEECH-89, 148-155.

[5] K.F. Lee (1989), "Speaker-independent phone recognition using Hidden Markov Models", IEEE Transactions ASSP 37, pp 1641-1648.

[6] J.P. Martens and L. Van Immerseel (1990). "An auditory model based on the analysis of envelope patterns", Proceedings ICASSP-90, 401-404.

[7] J.P. Martens and L. Depuydt (1990). "Broad phonetic classification and segmentation based on neural networks and dynamic programming", to appear in Speech Communications 10 (number 1).

[8] S. Roucos and M. Dunham (1987), A stochastic segment model for phoneme-based continuous speech recognition", Proceedings ICASSP-87, 73-76.

[9] T.J. Sejnowski and C.R. Rosenberg (1987), "Parallel networks that learn to pronounce English text", Complex Systems 1, 145-168.

Automatic Learning of a Production Rule System for Acoustic-Phonetic Decoding

M.-J. Caraty, C. Montacié and X. Rodet

Laforia - Université Paris 6, CNRS - URA 1095, 4 place Jussieu 75252 Paris Cedex 05. France

Abstract: Results are reported of experiments in the use of Charade, automatic learning of a production rule system, for acoustic-phonetic decoding. The Charade system is evaluated in terms of performance to classify phonetic macro-classes from generated production rules. As points of comparison, the same experiments are carried out with an usual classifier (i.e., Hamming Distance Nearest Neighbor) and a neural net based technique (i.e., Modified Hopfied Net). The results can be summarized as follows :
— For a given reasonable error rate, Charade classifier gives a higher accuracy rate than HDNN and performs as well as MHN.
— The generated production rules can be analysed and interpreted for knowledge acquisition.

Keywords: Acoustic-Phonetic Decoding - Automatic Learning - Production Rule System

1. Introduction

Knowledge Acquisition based on observation or experience is of major necessity as far as the performance of Automatic Recognition Systems is concerned. An automatic technique for generalization and learning of a production rule system from a training set of examples should allow the acquisition and the constitution of a consistent knowledge database [1, 2, 6, 7].

The Charade system [4] was designed to detect logical or statistical regularities existing in a set of examples and to generate a production rule system reflecting such regularities.The learning technique is mainly characterized by a principle of representation based on the Hilbert Cube taking advantage of mathematical properties and optimized exploration of the description space for the rules generation.

The results of the Charade automatic learning (Ch) will be presented for the classification of phonetic macro-classes on a reference database of continuous speech. As points of comparison, we chose two different classification techniques : Hamming Distance Nearest Neighbor (HDNN) [3] and a Modified Hopfield Net (MHN) [5].

2. Charade : Automatic Learning of a Production Rule System

In french, Charade stands for Hilbert Cubes Applied to Representation and Learning Based on Examples Described. From a description language, a set of axioms reflecting the language semantics and a set of examples expressed in that language, Charade generates a consistent production rule system.

NATO ASI Series, Vol. F 75
Speech Recognition and Understanding. Recent Advances
Edited by P. Laface and R. De Mori

2.1. Description and Representation Space of Examples

The description d(E) of an example E is a conjunction of descriptors : $d(E) = d_1 \& d_2 \& \ \& d_D$. Each descriptor d_i being an atomic proposition or the negation of an atomic proposition. With such a description of examples, the elementary operations in the field of learning (i.e., generalization and discrimination) are computed as simple conjunction of descriptors. For instance, the smallest generalization of two examples is the conjunction of common descriptors of these examples.

The representation space of the learning set (e.g., a set of n examples) is a Hilbert Cube : a unitary hypercube in an n dimensional space. In this representation space, -each axis is associated to an individual, -each vertex corresponds to a subset of individuals characterized by its smallest generalization, -the belonging of an individual to a vertex is defined by the projection of the vertex onto the corresponding axis, -edges of the cube are considered as inheritance relationships towards the origin allowing an optimization of the representation. Indeed, a vertex inherits of every vertex above it in the hierarchy of the cube.

2.2. Principles of Logical Rule Generation

The induction principle is the following : given two descriptors d_1 and d_2 then if all the examples of the learning set containing d_1 contain also d_2 then $d_1 \Rightarrow d_2$. To introduce the generation principle, let us consider the two following representation spaces :
— The Hilbert Cube of Examples (C_{Ex}), describing the set of subsets of the learning examples, ordered by set inclusion.
— The Hilbert Cube of Descriptors (C_{Des}), describing the set of descriptor conjonctions, ordered by logical implications.
Then, the principle for the generation of the rules will consist to define the link between these two ordering relations by an exploration of the Cube of Descriptors.

For the generation of a potential rule, two main functions are defined as follows :
— $\delta : C_{Ex} \longrightarrow C_{Des}$. For each vertex of C_{Ex} (i.e., a set of examples), δ associates the vertex of C_{Des} corresponding to the smallest generalization of the set.
— $\gamma : C_{Des} \longrightarrow C_{Ex}$. For each vertex of C_{Des} (i.e., a descriptors conjunction), γ associates the vertex of C_{Des} corresponding to the set of all the examples for which this descriptors conjunction appears.
The application $\beta = \delta \circ \gamma$ has the property to show up all the logical relations present in the learning set. Two other functions ω and τ are defined in order to eliminate redundancies related to inheritance relationship and implication transitivity.

With the previous function definitions we can obtain for each vertex V_{Des} of the Cube of Descriptors, a potential rule of the type : $V_{Des} \Rightarrow \tau(\omega(\beta(V_{Des})))$. For the generation of the complete rule system, an exhaustive exploration of the Cube of Descriptors is excluded. For feasability of the technique of generation, various limitation theorems for the exploration are used. Two basic theorems determine nullity criteria of vertices of the Cube of Descriptors (i.e., vertices for which $\tau \circ \omega \circ \beta$ is empty). The goal is to eliminate irrelevant vertices (i.e., the vertices who will not generate new rules). Other theorems are introduced as constraints. They are related to the desired properties of the rule-based system. For instance, the descriptors structuration, the noise factor, the terminal condition of the rules, etc... Fifteen parameters may be adjusted in the Charade system.

3. Application for Speech Processing

The application in the field of speech, consists in the classification of spectra according to phonetic macro-classes. Short time spectrum is an example in our experiment. Spectra are manually labeled from a continuous speech database. The description of an example is a conjunction of binary descriptors.

The selection of a parametric representation of acoustic data is an important task in the design of any speech experiment. Knowledge of audition leads us to look after the frequency content of acoustic signals and more precisely to look for spectral masses. For our experiment, we choose a binary description of spectral power balances. For each labeled phone, a signal frame of 32 ms duration is extracted. Channels on the power spectrum $\{E_k\}$ are computed through a triangular filter according to Mel frequency scale. Four analysis levels are considered. At level i, the frequency range is divided into 2^i channels. For any pair of contiguous channels (e.g., k and k+1) a descriptor is computed as the value of the logical expression $(E_k > E_{k+1})$ (cf. table 4). If frequency channels are narrow enough, such a parametrization allows for localisation of spectral masses such as formants.

4. Experiments

Concerning the classification of phonetic macro-classes, four oppositions are studied :

— Oral versus Nasal Vowels (i.e., [i, e, E, A, o, O, y, u, eu] vs [an, on, un])
— Voiced versus Unvoiced Fricatives (i.e., [v, z, j] vs [f, s, ch])
— Voiced versus Unvoiced Plosives (i.e., [b, d, g] vs [p, t, k])
— Glides versus Nasal Consonants (i.e., [l, r] vs [m, n])

The various classification techniques we tested (i.e., HDNN, MHN and Ch) were carried out on a reference database. The database we used is the Acoustic Corpus SYL which is a part of BDSONS supported by GRECO *Communication parlée*, CNRS, France. This corpus contains 192 diphone dedicated sentences of continuous speech per speaker.

4.1. Reference Systems

In order to evaluate the learning and the accuracy rate of the Charade system, we chose two reference systems : Hamming Distance Nearest Neighbor and a Modified Hopfield Net.

The HDNN classifier is based on the Hamming distance. This distance is the number of bits in the test which does not match the corresponding reference bits.

The Classical Hopfield Net is defined by N nodes containing hard limiter and binary input and output $\mu_i(t)$. D nodes represent the D descriptors of the parametrisation, E nodes coding the various classes are used for the classification (N=D+E). Each node is fed back to all other nodes via weights t_{ij} computed from M patterns $\{x^S\}$, (S = 1,.., M) with the formula (1).

$$t_{ij} = \sum_{S=1}^{M} (x_i^S * x_j^S) \quad \text{for } 1 \leq i \leq N,\ 1 \leq j \leq N,\ i \neq j\ , \ = 0 \quad \text{for } i = j. \tag{1}$$

The value of a node $\mu_i(t+1)$ is computed via a hard limiter F with the formula (2).

$$\mu_i(t+1) = F\left(\sum_{j=1}^{N} t_{ij} * \mu_i(t)\right) \quad \text{for } 1 \leq i \leq N, \text{ with } F(x) = -1 \text{ if } x < 0,\ = +1 \text{ if } x > 0. \tag{2}$$

The Classical Hopfield Net performances decrease when the number of classes (E) is less than 0.15 the number of nodes (N) in the net. In our experiment, the results obtained with this algorithm were quite insufficient (i.e., a recognition rate lower than 50%). Two modifications of this algorithm improve significantly this accuracy rate.

The *first* modification concerns the offset of the hard limiter F. The new one (S_i) is defined as the optimal offset node. From training patterns, for every node μ_i we calculate mean and standard deviation of positive and negative activations (i.e., m_{pi}, d_{pi}, m_{ni}, d_{ni}). With such notations, the offset is computed with the formula : $S_i = m_{ni} + d_{ni} * (m_{pi} - m_{ni}) / (d_{ni} + d_{pi})$.

The *second* modification concerns the decision principle. After convergence a supplementary step is computed without hard limiter. Decision is taken with the optimal activation formula (3).

$$\left(\left(\sum_{j=1}^{N} t_{ij} * \mu_i(t) \right) - S_i \right) / d_{pi} \qquad (3)$$

5. Results

For each experiment :

— The training set, the test set and the generated rules number are showed on the table 1.

— Three results areas (i.e., accuracy, error and rejection) corresponding to the various classifier tested (i.e., HDNN, MHN and Ch) are showed on the table 2. To take into account the bias of the test set (i.e., phonemes occurences), the results are computed as means corresponding to the various phonetic classes of the test set.

Oppositions	Oral vs Nasal Vowels		Voiced vs Unvoiced Fricatives		Voiced vs Unvoiced Plosives		Glides vs Nasal Consonants	
Training Set	54	54	54	54	54	54	54	54
Test Set	1209	294	192	219	232	231	264	145
Rules Number	13	15	3	2	7	4	6	9

Table 1. Learning characteristics

Macro-classes	accuracy rate			error rate			rejection rate		
	HDNN	MHN	Ch	HDNN	MHN	Ch	HDNN	MHN	Ch
Oral Vowels	78.5	87.7	67.9	21.5	12.3	4.8	0.0	0.0	27.3
Nasal Vowels	83.7	67.8	57.4	16.3	32.2	8.7	0.0	0.0	33.7
Voiced Fricatives	96.3	99.3	77.1	3.7	0.7	8.3	0.0	0.0	14.6
Unvoiced Fricatives	99.4	88.9	92.7	0.6	11.1	0.0	0.0	0.0	7.3
Voiced Plosives	87.1	85.6	77.1	12.9	4.4	4.3	0.0	0.0	18.6
Unvoiced Plosives	95.2	98.1	43.4	4.8	1.9	14.3	0.0	0.0	42.3
Glides	87.3	97.1	54.5	12.7	2.9	9.9	0.0	0.0	36.6
Nasal Consonants	96.4	81.5	81.1	3.6	18.5	7.5	0.0	0.0	11.4
Mean	**87.8**	**88.0**	**68.8**	**12.2**	**12.0**	**6.6**	**0.0**	**0.0**	**24.6**

Table 2. Classification results for HDNN, MHN and Ch

HDNN and MHN, used without any efficient rejection criteria, give the best accuracy rate and perform alike. However, their error rates (i.e., about 12%) are rather high. For acoustic-phonetic decoding, the goal is mainly to minimize error rates. Given a same reasonable error rate (e.g., ≈5%), an experiment shows Charade gives the highest accuracy rate (cf. table 3). This experiment consisted in using thresholds for rejection criteria (i.e., maximum distance for HDNN and minimum activation for MHN).

Classifier & Thresholds	**Ch**	HDNN [5]	**HDNN [4]**	HDNN [3]	MHN [2]	**MHN [3.5]**	MHN [3.75]
Accuracy Rate	**68.8**	76.6	**57.7**	28.1	81.4	**65.0**	58.2
Error Rate	**6.6**	9.9	**5.9**	1.8	7.3	**5.9**	4.4
Rejection Rate	**23.6**	13.5	**36.4**	70.1	11.3	**28.1**	37.4

Table 3. Classifier results for HDNN and MHN with different thresholds for rejection criteria

An important advantage of the Charade system lies in the ability to analyse and interpret the generated rules. During the test set classification step the analysis of rules activation frequency, per macro-classes, shows up the significant clustering rules. For instance, for Oral versus Nasal Vowels classification and with the indication of the most concluded phonemes, the found significant rules are the following :

— Four principal rules conclude on Oral Vowels :

R1. E[580-935] < E[935-1390] & E[1390- 2025] > E[2025-2950] => [A, E, e, y]

R2. E[62-312] > E[312-580] & E[580-935] < E[935-1390] => [e, i, y]

R3. E[62-580] < E[580-1390] & E[1390- 2025] > E[2025-2950] => [E, A, e, y]

R4. E[62-580] < E[580-1390] & E[62-312] > E[312-580] => [e,y]

— Two principal rules conclude on Nasal Vowels :

R1. E[312-580] < E[580-935] & E[1390-2025] < E[2025-2950] => [an,un]

R2. E[580-1390] > E[1390-2950] & E[312-437] < E[437-580] & E[748-935] > E[935-1142] => [an]

The rules can be interpreted in terms of cues (e.g., invariant primitives) relevant to features (e.g., Oral versus Nasal). Spectral masks of vowels, drawn from the same analysis carried out per phoneme, are shown to be coherent with phonetic knowledge. The masks of two well classified cardinal vowels (e.g., [A, i]) are shown with the indication of the theoretical first three formants for french language (cf. figure 1 & 2).

6. Conclusion

The results of the experiments show, for a given error rate, Charade performs as well as MHN and better than HDNN. In comparison with the usual classifiers, Charade offers various advantages :

— The generated production rules can be analysed and interpreted for knowledge acquisition.

— The rule system autocoherence (i.e., accuracy rate on the learning set) gives an information measure of the parametrisation in terms of invariance. It puts in question the validity of the description language for the considered problem.

— The principle of decision inherent in a rule system allows the integration of heterogenous data (e.g., information coming from various signal processing methods).

Acknowledgement : we are especially indebted to Jean-Gabriel Ganascia for providing us with the CHARADE system and for making helpful suggestions.

References

[1] N. CARBONELL, D. FOHR, J.-P. HATON, F. LONCHAMP, J.M. PIERREL : **An Expert System for the Automatic Reading of French Spectrograms.** Proc ICASSP, 42.8, 1984.

[2] R. DE MORI, A. GIORDANA, P. LAFACE, L. SAITTA : **An Expert System for Speech Decoding.** Proc. AAAI. 1982.

[3] R.G. GALLAGER : **Information Theory and Reliable Communication.** John Wiley & Sons, New York, 1968.

[4] J.-G. GANASCIA : **Agape et Charade : deux techniques d'apprentissage appliquées à la construction de bases de connaissance.** Thèse d'état, Université Paris-Sud, 1987.

[5] J.J. HOPFIELD : **Neural Networks and Physical Systems with the Emergent Collective Computational Abilities.** Proc Nat. Acad. Sci. USA vol 79, pp 2554-2558, 1982.

[6] H. MELONI, R. BULOT : **Signal Processing and Pattern Recognition in Prolog II for Acoustic-Phonetic Decoding of Continuous Speech.** Proc RFIA, pp 21-36, 1987.

[7] G. MERCIER, A. COZANNET, J. VAISSIERE : **Recognition of Speaker-Dependant Continuous Speech with Keal-Nevezh** NATO ASI Series, Vol F46 pp 459-463., 1988.

Niveau 1				Niveau 4			
d1	E[62.3 - 1390]	>	E[1390 - 7097]	d12	E[62.3 - 186.9]	>	E[186.9 - 311.5]
				d13	E[186.9 - 311.5]	>	E[311.5 - 436.7]
Niveau 2				d14	E[311.5 - 436.7]	>	E[436.7 - 580]
d2	E[62.3 - 580]	>	E[580 - 1390]	d15	E[436.7 - 580]	>	E[580 - 748]
d3	E[580 - 1390]	>	E[1390 - 2950]	d16	E[580 - 748]	>	E[748 - 935]
d4	E[1390 - 2950]	>	E[2950 - 7097]	d17	E[748 - 935]	>	E[935 - 1142]
				d18	E[935 - 1142]	>	E[1142 - 1390]
Niveau 3				d19	E[1142 - 1390]	>	E[1390 - 1678]
d5	E[62.3 - 311.5]	>	E[311.5 - 580]	d20	E[1390 - 1678]	>	E[1678 - 2025]
d6	E[311.5 - 580]	>	E[580 - 935]	d21	E[1678 - 2025]	>	E[2025 - 2439]
d7	E[580 - 935]	>	E[935 - 1390]	d22	E[2025 - 2439]	>	E[2439 - 2950]
d8	E[935 - 1390]	>	E[1390 - 2025]	d23	E[2439 - 2950]	>	E[2950 - 3595]
d9	E[1390 - 2025]	>	E[2025 - 2950]	d24	E[2950 - 3595]	>	E[3595 - 4455]
d10	E[2025 - 2950]	>	E[2950 - 4455]	d25	E[3595 - 4455]	>	E[4455 - 5622]
d11	E[2950 - 4455]	>	E[4455 - 7097]	d26	E[4455 - 5622]	>	E[5622 - 7097]

Table 4. 26 binary descriptors corresponding to 4 analysis levels

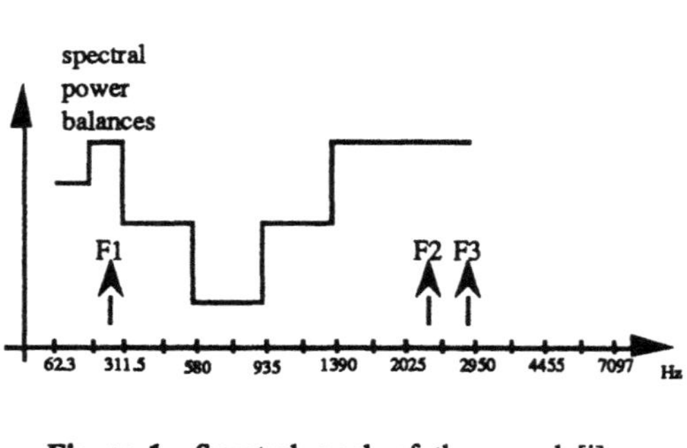

Figure 1. Spectral mask of the vowel [i]

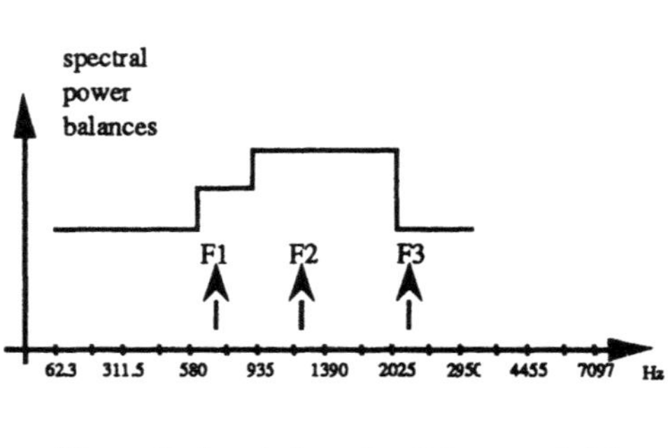

Figure 2. Spectral mask of the vowel [a]

Part 4

Stochastic Models for Language and Dialogue

STOCHASTIC GRAMMARS AND PATTERN RECOGNITION

Hermann Ney

Philips GmbH Forschungslaboratorium Aachen
D-5100 Aachen, Germany

SUMMARY: This paper presents a unifying framework of syntactic and statistical pattern recognition for one-dimensional observations and signals like speech. The syntactic constraints will be based upon stochastic extensions of the grammars in the Chomsky hierarchy. These extended stochastic grammars can be applied to both discrete and continuous observations. Neglecting the mathematical details and complications, we can convert a grammar of the Chomsky hierarchy to a stochastic grammar by attaching probabilities to the grammar rules and, for continuous observations, attaching probability density functions to the terminals of the grammar. In such a framework, a consistent integration of syntactic pattern recognition and statistical pattern recognition, which is typically based upon Bayes' decision rule for minimum error rate, can be achieved such that no error correction or postprocessing after the recognition phase is required. Efficient algorithms and closed-form solutions for the parsing and recognition problem will be presented for the following types of stochastic grammars: regular, linear and context-free. It will be shown how these techniques can be applied to the task of continuous speech recognition.

1. INTRODUCTION

In pattern recognition, it is usual to dichotomize between statistical approaches and syntactic approaches. The statistical approach is ideally suited for handling patterns which can be represented by continuous-valued measurements. These measurements are used to form a vector in a high-dimensional feature space. Typically the statistical approach is based on the Bayes' decision rule for minimum error rate [Duda, Hart 1973] and therefore also referred to as decision theoretic approach.

On the other hand, there are applications where the structure of patterns plays an important role and thus a formalism for handling pattern structures is required. Such syntactic approaches typically recourse to concepts from the theory of formal languages in the Chomsky hierarchy [Hopcroft, Ullman 1979; Revesz 1985] that were originally developed to describe the linguistic structure of natural language. For perfect input patterns, i.e. for symbol strings with no errors in them, there are two well-known and most powerful parsing algorithms that analyze the unknown structure of the input string, namely the Earley algorithm [Earley 1970] and the CYK algorithm which is named after its independent authors Cocke, Younger and Kasami [Aho, Hopcroft 1972, p.314].

There have been successful attempts to extend the structure analyzing algorithms to cope with non-perfect input patterns, i.e. strings with deletion, insertion and confusion of symbols [Aho, Peterson 1972; Lyon 1974; Tanaka, Fu 1978]. There are two textbooks [Fu 1982; Gonzalez, Thomason 1982] that are devoted to syntactic pattern recognition. The widely used approach is to extend the formal grammars to stochastic grammars for handling symbol errors such as deletions, insertions and confusions and to incorporate error recovering strategies into the parsing algorithms.

However there are a number of applications, where these extensions and modifications are not sufficient: the strings to be recognized may consist of continuous-valued vectors rather than discrete symbols from a finite alphabet, and the ambiguity inherent in the data may be extremely high.
An example of such an application is the recognition of continuous speech which is difficult for a number of reasons:

- the acoustic data themselves are highly variable;
- the speaking rate can vary in an extremely nonlinear way;

NATO ASI Series, Vol. F 75
Speech Recognition and Understanding. Recent Advances
Edited by P. Laface and R. De Mori

- there are no cues in the acoustic signal that mark the boundaries between words or sounds;
- there are often linguistic constraints on the set of 'legal' word sequences that have to be taken into account in the recognition phase.

Despite or because of these difficulties, at the same time, a lot of algorithms and experimental results are available for speech recognition. Therefore we will use the task of continuous speech recognition as application for the algorithms to be described.

The intention of this paper is to give a unifying view of the integration of syntactic and statistical pattern recognition and to present efficient algorithms for intergrated syntactic-statistical pattern recognition. The approach is holistic in the sense that a full integration of the structural rules and constraints into the statistical approach is achieved. The characteristics of the approach are:

- Three subsets of grammars in the Chomsky hierarchy will be considered simultaneously: regular, linear and context-free grammars.
- The stochastic grammars are extended to handle sequences of continuous-valued vectors.
- The Bayes' decision rule for minimum error rate is incorporated into the analysis.
- Consistent parsing and recognition strategies are described that do not require error recovering techniques. The algorithms result in closed-form solutions in the sense that no approximations are involved such that they work even in the case of highly imperfect and corrupted data such as speech signals.

The emphasis of the paper will be placed upon efficient search strategies and the corresponding architecture of the recognition system, in particular the question of how to handle the interdependence between elementary low-level recognition and the structural constraints as expressed by the grammar rules. As opposed to this decision or search problem, the modelling problem, i.e. the question of how to arrive at the structure of the models or in other words the grammar rules, will only be touched upon occasionally.

For stochastic grammars, it is helpful to distinguish three grammar types according to their generative power: regular, linear, context-free grammars. For each of these three stochastic grammar types, we have to study and design algorithms for the following problems, where the observations typically are continuous-valued and highly ambiguous:

1. The parsing problem (Viterbi scoring): given the grammar and a set of observations, what is the most likely parse tree that produced the observations ?
2. The membership problem (Baum-Welch scoring): given the grammar and a sequence of observations, what is the probability that the grammar produced the observations ?
3. The training problem: given the grammar rules and a set of observation sequences, how can the probabilities of the grammar rules be estimated such that the likelihood of the grammar producing the observations is maximum ?

The outline of the paper is as follows. After the review of the Bayes' decision rule for minimum error rate, we present the principles of formal grammars and the Chomsky hierarchy along with examples and a discussion of normal forms. We then move to stochastic grammars and extend them to cover strings of continuous-valued observations. In the next section, we present the algorithms for the parsing problem and the membership problem in the context of stochastic grammars and study their computational complexity. We then study the problem of training the unknown parameters of the grammar models. Finally we show how these techniques are applied to continuous speech recognition.

2. BAYES' DECISION RULE AND SPEECH RECOGNITION

From this paper's point of view, the primary function of a pattern recognition system is to arrive at decisions concerning the class membership of unknown patterns. In our case the patterns to be recognized are assumed to be a sequence of either discrete or continuous-valued observations. The system functions are depicted in Fig. 1. Often, in particular in image recognition applications, there is a preprocessing step that produces a sequence of so-called pattern primitives. Although some image recognition application may fit in the framework of this paper, it should be clear that truly two-dimensional applications cannot be handled by our approach because phrase structure grammars are not able to capture two-dimensional constraints and relationships.

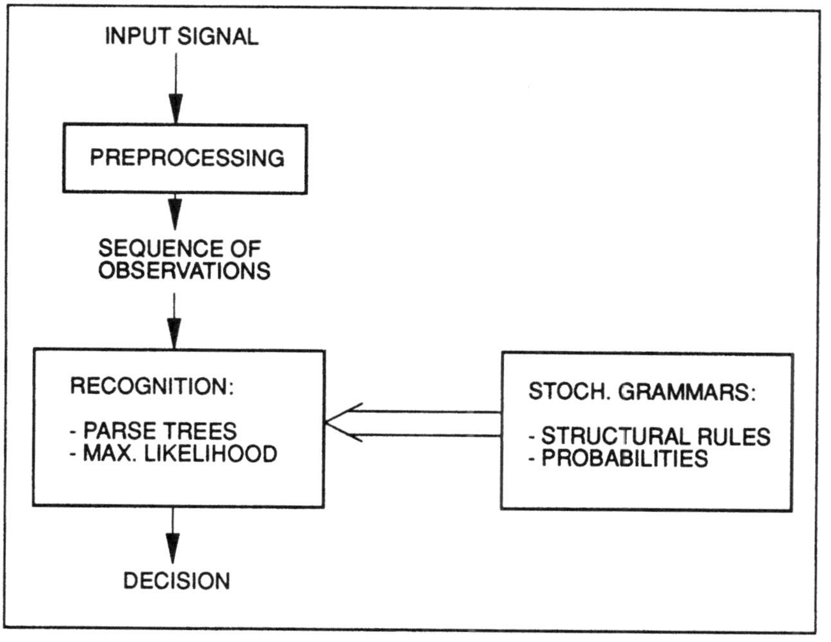

FIG.1. System architecture based on a stochastic grammar

In addition to determining the class membership of the unknown pattern, the task of the recognition system is also to analyze the structure of the input sequence and to make use of this knowledge in order to improve the recognition reliability. As shown in Fig. 1, there are two kinds of knowledge sources in the system: the grammar rules that describe the structure of the pattern string and probabilities associated with these rules.

The decision of the pattern recognition system will be based upon Bayes' decision rule for minimum error rate. Its application to speech recognition is illustrated in Fig. 2. The acoustic analysis produces a sequence of acoustic vectors that typically represent the short term intensity spectrum of the speech signal. There are three knowledge sources in the system: the inventory of phoneme models, the pronunciation lexicon and the language model. In order to minimize the probability of error according to Bayes' decision rule [Jelinek 1976], the task of a speech recognizer must be to determine that

sequence of words $w' = w[1:N]$ of unknown length N

that has most probably caused the

sequence of observations $x' = x[1:I]$ over time $i=1,...,I$.

For notational convenience, we use the apostrophe to distinguish sequences and single symbols. Thus we have for the unknown word sequence w_0':

$$w_0' = \text{argmax} \{ P(w'|x'): w' \}$$

$$= \text{argmax} \{ P(w')\, P(x'|w'): w' \}$$

The important advantage of the Bayes' decision rule is that it provides a consistent scoring criterion for a task like speech recognition, where reliable recognition at the acoustic-phonetic level is heavily error-prone and therefore high-level knowledge and constraints must be combined in a suitable way to improve the recognition reliability.

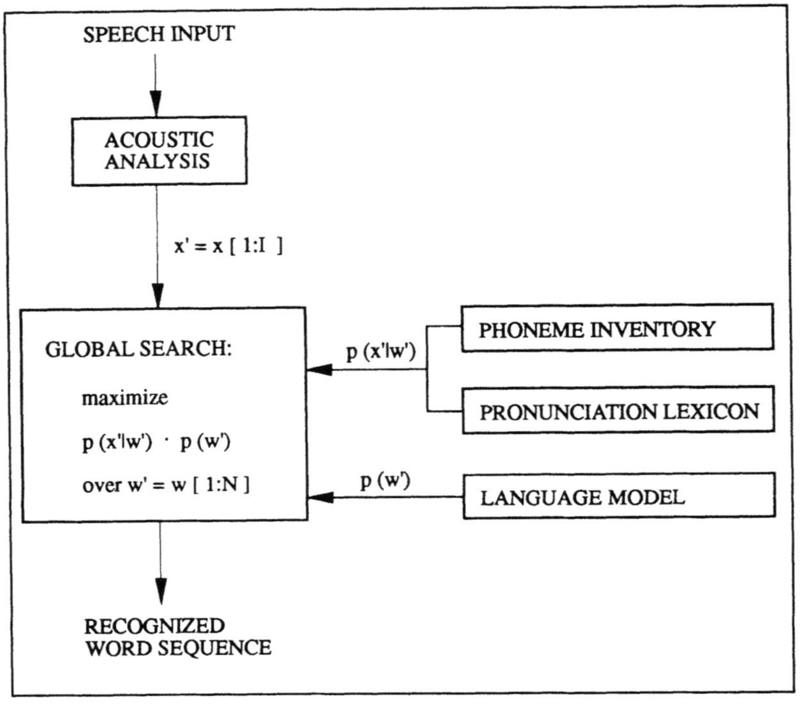

FIG.2. Application of Bayes' decision rule to speech recognition

The formalism of stochastic grammars can be used for various purposes:

- to compute the prior probability P(w') for any arbitrary word sequence; these prior probabilities should ideally capture constraints resulting from syntax, semantics and pragmatics;
- to compute the conditional probability P(x'|w') of a sequence x' of continuous observations, conditioned on word sequence w' or also a single word w ;
- to compute the product P(w') P(x'|w') and to optimize the Bayes' criterion. In all these cases, stochastic grammars will be utilized to model the redundancy due to the internal structure of the strings to be recognized.

The most prominent examples of stochastic grammars in speech recognition are the Hidden Markov models and the language models. Hidden Markov models are used to model the observations at the acoustic-phonetic level and speaking rate variations at the same time. From the grammar point of view, they are special cases of stochastic regular grammars whose nonterminals or states are identified with points on a time scale. The task of language models is to capture the inherent linguistic constraints in the spoken sentences to be recognized. In most cases, the constraints are expressed in terms of a finite state network or a stochastic n-gram model. Language models based on context-free grammars have been used only recently and only for small tasks, due to their computational complexity.

By studying these questions, we will see that stochastic grammars and their extensions for handling continuous-valued vector strings provide a common basis for various algorithms. For example, the dynamic programming based time alignment algorithm used in Hidden Markov models can be shown to be a parser for stochastic regular grammars; the dynamic programming based parsers for stochastic context-free grammars can be shown to be generalizations of the CYK parser in a bottom-up strategy and of the Earley parser in a top-down strategy.

3. FORMAL GRAMMARS: THE CHOMSKY HIERARCHY OF PHRASE STRUCTURE GRAMMARS

3.1 Definitions and Principles

In this section, we review the terminology and basic properties of formal grammars because they form the basis for the stochastic extensions of the nonprobabilistic formal grammars. In pattern recognition, the typical task is to recognize strings of symbols or observations. For the moment, we assume a finite set of discrete symbols from which the strings of the language can be built up. Later we will extend our approach and identify each symbol of the discrete alphabet with a pattern class, which has generated the random observation vector.

A phrase structure grammar is a mathematical system for defining a language of one-dimensional strings and assigning a structure to these strings. The key property of phrase structure grammars, or simply formal grammars, is that they provide an efficient means for defining a language by a finite number of specifications, although the number of strings or elements in the language may be unlimited.

A phrase structure grammar is defined as a fourtuple

$G = (N,T,S,F)$, where

1. N is a finite set of nonterminal symbols, also called variables or syntactic categories.
2. T is a finite set or alphabet of terminal symbols, disjoint from N.
3. S is a distinguished element in N called the sentence or start symbol.
4. F is a finite set of production or rewrite rules, where each rule maps a symbol string u' in $(N \cup T)'$ N $(N \cup T)'$ on a symbol string v' in $(N \cup T)'$. The apostrophe ' is used to denote the formation of the closure of a set, which is defined to be the set of all strings over the set.

The productions in F are written in the form

$$u' \to v',$$

where the symbol $\to$, of course, is not in $(N \cup T)$. According to the definition, u' must always be different from the empty string whereas v' may stand for the empty string. Productions are used to derive new strings from given ones by replacing a substring on the left-hand side of a rule by the right-hand side of the same rule. Given two words u' and v' in $(N \cup T)'$, we say that u' derives v' or v' is derivable from u', in symbols,

$$u' \Rightarrow v',$$

if there is a sequence of production rules such that the rewriting operations start with string u' and end with v'. A string a' in T ' is said to be generated by the grammar G, if a' can be derived by the start symbol S:

$$S \Rightarrow a'.$$

The language L(G) of a grammar G is the set of all strings generated by the grammar:

$$L(G) = \{a' \mid a' \text{ in } T',\ S \Rightarrow a'\}.$$

Hence we see that nonterminal symbols are used solely as intermediary or auxiliary symbols in the sequence of derivations starting with S. Along with the productions, they are abstractions in the sense that they do not exist as members of the language defined by the grammar.

In this paper, we use -more or less consistently- the following conventions regarding grammars. The capital letters A,B,C,D and S denote nonterminals. The lower case letters a,b,c represent terminals. The apostrophe symbol is consistently used to denote strings over a language, e.g. a' and A' are strings of terminals and nonterminals, respectively. The apostrophed lower case letters u', v' denote mixed strings of nonterminals and terminals. In most cases, the membership of the symbols will become clear anyway from the context. The length of a string u' is defined as the number of symbols and is denoted as |u'|. For erroneous discrete symbols or noisy continuous observations to be considered later on, we will use letters from the end of the alphabet, e.g. x, y, z.

3.2 Hierarchy of Phrase Structure Grammars

The phrase structure grammars can be classified on the format of the production rules. As we will see, this format reflects the generative power of a given grammar.

Let A, B be in N, u',v' in $(N \cup T)'$ and a',b' in T '. A phrase structure grammar is said to be

1. Unrestricted if there are no restrictions.
2. Context sensitive if each production is length-increasing (more precisely, not length-decreasing), i.e.

   $u' \rightarrow v'$, where $|u'| < |v'|+1$

3. Context-free if each production has exactly one nonterminal on the left-hand side, i.e.

   $A \rightarrow v'$

4. Linear if each production rule has exactly one nonterminal on the left-hand side and at most one nonterminal on the right-hand side, i.e.

   $A \rightarrow a'Bb'$ or $A \rightarrow a'$

5. Regular if all productions are either right-linear, i.e.

   $A \rightarrow a'B$ or $A \rightarrow a'$,

   or left-linear, i.e.

   $A \rightarrow Ba'$ or $A \rightarrow a'$.

These definitions are either identical or can be shown to be equivalent to the definitions in the standard textbooks [Hopcroft, Ullman 1979; Revesz 1985]. There might be, however, a difference as to whether the grammars allow the derivation of the empty string or not. If for some reasons the empty string is required as a member of the language, the original empty-free grammar can be modified by a special transformation [Revesz 1985, p.18,45; Thomason, Gonzalez 1982, p.30]. For pattern recognition purposes, the empty string is rarely useful because the grammar is intended to provide a structural description of the strings of the language.

Apart from the class of linear grammars, the above taxonomy is the Chomsky hierarchy. Each class is a proper subset of the preceding class, e.g. the linear grammars are a proper subset of the context-free grammars etc.

3.3 Normal Forms and Examples of Grammars

There are a lot of well-known transformations that are concerned with the format of the right-hand sides of the production rules. Examples of such normalizing transformations can be found in [Revesz 1985, p.46; Aho, Ullman 1972, p.165; Hopcroft, Ullman 1979, p.92]. The original grammar and the transformed grammar are equivalent in the sense that they define exactly the same language. These transformations can be obtained by fully automatic algorithms. The resulting form of the grammar, called normal form, is important because it imposes a certain structure on the format of production rules that simplifies the theoretical analysis of the rules and provides the starting point for the pattern recognition approach to be described. The normal forms used in this paper are summarized in Fig. 3. It is interesting to note that the characteristic properties of each grammar type can be found directly in its corresponding normal form.

GRAMMAR	NAME	NORMAL FORM
REGULAR	LEFT LINEAR	A → Ba A → a
LINEAR	-	A → aBb A → a | ab
CONTEXT-FREE	CHOMSKY	A → BC A → a
CONTEXT-SENS.	KURODA	AB → CD A → BC A → B A → a

FIG. 3. Normal forms of formal grammars

To illustrate the mathematical formalism, we try to give typical examples of each grammar. In general, the reader should realize that it may be extremely hard to determine what language is generated by the grammar. We will use examples with a small ('minimum') number of terminals and nonterminals in order to put the important differences in the generative power into evidence. The idea behind these examples is to illustrate the structures that can be captured by phrase structure grammars.

- REGULAR GRAMMAR.

Consider the grammar

$$G = (\{A,S\},\{a,b\},S,F), \qquad \text{where } F: \; S \rightarrow Sa \mid Ab \mid a \quad A \rightarrow Aa \mid Sb \mid b$$

The symbols | is used as an abbreviation for indicating alternative rules to be applied to the left-hand side. This grammar generates strings consisting of any number of a's and an even number of b's. There is the well-known theorem that states the equivalence between regular grammars and finite state networks or automata [Hopcroft, Ullman 1979]. For the grammar at hand, the equivalent finite state automaton is depicted in Fig. 4. Each state of the network is identified with a nonterminal of the grammar. Thus a rule A → Sb is translated as a transition from state S to state A, whereby terminal b is generated. There is an additional state in the network that serves as initial state and covers the rules of the type A →b. This rule is translated as a transition from the initial state to state A, whereby terminal b is generated. Other choices for setting up the one-to-one correspondence between regular grammars and finite state automata are possible as well. The choice we have taken here allows us to identify a state A with a beginning portion of a string in the sense that it is derived by the corresponding nonterminal A. By unfolding the finite state machine along the axis of the string positions, we obtain a so-called lattice or trellis that captures all possible state sequences through the network as shown in Fig. 1. Of course, it is clear that the path through the network could also be defined in the reverse order by starting from state S and ending in state $.

- LINEAR GRAMMAR.

Linear grammars are well suited to model symmetries in strings. Consider the grammar [Hopcroft, Ullman 1979, p.81]

$$G = (\{S\},\{a,b,c\},S,F), \qquad \text{where } F: \; S \rightarrow aSb \mid c$$

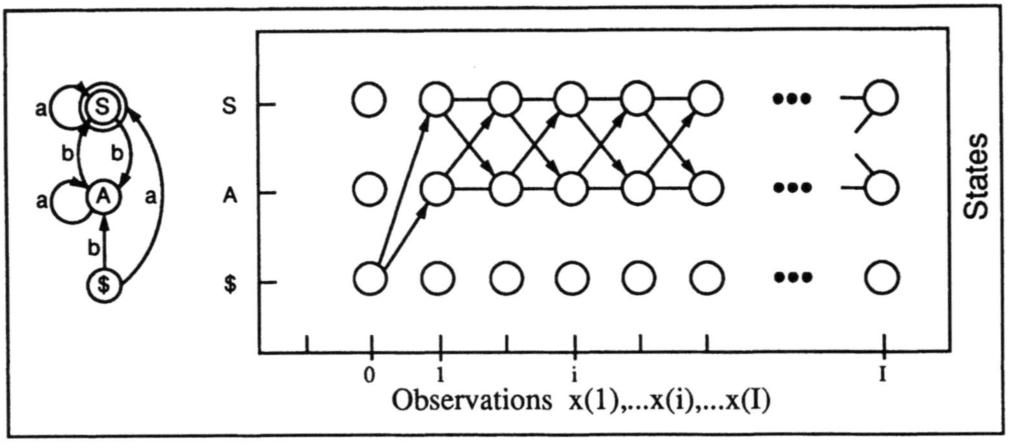

FIG. 4. Finite state automaton and trellis of a regular grammar

This grammar generates strings of balanced parentheses a^n c b^n around the symbol c, where each left parenthesis (a) has a matching right parenthesis (b). Note that this example of a linear grammar is rather simple: there is only one nonterminal, namely the start symbol.

- CONTEXT-FREE GRAMMAR.

  Consider the grammar

  $G = (\{B\},\{a,b\},S,F)$, where F: $B \rightarrow aBB \mid b$

Note that here the start symbol is denoted as B. As can be shown by induction, this grammar generates strings consisting of a's and b's, where the last symbol is b and the remaining portion of the string has an equal number of a's and b's. Thus in a certain sense, the grammar is able to count. It can be shown that no linear grammar exists for this language [Chomsky, Schuetzenberger 1963]. A rule of the type $B \rightarrow aBB$ plays a basic role in other grammars that count, e.g. a grammar that produces strings that consist of an equal number of a's and b's [Hopcroft, Ullman 1979, p.81].

A summary of typical applications and structures for the four types of grammars is given in Fig. 5. Note that there is not a unique relationship between a language and a grammar. For example, the two grammars

$G = (\{S\},\{a\},S,F)$, where F: $S \rightarrow SS \mid a$
$G = (\{S\},\{a\},S,F)$, where F: $S \rightarrow Sa \mid a$

describe the same language.

3.4 Parsing

So far we have considered the generative concept of a grammar, i.e. a mathematical formalism for generating formal languages. The analytic aspect is how we can decide whether a given string belongs to the language generated by a grammar and if so which derivation it has. A number of abstract mathematical devices, called automata, have been designed to perform this task for each of the grammars in the Chomsky hierarchy. These devices are [Hopcroft, Ullman 1979]

the finite-state automaton	for regular grammars,
the push-down automaton	for context-free grammars,
the linear-bounded automaton	for context sensitive grammars,
the Turing machine	for unrestricted languages.

GRAMMAR	STRUCTURES, APPLICATIONS
REGULAR	FINITE STATE NETWORK, LEXICAL ANALYZER, TEXT EDITOR
LINEAR	BALANCED PARANTHESES, MIRROR STRINGS
CONTEXT-FREE	ARITHM. EXPRESSIONS, COUNTING OF TWO SETS, PROGRAMMING LANGUAGES, RESTRICTED NATURAL LANGUAGE
CONTEXT-SENS.	STRING DOUBLING, COUNTING OF THREE SETS, UNRESTRICTED NATURAL LANGUAGE

FIG. 5. Examples of structures described by formal grammars

However, it is possible to study the analytic aspects of the grammars in a different way by considering techniques and algorithms that may be implemented on a computer or one of the above devices. An analytic algorithm that determines a derivation, i.e. a sequence of productions for a given string in a language L(G) with a known grammar G is usually called a parsing algorithm.

The above devices and most of the parsing algorithms have been designed for error-free observations and for nonprobabilistic grammars. They are not well suited for pattern recognition purposes because they are not able to cope with recognition errors in the string or with the probabilistic aspects of the grammar in an efficient way.

The parsing result is not always unique. A string in L(G) may have more than one derivation in the grammar G. Derivations that differ only in the order of rule applications are defined to be the same. Usually the notion of a leftmost derivation is introduced in order to force the derivation to proceed in an orderly fashion. A grammar is called ambiguous if it generates at least one terminal string in at least two different derivations. The most general case considered in this paper is the context-free grammar for which the sequence of derivation rules is usually represented by a so-called derivation or parse tree.

4. STOCHASTIC OR PROBABILISTIC GRAMMARS

4.1 Motivation and Principles

In this section, we will describe the probabilistic extensions of context-free grammars and their subsets. Up to now we have considered only the structural description for strings of a given language. This approach requires an extension for two reasons. First, all string patterns under consideration do not occur with the same frequency. In particular, certain structures in the strings may be more likely to occur than others. Therefore it could be useful to specify the frequencies with which the derivation occur in the language under consideration. These frequencies form some sort of a-priori knowledge about the strings in the language. The second reason, in particular for pattern recognition applications, is the unreliability in the recognition of what is usually called pattern primitives. These primitives are used as the terminals of the grammar. In practical situations, noise and distortions cause errors in the recognition of the primitives, and these errors have to be taken into account by the grammar. We will first stick to discrete observations as input data and then extend the approach towards continuous-valued observations. As in the nonprobabilistic case, the grammar gives a finite description of the underlying structures, i.e. of the probability distribution taken over all possible sequences of observations.

A meaningful generalization is the definition of a stochastic or probabilistic grammar. Each rule of the grammar

$$r: u' \to v'$$

is assigned a probability, in symbols:

$$p(r) = p(r{:}u' \to v'|u') = p(u' \to v'|u') = p(v'|u').$$

Depending on the situation, we will use each of theses three notations to denote the probability of a rule $r{:}u' \to v'$. This probability is interpreted as the conditional probability of producing string v', given nonempty string u'. Thus it is natural to consider the probabilities to be conditional probabilities and require them to be normalized:

$$\sum_{v'} p(u' \to v'|u') = 1.$$

A stochastic grammar for which all probabilities are normalized for all left-hand sides u' of productions is called normalized or proper [Gonzalez, Thomason 1982, p.180]. The training of these probabilities will be considered in a later section.

A derivation in a stochastic grammar is a sequence of stochastic events each of which is the application of a rule. A stochastic grammar is called (stochastically) unrestricted if the elementary steps of a derivation are mutually independent, i.e. independent of the order in which they are applied [Gonzalez, Thomason 1982, p.179]. In other words, the probability of applying any particular rule is independent of previously applied rules. In this paper, only unrestricted stochastic grammars are considered. Utilizing this property of a stochastic grammar, we can assign a probability to each derivation $r'{:}\ A \Rightarrow v'$ with $r' = r(1),...,r(m)$ by defining:

$$P(r'{:}A \Rightarrow v'|A) = P(r') = p(r(1)) \cdot ... \cdot p(r(m)).$$

In other words, given the derivation or parse tree, we assign a probability to each rule and thus obtain the probability of the whole derivation as the product of these individual probabilities. Now using the same scheme, we can write the probability of nonterminal A generating string v' by summing up over all possible derivations r' starting from A:

$$P(A \Rightarrow v'|A) = P(v'|A) = \sum_{r'} P(r'{:}A \Rightarrow v'|A)$$

Using the rule probabilities, each terminal string a' in the language L(G) generated by the grammar G can be assigned a probability as the conditional probability $P(S \Rightarrow a'|S)$ of the start symbol S generating string a':

$$P(S \Rightarrow a'|S) = \sum_{r'} P(r'{:}S \Rightarrow a'|S)$$

Now a natural way is to use the production probabilities $P(S \Rightarrow a'|S)$ of a string a' in a language L(G) to induce a probability measure on the language L(G). Therefore the question is important whether it is true that

$$\sum_{a'} P(S \Rightarrow a'|S) = 1$$

where we have used the convention that $P(S \Rightarrow a'|S)=0$ for terminal strings a' that are not in L(G). Such a grammar is called consistent. Consistency tests along with examples of inconsistent grammars are given in [Fu 1982, p.206] and [Thomason, Gonzalez 1982, p.188].

GRAMMAR	PROCESS DESCRIBED NAME	 RULE	ASSIGNED PROBABILITY
CONTEXT-FREE	STRUCTURE	$A \rightarrow BC$ $A \rightarrow a$	$p(A \rightarrow BC \mid A)$ $p(A \rightarrow a \mid A)$
	MEASUREMENT	$a \rightarrow x$	$p(x \mid a)$
REGULAR	STRUCTURE	$A \rightarrow Ba$ $A \rightarrow a$	$p(A \rightarrow Ba \mid A)$ $p(A \rightarrow a \mid A)$
	MEASUREMENT	$a \rightarrow x$	$p(x \mid a)$

FIG. 6. Summary of processes and their probabilities in the generation/recognition of continuous and/or noisy observations

4.2 Pattern Recognition: Noisy and Continuous Observations

What has been considered so far, is the standard type of stochastic grammars. In order to establish the link to pattern recognition, we assume that the observations, no matter whether discrete or continuous-valued, are random observations from the corresponding terminal class. Denoting the observations by x and the terminal class by a, we can write this measurement process as:

$$a \rightarrow x,$$

with the probability:

$$p(a \rightarrow x|a) = p(x|a)$$

For discrete observations, the quantity $p(x|a)$ is a probability distribution over the possible discrete observations x; for continuous observations, it is a probability density function. Note that the normalization properties are still satisfied, since both $p(A \rightarrow a|A)$ and $p(x|a)$ are normalized individually. As a result of this extension of the standard stochastic grammars, we have two kinds of rules or processes in the grammar which are summarized in Fig. 6 for the case of regular and context-free grammars.

Thus the conditional probability $P(A{:}S \Rightarrow x'|A)$ of observing or producing the measurement sequence x' given nonterminal A can be thought of as composed of two kinds of probabilities:

- the structural probabilities, reflecting prior knowledge about the structure of the possible sequences;
- the measurement probabilities, which are intrinsic to a true pattern recognition problem.

These measurement probabilities must therefore result in an inherent ambiguity at the level of the observations. This type of ambiguity must be distinguished from the ambiguity that arises from the grammar rules.

It is important to emphasize what we have achieved now by the extended formalism of stochastic grammars: a finite mechanism for describing probability (density) distributions over strings of observations with internal structures. The approach we have chosen for introducing noisy and/or continuous observations allows a clear distinction between the terminal symbols as class identifiers and the observations as the measured data. A different approach is to modify the grammar rules such that they cover all possible errors at the level of the observable terminals [Fu 1982, p.250]. Such an approach works only in the case of discrete observations.

5. PARSING AND MEMBERSHIP ALGORITHMS FOR STOCHASTIC GRAMMARS

5.1 Problem Specification

This section covers the central topic of this paper, namely efficient parsing and search algorithms for stochastic grammars. In the previous sections, stochastic grammars were introduced as means for generating strings of a language. In view of the general architecture shown in Fig 2., our goal in recognition is to compute the prior probabilities P(w') of a symbol string (or word sequence in the case of speech recognition) and the conditional probabilities P(x'|w'), where x' denotes the sequence of continuous observations. For the second case, there is a slight complication in that we would need a grammar for handling word concatenations and their effect on the acoustic observations. We will postpone a more thorough discussion to a later section and assume for this section that we have only single words: P(x'|w).

For a given string of observations, there are two problems to be considered, namely the parsing problem and the membership problem. In the parsing problem, we want to find the most likely parse tree that has produced the observations, i.e. we want to calculate the so-called Viterbi score:

$$\max_{r'} P(r':S \Rightarrow x'|S)$$

In contrast to that, the membership problem is to compute the probability that the grammar has generated the observations. Since there may be several parse trees, we have to sum up the probabilities of all these parse trees. The corresponding quantity is usually referred to as Baum-Welch score:

$$P(S \Rightarrow x'|S) = \sum_{r'} P(r':S \Rightarrow x'|S)$$

The terminology 'Viterbi score' and 'Baum-Welch score' arises from the framework of Markov processes [Levinson et al. 1983].

5.2 Parsing Problem: Viterbi Scoring

First we address the parsing problem. To perform the optimization, we consider the normal forms of the grammars and their interpretation as shown in Fig. 7.

Context sensitive grammars do not fit into the formalism to be developed and are excluded. We consider the three grammar types in parallel to put the unifying approach into evidence. For all three grammar types, we introduce a probability that nonterminal A derives a certain subportion of the whole input string of continuous-valued observations. We denote the time axis as i=0,...,I and identify the indices with boundaries rather than segments in order to simplify the notation. In addition, we use the notation x[j:i] for the observations between times j and i. For a given portion x[j:i] of the string and nonterminal A, we define the partial likelihood:

$$Q(j,i|A) := \max_{r'} P(r':A \Rightarrow x[j:i]|A)$$

This is the probability of the best hypothesis that observation portion x[j:i] is explained by nonterminal A. It should be noted that this definition captures only one, i.e. the most likely derivation. The triple argument in the score Q(j,i|A) is related to the parse chart or parse table used in the CYK [Aho, Hopcroft 1972, p.314] and Earley [1970] algorithms for the parsing of nonstochastic grammars. The partial likelihood Q(j,i|A) can be viewed as a sort of probabilistic extension of these chart tables.

Now using the grammar rules shown in Fig. 7, we can define recursion equations by building up portions of growing length using previously computed portions according to the grammar rules. The starting point for these recursions is provided by those grammar rules that explain elementary portions of the input sequence, i.e. portions of length 1 or 2.

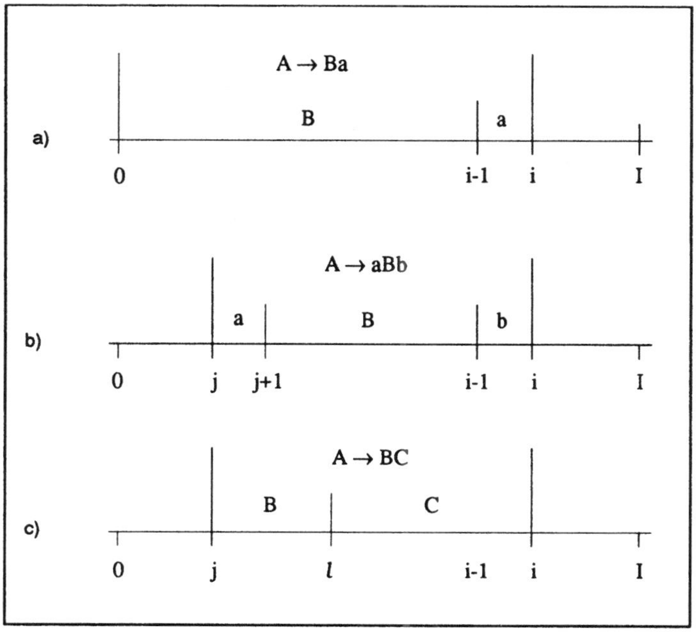

FIG. 7. Illustration of the recombination step for regular (a), linear (b) and context-free (c) grammars

- REGULAR GRAMMAR

The regular grammar is a special case because the nonterminals always explain a portion that starts with j=0 and thus only one position index i is required as shown in Fig. 7. According to the normal form, we have the initialization step:

$$Q(0,1|A) = \max \{ p(A \to a|A)\, p(x(1)|a): A \to a \}$$

and the recursive optimization:

$$Q(0,i|A) = \max \{ p(A \to Ba|A)\, p(x(i)|a)\, Q(0,i-1|B): A \to Ba \}$$

The final result of the recursion is then Q(0,I|S), i.e. the probability of the best derivations starting with the sentence symbol S. By looking at the trellis shown in Fig. 4, it becomes clear that the argument (time,nonterminal)=(i,A) can be identified with the gridpoints of the trellis.

- LINEAR GRAMMAR

For the linear grammar type, there are two ways of initializing according to whether we start with one or two elementary observations:

$$Q(i-1,i|A) = \max \{ p(A \to a|A)\, p(x(i)|a): A \to a \}$$
$$Q(i-2,i|A) = \max \{ p(A \to ab|A)\, p(x(i-1)|a)\, p(x(i)|b): A \to ab \}$$

In contrast to the regular grammar, the recursion must be carried out over both time indices j and i:

$$Q(j,i|A) = \max \{ p(A \to aBb|A)\, p(x(j+1)|a)\, p(x(i)|b)\, Q(j+1,i-1|B): A \to aBb \}$$

- CONTEXT-FREE GRAMMAR

For the most general case, the context-free grammars, we have as initialization step:

$$Q(i-1,i|A) = \max \{ p(A \to a|A)\, p(x(i)|a): A \to a \}$$

a)

```
FOR EACH NONTERMINAL A DO
    Q (0,i | A) = max { p(A → a | A) p(x(1) | a) : A → a }
FOR EACH POSITION i = 2,3,...,I DO
    FOR EACH NONTERMINAL A DO
        Q (0,i | A) = max { p(A →Ba | A) p(x(i) | a) Q (0,i-1 | B) : A → Ba }
FINAL SCORE: Q (0,I | S)
```

b)

```
FOR EACH POSITION i = 1,2,...I DO
    FOR EACH NONTERMINAL A DO
        Q (i-1,i | A) = max { p(A → a | A) p(x(i) | a) : A→ a }
        Q (i-2,i | A) = max { p(A → ab | A) p(x(i-1) | a) p(x(i) | b) : A → ab }
    FOR EACH POSITION j = i-3,i-4,...,0 DO
        FOR EACH NONTERMINAL A DO
            Q (j,i | A) = max { p(A →aBb | A) p(x(j+1) | a) ·
                                · p(x(i) | b) Q (j+1,i-1|B) :A →aBb }
FINAL SCORE: Q ( 0,I | S )
```

c)

```
FOR EACH POSITION i = 1,2,...I DO
    FOR EACH NONTERMINAL A DO
        Q (i-1,i | A) = max { p(A → a | A) p(x(i) | a) :A → a }
    FOR EACH POSITION j = i-2,i-3,...,0 DO
        FOR EACH NONTERMINAL A DO
            Q (j,i | A) = max { p(A →BC | A) Q (j,l |B) Q (l,i |C) : A → BC, j<l<i }
FINAL SCORE: Q (0,I | S)
```

FIG. 8. Parsing algorithms for a) regular grammars, b) linear grammars, c) context-free grammars

In the recursive optimization, we have to optimize with respect to two variables, the unknown rule $A \rightarrow BC$ and the unknown boundary l with $j<l<i$:

$$Q(j,i|A) \;=\; \max \{\; p(A \rightarrow BC|A)\; Q(j,l|B)\; Q(l,i|C):\; A \rightarrow BC,\; j<l<i \;\}$$

The two optimization steps can be rearranged and we have:

$$Q(j,i|A) \;=\; \max_{A \rightarrow BC} \;\{\; p(A \rightarrow BC|A) \;\max_{j<l<i}\; \{Q(j,l|B)\; Q(l,i|C)\} \;\}$$

When using the above recursion relations, we must arrange the order of operations in a suitable way. Fig. 8 shows the resulting algorithms for the regular, linear and context-free grammar. Each of the three algorithms proceeds strictly from left to right and builds up derivations of growing length such that each new hypothesis covers a larger portion of the input string by combining previously computed adjacent derivations of smaller portions. The computational complexity of the resulting algorithms can be summarized as follows:

regular grammar:	linear
linear grammar:	quadratic
context-free grammar:	cubic

In the nonprobabilistic case, the cubic complexity is a well-known fact for the Earley and CYK parsers; in addition, there are special techniques to reduce this complexity further down to a quadratic or even linear dependence for certain subsets of the context-free grammars, such as unambiguous grammars or LL(k) grammars [Aho, Ullman 1972, p.325,333]. Due to the ambiguity inherent in a pattern recognition task, these reductions are not possible in the stochastic case. Note that the presented algorithms work in a bottom-up manner: the algorithm computes all possible hypotheses even if some might not be needed. It is possible to add a top-down component [Paeseler 1988]. This is equivalent to Earley's approach [Earley 1970] which can be viewed as a top-down extension of the CYK algorithm after transformation to Chomsky normal form [Graham et al. 1976].

The type of recursive optimization we have presented is usually referred to as dynamic programming or 'divide and conquer' strategy. Dynamic programming is very well tailored to such a type of optimization problem where only a small number of choices or decisions have to be taken at each optimization step. By dynamic programming, the overall optimization problem is broken down into a sequence of subproblems that are easier to solve. In the problem under consideration, it results in filling in the table of Q(j,i|A) scores for all (j,i|A) arguments by recursively evaluating the recurrence relation step by step. Thus the algorithm creates a table of solutions to all subproblems that might ever occur. No preliminary decisions have to be taken. The dynamic programming algorithm permits a straightforward implementation without recourse to feedback or special strategies for correcting decision errors. The attractive property of dynamic programming is that the internal structure of the observations is explicitly taken into account by recombining hypotheses at exactly the same rate that new hypotheses are formed. In the case of stochastic grammars, the applicability of dynamic programming is due to the following properties:

- the grammar rules are length increasing;
- there is only a finite number of nonterminals and grammar rules;
- there is a unique correspondence between each subportion of the input string and the nonterminal deriving this subportion.

The last property is not given in the case of context sensitive grammars.

After the table of Q(j,i|A) values has been filled, the globally optimal parse tree is constructed by tracing back the decisions taken at each recursion step. In the case of context-free grammars, we also might be interested in knowing the boundaries between the segments. For this purpose, special techniques must be introduced to keep track of the decisions taken about the segment boundaries and the grammar rules at each recombination step. The full details of the algorithms are given in Fig. 9 for regular and context-free grammars.

It is important to emphasize the reduction in computational complexity that has been achieved by using the dynamic programming technique. Without any recombination, the computational complexity would have been exponential in the length of the input string. In fact, this is the complexity we would obtain if we used backtrack parsing [Aho, Ullman, p.299].

```
FOR EACH NONTERMINAL A: Q (0,1| A ) = 0
    FOR EACH RULE r: A→a DO
        X = p (A→a | A) p(x(1) | a)
        IF X > Q (0,1 | A) THEN RECOMBINATION:
            SCORE:    Q (0,1 | A) = X
            RULE:     F (0,1 | A) = r
FOR EACH POSITION  i = 2,3,...I  DO
    FOR EACH NONTERMINAL A:  Q (0,i | A) = 0
    FOR EACH RULE  r: A→Ba  DO
        X = p (A→Ba | A) p(x(i) | a)  Q (0,i-1 | B)
        IF X > Q (0,i | A) THEN RECOMBINATION:
            SCORE:    Q (0,1 | A) = X
            RULE:     F (0,1 | A) = r
```

FIG. 9a. Implementation of the parsing algorithm for regular grammars

5.3 Membership Problem: Baum-Welch Scoring

The Baum-Welch scoring can be considered to provide the probabilistic counterpart of the membership problem in the non-stochastic case: What is the probability that the grammar produced the string of observations?
In order to arrive at an efficient algorithm for computing this probability, we introduce the following quantity:

$$Q(j,i|A) := \sum_{r'} P(r':A \Rightarrow x[j:i]|A)$$

which is the probability that nonterminal A explains subportion x[j:i] of the input string. In contrast to the Viterbi criterion, we now have to sum up over all possible realizations, i.e. contributions resulting from each different derivation.

To simplify the formulas, we always assume:

$p(A \rightarrow u'|A)=0$ if there is no such rule.

The algorithms for an efficient evaluation are obtained in a way similar to the Viterbi case. The difference is that instead of retaining only the best derivation, we have to sum over all possible explanations according to the normal forms shown in Fig. 7.

For the regular grammars, we have the initialization step:

$$Q(0,1|A) = \sum_{A \rightarrow a} p(A \rightarrow a|A)\ p(x(1)|a)$$

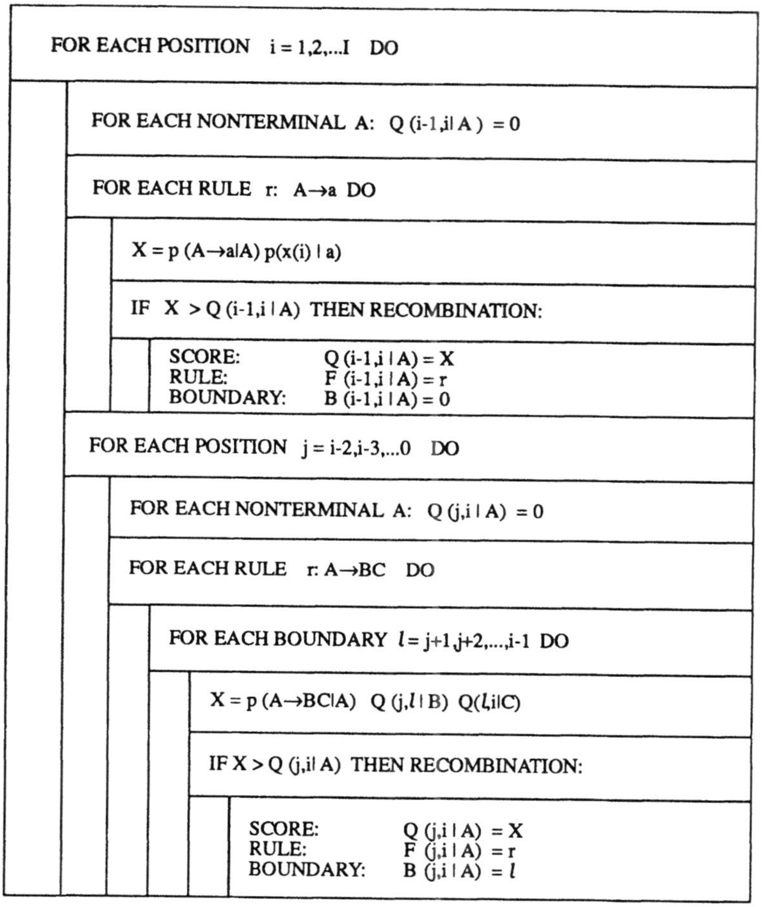

FIG. 9b. Implementation of the parsing algorithm for context-free grammars

The recursive relation is:

$$Q(0,i|A) = \sum_{A \to Ba} p(A \to Ba|A)\, p(x(i)|a)\, Q(0,i-1|B)$$

For the context-free grammars, we have as initialization step:

$$Q(i-1,i|A) = \sum_{A \to a} p(A \to a|A)\, p(x(i)|a)$$

In the recursions, we have to sum over all rules A→BC and all intermediate boundaries l, where j<l<i:

$$Q(j,i|A) = \sum_{A \to BC} \sum_{j<l<i} p(A \to BC|A)\, Q(j,l|B)\, Q(l,i|C)$$

By rearranging the sums, we obtain:

$$Q(j,i|A) = \sum_{A \to BC} p(A \to BC|A) \sum_{j<l<i} Q(j,l|B)\, Q(l,i|C)$$

This probabilistic extension of the CYK chart parser was given by Baker [1979]. It is straightforward to formulate the corresponding equations for the case of linear grammars. The resulting algorithms have the same computational complexity as in the case of the Viterbi scoring. Due to the problem formulation, there is nothing to be kept track of during the recursion.

6. TRAINING PROCEDURE

In this section, we derive a method for estimating the unknown parameters of a grammar, i.e. typically the rule probabilities and the parameters of the probability density functions describing the measurement process. We assume that the structure of the grammar is kept fixed, in other words the rules are given. Learning the grammar rules is usually referred to as grammatical inference [Fu 1982; Gonzalez, Thomason 1978] and is beyond the scope of this paper.

The training problem will be considered in the approximation of the Viterbi criterion. The advantage of this approach is its simplicity. The idea is similar to the approach used in clustering algorithms like K-MEANS or ISODATA where the cluster centres are not known, but their estimates can be improved by an iterative procedure [Duda, Hart 1973]. The experimental results indicate [Rabiner 1988] that there is no crucial loss in accuracy for the pattern recognition purposes. Training for the Baum-Welch criterion is covered by several authors [Baker 1975; Baker 1979; Jelinek 1976; Levinson et al. 1983; Rabiner 1988].

In general, even in the case of a detailed theoretical model there are a number of unknown model parameters that need adjusting. We use the following notation:

h' = the set of unknown model parameters;
r' = the parse trees or rule sequences;
x' = the total set of training sequences of observations.

To estimate the set of parameters h', we are given a sequence of training observations x' as obtained from all training samples and a script specifying the class membership of the training data, e.g. the word sequences that were spoken in the case of speech recognition. A standard statistical technique is to maximize the posterior probability over the unknown parameters h' [Duda, Hart 1973]. For practical computations, it is convenient to have an expression in terms of a prior probability and a conditional probability rather than a posterior probability. Assuming $P(x')$ to be constant, we can write the maximum a-posteriori estimate h'' as:

$$h'' = \text{argmax } \{ P(h',x'): h' \} = \text{argmax } \{ P(x'|h')\, P(h'): h' \}$$

Now the prior distribution $P(h')$ is assumed to be constant, and we obtain the maximum likelihood estimate:

$$h'' = \text{argmax } \{ P(x'|h'): h' \}$$

In order to derive an approximation to this rather complex training criterion, we use Viterbi scoring and thus obtain an explicit functional dependence of the maximum likelihood estimate on the parse trees r' and parameters h':

$$(h'',r'') = \text{argmax } \{ P(x'|r',h'): (r',h') \}$$

Due to the interdependence between the parse tree and the parameter estimates, it is not clear how to solve this problem in such a way that attaining the global optimum can be guaranteed. However there is an straightforward iterative procedure that can be shown to be locally convergent and that is sometimes referred to as Viterbi training [Brown et al. 1983; Jelinek 1976] or decision-directed learning [Duda, Hart 1973].

This iterative procedure is illustrated in Fig. 10. Each iteration consists of two steps. In the first step, we use the old estimates h'' (or some preliminary estimates in the first iteration) to determine the new parse trees r'':

$$r'' = \text{argmax } \{ P(x'|r',h'): r' \}$$

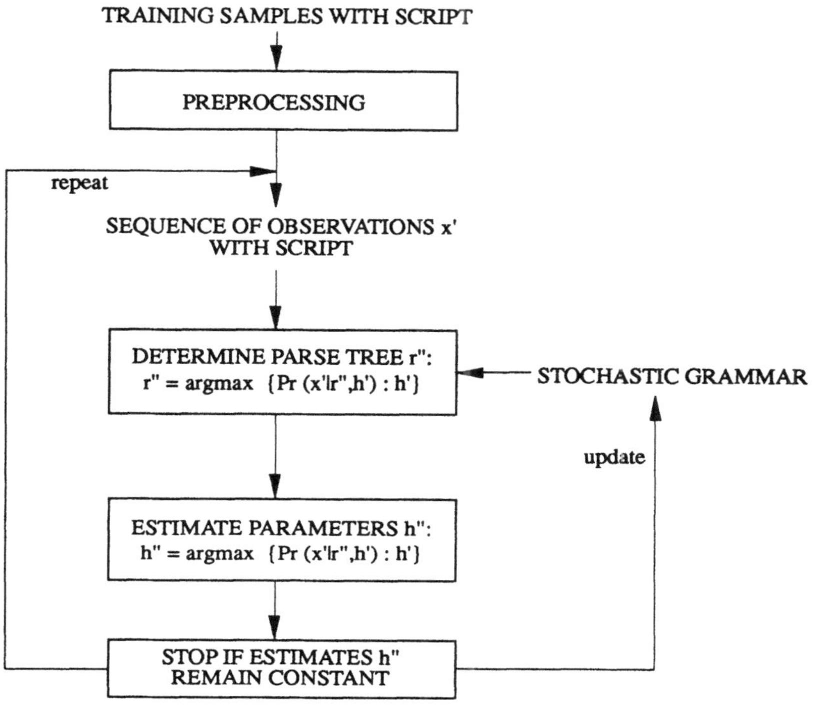

FIG. 10. Illustration of the training procedure in the Viterbi approximation

This registration parse tree can be determined by dynamic programming. In the second step, we keep the new parse trees r" fixed to reestimate the parameters h":

$$h'' = \text{argmax} \{ P(x'|r'',h'): h' \}$$

The iterations are repeated until the parameter estimates h" are constant or the differences are sufficiently small. The iterative process is bound to produce a convergent sequence of estimates since at each step of each iteration the probability score must either increase or remain constant and, on the other hand, the probability scores have an upper bound of one.

What varies from iteration to iteration, are the parse trees and thus the selection of observations that is used to refine the estimates of the model parameters. Note that rules with zero probability remain zero.

So far we have not considered how in the second step of the iteration the new parameters h' can be determined. This depends very much on the type of probability models used. Consider any type of regular, linear or context-free grammar with given rules and define the following counts:

N(A→u'): number of rule applications A→u'
N(A): total number of occurrences of A

The rule probabilities p(A→u'|A) are then estimated as relative frequencies:

$$\hat{p}(A \rightarrow u'|A) = \frac{N(A \rightarrow u')}{N(A)}$$

To estimate the unknown parameters of the probabilities of the measurement process, we have to collect the observations for each probability distribution using the parse trees and perform maximum likelihood estimation for the parameters of these distributions. For discrete models of the measurement process, of course, we again obtain the relative frequencies as estimates. For continuous observations, widely used models are the Gaussian and the Laplacian models for which maximum likelihood estimation produces the following estimates:

Gaussian model: - mean vector,
- covariance matrix;

Laplacian model: - median,
- absolute deviation.

Such a training approach has been found to be very efficient with respect to computation time and to produce results which are comparable to the forward-backward training procedure [Rabiner 1988]. In [Ney 1990], it is described how mixture densities can be efficiently trained using the Viterbi approximation and how this approach has been successfully applied to phoneme modelling for continuous speech recognition.

7. APPLICATIONS TO SPEECH RECOGNITION

7.1 Hidden Markov Models: Stochastic Regular Grammars for Acoustic Modelling

Hidden Markov Models (HMM) are a well established and successful technique in whole word and subword modelling [Baker 1975; Jelinek 1976; Levinson 1985; Rabiner 1990; Ney 1990]. We will show that they are a special case of stochastic regular grammars as they were introduced in Section 4.

Hidden Markov models are used to efficiently model the class conditional probabilities P(x[1:I]|w) of the observation sequence x[1:I] for a given word or subword class w. Fig. 11 shows a typical example of such a Hidden Markov model that may present either a whole word or a subword unit. When proceeding from one state to another state, the model produces or, in recognition mode, accepts an acoustic vector. Such a model can be interpreted as a special stochastic regular grammar. The states can be identified with the nonterminals of the grammar as was shown in Fig. 4. However, the standard interpretation is different from the convention used in this paper. Ultimately this is related to the question whether we start with the initial state of the finite state machine or with the terminal state, i.e. the state corresponding to the sentence symbol. It should be clear that both interpretations are equivalent if they are used in a consistent way.

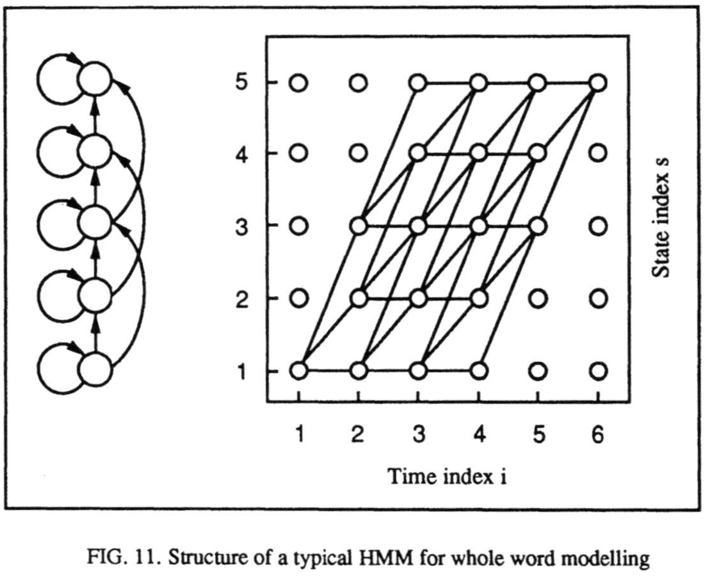

FIG. 11. Structure of a typical HMM for whole word modelling

In most cases, the probabilities are not conditioned on a grammar rule as a whole or in other words the transition, but only one of the two nonterminals (states) occurring in the normal form shown in Fig. 3. This type of HMM probability is referred to as state assigned probability as opposed to the general case of a transition assigned probability. In addition, the probabilities are conditioned on the state that has just been either left or reached. For more details on the relations between stochastic finite state automata and stochastic grammars, see [Casacuberta 1990]. Thus the probability attached to the two events

$A \rightarrow Ba$ and $a \rightarrow x$

is written as $p(A|B)\, p(x|A)$. Now since we have a linear arrangement of states, it is natural to identify the states with natural numbers $s=1,...,S$ rather than abstract nonterminal symbols. Thus the whole HMM is described by two types of probabilities:

$q(s|s')$: the transition probability of going from state s' to state s;

$p(x|s)$: the emission probability (density) of emitting/observing vector x when reaching state s.

As shown in Fig. 11, the different types of transitions (forward, skip, loop) allow for variations in the temporal development of the string of acoustic vectors, i.e. in the speaking rate. Unlike the acoustic vectors, the states as such cannot be observed and are hidden (hence Hidden Markov Model). The function of the states is to limit all the statistical dependencies both in emission and transition probabilities to the states.

In recognition mode, all combinations of times and states must be systematically considered as illustrated by the trellis shown in Fig. 11. The formulation of the Baum-Welch recursion for computing $P(x[1:I]|w)$ is as follows. Using an auxiliary quantity $Q(i;s)$ with time i and state s as arguments, we have:

$$Q(i;s) = \sum_{s'} \{q(s|s')\, p(x(i)|s)\, Q(i-1;s')\}$$

The summations are carried out over all states s' from which state s can be reached. The result is the value of the final gridpoint of the trellis:

$$P(x[1:I]|w) = Q(I;S)$$

The experimental results show that the Viterbi approximation results in the same recognition performance [Levinson et al. 1983]. The Viterbi recursion is obtained by replacing the sum by the maximum:

$$Q(i;s) = \max_{s'} \{q(s|s')\, p(x(i)|s)\, Q(i-1;s')\}$$

In lieu of summing over all the transitions going into state s, only the transition with the highest probability is considered. The Viterbi criterion has the advantage that is allows a further simplification of the recursion formula. By taking the negative logarithm of each quantity in the preceding equation, we obtain the following recursion:

$$\begin{aligned} D(i;s) &= \min_{s'} [t(s|s')+d(x(i)|s)+D(i-1;s')] \\ &= d(x(i)|s) + \min_{s'} [t(s|s')+D(i-1;s')] \end{aligned}$$

$D(i;s)$ is the so-called partial cumulative score that is computationally more convenient than the probabilities themselves. Often the time distortion penalties $t(s|s')$ are assumed to depend only on the type, i.e. the jump width of the transitions:

$$t(s|s') = t(s-s')$$

In addition we can interpret $d(x|s)$, i.e. the negative logarithm of the emission probability $p(x|s)$ as a distance measure, which is exactly correct in the case of a Gaussian model. The resulting type of dynamic programming is often referred to as dynamic time warping (DTW).

The same technique can also be used to compute a 'distance' between two symbol strings [Levenshtein 1966; Ackroyd 1980]. To model the three types of symbol errors, namely insertions, deletions and confusions, we have to identify them with suitable transitions in the HMM. This is straightforward for confusion errors, which are modelled by the emission probabilities, and the insertion errors, which can be modelled by loop transitions. The treatment of deletion errors is more tricky because this type of error violates the length increasing property of the corresponding grammar rules. However, instead of considering a deletion error in isolation, we can assign the deletion error to the subsequent symbol and model the error by a skip transition. It is clear that in order to capture pairs of deletion errors, we would have to introduce skip transitions over two states.

7.2 Integrating Language and Acoustic Grammars for Continuous Speech Recognition: Two-Level Algorithm and One-Pass Algorithm

So far we have not yet considered the problem of maximizing the product of the two probabilities P(x'|w') and P(w') as shown in Fig. 2. In a typical speech recognition application, we have two grammars as illustrated in Fig. 12. The acoustic grammar provides the conditional probabilities P(x'|w) for a given word w with known word boundaries. The language grammar provides the prior probabilities P(w'). There is no closed-form solution to this problem in the case of Baum-Welch scoring, and special search techniques are required such as stack decoding or best first [Jelinek 1976]. This problem is caused by the fact that for the Baum-Welch criterion we have to sum up over all possible realizations, i.e. both word boundaries and state sequences.

However the Viterbi criterion allows two distinctly different approaches, which result in a non-exponential complexity for the full search case. The two solutions are illustrated in Fig. 12. The left-hand side of Fig. 12 described the so-called two-level approach. In this approach, we have an explicit separation between parsing for the acoustic grammar and parsing for the language grammar. The interface between the two parsing procedures is a table of word hypotheses that explains all possible subportions of the input sequence. Thus in a full search strategy, this table will be rather huge. At the second level, the so-called phrase level, these word hypotheses are then used to form sentence hypotheses and find the most likely word sequence (in the Viterbi approximation) [Sakoe 1979; Ney 1984]. Since the input to this level are hypotheses depending on two time indices, namely beginning and end of each subportion, the required parsing strategy might be quite complex. Due to this reason and due to the size of the table of word hypotheses, this approach has been used only rarely so far: either the vocabulary is rather small [Sakoe 1979], or some segmentation of the original sequence of acoustic vectors has been performed beforehand. In most cases the language grammar is given in terms of a nonprobabilistic finite state network or a stochastic trigram or bigram grammar. There are only a few studies where the language grammar was a context-free grammar [Ney 1987; Paeseler 1988; Miller, Levinson 1988; Chow, Roukos 1989]. The additional problem here is that a full search algorithm has a computational complexity which is cubic in the length of the input sentence.

In connection with the architecture shown in Fig. 2, there is the interesting question whether it is possible to combine the two separate parsing algorithms for the acoustic grammar and for the language grammar into just one single parsing algorithm. Indeed this is possible. From the abstract point of view, what has to be done, is to expand these two grammars into one single grammar. If the language grammar is a context-free grammar and the acoustic grammar is a regular grammar (HMM), the resulting total grammar would be context-free. Not much is gained in this case because we still need hypotheses with two time indices in the search procedure [Ney 1987]. However there is the case when both the language grammar and the acoustic grammar are stochastic regular grammars. Instead of using the two-level approach that requires hypotheses depending on two time indices, we can combine the two grammars into one single grammar and use the corresponding parsing strategy with just one time index. This idea is the basis of the one-pass algorithm [Vintsyuk 1971; Bridle et al. 1982; Ney 1984].

The input utterance is known to be composed of words from a given vocabulary. Each word is modelled by an HMM, which may be obtained by concatenating subword units according to the pronunciation lexicon. This set of reference models defines the acoustic grammar. Now when inserting this acoustic grammar into the language grammar, we have to make copies of each acoustic word model in order to preserve the linguistic constraints expressed by the language grammar. These word copies are denoted by the index k=1,...,K. In the HMM and the corresponding trellis shown in Fig. 11, the states then have an additional index k, and the grid

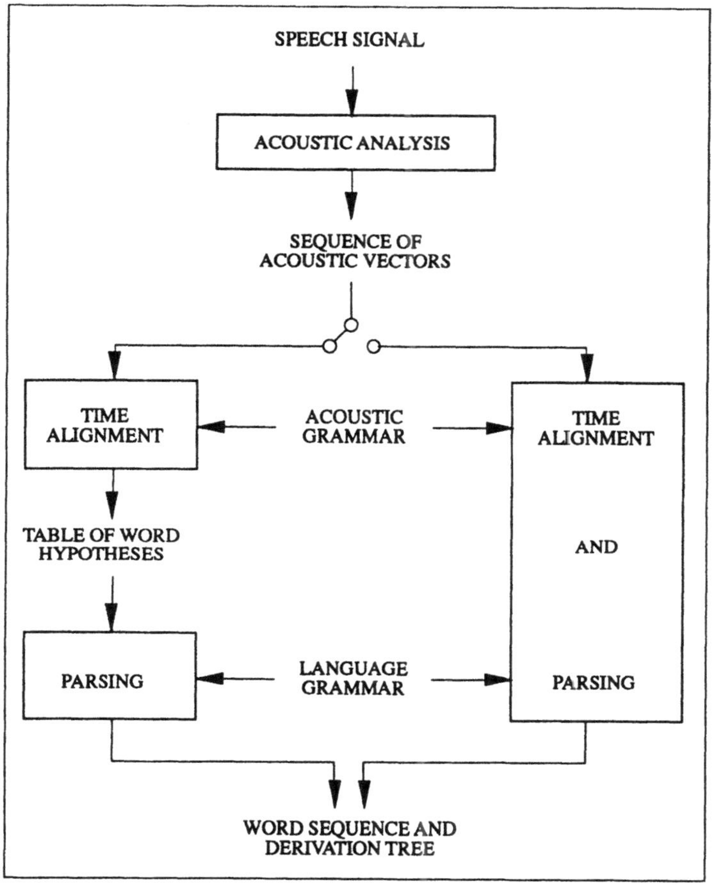

FIG. 12. Integration of an acoustic grammar and a language grammar into a recognition system

points of the trellis are defined by three indices: word (copy) index k, state index s and time index i. Thus we obtain a new 'super' trellis through which we have to find the best state sequence. The technique of dynamic programming can now be applied in the same way as above. We define a cumulative score D(i;s,k) as the partial sum over the local scores and time distortion penalties to grid point (i,s,k).
In the word interior, we have to perform the recombination over all reference models k=1,...,K and all states s=1,...,S(k):

$$D(i;s,k) = d(x(i)|s,k) + \min \{ D(i-1;s',k) + t(s|s',k): s' \}$$

In this recombination process, we include an additional artificial grid point with s'=0 in order to allow new words to be started up. This process of starting up new words must be controlled by the language model. Instead of the original stochastic regular grammar, we assume that the grammar has been transformed as follows. The grammar can be represented as finite state network as shown in Fig. 4. Each transition in this language network is uniquely identified by the word copy index k. Therefore the language constraints and probabilities are expressed by the following table:

C(k|k'): = negative logarithm of the probability that word copy k follows word copy k'.

Of course, instead of the table, any algorithm may be used to calculate these probabilities on demand. Thus after all gridpoints for a given time i have been processed, the recombination of hypotheses about potential word boundaries is carried out according to the equation:

$$D(i;0,k) = \min \{ D(i;S(k'),k') + C(k|k'): k' \}$$

In the next time frame i+1, this value is then used to start-up new words in the word interior recursion. The dynamic programming recursions build up hypotheses by proceeding along the time axis of the input signal from left to right and thus allow "on line" calculation.

More details about the one-pass algorithm are given in [Bridle et al. 1982, Ney 1984], in particular techniques for keeping track of the decisions taken at each recombination step. Techniques for reducing the full search space to a pruned search are given in [Bridle et al. 1982; Ney et al. 1987]. The one-pass algorithm provides a closed-form solution to the problem of handling the interdependence of a number of operations in continuous speech recognition, namely nonlinear time alignment, word boundary detection, language constraints and word identification. This idea of explicitly capturing this interdependence is the basis of the one-pass algorithm which is now used in a number of successful speech recognition systems.

8. SUMMARY

In this paper we have attempted to present a unification of statistical and syntactic pattern recognition with particular applications to speech recognition. Such a unifying view allowed us to clarify relations between a number of well-known, but apparently different techniques: stochastic grammars and languages, Hidden Markov Models, Viterbi algorithm, one-pass algorithm, two-level algorithm, Levenshtein distance, error-correcting parsers, CYK parser, Earley parser, bottom-up and top-down strategies.

ACKNOWLEDGMENT

This work was supported in part by the Federal German Ministery for Research and Technology (BMFT) under grant No. 413-5839-ITM 8401 and 8801. Only the author is responsible for the contents of this publication.

REFERENCES

[Ackroyd 1980] M.H. Ackroyd: "Isolated Word Recognition using the Weighted Levenshtein Distance", IEEE Trans. on Acoustics, Speech and Signal Proc., Vol. ASSP-28, pp.243-244, April 1980.

[Aho, Hopcroft 1972] A.V. Aho, J.D. Ullman: "The Theory of Parsing, Translation, and Compiling", Prentice-Hall, Englewood Cliffs, NJ, 1972.

[Aho, Peterson 1972] A.V. Aho, T.G. Peterson: "A Minimum Distance Error-Correcting Parser for Context-Free Languages", SIAM J. Comput., Vol. 4, pp. 305-312, Dec. 1972.

[Baker 1975] J.K. Baker: "Stochastic Modelling for Automatic Speech Understanding", in D.R. Reddy (ed.): 'Speech Recognition', Academic Press, New York, NY, pp. 521-541, 1975.

[Baker 1979] J.K. Baker: "Trainable Grammars for Speech Recognition", Speech Commun. Papers of 97th Meeting of ASA, pp. 547-550, June 1979.

[Bridle et al. 1982] J.S. Bridle, R.M. Chamberlain, M.D. Brown: "An Algorithm for Connected Word Recognition", Proc. IEEE Int. Conf. on Acoustics, Speech and Signal Proc., Paris, pp. 899-902, May 1982.

[Brown et al. 1983] P.F. Brown, C.H. Lee, J.C. Spohrer: "Bayesian Adaptation in Speech Recognition", Proc. IEEE Int. Conf. on Acoustics, Speech and Signal Proc., Boston, MA, pp.761-764, April 1983.

[Casacuberta 1990] F. Casacuberta: "Some Relations among Stochastic Finite State Networks Used in Automatic Speech Recognition", IEEE Trans. on Pattern Analysis and Machine Intelligence, Vol. PAMI-12, pp.691-695, July 1990.

[Chomsky, Schuetzenberger 1963] N. Chomsky, M.P. Schuetzenberger 1963: "The Algebraic Theory of Context-Free Languages", in P. Braffort, D. Hirschberg (eds.): 'Computer Programming and Formal Systems', North-Holland, Amsterdam, 1963.

[Chow, Roukos 1989] Y.L. Chow, S. Roukos: "Speech Understanding Using a Unification Grammar", Proc. IEEE Int. Conf. on Acoustics, Speech and Signal Proc., Glasgow, pp.727-730, May 1989.

[Duda, Hart 1973] R.O. Duda, P.E. Hart: "Pattern Classification and Scene Analysis", Wiley, New York, 1973.

[Earley 1970] J. Earley: "An Efficient Context-Free Parsing Algorithm", Comm. ACM, Vol. 13, pp. 94-102, Feb. 1970.

[Fu 1982] K.-S. Fu: "Syntactic Pattern Recognition and Applications", Prentice-Hall, Englewood-Cliffs, NJ, 1982.

[Gonzalez, Thomason 1982] R.C. Gonzalez, M.G. Thomason: "Syntactic Pattern Recognition", Addison-Wesley, Reading, MA, 1982.

[Graham et al. 1976] S.L. Graham, M.A. Harrison , W.L. Ruzzo: "On Line Context Free Language Recognition in less than Cubic Time", Proc. 8th Annual ACM Symposium on Theory of Computing, pp. 112-120, 1976.

[Hopcroft, Ullman 1979] J.E. Hopcroft, J.D. Ullman: "Introduction to Automata Theory, Languages, and Computation", Addison-Wesley, Reading, MA 1979.

[Jelinek 1976] F. Jelinek: "Continuous Speech Recognition by Statistical Methods", Proc. of the IEEE, pp.532-556, April 1976.

[Levenshtein 1966] V.I. Levenshtein: "Binary Codes Capable of Correcting Deletions, Insertions and Reversals", Sov. Phys. Dokl., Vol.10, pp.707-710, Feb. 1966.

[Levinson et al. 1983] S.E. Levinson, L.R. Rabiner, M.H. Sondhi: "An Introduction to the Application of the Theory of Probabilistic Functions of a Markov Process to Automatic Speech Recognition", The Bell System Technical Journal, Vol. 62, pp.1035-1073, April 1983.

[Levinson 1985] S.E. Levinson: "Structural Methods in Automatic Speech Recognition", Proc. IEEE, Vol. 73, pp. 1625-1650, Nov. 1985.

[Lyon 1974] G. Lyon: "Syntax-Directed Least-Errors Analysis for Context Free Languages", Comm. ACM, Vol. 17, pp.3-14, Jan. 1974.

[Miller, Levinson 1988] L.G. Miller, S.E. Levinson: "Syntactic Analysis for Large Vocabulary Speech Recognition Using a Context-Free Covering Grammar", Proc. IEEE Int. Conf. on Acoustics, Speech and Signal Proc., New York, NY, pp.271-274, April 1988.

[Ney 1984] H. Ney: "The Use of a One-Stage Dynamic Programming Algorithm for Connected Word Recognition", IEEE Trans. on Acoustics, Speech and Signal Proc., Vol. ASSP-32, pp.263-271, April 1984.

[Ney 1987] H. Ney: "Dynamic Programming Speech Recognition Using a Context Free Grammar", Proc. IEEE Int. Conf. on Acoustics, Speech and Signal Proc., Dallas, TX, pp.69-72, April 1987.

[Ney et al. 1987] H. Ney, D. Mergel, A. Noll, A. Paeseler: "A Data-Driven Organization of the Dynamic Programming Beam Search for Continuous Speech Recognition", Proc. IEEE Int. Conf. on Acoustics, Speech and Signal Proc., Dallas, TX, pp.833-836, April 1987.

[Ney 1990] H. Ney: "Acoustic Modelling of Phoneme Units for Continuous Speech Recognition", Proc. 5th European Signal Processing Conf., Barcelona, pp.65-72, Sep. 1990.

[Paeseler 1988] A. Paeseler: "Modification of Earley's Algorithm for Speech Recognition", Proc. NATO ASI, in H. Niemann, M. Lang, G. Sagerer (eds.): 'Recent Advances in Speech Understanding and Dialog Systems', Springer, Berlin, pp. 465-472, 1988.

[Rabiner 1988] L.R. Rabiner: "Mathematical Foundations of Hidden Markov Models", Proc. NATO ASI, in H. Niemann, M. Lang, G. Sagerer: 'Recent Advances in Speech Understanding and Dialog Systems', Springer, Berlin, pp. 183-205, 1988.

[Revesz 1985] G.E. Revesz: "Introduction to Formal Languages", McGraw-Hill, New York, 1985.

[Sakoe 1979] H. Sakoe: "Two-Level DP-Matching - A Dynamic Programming Based Pattern Matching Algorithm for Connected Word Recognition", IEEE Trans. on Acoustics, Speech and Signal Proc., Vol. ASSP-27, pp.588-595, Dec. 1979.

[Tanaka, Fu 1978] E. Tanaka, K.-S. Fu: "Error-Correcting Parsers for Formal Languages", IEEE Trans. on Computers, Vol. C-27, pp.605-616, July 1978.

[Vintsyuk 1971] T.K. Vintsyuk: "Element-wise Recognition of Continuous Speech Composed of Words from a Specified Dictionary", Kibernetika, Vol.7, pp.133-143, Mar.-Apr. 1971.

Basic Methods of Probabilistic Context Free Grammars[1]

F. Jelinek, J.D.Lafferty, and R. L. Mercer

Continuous Speech Recognition Group, IBM Research, T.J. Watson Research Center,
Yorktown Heights, NY 10598

Abstract: In automatic speech recognition, language models can be represented by Probabilistic Context Free Grammars (PCFGs). In this lecture we review some known algorithms which handle PCFGs; in particular an algorithm for the computation of the total probability that a PCFG generates a given sentence (Inside), an algorithm for finding the most probable parse tree (Viterbi), and an algorithm for the estimation of the probabilities of the rewriting rules of a PCFG given a corpus (Inside-Outside). Moreover, we introduce the Left-to-Right Inside algorithm, which computes the probability that successive applications of the grammar rewriting rules (beginning with the sentence start symbol s) produce a word string whose initial substring is a given one.

Keywords: Probabilistic Context Free Grammar, Inside-Outside Algorithm, Language Model for Speech Recognition.

1 Introduction

Language models improve automatic speech recognition providing information to reduce the search space during processing. Recognition performance depends on the language perplexity, which can be thought of as the average number of choices per word decoded. Furthermore, as the acoustic level returns a score (a probability if Hidden Markov Models are used) for each word tested, a probabilistic language model fits naturally into the context of uncertainty created by pattern matching [1]. A probabilistic language model allows every string of words to be scored with a generation probability. In the past, N-gram models have been proposed, but they do not capture the whole structure of a sentence, and more powerful formalisms are needed.

Context Free Grammars (CFGs) are now widely accepted as a reasonable model for the generation of natural language corpora but, due to the intrinsic difficulty of this task, CFGs that

---

[1]These notes have been assembled by Anna Corazza and Roberto Gretter (IRST, Trento, Italy).

NATO ASI Series, Vol. F 75
Speech Recognition and Understanding. Recent Advances
Edited by P. Laface and R. De Mori

provide good coverage are highly ambiguous. Probabilistic Context Free Grammars (PCFGs) allow both selection of the most probable parse as the "correct" one [3, 5] and considerably faster parsing by pruning of low probability subtrees. Moreover, the entropy of a given grammar gives an objective score of its capability in reducing the search space in speech recognition.

(s - 1)	s	→	np	vp	P(1|s)	= .5
(s - 2)	s	→	s	pp	P(2|s)	= .25
(s - 3)	s	→	v	np	P(3|s)	= .05
(s - 4)	s	→	v	pp	P(4|s)	= .1
(s - 5)	s	→	flies		P(5|s)	= .04
(s - 6)	s	→	like		P(6|s)	= .04
(s - 7)	s	→	time		P(7|s)	= .02
(vp - 1)	vp	→	v	np	P(1|vp)	= .2
(vp - 2)	vp	→	v	pp	P(2|vp)	= .4
(vp - 3)	vp	→	flies		P(3|vp)	= .16
(vp - 4)	vp	→	like		P(4|vp)	= .16
(vp - 5)	vp	→	time		P(5|vp)	= .08
(np - 1)	np	→	det	n	P(1|np)	= .4
(np - 2)	np	→	n	n	P(2|np)	= .1
(np - 3)	np	→	np	pp	P(3|np)	= .1
(np - 4)	np	→	arrow		P(4|np)	= .16
(np - 5)	np	→	flies		P(5|np)	= .08
(np - 6)	np	→	time		P(6|np)	= .16
(pp - 1)	pp	→	p	np	P(1|pp)	= 1.0
(n - 1)	n	→	arrow		P(1|n)	= .4
(n - 2)	n	→	flies		P(2|n)	= .2
(n - 3)	n	→	time		P(3|n)	= .4
(v - 1)	v	→	flies		P(1|v)	= .4
(v - 2)	v	→	like		P(2|v)	= .4
(v - 3)	v	→	time		P(3|v)	= .2
(p - 1)	p	→	like		P(1|p)	= 1.0
(det - 1)	det	→	an		P(1|det)	= 1.0

Table 1.1. A probabilistic context-free grammar in Chomsky normal form

These arguments substantiate the introduction of a PCFG as language model. The set of terminal symbols of this grammar should be given by the language vocabulary $\mathcal{V} = \{ v_1, v_2, \ldots, v_N\}$, while the phrase markers adopted for syntax analysis, $\mathcal{G} = g_1, g_2, \ldots, g_M$, compose the set of non-terminals; a special symbol $s \in \mathcal{G}$, which indicates the whole sentence, is the start symbol of the grammar. It is not a restriction to consider the grammar in Chomsky Normal Form (CNF), such that there are only two kinds of productions: the non-terminal ones, $H \rightarrow G_1G_2$, and the terminal ones, $H \rightarrow v$, where H, G_1, G_2 are non-terminal symbols, while v is a terminal symbol. A probability, called $P(H \rightarrow G_1G_2)$ or $P(H \rightarrow v)$, is associated to each production. The following normalization condition must hold for every $H \in \mathcal{G}$:

$$\sum_{G_1G_2 \in \mathcal{G}} P(H \rightarrow G_1G_2) + \sum_{v \in \mathcal{V}} P(H \rightarrow v) = 1. \quad (1)$$

As an example we will consider the PCFG presented in Table (1.1) and the five possible parses of the sentence *"Time flies like an arrow"* shown in Figure (1.1).

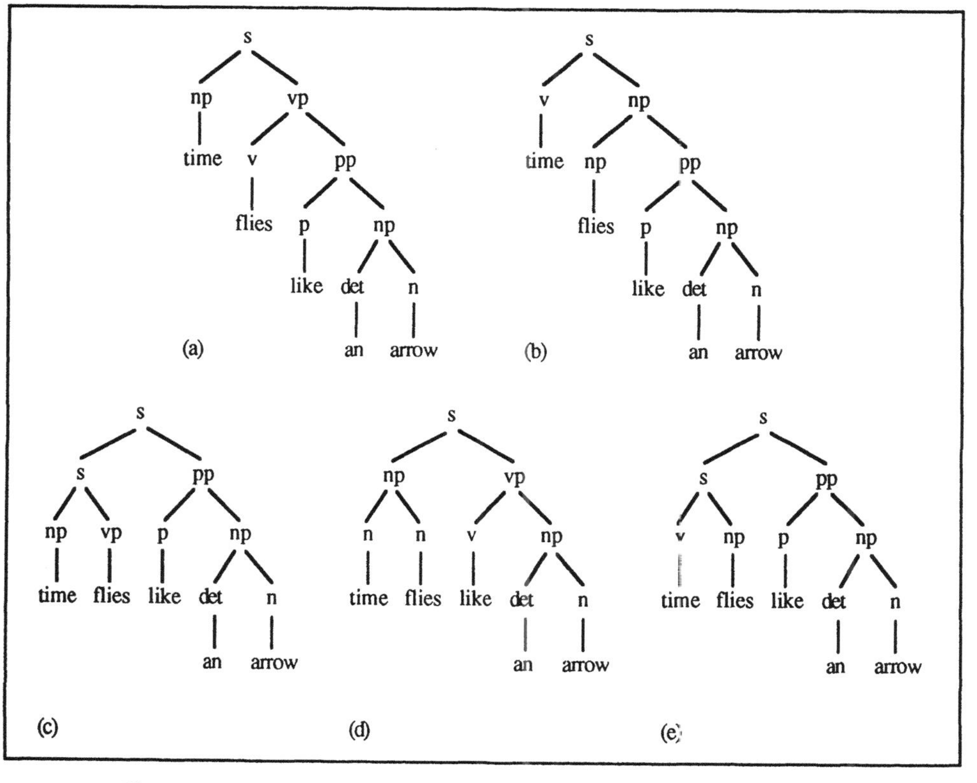

Figure 1.1. The five possible parses of the sentence *"Time flies like an arrow"*

In this probabilistic framework, four basic problems have to be dealt with:

1. finding the probability $P(s \rightarrow w_1w_2...w_K)$ that the grammar generates sentence $w_1w_2...w_K$ in any way whatsoever; this is the goal of the Inside algorithm, presented in Section 2;
2. given an arbitrary sentence $w_1w_2...w_K$, finding the most probable production process (parse) generating the sentence; the Viterbi algorithm, briefly treated in Section 4, can be used;
3. given an initial substring $w_1w_2...w_n$ of any length n, finding the probability that it was generated by the grammar; the Left-to-Right Inside algorithm is presented in Section 5;
4. the *probability estimation* problem: finding the maximum likelihood production probabilities $P(H \rightarrow G_1G_2)$ and $P(H \rightarrow v)$ accounting for the generation of given (large) training text; see the Inside-Outside algorithm [2, 7] in Section 3.

2 Inside algorithm

An important question is: with what probability does a given PCFG generate a given sentence? In principle the answer is obvious. The desired probability, which will be indicated as $P(s \overset{*}{\rightarrow} w_1w_2...w_K)$, is equal to the sum of the probabilities of all possible trees generating the sentence. Its computation will be based on probabilities $P(H \overset{*}{\rightarrow} w_i... w_{i+n})$, for all $H \in \mathcal{G}$, $i \geq 1$, $n \geq 0$. It is easy to demonstrate that, when context-free grammars are employed, the probability of a tree having the topmost production $H \rightarrow G_1G_2$ equals the product of the probabilities of the two subtrees having root node G_1 and G_2 times $P(H \rightarrow G_1G_2)$. In Figure 2.1 one of the possible parse trees for the sentence *"Time flies like an arrow"*, along with its probability, is shown. This is the key idea on which John Cocke based his algorithm (CYK) [9].

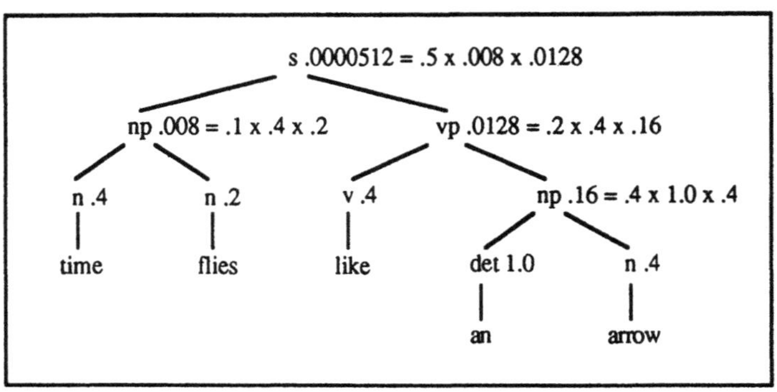

Figure 2.1. Probabilities of a derivation tree

An algorithm exists to compute $P(H \stackrel{*}{\rightarrow} w_i \ldots w_{i+n})$: the Inside algorithm, whose name will be justified when we discuss estimation of production probabilities. The data structure adopted by the Inside algorithm is the same parse triangle used by the CYK algorithm. The problem is simplified if the grammar is in CNF, but a complete generalization is possible.

The basis of the Inside algorithm is given by the following equation:

$$\begin{aligned} P(H \stackrel{*}{\rightarrow} w_i \ldots w_{i+n}) = \sum_{G_1G_2} P(H \rightarrow G_1G_2)[\, & P(G_1 \stackrel{*}{\rightarrow} w_i)P(G_2 \stackrel{*}{\rightarrow} w_{i+1} \ldots w_{i+n}) + \\ & + P(G_1 \stackrel{*}{\rightarrow} w_i w_{i+1})\, P(G_2 \stackrel{*}{\rightarrow} w_{i+2} \ldots w_{i+n}) + \ldots + \\ & + P(G_1 \stackrel{*}{\rightarrow} w_i \ldots w_{i+n-1})P(G_2 \stackrel{*}{\rightarrow} w_{i+n})]. \end{aligned} \qquad (2)$$

Equation (2) can be applied iteratively for $n=1, 2, \ldots$.

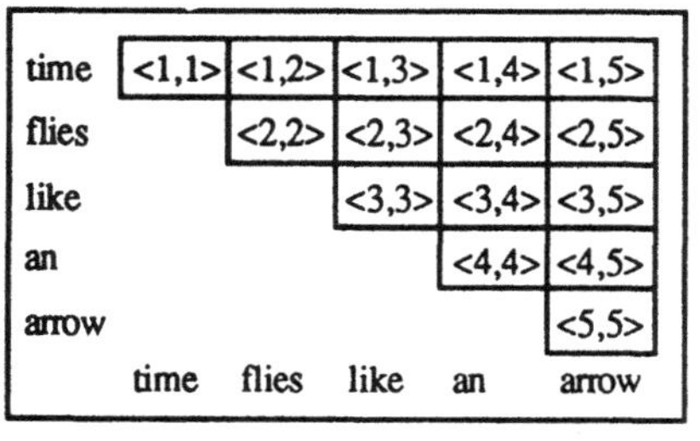

Figure 2.2. The parse triangle

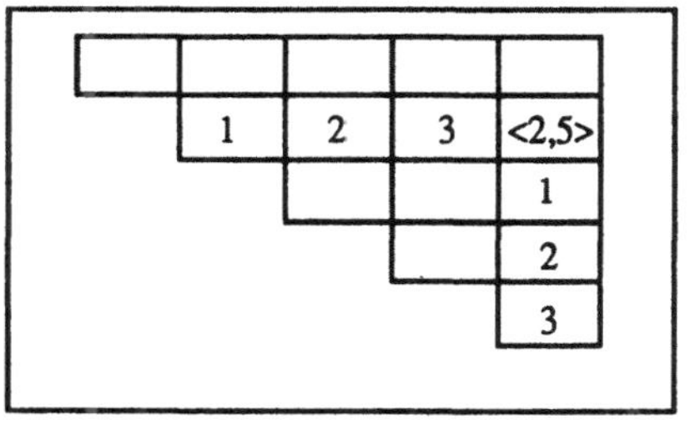

Figure 2.3. Contributions to cell <2,5>

An efficient procedure to apply this computation is based on the parse triangle represented in Figure 2.2; each box corresponds to a particular substring of $w_1 \ldots w_K$: $<i,j>$ stands for the substring $w_i \ldots w_j$. The order with which the triangle has to be filled is suggested by Equation (2). In fact, the computation of the probability that a given non-terminal H spans the substring corresponding to a particular box $<i,j>$ (i.e. $w_i \ldots w_j$) only involves the probabilities that every non-terminal spans the substrings corresponding to the boxes that are below and on the left of the box $<i,j>$. An example of this property is presented in Figure 2.3: for the computation of the probability referred to by the string $<2,5>$ it is necessary to compute the pairs of probabilities corresponding to the boxes indicated with *1, 2, 3*:

$$\begin{aligned} P(H \stackrel{*}{\rightarrow} w_2w_3w_4w_5) = \sum_{G_1G_2} P(H \rightarrow G_1G_2)[& P(G_1 \stackrel{*}{\rightarrow} w_2)\, P(G_2 \stackrel{*}{\rightarrow} w_3w_4w_5) + \\ & + P(G_1 \stackrel{*}{\rightarrow} w_2w_3)\, P(G_2 \stackrel{*}{\rightarrow} w_4w_5) + P(G_1 \stackrel{*}{\rightarrow} w_2w_3w_4)\, P(G_2 \stackrel{*}{\rightarrow} w_5)]. \end{aligned} \qquad (3)$$

Thus the triangle will be filled one diagonal after the other beginning with the longest one and ending with the vertex. This order reflects the fact that the Inside algorithm is bottom-up.

Figure 2.4 gives the table produced by applying the Inside algorithm to the sentence *"Time flies like an arrow"*.

	time	flies	like	an	arrow
time	v->time n->time np->time vp->time s->time	np->n<1,1>n<2,2> vp->v<1,1>np<2,2> s->v<1,1>np<2,2> s->np<1,1>vp<2,2>	s->np<1,2>vp<3,3>		np->np<1,2>pp<3,5> vp->v<1,1>np<2,5> s->s<1,2>pp<3,5> s->np<1,2>vp<3,5> s->v<1,1>np<2,5> s->np<1,1>vp<2,5>
flies		v->flies n->flies np->flies vp->flies s->flies	s->np<2,2>vp<3,3>		np->np<2,2>pp<3,5> vp->v<2,2>pp<3,5> s->v<2,2>pp<3,5> s->s<2,2>pp<3,5> s->np<2,2>vp<3,5>
like			p->like v->like vp->like s->like		pp->p<3,3>np<4,5> vp->v<3,3>np<4,5> s->v<3,3>np<4,5>
an				det->an	np->det<4,4>n<5,5>
arrow					n->arrow np->arrow

Figure 2.4. The complete parse triangle for the sentence *"Time flies like an arrow"*

3 Inside-Outside algorithm

The application of probabilistic grammars requires a correct estimation of the rewriting rule probabilities. In this section we will sketch the Inside-Outside (I/O) algorithm that extracts these unknown parameters from a training corpus. A solid mathematical treatment of the problem is possible, proving the properties of the I/O procedure and describing the circumstances under which it converges [2, 6]. The I/O algorithm is iterative and is a special case of the Expectation-Maximization class of algorithms [4]. As such it has the property that at each iteration, it augments the derivation probability of the training corpus.

A more intuitive way to estimate these quantities would be by counting the number of occurrences of each production in the corpus. An algorithm could be:

1. Find the parse for each sentence.

2. For each use of rule $H \rightarrow G_1G_2$, add 1 to the corresponding counter:

$$C'(H \rightarrow G_1G_2) = C(H \rightarrow G_1G_2) + 1. \tag{4}$$

3. Estimate probabilities:

$$\hat{P}(H \rightarrow G_1G_2) = \frac{C(H \rightarrow G_1G_2)}{C(H)} \tag{5}$$

where $C(H)$ is the sum of the counters associated to all the rules rewriting H.

Of course, the problem is that if the corpus is large then there is no practical way to employ human experts to determine the correct parses for all its sentences. However, if we can evaluate the probability of each parse, we can think of this probability as the degree of confidence we have in this parse. Then for each use of a rule, we could augment the corresponding counter by the confidence in a particular parse. This would result in the following more refined approach:

1. Find all parses $\pi_1, \pi_2, \ldots, \pi_k$ of each sentence, and their probabilities $P(\pi_i)$. Note that

$$\sum_{i=1}^{k} P(\pi_i) = 1. \tag{6}$$

2. For each use of rule $H \rightarrow G_1G_2$ in parse π_i, add $P(\pi_i)$ to the corresponding counter:

$$C'(H \rightarrow G_1G_2) = C(H \rightarrow G_1G_2) + P(\pi_i). \tag{7}$$

3. As before, estimate probabilities:

$$\hat{P}(H \rightarrow G_1G_2) = \frac{C(H \rightarrow G_1G_2)}{C(H)} \tag{8}$$

where $C(H)$ is the sum of the counters associated to all the rules rewriting H.

This approach presents an evident main problem: the probabilities $P(\pi_i)$ that must be summed in the counters depend on the probabilities to be estimated. It is reasonable to apply this procedure starting from an initial guess of the rules probabilities and improving the estimation in a loop.

This is the basic framework of the Inside-Outside algorithm. The following procedure based on the CYK parse triangle is equivalent to the latter, but allows the computational complexity to be kept low, which is critical because in practical cases the number of parses of a given sentence is high.

For each box $<i, i+n>$ we define $P(H \rightarrow G_1G_2;<i,i+n>)$ as the sum of $P(\pi_i)$ for all parses that use $H \rightarrow G_1G_2$ to produce the substring $w_i \ldots w_n$; so it can be written:

$$P(H \rightarrow G_1G_2; <i,i+n>) = \frac{\sum_{j=1}^{n} P(H<i,i+n> \rightarrow G_1<i,i+j-1>G_2<i+j,i+n>; s \overset{*}{\rightarrow} w_1 \ldots w_K)}{P(s \overset{*}{\rightarrow} w_1 \ldots w_K)} \tag{9}$$

where

$$P(H<i,i+n> \rightarrow G_1<i,i+j-1>G_2<i+j,i+n> ; s \overset{*}{\rightarrow} w_1 \ldots w_K) =$$

$$= P(s \overset{*}{\rightarrow} w_1 \ldots w_{i-1}\, H\, w_{i+n+1} \ldots w_K) \times$$

$$\times P(H<i,i+n> \rightarrow G_1<i,i+j-1>G_2<i+j,i+n> ; s \overset{*}{\rightarrow} w_1 \ldots w_K \mid s \overset{*}{\rightarrow} w_1 \ldots w_{i-1}\, H\, w_{i+n+1} \ldots w_K) =$$

$$= P(s \overset{*}{\rightarrow} w_1 \ldots w_{i-1}\, H\, w_{i+n+1} \ldots w_K) P(H \rightarrow G_1G_2) P(G_1 \overset{*}{\rightarrow} w_i \ldots w_{i+j-1}) P(G_2 \overset{*}{\rightarrow} w_{i+j} \ldots w_{i+n}). \tag{10}$$

Denoting

$$P(i-1>H<i+n+1) = P(s \overset{*}{\rightarrow} w_1 \ldots w_{i-1}\, H\, w_{i+n+1} \ldots w_K) \tag{11}$$

$$P(G_1<i,i+j-1>) = P(G_1 \overset{*}{\rightarrow} w_i \ldots w_{i+j-1}) \tag{12}$$

$$P(G_2<i+j,i+n>) = P(G_2 \overset{*}{\rightarrow} w_{i+j} \ldots w_{i+n}) \tag{13}$$

one obtains:

$$P(H \rightarrow G_1G_2; <i,i+n>) = \frac{\sum_{j=1}^{n} P(i-1>H<i+n+1)\, P(G_1<i,i+j-1>)\, P(G_2<i+j,i+n>)\, P(H \rightarrow G_1G_2)}{P(s<1,K>)} \tag{14}$$

The procedure presented in the Inside algorithm can be applied to compute the probabilities $P(H<i,i+n>)$ (inside probabilities). In the following an analogous procedure is given for the computation of outside probabilities ($P(i-1>H<i+n+1)$). Figure 3.1(a) and (b) give a bipartition of the event $i-1>H<i+n+1$, so that a recursive way to write the outside probability is:

$$P(i-1>H<i+n+1) =$$

$$\sum_G P(F\rightarrow GH) \sum_{h=1}^{i-1} P(i-h-1>F<i+n+1)P(G<i-h,i-1>) +$$

$$\sum_E P(F\rightarrow HE) \sum_{m=1}^{K-i-n} P(i-1>F<i+n+m+1)P(E<i+n+1,i+n+m>) \qquad (15)$$

Note that the procedure is analogous to that of the Inside algorithm, but it refers to the generation of strings larger than the one examined. From Equation (15) one can see that the computation of the outside probability corresponding to a given box $<i,j>$ only involves boxes above and on the right of $<i,j>$ (see Figure 3.2).

Since sentences belonging to the language generated by the grammar have s as starting symbol, we begin the computation of outside probabilities with the condition $P(0>s<K+1) = 1$ in box $<1,K>$. This done, the diagonal immediately below the top right-hand box $<1,K>$ can be completed. Proceeding top-down in this way, the whole parse triangle can be filled.

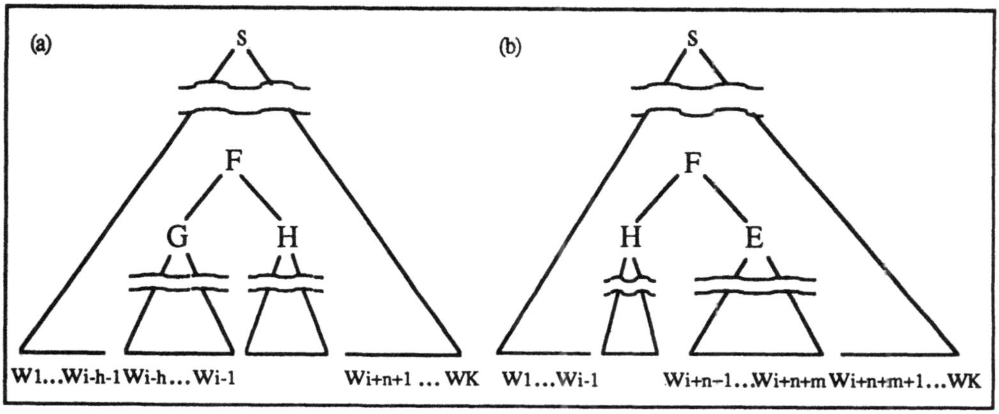

Figure 3.1. The two ways in which the non-terminal H can be generated during the Outside process

In conclusion, the inner loop of the Inside-Outside algorithm becomes the following:

1. Calculate the inside probabilities $P(G<i, i+n>)$ (Inside algorithm).
2. Calculate the outside probabilities $P(i>H<i+n)$ (Outside algorithm).
3. For each box, add to the counter $C(H\rightarrow FG)$ the quantity:

$$\frac{P(H\rightarrow FG)\, P(i-1>H<i+n+1) \sum_{j=1}^{n} (P(F<i,i+j-1>)\, P(G<i+j,i+n>))}{P(s<1,K>)} \qquad (16)$$

4. Re-estimate the rule probabilities by

$$\hat{P}(H \rightarrow FG) = \frac{C(H \rightarrow FG)}{C(H)}. \tag{17}$$

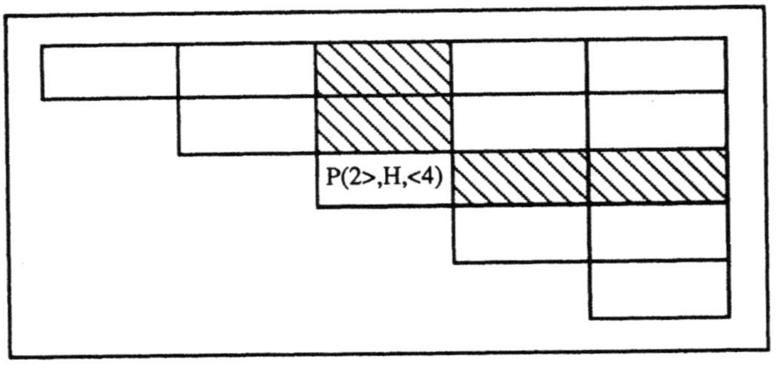

Figure 3.2. Boxes needed to compute P(2>,H,<4)

4 Viterbi algorithm

When the interpretation of an input sentence is requested, it is useful to find its correct parse. In the probabilistic framework introduced, the "correct" parse is a high probability parse. It is therefore of interest to have an algorithm that finds the most probable parse.

Let T be a parse tree of a given sentence, and let U be a subtree of T having H as root label and spanning $<i,j>$.If a subtree $U' \neq U$ exists, with the same root H and spanning the same substring $<i,j>$, the parse tree T' is obtained from T replacing U by U'. Viterbi algorithm is based on the key observation that if $P(U) > P(U')$, then $P(T) > P(T')$.

This note allows the procedure described for the Inside algorithm to be adapted to the problem of finding the most probable parse of a given sentence. The only difference is that in each box, for every non-terminal H, among all the possible subtree probabilities, only the largest one has to be retained, instead of their sum.

5 Left-to-Right Inside algorithm

The most popular approaches to speech recognition are left-to-right, so that words are examined following their temporal order. Probabilistic language models can help recognition by providing some hypotheses for the next word. Assuming that successive sentences are independent of each other and that w_1 is the first word of the current sentence, these hypotheses can be scored by the following probability:

$$P(w_{n+1} = v \mid w_1 w_2 \ldots w_n) = \frac{P_L^*(s \Rightarrow w_1 w_2 \ldots w_n v)}{P_L^*(s \Rightarrow w_1 w_2 \ldots w_n)} \tag{18}$$

where v is the word hypothesis to be scored. In Equation (18) $P_L^*(s \Rightarrow w_1 w_2 \ldots w_n)$ indicates the probability that the grammar generates a tree whose left-most terminals are $w_1 w_2 \ldots w_n$:

$$P_L^*(s \Rightarrow w_1 w_2 \ldots w_n) = P(s \overset{*}{\rightarrow} w_1 w_2 \ldots w_n) + \sum_{v_1 \in \mathcal{V}} P(s \overset{*}{\rightarrow} w_1 w_2 \ldots w_n v_1) + \\ + \sum_{v_1 v_2 \in \mathcal{V}} P(s \overset{*}{\rightarrow} w_1 w_2 \ldots w_n v_1 v_2) + \ldots ; \tag{19}$$

Thus $P_L^*(s \Rightarrow w_1 w_2 \ldots w_n)$ is the sum of the probabilities of all trees whose yield begins with $w_1 w_2 \ldots w_n$ and has length n, $n+1$, ... As we will see later, its computation is based on the following probabilities:

• $P(H \rightarrow w_i w_{i+1} \ldots w_{i+n}) = P(H\langle i,i+n \rangle)$ inside probabilities: it is the probability that the production process starting with the non-terminal H results in exactly the word sequence $w_i w_{i+1} \ldots w_{i+n}$;

• $P_L^*(H \Rightarrow w_i w_{i+1} \ldots w_{i+n})$ left-corner string production probabilities: it denotes the sum of the probabilities of all subtrees with root node H resulting in word strings whose initial subsequence is $w_i w_{i+1} \ldots w_{i+n}$;

• $P_L(H \rightarrow G) = \sum_F P(H \rightarrow GF)$ probability that H is rewritten by a pair whose left-hand element is G.

We will present an algorithm to compute $P_L^*(H \Rightarrow w_i w_{i+1} \ldots w_{i+n})$ iteratively. For $H=s$ and $i=1$, this is the quantity of interest. The probabilities $P_L^*(H \Rightarrow w_i)$ will turn out to be basic to this algorithm. In fact, we will bootstrap from them to the probabilities $P_L^*(H \Rightarrow w_i w_{i+1} \ldots w_{i+n})$ just as the Inside algorithm (Section 2) bootstraps from probabilities $P(H\langle i,i \rangle)$ to probabilities $P(H\langle w_i w_{i+1} \ldots w_{i+n} \rangle)$. For its computation the partial trees represented in Figure (5.1) have to be considered:

$$P_L^*(H \Rightarrow w_i) = P(H \rightarrow w_i) + \sum_G P_L(H \rightarrow G) P(G \rightarrow w_i) + \\ + \sum_{A_1} P_L(H \rightarrow A_1) \sum_G P_L(A_1 \rightarrow G) P(G \rightarrow w_i) + \\ + \sum_{A_1} P_L(H \rightarrow A_1) \sum_{A_2} P_L(A_1 \rightarrow A_2) \sum_G P_L(A_2 \rightarrow G) P(G \rightarrow w_i) + \ldots =$$

$$= P(H\rightarrow w_i) + \sum_{G} P(G\rightarrow w_i)\ [$$

$$P_L(H\rightarrow G) + \sum_{A_1} P_L(H\rightarrow A_1)P_L(A_1\rightarrow G) + \sum_{A_1A_2} P_L(H\rightarrow A_1)P_L(A_1\rightarrow A_2)P_L(A_2\rightarrow G) + \ldots] =$$

$$= P(H\rightarrow w_i) + \sum_{G} Q_L(H\Rightarrow G)P(G\rightarrow w_i) \qquad (20)$$

where the quantity $Q_L(H\Rightarrow G)$ (see Figure 5.2) is defined by:

$$Q_L(H\Rightarrow G) = P_L(H\rightarrow G) + \sum_{A_1} P_L(H\rightarrow A_1)P_L(A_1\rightarrow G) +$$
$$\sum_{A_1A_2} P_L(H\rightarrow A_1)P_L(A_1\rightarrow A_2)P_L(A_2\rightarrow G) + \ldots +$$
$$\sum_{A_1\ldots A_k} P_L(H\rightarrow A_1)P_L(A_1\rightarrow A_2)\ldots P_L(A_{k-1}\rightarrow A_k)P_L(A_k\rightarrow G) + \ldots \qquad (21)$$

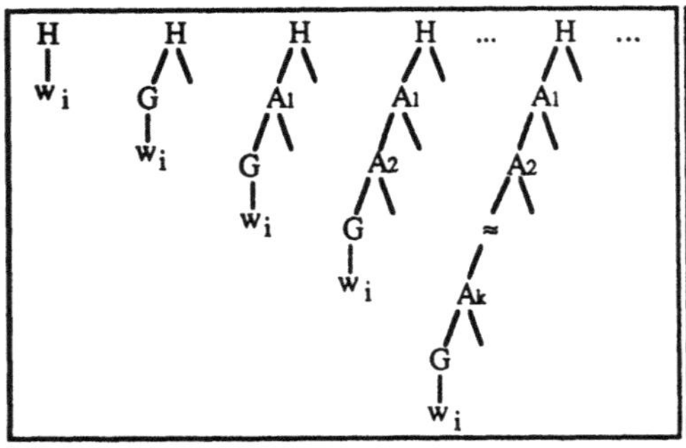

Figure 5.1. Terms of the equation (20)

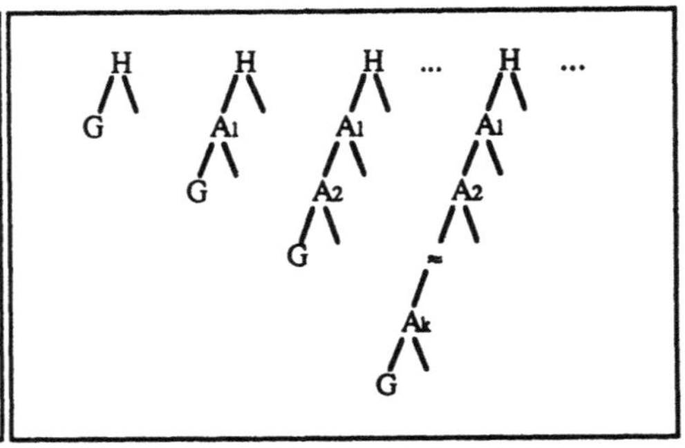

Figure 5.2. Terms of the equation (21)

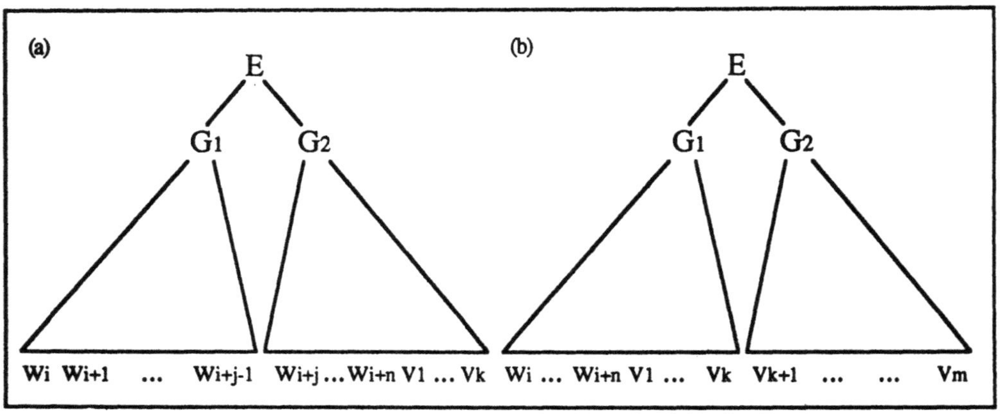

Figure 5.3. Two cases for equation (22)

Note that $Q_L(H \Rightarrow G)$ is the sum of probabilities of all trees with root node H that produce G as the leftmost (first) non-terminal. It is not in general a probability because the same tree may be counted twice or more.

Based on the decomposition shown in Figure (5.3), if the string length is greater than 1 we get the following equation:

$$P_L^*(E \Rightarrow w_i w_{i+1} \ldots w_{i+n}) = \sum_{G_1 G_2} P(E \rightarrow G_1 G_2)\, R_n(G_1,G_2) + \sum_{G_1 G_2} P(E \rightarrow G_1 G_2)\, P_L^*(G_1 \Rightarrow w_i w_{i+1} \ldots w_{i+n}) =$$
$$= \sum_{G_1 G_2} P(E \rightarrow G_1 G_2)\, R_n(G_1,G_2) + \sum_{A_1} P_L(E \rightarrow A_1)\, P_L^*(A_1 \Rightarrow w_i w_{i+1} \ldots w_{i+n}) \quad (22)$$

where

$$R_n(G_1,G_2) = \sum_{j=1}^{n} P(G_1 \stackrel{*}{\rightarrow} w_i \ldots w_{i+j-1})\, P_L^*(G_2 \Rightarrow w_{i+j} \ldots w_{i+n}). \quad (23)$$

Note that in the second summation the variable G_1 has been renamed as A_1, and the definition of $P_L(H \rightarrow G)$ has been used. Renaming E in (22) as A_1, and A_1 as A_2, and substituting the obtained expression $P_L^*(A_1 \Rightarrow w_i w_{i+1} \ldots w_{i+n})$ into (22), we obtain:

$$P_L^*(E \Rightarrow w_i w_{i+1} \ldots w_{i+n}) =$$
$$= \sum_{G_1 G_2} [\, P(E \rightarrow G_1 G_2) + \sum_{A_1} P_L(E \rightarrow A_1) P(A_1 \rightarrow G_1 G_2)\,] \times R_n(G_1,G_2) +$$
$$\sum_{A_1 A_2} P_L(E \rightarrow A_1) P_L(A_1 \rightarrow A_2) P_L^*(A_2 \Rightarrow w_i w_{i+1} \ldots w_{i+n}). \quad (24)$$

Next, renaming E in (22) as A_2, and A_1 as A_3, and substituting the result into (24), we get:

$$P_L^*(E \Rightarrow w_i w_{i+1} \ldots w_{i+n}) =$$
$$= \sum_{G_1 G_2} [\, P(E \rightarrow G_1 G_2) + \sum_{A_1} P_L(E \rightarrow A_1) P(A_1 \rightarrow G_1 G_2) +$$
$$+ \sum_{A_1 A_2} P_L(E \rightarrow A_1)\, P_L(A_1 \rightarrow A_2)\, P(A_2 \rightarrow G_1 G_2)] \times R_n(G_1,G_2) +$$
$$+ \sum_{A_1 A_2 A_3} P_L(E \rightarrow A_1) P_L(A_1 \rightarrow A_2) P_L(A_2 \rightarrow A_3) P_L^*(A_3 \Rightarrow w_i w_{i+1} \ldots w_{i+n}). \quad (25)$$

The pattern is now clear. Since

$$\sum_{A_1 \ldots A_k} P_L(E \rightarrow A_1) \ldots P_L(A_{k-1} \rightarrow A_k)\, P_L^*(A_k \Rightarrow w_i w_{i+1} \ldots w_{i+n}) \quad (26)$$

tends to 0 as k grows without limit, then using successive re-substitutions, and defining the function (see Figure 5.4)

$$Q_L(H \Rightarrow G_1G_2) = P(H \rightarrow G_1G_2) + \sum_A Q_L(H \Rightarrow A)P(A \rightarrow G_1G_2) \tag{27}$$

we get the final formula:

$$P_L^*(H \Rightarrow w_i w_{i+1} \ldots w_{i+n}) = \sum_{G_1G_2} Q_L(H \Rightarrow G_1G_2) \times R_n(G_1,G_2) =$$

$$= \sum_{G_1G_2} Q_L(H \Rightarrow G_1G_2) \sum_{j=1}^{n} P(G_1 \overset{*}{\rightarrow} w_i \ldots w_{i+j-1}) \, P_L^*(G_2 \Rightarrow w_{i+j} \ldots w_{i+n}). \tag{28}$$

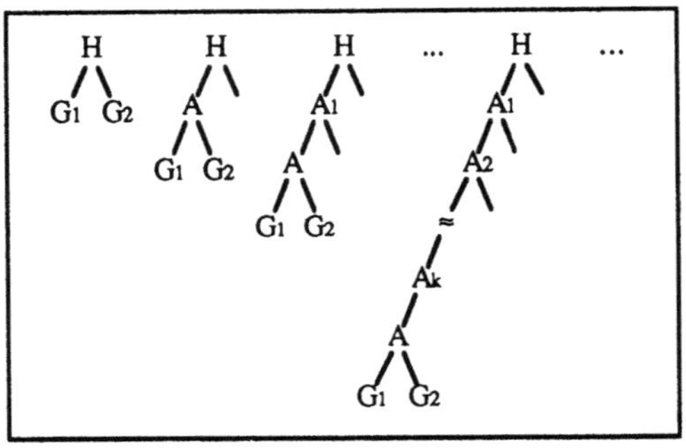

Figure 5.4. Terms of the equation (27)

Note that Equation (28) is very similar to Equation (2), used for the Inside algorithm. This fact suggests the use of an algorithm similar to the Inside one (compare the right pattern of Figure 5.5 with the left one used for the Inside algorithm).

1	6	10	13	15
	2	7	11	14
		3	8	12
			4	9
				5

1	3	6	10	15
	2	5	9	14
		4	8	13
			7	12
				11

Figure 5.5. Patterns for the computation of the parse triangle cells in the normal case (completely bottom-up) and in the left-to-right case

The final problem to solve is how to compute $Q_L(H \Rightarrow G)$. If the matrices $\mathbf{Q}$ and $\mathcal{P}$ are defined with elements:

$$\mathbf{Q}[H,G] = Q_L(H \Rightarrow G)$$
$$\mathcal{P}[H,G] = P_L(H \rightarrow G) \tag{29}$$

the following matrix equation can be written

$$\mathbf{Q} = \mathcal{P} + \mathcal{P}^2 + \mathcal{P}^3 + \dots, \tag{30}$$

and, multiplying both sides by $\mathcal{P}$:

$$\mathbf{Q}\,\mathcal{P} = \mathcal{P}^2 + \mathcal{P}^3 + \mathcal{P}^4 + \dots . \tag{31}$$

From Equations (30) and (31):

$$\mathbf{Q}\,[\mathcal{U} - \mathcal{P}] = \mathcal{P} \tag{32}$$

where $\mathcal{U}$ is the unit diagonal matrix; finally

$$\mathbf{Q} = \mathcal{P}\,[\mathcal{U} - \mathcal{P}]^{-1} \tag{33}$$

where the inverse of $[\mathcal{U} - \mathcal{P}]$ exists because $\mathcal{P}$ is a stochastic matrix. Note that the computation of $\mathcal{P}$ and $\mathbf{Q}$ does not depend on the input string, and it can be performed off-line.

In conclusion, the algorithm can be summarized in the following steps:

1. Compute off-line

$$\mathbf{Q} = \mathcal{P}\,[\mathcal{U} - \mathcal{P}]^{-1}$$

and therefore

$$P_L^*(H \Rightarrow v) = P(H \rightarrow v) + \sum_G Q_L(H \Rightarrow G)P(G \rightarrow v)$$
$$Q_L(H \Rightarrow G_1G_2) = P(H \rightarrow G_1G_2) + \sum_A Q_L(H \Rightarrow A)P(A \rightarrow G_1G_2)$$

2. Assuming that $P_L^*(s \Rightarrow w_1w_2 \dots w_n)$ has been computed,
   - A. Compute $P(G \overset{*}{\rightarrow} w_n)$, $P(G \overset{*}{\rightarrow} w_{n-1}w_n)$, ..., $P(G \overset{*}{\rightarrow} w_1 \dots w_n)$ for all non-terminals G.
   - B. Compute $P_L^*(H \Rightarrow w_{n+1})$, $P_L^*(H \Rightarrow w_nw_{n+1})$, ..., $P_L^*(H \Rightarrow w_2 \dots w_{n+1})$, for all non-terminals H.-
   - C. Compute $P_L^*(s \Rightarrow w_1w_2 \dots w_{n+1})$.

6 Conclusions

The goal of this work was to introduce probabilistic context-free grammars as a language model for speech recognition. For this application, some algorithms are needed:

1. in Section 2 an algorithm is given to compute the probability that a PCFG generates a given sentence;
2. Section 5 presents an algorithm to compute the probability that a given substring is an initial part of a sentence generated by the PCFG;
3. when the best parse of a sentence is requested, the Viterbi algorithm can be applied (Section 4);
4. the use of a probabilistic grammar is realistic if an algorithm is given for the rules probabilities estimation from a training text (Section 3).

We assumed the existence of an underlying CFG. How is it to be procured? Is the use of expert grammarians the only way? If rule probabilities can be extracted from a large text, why not the entire grammar [7]? What if one were to (a) determine a set of non-terminals $\mathcal{G}$, (b) allow initially all the conceivable rules $F \rightarrow GH$, and (c) estimate their probabilities by the I-O algorithm, hoping that the estimation process itself would identify the unnecessary productions by giving them zero probabilities? This outcome is obviously too good to be attainable. In any practical approach, the initial set of productions must be pared down, and the probabilities of the remaining productions must be given reasonable initial values [8].

References

1. Bahl L.R., Jelinek F., and Mercer R.L.: A Maximum Likelihood Approach to Continuous Speech Recognition, IEEE Transactions on Pattern Analysis and Machine Intelligence, vol. PAMI-5, No 2, pp 1798-190, March 1983.
2. Baker J.K.: Trainable Grammars for Speech Recognition, Proceedings of the Spring Conference of the Acoustical Society of America, 1979.
3. Brown P., Jelinek F., Lafferty J., and Mercer R.L., to appear.
4. Dempster A.P., Laird N.M., and Rubin D.B.: Maximum Likelihood from Incomplete Data via the EM Algorithm, Journal of the Royal Statistical Society (B), 39, pp 1-38
5. Fujisaki T., Jelinek F., Cocke J., Black E., and Nishino T.: Probabilistic Parsing Method for Sentence Disambiguation, International Workshop on Parsing Technologies, pp 85-94, Pittsburgh, PA, August 1989.
6. Jelinek F.: Markov Source Modeling of Text Generation, The Impact of Processing Techniques on Communications, J.K. Skwirzinski ed., published by Nijhoff, Dordrecht, 1985.
7. Lari K., Young S. Y.: The Estimation of Stochastic Context-Free Grammars using the Inside-Outside Algorithm, Computer Speech and Language, Vol 4, No 1, 1990, pp 35-56.
8. Sharman R.A., Jelinek F., and Mercer R.L.: Generating a Grammar for Statistical Training, Proceedings of the IBM Conference on Natural Language Processing, October 1988, pp 81-89, NY.
9. Younger D.H.: Recognition and Parsing of Context-Free Languages in Time N**3, Information and Control, 10, 1967, pp 1980-208

A Probabilistic Approach to Person-Robot Dialogue

Renato De Mori[1,2], Jacqueline Bourdeau[2], and Roland Kuhn[1]

[1] McGill University School of Computer Science, Montreal, Canada H3A 2K6
[2] Centre de Recherche en Informatique de Montreal, Montreal, Canada H3T 1P1

Abstract

This paper surveys current work on systems that support spoken man-machine dialogue. Subsequently, we describe our current research on a natural language interface for robot control, indicating how our system uses probability scores to evaluate and combine hypotheses generated at different conceptual levels.

Keywords

oral dialogue - speech recognition - person-machine dialogue - natural language interface - probabilistic approach

1 Introduction

Oral dialogue (OD) means a spoken conversation with initiative shared between the user and computer. OD focuses on spoken communication on both sides of the conversation, either in a system whose sole mode of interaction is speech, or combined with visual information, editing, pointing or other direct manipulation. The concept is being explored, and experimental work is underway with dialogue systems which make use of oral dialogue.

Dialogue systems do not imply a specific type of input/output interface, whether spoken or written. They may accept a combination of the two, where each plays a role. Natural Language (NL) techniques are used to complement, confirm, validate or correct the spoken input. Since the dialogue may include information from another source of

NATO ASI Series, Vol. F 75
Speech Recognition and Understanding. Recent Advances
Edited by P. Laface and R. De Mori

input, such as visual information, dialogue systems often imply multimodal interaction, and have to be designed and managed with this in mind.

Oral dialogue systems do not imply any particular type of processing. They may employ HMMs, parsing techniques, other techniques, or combinations of these. The current trend is to find a complementary role for different techniques, involving much methodological work on such questions.

Oral Dialogue has a strong, two-way relationship with Speech Recognition (SR). SR research considers a large variety of topics, including studies in signal processing and connectionist methods. The goals are to improve the recognition performance rate, and to solve problems posed by multiple speakers and large vocabularies. Areas addressed in human-machine communication include speed, performance, and accuracy, as well as exploration of the application of SR to real world situations. SR techniques form the basis for OD; however, knowledge gained through dialogue system research can lead to improvement of SR performance.

Oral dialogue design can be conducted using different approaches: modelling human-human dialogue, simulating a machine, or by directly building a working prototype in order to test and subsequently refine a design. One major design issue is to find optimal trade-offs between user-friendliness and robustness, between user freedom and performance.

Oral dialogues are different from other kinds of NL phenomena in that they have a structure, are goal-oriented, and the choice of words is constrained. Dialogue understanding is more tractable than spontaneous speech since the concepts and the words are well defined, the domain is restricted, and the goal is well known. Each phrase has a specific function in the conversation, such as information gathering, answer, interpretation, etc.

2 Overview of Person-Machine Dialogue Systems

Since the 70's, research work has progressively evolved from speech recognition to speech understanding, leading researchers to consider the NL understanding approach and techniques. This is how SR researchers started to explore problems common to speech and language processing, and to consider parsing techniques and the linguistic approach with its classical dimensions: syntactic, semantic, lexical and pragmatic. Some of them [3,12] refer to Searle's Speech Act Theory. Most people working on the integration of SR and NL language understanding techniques have a background in SR, and are interested in dialogue constraints. The use of dialogue constraints varies from correcting SR errors to selecting word candidates and reducing the search space.

Artificial Intelligence (AI) also plays a role in this evolution. Some researchers simply apply established techniques; others use a more global AI conceptual approach. Most recently, researchers have begun to use knowledge acquisition techniques for dialogue systems rather than relying only on traditional knowledge-based techniques.

As much as the desire to apply SR technology, it seems that it is the research on applications that led the researchers to work directly on dialogue systems. These applications include machine translation [16], telephone services [12], database inquiry [2,38], office work automation [1,25], telerobotics [4], air traffic control [18]. Design issues are raised for driving voice-driven dialogues [11,25]. Conceptual issues are suggested for multimodal dialogue [33,1], and intelligent dialogue [7,8], aiming at, ideally, their integration.

Towards the end of the 80's, the domains of natural language processing and speech understanding began to intersect, and dialogue systems became a new research field.

Below is a description of the activities of the research teams whose research focuses on oral dialogue systems, as well as their particular approaches and some problems that arise out of this new research field.

2.1 Speech Group, Medialab, MIT, USA

Medialab Speech Group's projects are to some extent influenced by the industrial orientation of this lab, as well as by the futuristic vision of its head, N. Negroponte. In his view, speech could become the dominant mode of interaction because it is frequently the most available communication channel, and because it allows the user to do remote work. The two goals of the group's approach are: 1) to integrate speech and language understanding, 2) to consider speech as one of several concurrent channels, across which a message content can be redundant. Bolt's research objective [1] is to obtain an integrated multimodal interface, where the design issue is to orchestrate the different modes, in order to obtain improved usability, performance and robustness. Principles for such orchestration may include: reducing the user's mental load, and obtaining summation and redundancy of information through the use of several modes.

2.2 Laboratory for Computer Science, MIT, USA

Spoken language understanding is part of the program of the Spoken Language Systems Group, as an extension of the SUMMIT SR system developed there. TINA is a natural language system developed by Seneff [28] specifically for speech understanding, in order to be integrated into SUMMIT. The system works with context-free grammars, ATNs, and lexical functional grammars. In converting context-free grammar rules into a probabilistic network structure, the grammar includes arc probabilities reflecting the frequency of occurrence of the syntactic structures. These probabilities are trained automatically from a set of parsed sentences; training can happen only if parsing was successful.

Zue and Seneff [42] developed VOYAGER as an application in order to test the two components, SUMMIT and TINA, connected in a serial manner. SUMMIT converts the speech signal into word hypotheses, and TINA gives an interpretation, which is then transformed into query functions. A back-end is responsible for response generation. Semantics was incorporated into TINA's parse tree. VOYAGER gives directional assistance for the Cambridge area. An experiment was conducted to test components and overall system response. The system could not parse 66.1% of the sentences, due to recognition errors, unknown words, or unseen linguistic structures. Experts found that 97% of the functions generated were correct. Recognition rate can be improved by incorporating context-dependent phoneme models.

2.3 Carnegie-Mellon University, USA

Under the direction of R. Reddy, work is being done by K. F. Lee and others on SR techniques; Ward focuses on speech understanding while Rudnicky concentrates on dialogue system design and experimentation. S. R. Young, in collaboration with Ward and Hauptmann, developed a system using dialogue context in SR [39,40]. The MINDS system generates expectations in order to predict the upcoming dialogue, then organizes constraints into hierarchical sets called "layers" to reduce its search space. Each prediction is then translated into a grammar usable by the speech recognizer. MINDS derives predictions from pragmatic knowledge sources such as dialogue structure, task semantics, general world knowledge, and user knowledge. Instead of using dialogue constraints to correct recognition errors or to select word candidates, MINDS uses layered contextual constraints to circumscribe dynamically the search space for words in a speech signal. It guides the recognition process from the very beginning. Experiments with CMU's recognizer, SPHINX, showed effective reduction of search space, as well as improved recognition performance. Performance is higher with users whose behaviour is consistent.

Ward [36] examines problems specific to spontaneous speech, and solutions that make SR systems robust enough to be able to handle phenomena such as filled pauses, restarts, interjections, ellipsis and ungrammatical constructions, problems common to SR and NL. A combination of techniques from both fields is developed, namely modelling accoustic properties (instead of words), with a flexible control structure. Grammar is used as a local constraint to group words into phrases, while global constraints from semantics combine the sequence of meanings into a defined action. A system called PHOENIX has been developed using the HMM word models of SPHINX.

Rudnicky [25] raises different design issues, beginning their investigation with task language analysis. To him, constraining the language means modelling the task language by user observation, in order to get a language as natural as possible; he calls it a "habitable" language. Fluent interaction is the second design objective, which implies modelling the dynamics of dialogue, as well as user characteristics. Error repair facilities and allowing the user to define new words are other design features which make the whole environment habitable. Experiments conducted on a financial planning task, with a voice-driven interface to the SPHINX system [25], comparing voice only, voice and keyboard, and keyboard only interfaces, demonstrated good performance in terms of task completion. However, subjects using voice took more time, this being due to two SR limitations: lack of speed and of accuracy.

2.4 Cambridge University

Work done by S.J. Young with Proctor [38] at Cambridge University, within an Alvey project called VODIS (Voice Operated Database Inquiry System), deals with design and implementation issues. The main focus is on intelligent dialogue control, in order to compensate better for SR errors. In the view of the authors, using a finite state transition network would rapidly become combinatorially too large, i.e. intractable. Their design methodology is to build a dialogue controller using an object-oriented framework, in which objects contain information needed to complete a transaction. With the inheritance mechanism of frames, dialogue design is data-driven, and is realized through successive refinement. As experiment shows, this methodology is valid and also easier to test, since each topic can be tested individually.

2.5 CNET-IRISA, France

Similarly to MIT and CMU, researchers at CNET-IRISA started from their SR system, KEAL, to explore understanding and dialogue systems. Siroux and Gillet [29] first introduced syntactic, semantic and pragmatic knowledge to implement dialogue function as an interface to KEAL. A context-free grammar takes into account semantic features, which are then interpreted to represent information relevant to the subsequent dialog. The dialogue manager instantiates a model graph, makes predictions, and controls the question-answer type of oral exchange. Experimental work was done in the PALABRE ESPRIT project by Guyomard and Siroux [12] on a telephone service application, the French general information service called "Yellow Pages". Dialogue modelling was performed after observation, in order to structure and constrain a cooperative type of dialogue. Dialogue features include cooperation, follow-on dialogues, use of implicit utterances, and instructions given to users. Experimentation showed the validity of this realistic approach in terms of cooperation and prediction; however, predictive power is a detriment to user's freedom. Linguistic expertise facilitates the task of recognition by reducing the number of utterances and their length. New requests issued from follow-on dialogue can be classified into three categories: slidings, mutations, and reproductions.

2.6 CRIN-INRIA, France

At CRIN-INRIA, Pierrel and Haton work on speech understanding; Haton's work on labeling is part of the SAM ESPRIT project. Carbonell and Pierrel [4] have been working since the beginning of the 80's on dialogue systems design. Their approach is knowledge-based, and they propose an architecture based on four knowledge sources, accessed independently by the dialogue component, giving as much modularity as possible. These knowledge sources are: acoustic-phonetic, prosodic, syntactic-semantic and lexical. The dialogue controller builds up a dialogue history out of dynamic information on the dialogue process. Knowledge about dialogue structures and strategies is represented by plans, which are then selected by the dialogue manager according to dialog progress. Souvay et al. [30] have applied an implementation of this design to the oral command of a Sonar system. Voice input was introduced to replace a complex input system composed of keyboard and menus, trackball and switches. Keyboard and mouse could be used in case of deficiencies in the voice input. Results are positive in terms of usability and performance: operators can keep looking at the screen and concentrate a command in a few words. This, with attendant reduction in mental overload and overall workload, allows operators to work faster.

2.7 LIMSI, France

At LIMSI, Matrouf et al. [18,19] have developed a task-oriented and knowledge-based approach to build a dialogue system for air traffic controller training. After having analyzed and categorized the corpus of messages used in real life, they have deduced rules and criteria, and built a model based on the theory of frames. The dialogue system consists of four parts: message analysis, validity control, correction and interpretation. Knowledge sources are vocabulary, phraseology, task model, and dialogue history. The focus is on the representation of the language model in order to reduce the search space. Pragmatic knowledge is used to make predictions which dynamically limit the recognition search space. The dialogue module is made of a network of frames. After message analysis, other modules are invoked for error detection and correction, message generation, updating of the dialogue network, and prediction for the next message. Predictions are made from current context and dialogue history, using a probabilistic syntax, in which the probability of the occurence of a concept is given a weight. Prediction is then propagated to the arc probabilities of the automaton corresponding to the concept. For experimentation, the AMADEUS SR system made at LIMSI was adapted in order to use transition probabilities in the recognition process. Results indicated that the use of probabilities and semantic knowledge increased the message recognition accuracy to 96%, and that dynamic probabilities provided an additional 0.5% improvement. Combined use of dynamic probabilities and semantic knowledge-based dialogue strategies gave a performance of 96.5% at the sentence level.

2.8 Advanced Telecommunications Research Institute, Japan

Integration of language and speech processing is extensively researched at ATR, as related to machine translation of telephone dialogues [16]. The goal is an automatic telephone interpretation system that will transform a spoken dialogue from the speaker's language to that of the listener automatically and simultaneously. Such spoken dialogues include elliptic and anaphoric expressions, as well as syntactically ill-formed expressions, all of which must be taken into account. A more limited goal is to handle task-oriented areas, where domain knowledge is expected to disambiguate the spoken language. Linguistically-based constraints of syntactic, semantic, and pragmatic information are utilized to narrow the number of word candidates. A method is developed that reduces the number of candidates presented by the SR component. In this method, the relationship of dependency is used for the analysis of spoken Japanese utterances. The current approach to analyzing Japanese dialogues is based on a lexico-syntactic grammar framework in terms of feature structures.

2.9 Bolt, Beranek and Newman, USA

At BBN, Schwartz and Makhoul have developed a SR system called BYBLOS using HMMs with context-dependent phoneme modelling, and are working on parallel implementation of the system. Roucos [24] developed a spoken language system, adding NL sources to BYBLOS: a unification grammar for syntax, a higher-order intensional logic language for semantics, and a Montague Grammar style framework for interfacing syntax and semantics. HMMs are used to determine a word lattice; then, a lattice parser finds all possible word sequences. The semantic interpreter determines the highest scoring meaningful ones, and interprets them. Experimental results with the DARPA Task Resource Management database demonstrate an improvement of accuracy with syntax from 71% to 87%, and, with both syntax and semantics, to 92%.

2.10 SRI, USA

SRI's Speech Research Group, under Murveit and Bernstein, aims at creating real-time systems for interactive problem-solving, by integrating SR and NL understanding. The key idea is to develop a parser that incrementally generates a state transition network to replace a finite state grammar in their HMM-based SR system [21]. Finite state constraints are used in order to permit integration into SRI's SR system, DECIPHER. The approach is to generate a dynamic grammar network that represents NL knowledge in a state transition network; only portions of the network within a pruned search are expanded. This approach reduces computation time while maintaining a very tractable recognition search space.

2.11 University of Erlangen, Federal Republic of Germany

At the University of Erlangen, Niemann and collaborators have developed one of the earliest speech understanding and dialogue systems, EVAR [22], where linguistic knowledge is used to interpret a sentence and discover the user's intent. Semantic and pragmatic knowledge are combined to filter hypotheses at different steps of the understanding process. Ehrlich and Niemann [9] use semantic and pragmatic knowledge to reduce the number of hypotheses by analyzing complex syntactic constituents. Their approach to semantic analysis is based on the caseframe theory and deep case theory. Results from experiments show that the number of hypotheses could be reduced by half through this filtering. Sagerer's approach to SR systems [27] is knowledge-based, using semantic networks.

This concludes our survey of leading research activity in Speech Research Groups. Despite the diversity of the work described, one may discern certain common themes. All of the groups aim to enhance recognition by incorporating a degree of understanding. In order to achieve this, they look for ways to make full use of their existing SR systems, adding linguistic processing techniques, while not reducing performance or robustness. The most common approach is to introduce NL constraints and/or AI techniques; much methodological work is being done on experimenting with different techniques. The NL techniques vary depending on the SR methodology, on the type of architecture of the existing SR system, and also on the overall orientation of the lab. The role of the understanding component, as compared with that for recognition, varies from reducing errors, or correcting them, to reducing the search space. Experiments performed with these spoken language systems, either on standard speech databases, or on new applications, show an improved recognition capacity. In some cases, limitations are reported to be due to the basic SR system.

In addition to the research programs in speech labs, one should be aware of other work related to OD, either on linguistic and communication problems in natural dialogue by Bunt [2,3], user modelling by Wahlster [35], or dialogue ergonomics by Waterworth [37]. We briefly introduce conceptual work conducted at DCIEM (Defense and Civil Institute of Environmental Medicine) by Taylor, and by Edwards and Mason. Taylor [31,32] proposes the Layered Protocol Model for analyzing and designing human-computer interaction. This view is inspired by psychological principles and also by previous work on ISO-OSI protocols. Taylor suggests the metaphor of computer-computer communication to describe human-computer dialog, and the Layered Protocol Model as a structure to design dialogue systems. In this view, multimodal dialogue consists of multiple dialogs conducted over multiplexed channels, each having its own coding and feedback features. Redundancy is necessary, as is feedback without delay, in order to obtain effective communication without misunderstanding. Edwards and Mason [7,8,17] focus on the aspect of intelligence in dialogue and propose a goal-oriented dialogue design model.

Psychological views like these, as well as linguistic ones, must be taken into account in OD systems design and evaluation, to make sure that psychological principles are respected, and in order to be able to discriminate between computing problems and other kinds of problems e.g. linguistic, psychological, and so on.

3 Person-Robot Dialogue

This section reports on progress in the construction of a system that supports spoken man-machine dialogue. A natural language interface for robot control has been designed. Such a system must meet the following, extremely demanding, criteria:

1. It must have access to a model of the environment, so that sentences or phrases that refer to the environment can be disambiguated
2. It must have a model of the robot's capabilities, so that at any point in the dialogue it knows what the robot is currently doing, what it is currently capable of doing, and what it has done in the recent past (note that this kind of state knowledge is only useful in the presence of the environment model just mentioned)
3. It must have a dialogue model, enabling it to manage a dialogue with the user - that is, it must be capable of pursuing its own conversational goals (the primary goal will usually be that of determining exactly what the user wants the robot to do) while responding flexibly to conversational initiatives by the user
4. At lower levels of the linguistic hierarchy, it must recognize sequences of words uttered by the user and convert them into semantic tokens which are passed to higher levels
5. It must permit information transfer between system components, so that information gained by one of the components facilitates the task of others.

The last point is particularly important. In our system, information passed from one component to another is often expressed in terms of **probabilities**. For instance, consider the way in which the speech recognition component interacts with the other components. The most urgent problem in large-vocabulary speech recognition is constraining the size of the search space. Thus, high level knowledge about the environment, semantics, discourse structure, user goals, and dialogue history is applied during spoken man-machine interaction to provide constraints for the speech recognition component. Since this component uses probabilistic models to evaluate competing acoustic hypotheses, the higher-level components provide it with probabilistic predictions about what will be said next. This uniform representation enables the speech recognition component to combine information from different levels in a consistent way. It may be the case that there is more than one plausible semantic interpretation of what the user said - if so, each of the interpretations will be passed up to higher levels of the system, along with the set of corresponding probabilities. Thus both the input and the output of this component are expressed probabilistically.

From a software engineering point of view, this approach leads to a transparent system, with a modular internal structure. There is no need of complicated protocols to decide how to assess information provided by other components - competing hypotheses can be scored in terms of their probabilities, allowing a best-first search to take place. If the history of speech recognition is any indication, a probabilistic system is also likely to be extremely robust.

4 Robot Behaviour and the Structure of Person-Robot Dialogue

Consider the circumstances in which person-robot dialogue will take place. As stated above, the system managing the natural-language robot-user dialogue requires three models: an **environment model**, a **robot state model**, and a **dialogue model**.

The **environment model** contains the system's knowledge about the aspects of the physical world that are relevant to its current task. Some of this knowledge will be derived from the robot's sensors, some of it perhaps directly entered into the environment model by the robot's designers, and some of it (as we will see) obtained from other humans through the mediation of the dialogue model. This model is expressed in an object-oriented formalism; the knowledge contained in the other two models is procedural, and is expressed in a Rule Description Language (RDL).

The **robot state model** contains the system's knowledge about what the robot can do and what it is doing now. In other words, it includes information about the states the robot can potentially be in, the nature of the transitions between states, and the robot's current state. It can be represented by an Augmented Transition Network Diagram that corresponds to a set of rules in the RDL.

The accompanying diagram (Figure 1) shows an example of an ATND. Most of the states S_i are associated with a Dialog Segment Manager D_i. Transitions between states are triggered by the setting of associated conditions to "true"; a condition may be set to "true" by a DSM (as in the example), or because of an observation made by the robot about its environment. In the example, state S_0 is the waiting state. The robot undergoes a transition to state S_1 when the user indicates that he wants the robot to undertake a task. In D_1, associated with S_1, the robot obtains a general description of this task from the user. In the example, there are two types of tasks: "Navigation" and "Manipulation". The corresponding states, S_2 and S_3, are each associated with DSMs that obtain more precise definitions of the task from the user.

In this schema, each Dialog Segment Manager D_i contains the system's knowledge about how to carry out a dialogue of a particular kind. This knowledge has two components: knowledge about what the user is likely to say next, and knowledge about how to fulfill the robot's dialogue goals by asking questions or providing the user with needed information. The former consists of a set of conceptual expectations, C_i. Each element of C_i is a different hypothesis about the semantic content of what the user is about to say; each conceptual hypothesis has an a priori probability. The conceptual hypotheses are implemented as frames. Each is associated with a **Conceptual Hypothesis Grammar** that will work in association with the speech recognition component to change the probability of that hypothesis and fill in its slots. Thus, as the user speaks, certain conceptual hypotheses which seem to match the input well are assigned higher probabilities, and their slots are filled in.

The different DSMs employ similar conversational strategies for extracting information from their human interlocutors in order to fill in slots for these active conceptual hypotheses, and to determine which was intended by the user. When the Dialog Segment terminates, the DSM will usually pass only the best scored conceptual hypothesis to working memory (unless the DS was terminated prematurely and will be resumed later).

Figure 1.

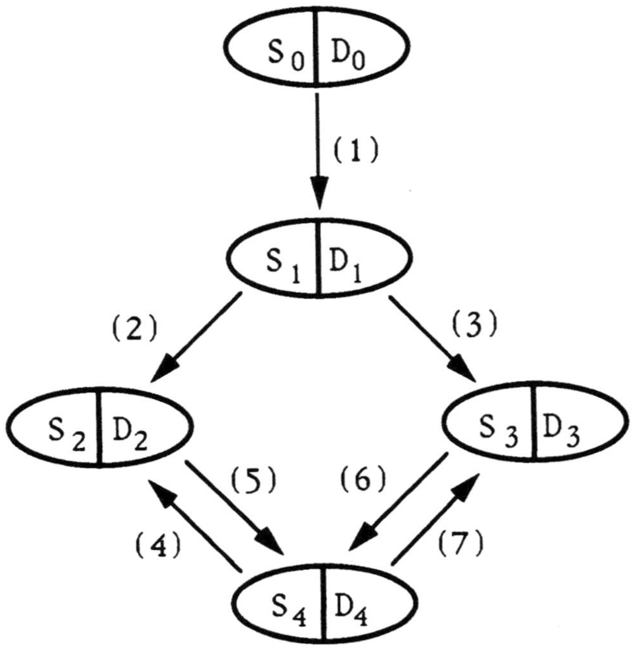

<u>Key</u>

c = condition, a = action, o = operation

<u>Arcs</u>

(1) c1 = recognized intention to define a task

(2) c2 = user has given NAVIGATION goal
o = push NAVIGATION

(3) c3 = user has given MANIPULATION goal
o = push MANIPULATION

(4) c = user resumes or initiates NAVIGATION goal
o = push NAVIGATION

(5) c = NAGIVATION complete or user interrupt occurs

(6) c = MANIPULATION complete or user interrupt occurs

(7) c = user resumes or initiates MANIPULATION goal
o = push MANIPULATION

Figure 2. Dialog Segment D_0

```
(start

 c1 := FALSE
 CHG := {Yes_no_answer}
 sentence := "Would you like me to carry out a task?"
 prepare_sentence(sentence)

 repeat

    output_sentence(sentence)
    activate(CHG)
    reply := speech_recognize(CHG)
    dialog_interpreter(reply,sentence,CHG,c1)
    prepare_sentence(sentence)

 until c1 = TRUE

 new_robot_status(S1)
)
```

Figure 3. Dialog Segment D_1

```
(start

 c2 := FALSE
 c3 := FALSE
 CHG := {Navigation_goal,Manipulation_goal}
 sentence := "What kind of task - navigation or manipulation?"
 prepare_sentence(sentence)

 repeat

    output_sentence(sentence)
    activate(CHG)
    reply := speech_recognize(CHG)
    dialog_interpreter(reply,sentence,CHG,c1,c2)
    prepare_sentence(sentence)

 until ((c1 = TRUE) or (c2 = TRUE))

 if (c1 = TRUE) then new_robot_status(S2)
 else new_robot_status(S3)
)
```

Figure 2 shows the high-level code for D_0 (the DS associated with the resting state S_0 shown in Figure 1) as it will appear in the RDL. Similarly, Figure 3 gives the high-level code for D_1.

In both cases, "speech-recognize" assesses the user's spoken input with the aid of the information in CHG and returns "reply", which is usually a single conceptual hypothesis, but may instead be a list of CHs, along with their probabilities. The procedure "dialog-interpreter" analyses "reply" on the basis of the current goals. If it decides the information is incomplete, it may formulate further speech output. On the other hand, if there is enough information, it will set a condition flag to "TRUE", so that the system undergoes transition to a new state.

In addition to local DSMs of the type just discussed, there is a global Dialog Manager which handles interrupts and resumes. An interrupt might occur because the user decides to change the topic, or because of an event in the environment. In either case, a new DS may be entered. The global DM knows which frames needed to complete a task have been only partially filled out; it ensures that these are resumed when the current DS is over.

It should now be clear how the dialogue model relates to the environment model and the robot state model. As the robot undergoes state transitions, various actions are triggered. Some of these are physical actions involving movement in space; others, triggered by new information provided by the robot's sensors, are updates of the environment model. A third kind of action employs the dialogue model.

This kind of action occurs when, for instance, the robot asks the human for further commands, or explains an obstacle it has encountered and asks for guidance. It may also occur in response to an initiative by the user. In each case, a particular DSM will be invoked. Which one is chosen depends to a large extent on the goals the robot is currently trying to satisfy. These goals involve information that the robot needs to carry out its physical tasks, and can be broken into two categories: either the robot needs to be given a command, or it has been blocked in the performance of a task by inadequate information about the environment. When its goals have been satisfied, the robot will try to terminate the Dialog Segment unless the human provides indications that he or she has relevant information to add.

A successful dialogue may lead to a robot state transition (the robot is starting a new task), to an update of the environment model, or both. Note that in the course of a dialogue, the system is highly goal-oriented: it does its best to extract needed information from the user, asking as many questions as necessary to do so. This feature of our dialogue model is shared with many natural language systems, especially natural language systems with a speech recognition component. If the user changes topic, the system will deal with the new topic, but it will eventually revert to the old topic as long as it still has unsatisfied conversational goals that have not been rendered irrelevant by new information.

5 The Conceptual Hypothesis Grammar

The CHG associated with each conceptual hypothesis is a crucial part of the dialogue model. It has three main functions: to provide the speech recognition component with constraints on its search through the space of possible words, to evaluate the probability score of the hypothesis based on information provided by the speech recognition component, and to fill in slots in the hypothesis if this probability score is sufficiently high.

The CHG is a two-level stochastic grammar. Hypothesis H may be realised in a number of different ways at the grammatical level. For instance, consider the number of different ways of saying "Yes": "Certainly", "I would like that very much", "I guess so", and so on. These realisations differ widely in grammatical structure; we will name them $G_1, G_2, ..., G_n$.

A particular grammatical structure G_i can in turn be parsed into a number of constituents:

$$G_i = < g_{1i}, g_{2i}, ..., g_{ki} > . \tag{1}$$

Finally, each one of these constituents g_{ji} yields a terminal. A terminal in a CHG could be a single word or a string of words, since certain multiword phrases in English recur so often that it makes sense to treat them as vocabulary items with an associated grammatical category. We will use the word "phrase" and the symbol p to refer to a terminal, with the understanding that this could represent either a single word or a string of words. The category g might yield a number of different terminals p. For instance, there may be a category "request-start" which includes such phrases as "please", "would you", "what I want is for you to", etc.

The CHG for a given hypothesis H assigns a probability $P(G_i|H)$ to each grammatical structure G_i. Now, suppose that the speech recognition component is evaluating the hypothesis that the speaker's last utterance was

$$U = < p_1, p_2, ..., p_k > . \tag{2}$$

Assume that p_1 is of grammatical category g_1, p_2 is of category g_2, and so on, and that this sequence of g's has the structure

$$G = < g_1, g_2, ..., g_k > . \tag{3}$$

Clearly, we have

$$P(U|H) = P(G|H)P(p_1|g_1)P(p_2|g_2)...P(p_k|g_k) \tag{4}$$

Now, consider the probabilities $P(p_i|g_i)$. We can divide these probabilities into two classes: those that depend on the circumstances of the dialogue (such as dialogue history and the environment) and those that do not. The first class contains slot-filling phrases and the second contains all other phrases, with some exceptions (these will be phrases that do not fill slots but are still influenced by dialogue history). For instance, the choice of p_1, p_3, and p_6 might be dialogue-dependent. Let $P_{dep} = \prod_i P(p_i|g_i)$ for all such dialogue-dependent p_is and g_is, and let $P_{ind} = \prod_j P(p_j|g_j)$ for the rest. Then we have

$$P(U|H) = P(G|H)P_{dep}P_{ind}. \tag{5}$$

Thus, P_{dep} is the component of the CHG that is most important for slot-filling. Consider a particular factor of this probability, $P(p_i|g_i)$. This gives the probability that, among type g_i phrases, phrase p_i will be uttered by the user. In our model, this probability is given as a weighted average of three components, $P_S(p_i|g_i)$, $P_K(p_i|g_i)$, and $P_C(p_i|g_i)$.

Here subscript S stands for "static", K stands for "knowledge", and C stands for "cache". The formula is:

$$P(p_i|g_i) = \lambda_1 P_S + \lambda_2 P_K + \lambda_3 P_C \tag{6}$$

where $\lambda_1 + \lambda_2 + \lambda_3 = 1$.

The static component is an a priori probability derived from the frequency of p_i among type g_i phrases in the past. The knowledge component reflects information about the robot's current situation, including relevant facts about the environment. For instance, if a noun phrase of type OBJECT is being considered, we would assign higher probabilities to phrases describing objects in the robot's vicinity. Finally, the cache component reflects phrase associations.

The idea underlying this component is that the occurrence of a given phrase may be used to predict the occurrence of other phrases; it can be formalized as follows. Suppose that the Viterbi algorithm or some other probabilistic algorithm has identified the most likely interpretation of the acoustic signals up to now as the sequence $p_1, p_2, ..., p_n$, where each phrase p_i is associated with a probability $P(p_i)$ derived from both acoustic and higher-level evidence. We are trying to identify phrase p_{n+1}. We consider only the L-sized window given by $p_{n-L+1}, p_{n-L+2}, ..., p_n$, discarding earlier information. Within this window, some phrases are more likely to have been correctly identified than others; $P(p_i)$ indicates the strength of the evidence that p_i was correctly identified.

Now, define $Q(p, p_j)$ as the strength of the likelihood that phrase p will occur, given that phrase p_j occurred among the last L phrases. This can be calculated from a training text by setting up a counter matrix $c_x(p_i, p_j)$ for each DS x, which is incremented every time the phrase p_i occurs while p_j is contained in the last L phrases in the training text. Q for DS x can be defined as

$$Q_x(p_i, p_j) = c_x(p_i, p_j)/(\sum_i c_x(p_i, p_j)). \tag{7}$$

Finally, we define a function

$$f_L(p, < p_{n-L+1}, ..., p_n >) = \sum_{j=1}^{L} P(p_{n-L+j})Q(p, p_{N-L+j}). \tag{8}$$

This function (which requires a normalization constant not given here to become a true probability) uses each of the previously recognized phrases within the L-sized window to estimate the probability that phrase p will occur; it assigns higher weight to evidence from those previously recognized phrases considered to have a high probability of being correct.

Thus, the occurrence of a phrase carries with it increased probabilities for other phrases which have often followed it within a window of size L while a certain topic is being talked about by the user. Even if a few phrases are inserted or deleted, the statistical association between two phrases yields information (within the limit imposed by the window size L). For example, suppose L is 5 and the training corpus includes several occurrences of strings where phrase p_a is followed by a gap of one or two other phrases followed by p_b. Suppose that during recognition, p_a is recognized with high estimated probability; p_b is then assigned high probability of occurring after p_a with a gap of one or two other phrases, but in addition it is assigned high probability of occurring with a zero gap, a gap of three, and a gap of four.

The Dialog Segment Manager that employs the CHG interacts directly with the speech recognition component, each guiding the other in its search through the space of hypotheses. The speech recognition component provides the DSM with a list of the currently hypothesized words, each accompanied by its corresponding acoustic probability. In turn, the DSM activates a list of words in the vocabulary considered most likely to follow the current word on the basis of the CHG. When the user's utterance has terminated, the DSM may retain several alternative CHs whose probability is sufficiently high to have survived the selection process. In that case, the DSM will probably ask further questions to eliminate all but one of the CHs.

In theory, the probability of a CH should be the sum of the probabilities of all parsings of the user utterance that yield that CH. For reasons of computational efficiency, it seems likely that most low-probability parsings will never be carried out, so the score associated with a CH will be the probability estimate of the highest-probability parsing with that interpretation.

6 Conclusions

The difficulties of the task faced by a system for spoken man-machine dialogue are well-known. They include not only the existence of multiple word candidates at each location in time, mentioned above, but repetition of syllables or of whole words, deletion of syllables or of whole words, and insertion of interjections like "ah" or "um" or of words or phrases not recognised by the system. The system must therefore be robust enough to generate a semantic token even when semantically unimportant parts of the phrase have been distorted by repetition, deletion, or insertion.

As the research surveyed in the first two sections indicates, this can be achieved only if there are strong constraints on the space of hypotheses examined by the speech recognition component. The model described above contains a hierarchy of constraints, each given probabilistically. At the uppermost level, the system uses its knowledge of dialogue structure to guess what topic the user is about to address; furthermore, it seeks to guide the user's responses through its own voice output (without being excessively domineering). Given a topic, the system can estimate how the user is likely to structure his response. Finally, at the phrase level, the system uses three kinds of constraints: an a priority probability distribution for phrases derived from past phrase frequencies, a probability distribution based on word association frequencies, and a probability distribution based on the current situation of the robot.

This model reflects some of the key features of natural language. It is clear that the use of language by human beings is very dependent on context: on the physical environment, on the recent experiences of the speaker, on the mood of the speaker, on what has been said by the speaker's interlocutors, and so on. One could sum up the effects of all these factors by saying that they determine what is "salient" to the speaker and the listeners at a given time and place. This is one of the most important problems in computational linguistics, and it manifests itself at many different levels - for instance, anaphor is one aspect of it. The model described above should give the system a method for estimating when certain words or phrases are likely to be spoken - in other words, when they are salient.

Thus, the dialogue model in general and the estimated probabilities for words and phrases in particular will respond as dynamically as possible to the recent history of the robot and to its physical environment. For instance, when the robot's vision system observes a physical obstacle blocking its path, the associated words and concepts will be assigned higher probabilities. In addition, the user's most recent use of words and phrases will be taken into account.

The model requires the assignment of a large number of probabilities. As far as possible, these probabilities will be obtained empirically from studies of transcripts of robot-user sessions. Thus design of the system will be iterative - as experience with the system grows, the probability estimates should get more and more accurate.

We have described a model for the incorporation of semantic and pragmatic information in a speech recognition system via explicitly specified probabilities. Implementation of the model will, we hope, yield performance improvements similar to those obtained by probabilistic approaches at other levels of the speech recognition hierarchy. As the use of voice-activated devices spreads to non-expert users, this kind of model is likely to receive more and more attention.

7 Acknowledgements

Acknowledgements to Charles Snow for attentive proofreading, to Diane Goupil for helpful comments, and to Henriette Bouchard and Sylvie Tellier for editing the bibliography.

8 Bibliography

1. Richard A. Bolt. "The integrated multi-modal interface" in The Transactions of the Institute of Electronics, Information and Communication Engineering, Vol. J70-D, no 11, pp. 2017-2025 (1987)

2. Harry Bunt. "On-line Interpretation in Speech Understanding and Dialogue Systems" in Recent Advances in Speech Understanding and Dialog Systems, NATO ASI Series, Series F; Computer and Systems Sciences, Vol. 46, pp. 348-395 Springer-Verlag (1988)

3. H.C. Bunt. "Information Dialogues as Communicative Action in Relation to Partner Modeling and Information Processing" in The Structure of Multimodal Dialogue, M.M. Taylor, F. Néel and D.G. Bouwhuis (Editors) Elsevier Science Publishers B.V. (1989)

4. N. Carbonell and J.M. Pierrel. "Task-oriented dialogue processing in human-computer voice communication" in Recent Advances in Speech Understanding and Dialog Systems, NATO ASI Series, Series F; Computer and Systems Sciences, Vol. 46, pp. 491-496 Springer-Verlag (1988)

5. N. Carbonnell and J.M. Pierrel. "Architecture and Knowledge Sources in a Human-Computer Oral Dialogue System" in The Structure of Multimodal Dialogue, pp. 417-429, M.M. Taylor, F. Néel and D.G. Bouwhuis (Editors) Elsevier Science Publishers B.V. (1989)

6. Guy Deville. "Modélisation d'énoncés finalisés dans un système de dialogue oral homme-machine" in XVIIIèmes Journées d'études sur la parole, pp. 242-247 (Conference held at Université de Montréal, May 28-31, 1990)

7. Jack L. Edwards and James A. Mason. "Toward intelligent dialogue with ISIS" in International Journal of Man-Machine Studies, 28, pp. 309-342 Academic Press (1988)

8. J.L. Edwards and J.A. Mason. "The Structure of Intelligence in Dialogue" in The Structure of Multimodal Dialogue, M.M. Taylor, F. Néel and D.G. Bouwhuis (Editors) Elsevier Science Publishers B.V. (1989)

9. U. Ehrlich and H. Niemann. "Using Semantic and Pragmatic Knowledge for the Interpretation of Syntactic Constituents" in Recent Advances in Speech Understanding and Dialog Systems, NATO ASI Series, Series F; Computer and Systems Sciences, Vol. 46, pp. 485-490 Springer-Verlag (1988)

10. P. Falzon. "Ergonomie cognitive du dialogue". 175 p. Presses Universitaires de Grenoble (1989)

11. M. Guyomard and J. Siroux. "Experimentation in the specification of an oral dialogue" in Recent Advances in Speech Understanding and Dialog Systems, NATO ASI Series, Series F; Computer and Systems Sciences, Vol. 46, pp. 497-501 Springer-Verlag (1988)

12. M. Guyomard, J. Siroux and A. Cozannet. "Le rôle du dialogue pour la reconnaissance de parole: Le cas du système pages jaunes" in XVIIIèmes Journées d'études sur la parole, pp.322-326 (Conference held at Université de Montréal, May 28-31, 1990)

13. J. Klein and J.M. Pierrel. "Interprétation d'expressions complexes dans un système de dialogue homme-machine" in XVIIIèmes Journées d'études sur la parole, pp. 248-252 (Conference held at Université de Montréal, May 28-31, 1990)

14. Roland Kuhn and Renato De Mori. "A Cache-Based Natural Language Model for Speech Recognition" in IEEE Transactions on Pattern Analysis and Machine Intelligence, Vol. 12, no 6, pp. 570-583 (June 1990)

15. S. Kunzmann, T. Kuhn and H. Niemann. "An experimental environment for the generation and verification of word hypotheses in continuous speech" in Speech Communication 7, pp. 381-388 North-Holland (1988)

16. Akira Kurematsu. "An overview of ATR basic research into telephone interpretation", Keynote Speech in ATR Symposium on basic research for telephone interpretation, Kyoto, Japan, December 11-12, 1989, 14 p.

17. James A. Mason and Jack L. Edwards. "Surveying projects on intelligent dialogue" in International Journal of Man Machine Studies, 28, pp. 259-307 Academic Press (1988)

18. K. Matrouf et al. "Adapting probability-transitions in DP matching process for an oral task-oriented dialogue" in Proceedings of ICASSP 90. New- Jersey : IEEE, 1990 : pp. 569-571 (Conference held at: New Mexico, April 3-6, 1990)

19. Karim Matrouf, Françoise Néel and Joseph Mariani. "Système de dialogue orienté par la tâche: une application en aéronautique" in Journal Accoustique 2, pp. 85-93 (mars 1989)

20. Robert Moore, Fernando Pereira and Hy Murveit. "Integrating Speech and Natural-Language Processing" in Proceedings of the Speech and Natural Language Workshop. California : Morgan Kaufmann, pp. 243-247. Sponsored by DARPA/ISTO. (Conference held at: Philadelphia, Pennsylvania, February 21-23, 1989)

21. Hy Murveit and Robert Moore. "Integrating Natural Language Constraints into HMM-based Speech Recognition" in Proceedings of ICASSP 90. New-Jersey : IEEE 1990 : pp. 573-576. (Conference held at: New Mexico, April 3- 6, 1990)

22. H. Niemann et al. "The Speech Understanding and Dialog System EVAR" in New Systems and Architectures for Automatic Speech Recognition and Synthesis, R. de Mori and C.Y. Suen (eds). NATO ASI Series F16, Berlin, pp. 271-302

23. J.M. Pierrel. "Dialogue oral Homme-Machine: connaissances linguistiques, stratégies et architectures des systèmes", Hermès, Paris (1987)

24. Salim Roukos. "Integrating Speech and Natural Language" in Proceedings of the Speech and Natural Language Workshop. California : Morgan Kaufmann, pp. 243-247. Sponsored by DARPA/ISTO. (Conference held at: Philadelphia, Pennsylvania, February 21-23, 1989)

25. Alexander I. Rudnicky. "The Design of Voice-Driven Interfaces" in Proceedings of the Speech and Natural Language Workshop. California : Morgan Kaufmann, pp. 120-124. Sponsored by DARPA/ISTO. (Conference held at: Philadelphia, Pennsylvania, February 21-23, 1989)

26. Alexander I. Rudnicky, Michelle Sakamoto and Joseph H. Polifroni. "Spoken Language Interaction in a Goal-Directed Task" in Proceedings of ICASSP 90. New-Jersey : IEEE, 1990 : pp. 45-48 (Conference held at: New Mexico, April 3-6, 1990)

27. G. Sagerer and F. Kummert. "Knowledge Based Systems for Speech Understanding" in Recent Advances in Speech Understanding and Dialog Systems, NATO ASI Series, Series F; Computer and Systems Sciences, Vol. 46, pp. 421-458 Springer-Verlag (1988)

28. Stephanie Seneff. "TINA: A probabilistic syntactic parser for speech understanding systems" in Proceedings of the Speech and Natural Language Workshop. California : Morgan Kaufmann, pp. 168-178. Sponsored by DARPA/ISTO. (Conference held at: Philadelphia, Pennsylvania, February 21-23, 1989)

29. Jacques Siroux and Dominique Gillet. "A system for man-machine communication using speech" in Speech Communication, Vol. 4, no 4, pp. 289-315 North-Holland (December 1985)

30. Gilles Souvay, Jean-Marie Pierrel, Evelyne Gallais and Pierre Alinat. "Diapason: un système de DIAlogue Pour la commande orAle d'une console SONar" in XVIIIèmes Journées d'études sur la parole, pp. 327-331 (Conference held at Université de Montréal, May 28-31, 1990)

31. M.M. Taylor. "Layered protocols for computer-human dialogue. I: Principles" in International Journal of Man-Machine Studies, 28, pp. 175-218 Academic Press (1988)

32. M.M. Taylor. "Layered protocols for computer-human dialogue. II: some practical issues" in International Journal of Man-Machine Studies, 28, pp. 219-257 Academic Press (1988)

33. M.M. Taylor, F. Néel and D.G. Bouwhuis (Editors). "The Structure of Multimodal Dialogue" Elsevier Science Publishers B.V. (1989)

34. G. Thurmair. "Semantic Processing in Speech Understanding" in Recent Advances in Speech Understanding and Dialog Systems, NATO ASI Series, Series F; Computer and Systems Sciences, Vol. 46, pp.397-419 Springer- Verlag (1988)

35. Wolfgang Wahlster and Alfred Kobsa. "User Models in Dialog Systems", Technical Report No. 30, Universitat des Saarlandes, January 1988.

36. Wayne Ward. "Understanding Spontaneous Speech" in Proceedings of the Speech and Natural Language Workshop. California : Morgan Kaufmann, pp. 137-141. Sponsored by DARPA/ISTO. (Conference held at: Philadelphia, Pennsylvania, February 21-23, 1989)

37. J. Waterworth. "Interactive Strategies for Conversational Computer Systems" in The Structure of Multimodal Dialogue, M.M. Taylor et al. eds, North-Holland (1988)

38. S.J. Young and C.E. Proctor. "The design and implementation of dialogue control in voice operated database inquiry systems" in Computer Speech and Language, Vol.3, no 4 , pp. 329-353 Academic Press (October 1989)

39. Sheryl R. Young, Wayne H. Ward and Alexander G. Hauptmann. "Layering predictions: Flexible use of dialog expectation in speech recognition" in Proceedings of the Eleventh International Joint Conference on Artificial Intelligence, Vol. 2 (Conference held in Detroit, Michigan, August 20-25, 1989)

40. Sheryl R. Young. "The Minds System: Using Context and Dialog to Enhance Speech Recognition" in Proceedings of the Speech and Natural Language Workshop. California : Morgan Kaufmann, pp. 131-136. Sponsored by DARPA/ISTO. (Conference held at: Philadelphia, Pennsylvania, February 21-23, 1989)

41. S.R. Young et al. "High Level Knowledge Sources in Usable Speech Recognition Systems" in Communications of the ACM, Vol. 32, no 2, (1989)

42. Victor Zue and al. "The Voyager Speech Understanding System: Preliminary Development and Evaluation" in Proceedings of ICASSP 90. New Jersey : IEEE, 1990 : pp. 73-76. (Conference held at: New Mexico, April 3-6, 1990)

Experimenting Text Creation by Natural-Language, Large-Vocabulary Speech Recognition

P. Alto, M. Brandetti, M. Ferretti, G. Maltese, F. Mancini, A. Mazza, S. Scarci, G. Vitillaro

IBM Italy Rome Scientific Center
via Giorgione 159, 00147 ROME (Italy)

1. Introduction

In the last years the probabilistic approach to speech recognition has allowed the development of high-performances large-vocabulary speech recognition systems [1] [2]. At the IBM Rome Scientific Center a speech-recognition prototype for the Italian language, based on this approach, has been built. The prototype is able to recognize in real time natural-language sentences built using a vocabulary containing up to 20000 words. [4]. Once and for all the user has to perform an **acoustic training phase** (about 20 minutes long), during which he is required to utter a predefined text. Words must be uttered inserting small pauses (a few centiseconds), between them. The prototype architecture is based on a personal computer equipped with special hardware. The first system we developed was aimed at a business and finance lexicon. Many laboratory tests have shown the effectiveness of the prototype as a tool to create texts by voice. After a first phase during which in-house experiments were carried on [5], the need arose to test the system in real work environments and for different applications. Two applications were considered: the dictation of radiological reports and of insurance company documents. Due to their characteristics, these applications seemed to be very well suited for our purposes. Since the vocabulary of the recognizer must be predefined, we had to adapt the system to the lexicon required by the new applications. The paper describes the techniques developed to efficiently adapt the basic component of the recognizer: the acoustic and language models. The results obtained experimenting automatic text dictation during real work are also presented.

2. System structure

To understand the steps necessary to perform the system adaptation a brief outline of its basic structure is needed. In the probabilistic approach to speech recognition we look for the sequence of words $\overline{W}$ which has the highest probability given the acoustic information $\overline{A}$ extracted from the observed signal [1]. In our case the acoustic signal is a sequence of **acoustic labels** extracted from the signal every centisecond and representing the energy content of the signal in 20 frequency bands.
Applying the Bayes theorem we can write:

$$P(\overline{W}|\overline{A}) = \frac{P(\overline{A}|\overline{W})P(\overline{W})}{P(\overline{A})} \tag{1}$$

where $P(\overline{A}|\overline{W})$ is the probability that the sequence of words $\overline{W}$ will produce the sequence of acoustic information $\overline{A}$. $P(\overline{W})$ is the *a priori* probability of the sequence of words $\overline{W}$. $P(\overline{A})$ is the probability of the sequence of acoustic information $\overline{A}$. We seek the maximum of the above expression with respect to $\overline{W}$. We can ignore $P(\overline{A})$ because it does not depend on $\overline{W}$. Therefore we need to maximize the numerator of the expression (1).
As a consequence of equation (1), the recognition task can be viewed as composed by four components:

1. an **acoustic processor** to extract the acoustic information $\overline{A}$ from the speech signal;

NATO ASI Series, Vol. F 75
Speech Recognition and Understanding. Recent Advances
Edited by P. Laface and R. De Mori

2. an **acoustic model** to compute the probability $P(\bar{A} \mid \bar{W})$;
3. a **language model** to compute the a-priori probability $P(\bar{W})$;
4. an efficient **search strategy**, to find, among all possible sequences of words $\bar{W}$, the most probable one.

While the signal processing stage and the search strategy can be considered as vocabulary-independent, both the acoustic and the language model depend on the application which the system is aimed for, so they must be adapted to the lexicon of the new application. In the next paragraphs we will present the structure of the models and the techniques employed to adapt them.

Acoustic Model

The acoustic model task is to compute $P(\bar{A} \mid \bar{W})$. The approach currently employed for the acoustic model is based on **hidden Markov models** that are finite state automata [3]. For every time slice the model takes a transition from the current state to one of the allowed states (the transition can also produce no state change). For each transition an acoustic label is produced [1]. Both the transitions and the label emission occur according two probability distributions which depend on the current state only. These models are called *hidden* because it is only possible to observe the sequence of acoustic symbols produced, while the sequence of states is unknown.

Each word belonging to the vocabulary is represented by a different model. To allow speech recognition for a very large vocabulary, the basic speech units modeled by the Markov sources are associated to the basic sounds of the language. So an alphabet of acoustic units must be defined to represent the basic sounds of the language. Examples of acoustic units used for speech recognition are: syllables, diphones, phones. We choose the phone as phonetic unit. The basic sounds of the Italian language were described by a set of 56 phonetic units [4]. For each phonetic unit a Markov model representing its pronunciation has been created. In our system all the phonetic units have the same topological structure. The distinction between different sounds is left entirely to the probability distributions, namely, to the *parameters* of the model. The computation of the parameters is accomplished during the acoustic training phase employing the predefined text uttered by the user. The model for each word is built by the concatenation of the Markov sources corresponding to the string of phonetic units forming its pronunciation.

Language Model

The task of the language model is to compute $P(\bar{W})$. The computation is performed in the following way:

$$P(\bar{W}) = \prod_{l=1}^{N} P(w_l \mid w_{l-1}, w_{l-2}) \tag{4}$$

This is the so-called **trigram language model**[2] : it contains the approximation of considering equivalent all the sentences ending with the same couple of words $w_{l-1}w_{l-2}$. The number of possible trigrams is so large that is practically impossible to collect the amount of data needed to estimate the probability of each of them. To overcome the problem, an interpolation between different probability distributions is performed. Three different distributions are computed for trigrams, bigrams and unigrams. The probability of word w_3 given the words w_1 and w_2 is estimated as follows:

$$P(w_3 \mid w_1 w_2) =$$
$$\lambda_3 \frac{C(w_1 w_2 w_3)}{C(w_1 w_2)} + \lambda_2 \frac{C(w_2 w_3)}{C(w_2)} + \lambda_1 \frac{C(w_3)}{N} + \lambda_0 \frac{1}{V}$$

where $C(w_1, \ldots, w_n)$ is the number of times the word string $w_1, \ldots, w_n$ was observed in the training data and V is the number of words in the vocabulary. The λ coefficients are estimated using the *expectation-*

maximization algorithm [9]. The trigram language model is an effective tool to represent the linguistic constraints for speech recognition purposes; on the other hand it requires a large amount of training data. The larger is the training corpus, the higher is the recognition rate [10].
Perplexity is a typical measure of the predictive power of a language model [11]. It estimates the average number of words that are considered equiprobable by the model. Of course, if the language model is not used, the perplexity value equals the vocabulary size.

Vocabulary definition

In our system the vocabulary is predefined; this means that the choice of the vocabulary is a crucial step for the system performances.
In the first version of our prototype a vocabulary containing more than 20000 words was used. It is aimed at the dictation of economy and finance reports. [5]. The 20000 words were chosen as the most frequent ones in a **corpus** containing 44 millions of words composed by articles from the most important Italian economy and finance newspaper (*Il Sole 24 Ore*), articles from an economy and finance newsmagazine (*Il Mondo*), and press agency news. The coverage of this vocabulary computed on a disjoint corpus was 96.5%.

3. System adaptation

To build a speech recognition system for a new application, we need:

- to define the vocabulary;
- to adapt the acoustic and language models.

While the adaptation of acoustic and language models can be automatized, the choice of vocabulary depends on factors like the lexicon of the application and the size of the available text. Thus in the next paragraphs, it will be described the tools that have been created to make the required adaptations easy and quick.

Acoustic model adaptation

We have just seen that each word is described by the concatenation of the Markov models of phonetic units forming its pronunciation. According to the chosen technique, the first step in adapting the system to new application, is to perform the phonetic transcription of all the words in the new vocabulary. To limit the number of words that must be transcribed, a large database was built containing all the words and the phonetic transcriptions used in previous vocabularies. By using the database it is possible to find all the words for which a new phonetic transcription must be supplied.
Usually, the phonetic transcription is performed manually. It is a very expensive process and for large vocabularies the transcriptions could contain errors. We tried to make the phonetic transcription process as automatic as possible. The systems that have been proposed to solve the problem of automatic phonetic transcription are based on rules [6] [7] or on automatic learning from training data [8]. Actually, these systems cannot provide the accuracy required for automatic speech recognition. This is due both to the complexity of the problem and to the difficulty to describe all the possibilities with a limited set of rules. We employed a different technique from the mentioned ones. We separated phonotactical knowledge (well described by a limited set of rules) from lexical knowledge (based on experience and not suitable for a formal description). Given the string representing the orthographic form of the word our system produces a set of phonetic transcriptions for that word, which are the ones that can be obtained applying our set of rules for the grapheme-to-phoneme translation. The user can choose manually the correct transcription on the basis of his lexical knowledge. Grapheme-to-phoneme translation for the Italian language has a relatively low uncertainty. A set of 78 rules allows to describe all the ambiguities. Each rule consists of a left part and a right part. The left part consists of a grapheme string and its (possibly empty) left and right graphemic contexts; the right part consists of the set of possible phonetic transcriptions for the grapheme string. The

set of transcriptions produced applying this set of rules is then pruned by means of a set of global rules (which, for example, reject all the transcriptions which do not have one and just one stressed vowel). The right phonetic transcription always belongs to the resulting set. The average number of phonetic transcriptions per word is 5. Using this method it was possible to adapt rapidly the recognizer to the new application. The quality of the produced transcriptions was at least equal to a completely manual phonetic transcription.

Language model adaptation

The language model is built following the previously described technique. We have created a software structure that takes a dictionary and a *corpus* as a starting point and give the probability estimations as output:

$$P(w_3 \mid w_1 w_2)$$

for any word w_3, given the context w_1, w_2.

4. In-house test for the new application

Before experimenting the speech-recognizer in a real environment we needed to perform in-house tests to assess the recognition rate of the system when used to dictate pre-defined texts. To perform the experiment a text containing phrases peculiar to the application must be created; it must be dictated by several different speakers. To make a meaningful test it is important that the text contains all the phonetic units used to build the acoustic model in phonetic contexts typical of the application.
Usually, the text is built manually trying to represent a large number of different contexts using the smallest number of sentences. To avoid this manual process a procedure has been built to prepare the test automatically, so it is possible to select efficiently a set of sentences containing phonetic contexts typical of the application and suitable to assess the accuracy of the recognizer.

This preliminary study on the usage of a voice-activated text editor indicated that large-vocabulary speech recognition can offer a very competitive alternative to traditional text entry. It is likely to be well accepted even by users who have a large experience in keyboard text entry. The results (error rate 2-3%) show that the system could be a revolutionary tool, for example, for the office of the future, but poses unprecedent human factors problems. These experiments have been very useful for us, because they have suggested, for example, improvements to the system interface and the need of some special keys to make easy the revision and the correction phases.

5. Real-work applications

Two distinct application areas were considered: medical report dictation and office (insurance company) documents and memos creation. We will describe separately how the vocabulary of each application was built.
The economy and finance lexicon is quite different from the lexicon required to dictate radiological reports, while it has some similarities with respect to the lexicon used in the insurance company reports.
In the first case the vocabulary was selected by using a corpus containing only radiological reports; in the second case the vocabulary was built taking the economy-finance vocabulary as a starting point.

Radiological Reports Vocabulary

The available corpus contained about 5 million words collected in four different hospitals. The hospital where the experiment was performed provided us with a corpus containing 50000 words
The first problem was the choice of the vocabulary size. We adopted the criterion of analyzing the variation of the coverage with respect to the vocabulary size (figure 1).

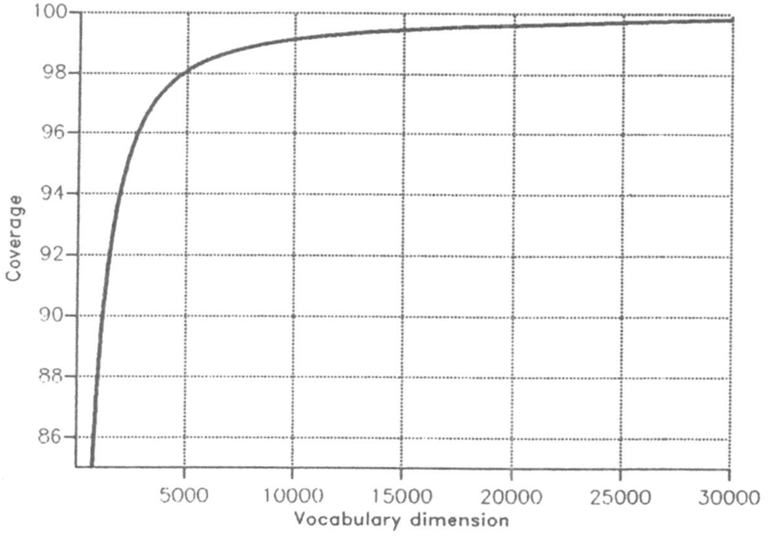

Figure 1. Coverage of corpus as function of vocabulary size

A vocabulary containing 5000 words seemed to us a reasonable trade-off between the need to reach high coverage and the need to have enough data to estimate the language model parameters.
One of the main characteristics found in this kind of lexicon was the presence of a set of words peculiar to the report dictation process at each location. To make the recognizer well suited to the needs of the experimenter all the 3200 different words found in corpus provided by the hospital where the experiment was performed, were included in the vocabulary. The vocabulary was completed adding 1900 words which were the most frequent ones in the other corpora not included in the previous list of 3200 words. The resulting vocabulary had a 100% coverage on reports from the experiment location and 97.5% coverage on other reports.

Insurance Company Reports Vocabulary

The selection of the words to be used in this vocabulary was performed by analyzing a corpus containing 1.5 million words provided by the insurance company. There was a certain similarity between this lexicon and the economy and finance one: the coverage obtained by economy and finance vocabulary (20000 words) on insurance corpus was 95.2%. The 15000 most frequent words of economy and finance vocabulary were selected and the 3100 most frequent words found in insurance corpus were added. The resulting vocabulary showed a 99% coverage on insurance texts and a 95.7% coverage on economy and finance ones.

Language models informations

Table 1. Language models characteristics		
Parameter	**Radiology Reports**	**Insurance Reports**
Vocabulary size	5100	18100
Training data	4.8	1.5 + 40
Millions of distinct trigrams	0.62	3.4
Millions of distinct bigrams	0.19	2.3
Perplexity	38	18

The radiology reports language model was trained using 4.8 million words. For the insurance company reports a corpus containing 1.5 million words typical of the application was merged with 40 million words extracted from the economy and finance corpus. In the table the most significant parameters of the language models are reported.

6. Results

In the following paragraph the results obtained experimenting the two prototypes during real work are reported.

Radiological Reports Dictation

Four doctors have used the recognizer during their every-day work to prepare the reports to be delivered to the patients. No one had any difficulty in inserting short pauses between words. The doctors dictated 150 reports containing more than 12000 words. The vocabulary coverage for the dictated reports was 98.6%. Table 2 reports the results of the experiment.

Table 2. Radiological reports dictation			
Speaker	**Number of reports**	**Error rate**	**Speaker's errors**
sp1	25	1.7%	1.4%
sp2	14	2.0%	1.2%
sp3	76	3.5%	0.9%
sp4	35	5.0%	3.2%

The error rate is referred to the number of errors done by the recognizer without taking into account errors due to words not included in the vocabulary. The speaker's error rate is referred to the errors due to misuse of the recognizer (wrong commands, absence of pause between words, etc.). The global error rate can be computed summing the numbers in the third and fourth column of the table. We can see that the recognizer's performances range from 90.5% to 95.6% of accuracy.

Insurance Company Reports Dictation

The experimentation was carried on by five different users who dictated more than 8000 words. The vocabulary coverage on the dictated text was about 99%. Table 2 reports the results obtained in this case.

Table 3. Insurance company reports dictation

Speaker	Error rate	Speaker's error rate
sp1	1.9%	1.0%
sp2	1.4%	0.5%
sp3	14.0%	2.0%
sp4	7.4%	1.5%
sp5	2.4%	0.8%

References

[1] L.R. Bahl, F. Jelinck, R.L. Mercer, **A Maximum Likelihood Approach to Continuous Speech Recognition,** *IEEE Trans. on Pattern Analysis and Machine Intelligence,* vol. PAMI-5, no. 2, 1983, pp. 179-190.

[2] F. Jelinek, **The development of an experimental discrete dictation recognizer,** *Proceedings IEEE,* vol. 73, no. 11, November 1985, pp. 1616-1624.

[3] L.R. Bahl, P.F. Brown, P.V. De Souza, R.L. Mercer, M.A. Picheny, **Acoustic Markov Models Used in the Tangora Speech Recognition System,** *Proc. IEEE International Conference on Acoustics, Speech and Signal Processing,*

[4] P. D'Orta, M. Ferretti, A. Martelli, S. Melecrinis, S. Scarci, G. Volpi, **Large-Vocabulary Speech Recognition: a System for the Italian Language,** *IBM Journal of Research and Development,* Vol. 32, No. 2, March 1988, pp.217-226.

[5] P. Alto, M. Brandetti, M. Ferretti, G. Maltese, S. Scarci, **Experimenting Natural-Language Dictation with a 20000-Word Speech Recognizer,** *IEEE CompEuro 89,* Hamburg, May 8-12, 1989, pp. 2-78 - 2-81.

[6] R. Carlson, B. Granstroem, **A Text-to-Speech System Based Entirely on Rules,** *Proc. IEEE International Conference on Acoustics, Speech and Signal Processing,* Philadelphia, PA, April 1976.

[7] D. H. Klatt, **Structure of a Phonological Rule Component for Synthesis-by-Rule Program** *IEEE Trans. on Acoust., Speech and Sig. Proc.,* vol. ASSP-24, no. 5, 1976, pp. 391-398.

[8] T. J. Sejnowski, C. R. Rosenberg, **Parallel Networks that Learn to Pronounce English Text,** *Complex Systems,* 1 (1987), pp. 145-168.

[9] F. Jelinek, R.L. Mercer, **Interpolated Estimation of Markov Source Parameters from Sparse Data,** *in "Pattern Recognition in Practice", E.L.Gelsema and L.N. Kanal, Ed.,* North-Holland, New York, 1980, pp. 381-387.

[10] M. Ferretti, G. Maltese, S. Scarci, **Measures of Language Model and Acoustic Information in Probabilistic Speech Recognition,** *Eurospeech 89,* Paris, September 1989, pp. 473-476.

[11] F. Jelinek, R.L. Mercer, L.R. Bahl, J.K. Baker, **Perplexity - a Measure of Difficulty of Speech Recognition Tasks,** *94th Meeting Acoustical Society of America,* Miami Beach, December 1977.

DUALGRAM: An Efficient Method for Representing Limited-Domain Language Models

C.A. Matheson, J.C. Foster, A.W. Black, and I.A. Nairn

Centre for Speech Technology Research (CSTR), University of Edinburgh, 80 South Bridge, Edinburgh EH1 1HN, Scotland [1]

Abstract: This paper describes DUALGRAM, a language model developed at CSTR to cope with certain problems arising from applications of speech recognition systems in limited domains. The essential feature of the model is the use of two finite state mechanisms running in parallel; one is a standard word Bigram, while the other is a tag Finite-State Machine which handles sequences of syntactic categories. Some advantages of the model are discussed in comparison to other approaches, and some results are presented on using the various models to describe unseen material.

Keywords: Language Modelling, Finite-State, Perplexity

1. Introduction

The main disadvantage with the successful speech recognition language models such as Bigrams and Trigrams (see e.g. [2]) is that they require very large amounts of data in order to produce reliable results. The actual amount required to produce useful statistics depends, of course, on the size of the vocabulary, but typically tens and often hundreds of millions of words of corpus are needed. For many possible applications of speech technology gathering this amount of data is extremely difficult and possibly prohibitively expensive. DUALGRAM represents an attempt to extract more accurate language models from the data, hence reducing the amount of material required.

There are two main criteria which a language model has to satisfy in speech applications. Firstly, the perplexity of the model should be as low as possible in order to increase the accuracy of the system. Some details on the measurement of this parameter in terms of branching factors are provided below. Secondly, the model should attempt to generate as much as possible of the notional language of the domain. It will normally be the case that the corpus, whatever its size, will not contain all the possible utterances in the domain. The language model must therefore be able to generalise from the corpus to the broader application language. It is arguable that a third criterion should be included: efficiency.

[1] This work was supported by the Information Engineering Directorate/Science and Engineering Research Council as part of the IED/SERC Large Scale Integrated Speech Technology Demonstrator Project (SERC grants D/29604, D/29611, D/29628, F/10309, F/10316) in collaboration with Marconi Speech and Information Systems and Loughborough University of Technology.

NATO ASI Series, Vol. F 75
Speech Recognition and Understanding. Recent Advances
Edited by P. Laface and R. De Mori

We assume here that the essentially finite-state characteristics of DUALGRAM serve to satisfy this requirement.

The first two desiderata pull in opposite directions; typically, as the domain of application is widened, the perplexity of the model increases. The DUALGRAM model thus seeks to limit the amount of uncertainty which results from extending the domain. Again, some measures of the success of the model in this area are provided below. Little detail is given on the use of the models in actual recognition runs; it is assumed here that the figures for perplexity which are included provide a more accurate measurement of this parameter.

2. Background

The model has been constructed for use in a speech recognition system for the IED/SERC Large Scale Integrated Speech Technology Demonstrator Project. The domain is Cytopathology, and the task is spoken text input. We have collected a corpus of some 5,000 reports from the Department of Pathology in the University of Edinburgh; although this represents almost a full year's output from the laboratories, these reports are quite short and the total usable corpus amounts to just over 10,000 utterances containing around 100,000 words. The relevant lexicon consists of 2,899 headwords.

In the tests which are reported below, an attempt is made to measure the success of language models in handling unseen data. This is assumed to provide a measure of the worth of the model as a representation of the language of the chosen domain. In practice, this measure was achieved by splitting the corpus in two and by further splitting the second half into four equal-sized sections. Tests were conducted by building a language model using the first half of the corpus and running it over the first quarter of the second half. A model was then constructed using the first half plus the first quarter of the second half; the new grammar was tested on the second quarter, and so on.

A total of 170 utterances were chosen for testing the recognition system and for calculating perplexity. The choice was random, except for the restriction that the strings should be less than 13 words long; this restriction was imposed by early versions of the acoustic front end. In calculating perplexity, the measure used is the geometric mean of the test set branching factor. The measure does not take account of probabilities as it is assumed here that these will not affect the necessary comparisons between models which are the main subject of this paper. In most cases, the addition of probabilities is a straightforward task; it is arguable, however, that this is more complicated in the case of DUALGRAM, and hence that simple branching rate comparisons are more useful.

3. DUALGRAM

The basis of DUALGRAM is very simple. The model is in essentially two halves; the first is a standard Bigram in which each lexical item specifies which words can follow it. As mentioned above, the model which is reported here does not include statistical information and hence is more accurately termed a followset grammar. The other half of DUALGRAM handles syntactic information, and there are a number of possible ways in which this representation can be generated. One possibility is to construct a finite-state machine (FSM) from a context-free grammar as described in [1]. If the grammar is small, this may be practicable. An alternative is to build FSMs directly from (tagged) corpora, and this method has been used in the models described in this paper. If the choice of tags is very general, then tagging a corpus can be done more or less automatically; it is noticeable that in constrained domains, with limited amounts of material, the problems arising from word class ambiguity are not great. Subsequently, constructing a tagged corpus is relatively straightforward.

In order to make this discussion concrete, let us assume that the corpus contains only the utterances below:

LARGE BLOOD CELLS ARE PRESENT
BLOOD IS PRESENT

The Bigram followsets will therefore contain the following information (the symbols *SS* and *ES* are used to mark the start and end of utterances respectively):

```
(IS (PRESENT))
(PRESENT (ES))
(ARE (PRESENT))
(CELLS (ARE))
(BLOOD (IS CELLS))
(LARGE (BLOOD))
(SS (BLOOD LARGE))
```

The syntactic side of the model records very broad facts about possible sequences of categories. The two strings of words above will produce the following FSM:

```
(0 NIL (ADJ 3) (NOUN 5))
(1 T)
(2 NIL (ADJ 1))
(3 NIL (NOUN 4))
(4 NIL (NOUN 5))
(5 NIL (VERB 2))
(6 NIL (VERB 3))
```

Each state records an identification index, a flag recording whether or not it is a terminal state, and lists of possible categories and following states. DUALGRAM uses the Bigram and this FSM in parallel; when recognising, strings of words effectively have to be acceptable to both models. In comparison to a syntactic model alone, this clearly reduces the number of possible words which can be hypothesised by the language model as specific lexical information is being included in addition to syntactic categorisation. However, the model is also more restrictive than a Bigram alone. Note, for example, that the Bigram model will accept an utterance such as **large blood is present* while DUALGRAM will reject this as the sequence of syntactic categories is not legal. Clearly, this is an unrealistic example; the sequence ADJ NOUN VERB ADJ would normally be perfectly well-formed. However, in practice, DUALGRAM does impose reasonable restrictions of this kind on lexical sequences as the figures which are provided in section 4 indicate.

Before looking at the results, there is another possible complication to DUALGRAM which should be described. The model was extended to allow for certain function words to be effectively ignored in the Bigram. This modification was introduced in order to permit certain kinds of generalisations from existing data; for example, wherever numbers occur in the corpus, as is common in NPs such as *eight millilitres of cloudy fluid*, it is clearly the case that many numbers can be substituted for *eight*. Similarly, it is arguable that determiners such as *the, this, each*, and so on, are mutually interchangeable. Such variation is not possible in any model which is based directly on lexical items and in order to capture this intuition, the notion of a "jumpset" was introduced to DUALGRAM. In this model, certain function words are specified as belonging to the jumpset. These are then effectively extracted from the corpus before the Bigram is generated; thus, for instance, where the corpus contains an NP such as *a large deposit of urothelial cells* and the jumpset consists of determiners and prepositions, the information in the Bigram will be as follows:

```
(SS (LARGE))
(LARGE (DEPOSIT))
(DEPOSIT (UROTHELIAL))
(UROTHELIAL (CELLS))
(CELLS (ES))
```

Recognition with this model proceeds as before except that when the syntactic FSM hypothesises a determiner or preposition, the lexical Bigram is not traversed. Hence at the beginning of the NP above, the FSM proposes any determiner and the Bigram imposes no restriction; this will therefore allow NPs such as *the large deposit of urothelial cells* to be recognised also. This model has been built and tested and the results are reported below.

Finally, it was felt that building the syntactic FSM directly from the corpus is overly restrictive and results in too many rejected strings. Another model was therefore tested which uses a 4-Gram of syntactic tags in tandem with the Bigram; the operation of the resulting DUALGRAM is precisely the same as described above.

4. Results

This section tabulates the results of running the language models proposed in the preceding sections. In each table, as outlined above, the corpus section refers to the parts of the second half of the corpus which were used as test data. The perplexity grows as the models become larger when the new data is added to the existing material. Firstly, the simple Bigram:

Corpus Section	Percentage of Unseen Data Recognised	Perplexity
1	33.7	58.3
2	41.1	67.8
3	45.9	75.9
4	49.6	82.5

Table I. Bigram model results

The main points to note are the final figures (i.e. for corpus section 4); the largest model has around a 50% chance of recognising unseen data with a branching factor of 82.5.

Similar tests were done using a Trigram:

Corpus Section	Percentage of Unseen Data Recognised	Perplexity
1	15.5	17.6
2	20.3	19.7
3	22.4	21.2
4	23.7	22.6

Table II. Trigram model results

The final Trigram model thus has a slightly less than one in four chance of describing unseen input and a significantly lower perplexity than the equivalent Bigram. The following results were obtained using a DUALGRAM in which the syntactic FSM is a 4-Gram:

Corpus Section	Percentage of Unseen Data Recognised	Perplexity
1	32.1	46.8
2	40.0	54.7
3	44.0	59.3
4	47.9	65.7

Table III. DualGram results with 4-Gram FSM and no jumpset

The results indicate that this DUALGRAM, as would be expected, falls between Bigrams and Trigrams. However, the model is much closer to a Bigram both in terms of unseen material described (finally 47.9% in comparison to 49.6% for the Bigram) and perplexity (65.7 as against 82.5). The reduction in perplexity, however, could be a significant factor in recognition accuracy, and it is arguable that the slightly lower chance of handling unseen material is acceptable. Two final results are presented below, in this case for DUALGRAMs in which the syntactic FSM is constructed from whole utterances. Firstly, with no jumpset:

Corpus Section	Percentage of Unseen Data Recognised	Perplexity
1	14.5	40.4
2	22.0	48.8
3	21.9	55.7
4	26.5	61.4

Table IV. DualGram results with full string FSM and no jumpset

These results are arguably the worst so far; the final model accepts only a little more unseen input than the Trigram (26.5% as against 23.7%) with a significantly worse perplexity (61.4 as against 22.6). However, the main reason for including it here is to provide comparison with a DUALGRAM containing a jumpset:

Corpus Section	Percentage of Unseen Data Recognised	Perplexity
1	18.0	104.6
2	23.4	118.5
3	23.2	130.7
4	27.2	140.6

Table V. DualGram results with full string FSM and jumpset

Perhaps surprisingly, this last model only marginally increases the coverage of unseen data at the expense of a large perplexity. It is possible that this is largely an artefact of the Cytopathology domain in which, for example, most of the cardinal numbers are actually found in the necessary environments; in other words, small amounts of corpus do contain instances of *two millilitres, three millilitres*, and so on.

5. Conclusions

We conclude that the DUALGRAM model which uses a 4-Gram FSM is a potentially useful device. It would be possible, of course, to extend the FSM in order to introduce further restrictions if required. More work is necessary if the DUALGRAM with jumpsets is to be usable. It would be possible to partition function words further into those which are not predicted in the Bigram and those which are and to stipulate that the latter set do not themselves predict subsequent lexical items. Thus, for instance, processing the NP *a deposit of cells* would result in a model in which *deposit* predicts both *of* and *cells*. This should reduce the perplexity while retaining the intuition that function words do not provide reliable information on subsequent lexical items.

Not included here is an analysis of the rates at which the successful description of unseen material increases. This study will provide further information on the usefulness of each model. Finally, we intend to add a statistical dimension to the DUALGRAM model as soon as possible as it is arguable, though unlikely, that the results of the comparisons will alter when probabilities are taken into account.

References

1. Black, A.W.: Finite State Machines from Features Grammars. Proceedings of International Workshop on Parsing Technologies, Carnegie Mellon University, Pittsburgh, PA. (1989)

2. Senders, W., Kugler, M., and Boves, L.: Simultaneous optimisation of several variables in a probabilistic language model. Proceedings of Eurospeech 89, Paris, Sept 1989. Vol II, pp.63-66.

Strategies for Speech Recognition and Understanding using Layered Protocols

M. Martin Taylor and Joyce van de Vegte

Defence and Civil Institute of Environmental Medicine, 1133 Sheppard Ave W., P.O. Box 2000, North York, Ontario, Canada M3M 3B9

Abstract

The theory of Layered Protocols, developed for more general description of intelligent communication and design of human-computer interfaces, is proposed as a framework for integrating syntactic, semantic, and pragmatic information of various types in a speech understanding system. A 9-element protocol structure is described, and the interaction of protocols in a processing network is suggested as a mechanism for understanding acoustic messages. Hidden Markov Models are a special case of the proposed structure.

Keywords: communication protocol, information fusion, processing framework

Introduction

We present a conceptual framework (called *Layered Protocols* or LP) within which different sources of information about the data and context of speech can be identified and integrated for use in speech recognition and understanding. LP theory was developed for different purposes: the description of communication between intelligent entities, and the design of human-computer interfaces. It requires no modification, however, to be applicable in integrating the information available in contextually-situated speech.

Early versions of LP theory, cast in terms of human-computer interaction, have been presented in [4–6]. Here we use some more recently developed features of the theory, but the cited papers should be consulted for more detail of the main theory.

Speech Recognition

Syntax-based approaches to speech recognition are based on the idea that a set of rules must exist that determine the permissible patterns of speech elements. But in 2000 years scholars have failed to agree on such a set of rules even for word patterns in a single natural language. Therefore, syntax-based approaches seem unlikely by themselves to provide human-like performance in understanding natural dialogue. They may be more useful when applied to smaller speech elements such as phones than to larger ones such as words.

Although syntactic rule sets cannot be taken as definitive descriptions of what can and cannot be said, nevertheless they do describe most of what actually is said. But often, especially in spoken communication, apparently obligatory items such as subjects, verbs, or objects may not occur, if the talker assumes that the "omitted" elements would be apparent to the intended listener. Deviations from syntactic "rules" usually follow meta-rules that have conventional interpretations. A complete speech understanding system would need to intepret every nuance that the talker incorporated in the acoustic signal. Such a system would have to

NATO ASI Series, Vol. F 75
Speech Recognition and Understanding. Recent Advances
Edited by P. Laface and R. De Mori

incorporate sophisticated models of the real-world situation, the task, the dialogue, and the partner (including the partner's corresponding models) [1].

Stochastic variation is the underlying theme of most of the best present-day speech recognizers. A word can be recognized because the acoustic data are statistically more likely to come from an intent to produce that word than any other word known to the recognizer. The words "statistically likely" cover the enormous range of variation among the stochastic techniques, from simple spectral template-matching to Dynamic Time Warping, Hidden Markov Modelling, and cache-based perplexity reduction [3]. Stochastic methods may substitute for ignorance of the actual rules, or there may be intrinsic stochastic variation in the phenomena of speech. But results to date suggest that it is better to apply stochastic methods than to use inadequate rules.

There is no clear boundary between syntactic and stochastic recognition techniques. One recognizer may use Markov modelling to determine a probability distribution for the next word, while using linguistically-based phonotactic rules to obtain probable phoneme sequences from the acoustic input. Another may use spectral template matching to provide probabilities for demisyllables but a prescribed syntax to specify permissible word sequences. In speech understanding systems, also, syntactic and stochastic approaches may be combined. In the LP approach, both syntactic and stochastic elements are potentially present at each stage of transforming the acoustic input into an internal knowledge representation.

Overview of Layered Protocol Theory

Layers

LP theory has two main constructs: the Layer, and the Protocol. Protocols allow the conceptual coordination of syntactic and stochastic aspects of recognition and understanding. Layers provide the skeleton of the structure, as shown in Figure 1. One partner (A) wants to get some information to the other (B), but since they do not have telepathic communication, the information must be encoded into a message that passes by means of an interface between the two. A's initial conceptual encoding must be recoded in several stages or levels of abstraction, perhaps proposition, phrase, word, and phoneme, before it can be transcribed into the mechanics of voice production and thence to an acoustic wave that can be intercepted by B's ear.

There are psychological and mathematical reasons why the encoding should be layered, depending in large part on the ability of B to provide timely feedback at the different levels of abstraction [6]. Delay in feedback systems is an invitation to instability. To reduce the likelihood of instability, the error-reducing redundancy in coding is partitioned, some aspects covering interactions over long stretches of the speech, some only short stretches. The topic

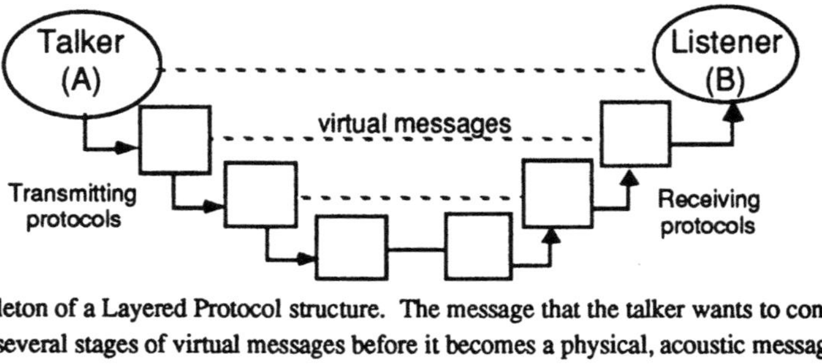

Figure 1. The skeleton of a Layered Protocol structure. The message that the talker wants to convey must be encoded through several stages of virtual messages before it becomes a physical, acoustic message

of conversation induces long-range redundancy, phrase syntax induces mid-range, and phonotactic constraints induce short range. Feedback can usually be provided at each higher layer rapidly enough to ensure that instability at that layer is unlikely. When feedback is unduly delayed, communication problems often arise.

Each layer in a layered protocol represents a recoding of the main message, but the form of this recoding may not be evident to the casual observer. A phoneme is obviously different from a word, and an acoustic vibration is different from a phoneme, but it is not always so clear that a speech act is different from a phrase, since both are expressed in the same words. In the LP analysis, however, the difference between speech act and phrase is as distinct as that between word and phoneme.

Protocols and Protocol Nodes

The second main construct of LP theory is the Protocol, which can be considered as both process and structure. As a process, a protocol includes the patterns in which elements of a one-way virtual message can be arranged—its syntax—and the ways in which the partners are likely to respond to each other—the General Protocol Grammar (GPG) [5].

The structure that executes the protocol process is called a *Protocol Node*, of which there are two complementary kinds: transmitting and receiving. A transmitting node encodes a higher-level virtual message into a less abstract virtual message, whereas a receiving node inverts the process. A passive speech understanding system consists of a network of receiving nodes.

Protocol Nodes are nodes in a network of potential and actual mutual support. In any specific recoding hierarchy, only a subset of the potential support links is used. The pattern of support active at any given moment is the called the Thread network. The part of the Thread network that supports a particular high-level message is called the Thread for that message. The Thread is crucial in analyzing the integration of information for speech recognition and understanding. It contains much of the dynamically-varying information that affects what is spoken.

Structure of a Protocol Node

The information required for executing a protocol has three time-scales: general capability (how to perform the protocol), dialogue-related (the situational context and the history and predicted future of the dialogue), and message-related (what has been received relating to this particular message, and both parties' beliefs about its reception). We find it convenient to treat the protocol node as if it consisted of three sections, corresponding to these three time scales. We call these three sections the Base, Thread, and Active Nodes. Thread Nodes are connected into the Thread Network, which supports a particular dialogue (or rather, a particular high-level message whose transmission involves a dialogue).

The three sections of a protocol node form one conceptual breakdown of the node. We also make an orthogonal breakdown of the protocol node (and thus each of its sections) into three conceptual components. The two most obvious components are the coder and the decoder, which transform higher-level messages into lower-level ones, and *vice versa*. But effective coding and decoding demand models of the partner's corresponding decoding and encoding processes and states. The protocol node therefore has three components: coder, decoder, and model. Each component has aspects of capability, dialogue-related knowledge, and message-related knowledge, and therefore each component can be seen as separated into

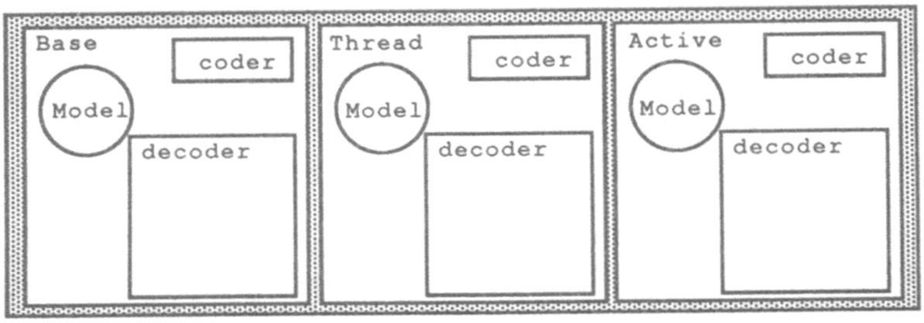

Figure 2. The 9-element structure of a protocol node. The Base node contains the static capabilities and data used by the protocol. The Thread node contains dynamic information that changes slowly relative to the duration of a single message, including dialogue history and real-world pragmatic information. The Active node contains everything that pertains only to the message currently being coded or decoded

the three sections: Base, Thread, and Active, making nine conceptual components in all (Figure 2).

An HMM uses only two or three of the nine components. The hidden Markov process may transform spectral template sequences into phoneme probabilities for the visible process. In the LP view, the decoder of the lower protocol contains a set of Markov process descriptions in its Base node, and so may the decoder of the higher protocol, as shown in Figure 3. Incoming data are compared against possible initial spectral templates, with their associated *a priori* probabilities. The probabilities of the visible elements in the higher process also come into play, as each of the lower, hidden, processes is associated with one of the higher elements.

In the HMM, all the probability distributions are fixed over the duration of the dialogue, although there are dynamic shifts that depend on the phonetic segments or sequences analyzed. These shifts express one level of the long-range interactions implied by the layered coding. But, for better speech recognition, other information might be at hand that could also change the probabilities, and the 9-element depiction of the protocol node can help to determine where such information should be applied.

In natural human conversation, stress is a signal as to whether a word represents a new or old topic—whether it is the *given* or the *new* part of a sentence. Words found in a cache such as that proposed by Kuhn and DeMori [3] are likely to be given rather than new, or to be function words in a recently used syntactic structure. Hence cached words are likely to be destressed, which would change their spectral and temporal character. In the LP theory, this interaction implies some information sharing in each direction across levels. If neutralized vowels or levelled pitch structures are detected at the spectral level, a marker *destressed* should probably be passed to the word level, inducing an increase in the probability that the word would be found in the Thread decoder cache. Conversely, the existence of a word in the cache should increase the probability assigned to the relevant spectral Markovian sequences for the word, and at the same time bias the spectral distribution probabilities toward neutralized phonemes.

Such information sharing may seem at first sight to be contrary to the segregation implied by the Layered Protocol structure, but in fact it is not. The Models are all seen as views on a kind of blackboard, in the manner of Hearsay [2]. A Model in any of the three nodes of a particular protocol represents just those aspects of the blackboard that are useful in the protocol. In particular, the Thread model contains the relevant dynamic states that change more slowly than the messages handled by its protocol. For example, the spectral-level Thread model could contain *p(word A(destressed))* and possibly *p(word B|word A)*, which are

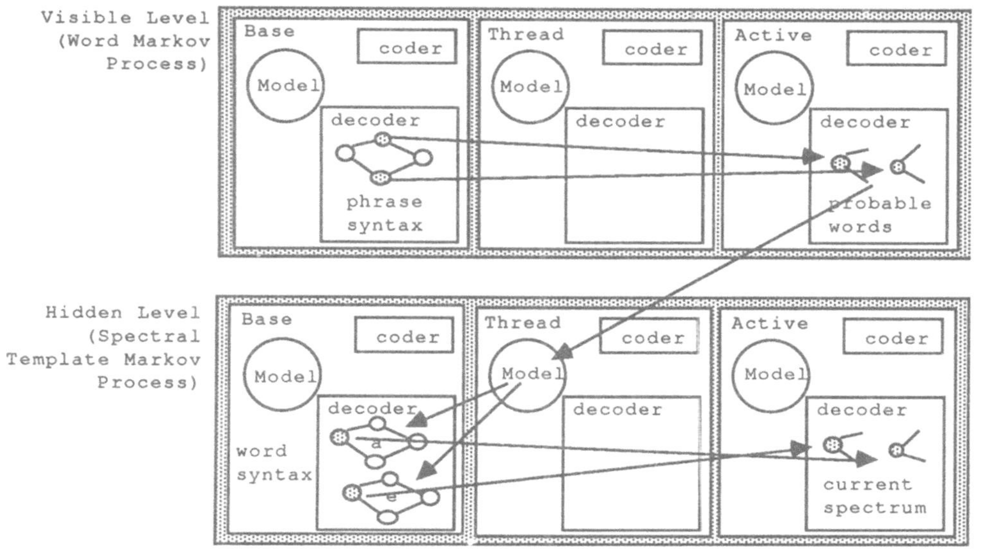

Figure 3. Hidden Markov Model as a pair of Layered Protocol nodes. The expected words in the upper protocol form part of the Thread model of the lower protocol, and serve to select the appropriate Markov processes for decoding the spectral sequence that arrives

expectations that might be contained in the supported word-level decoder. Depending on the extent to which the decoding was data-driven or expectation-driven, these probability estimates might have a small or a large influence on its result. In situations of low SNR they would have a large influence, but at high SNR their influence would be less.

When we look at speech understanding rather than speech recognition, the LP structures are the same. Only the content of the various protocol elements changes. Markovian processes become subservient to processes that we would attribute to the protocol Models, in particular models of the partner, models of the task situation, models of the behaviour of real-world objects and processes, and so forth. Decoded messages often change the contents of these models, which are pragmatic views of the world. If, pragmatically, the listeners knows that the time is near sunset, and the present sky is an object of discourse, then the probability of the word "red" is enhanced.

Syntax, Semantics, and Pragmatics

We propose that "syntax," "semantics," and "pragmatics" are aspects of every protocol, rather than of language as a whole. Seen this way, they maintain their original intent while acquiring clear and consistently separable meanings. We define "syntax" as the set of permitted patterns of elements in messages produced by or recognized by a protocol. If the patterns have associated probabilities, the probabilities also form part of the syntax for the protocol. Phonotactic rules are in this view an aspect of phonetic syntax, and rhetoric rules are an aspect of the syntax of argument. "Semantics" applies to the set of ways in which a particular higher-level message can be encoded. For example, the concept of *ship* can be encoded as "ship," "schooner," "she," "the old hulk," "Betsy II," and so forth. "Semantics" also means the inverse: the set of higher-level messages that can be encoded using a particular lower-level message as part of the encoding. "Ship" can represent a floating hotel, a seagoing gun platform, the act of sending materials to a destination, to go aboard a boat, or a boat taking on water over the side. "Syntax" and "Semantics" are both attributes of the Base coder or decoder of a protocol.

"Pragmatics" applies to the state of the Thread model that affects the selection among syntactic or semantic possibilities. Since many parts of the Thread model have to do with the "real world," pragmatics at the higher protocol levels has to do with whether the message makes sense in the present situational context, as it does in customary usage. But the LP view of pragmaticst goes beyond just the real world situation, inasmuch as the Thread model includes the necessary modelling of the dialogue state and of the partner (including the partner's view of the real world, the dialogue and of oneself).

In addition to these three concepts, a fourth, normally more humble, linguistic concept is elevated to a parallel status: *anaphora*, which refers to information about the recent use of each protocol. The most recent message element of each type distinguished by the syntax is held in the Thread node, in the coder or decoder depending on whether the protocol is transmitting or receiving. Kuhn and DeMori's [3] cache provides this kind of anaphoric referent in the Thread decoder of the word-level protocol.

Anaphora are not limited to words. They become more common at higher protocol levels, and can be a key to understanding the intentions of a cooperative talker. One does not normally repeat information one has just provided, unless one wishes to suggest that the listener failed to appreciate it. One assumes that it remains in the listener's focus.

Conclusion

We have tried to present enough of an overview of the Layered Protocol model to permit the reader to judge how and where specific kinds of information can be used in speech recognition and understanding. How the information is integrated depends to a large extent on the mechanisms used for individual decoders. But the use of the LP structure may help designers to determine what information they can profitably use, and what they can not or should not use. We feel that this structure will prove most valuable when speech is used in an interactive interface, but it may prove helpful even in more abstract studies of how speech can be recognized and understood.

References

1. Edwards, J. L. and Mason, J. A. Toward intelligent dialogue with ISIS. Int J. Man-Machine Studies, 28, 309–342 (1988)
2. Erman, L. D, Hayes-Roth, F., Lesser, V.R., and Reddy, D. R.: The Hearsay-II speech-understanding system: Integrating knowledge to resolve uncertainty. ACM Computing Surveys, 12, 213–253 (1980)
3. Kuhn, R. and DeMori, R. A cache-based natural language model for speech recognition. IEEE PAMI, (In Press, June 1990)
4. Taylor, M. M. Layered Protocols for computer-human dialogue. I: Principles. Int. J. Man-Machine Studies, 28, 175–218 (1988a)
5. Taylor, M. M. Layered Protocols for computer-human dialogue. II: some practical issues. Int. J. Man-Machine Studies, 28, 219–257 (1988b)
6. Taylor, M. M. Response timing in layered protocols: a cybernetic view of natural dialogue. In: The Structure of Multimodal Dialogue (M.M. Taylor, F. Néel, and D. G. Bouwhuis, eds.), Amsterdam: Elsevier Science Publishers B.V. (North Holland) (1989)

Part 5

Understanding and Dialogue Systems

TINA: A Probabilistic Syntactic Parser for Speech Understanding Systems[1,2]

Stephanie Seneff

Spoken Language Systems Group, Laboratory for Computer Science, Massachusetts Institute of Technology, Cambridge, Massachusetts 02139

Abstract: A new natural language system, TINA, has been developed for applications involving speech understanding tasks, which integrates key ideas from context free grammars, Augmented Transition Networks (ATN's) [12], and Lexical Functional Grammars (LFG's) [2]. The parser uses a best-first search strategy, with probability assignments on all arcs obtained automatically from a set of example sentences. An initial context-free grammar, derived from the example sentences, is first converted to a probabilistic network structure. Control includes both top-down and bottom-up cycles, and key parameters are passed among nodes to deal with long-distance movement and agreement constraints. The probabilities provide a natural mechanism for exploring more common grammatical constructions first. Arc probabilities also reduced test-set perplexity by nearly an order of magnitude. Included is a new strategy for dealing with movement, which can handle efficiently nested and chained gaps, and rejects crossed gaps.

Keywords: Natural Language, Speech Understanding, Parsing

Introduction

Most syntactic parsers have been designed with the assumption that the input word stream is deterministic: i.e., at any given point in the parse tree it is known with certainty what the next word is. As a consequence, these parsers generally cannot be used effectively, if at all, to provide syntax-directed constraint in the speech recognition component of a spoken language system. In a fully integrated system, the recognizer component should only be allowed to propose partial word sequences that the natural language component can interpret. Any word sequences that are syntactically or semantically anomalous should probably be pruned prior to the acoustic match, rather than examined for approval in a verification mode. To operate in such a fully integrated mode, a parser should have the capability of considering a multitude of hypotheses simultaneously. The control strategy should have a sense of which of these hypotheses, considering both linguistic and acoustic evidence, is most likely to be correct at any given instant in time, and to pursue that hypothesis only incrementally before reexamining the evidence. The linguistic evidence

[1]This paper first appeared in the proceedings of the first DARPA Speech and Natural Language Workshop, held in Philadelphia, PA, 21-23 February, 1989.

[2]This research was supported by DARPA under Contract N00039-85-C-0254, monitored through Naval Electronic Systems Command.

NATO ASI Series, Vol. F 75
Speech Recognition and Understanding. Recent Advances
Edited by P. Laface and R. De Mori

should include probability assignments on proposed hypotheses; otherwise the perplexity of the task is much too high for practical recognition applications.

This paper describes a natural language system, TINA, which addresses many of these issues. The grammar is constructed by converting a set of context-free rewrite rules to a form that merges common elements on the right-hand side (RHS) of all rules sharing the same left-hand side (LHS). Elements on the LHS become parent nodes in a family tree. Through example sentences, they acquire knowledge of who their children are and how they can interconnect. Such a transformation permits considerable structure sharing among the rules, as is done in typical shift-reduce parsers [11] . Probabilities are established on arcs connecting pairs of right siblings rather than on rule productions. We believe this is a more reasonable way to use probabilities in a grammar. Context-dependent constraints to deal with agreement and gaps are realized by passing features and flags among immediate relatives.

General Description

TINA is basically a context-free grammar, implemented by expansion at run-time into a network structure, and augmented with flags/parameters that activate certain filtering operations. The grammar is built from a set of training sentences, using a bootstrapping procedure. Initially, each sentence is translated by hand into a list of the rules invoked to parse it. After the grammar has built up a substantial knowledge of the language, many new sentences can be parsed automatically, or with minimal intervention to add a few new rules incrementally. The arc probabilities can be incrementally updated after the successful parse of each new sentence.

Figure 1 illustrates the process of converting a set of context-free rules to a probabilistic network structure of *grammar nodes*. All rules with the same LHS are combined to form a structure describing possible interconnections among children of a parent node associated with the left-hand category. A probability matrix connecting each possible child with each other child is constructed by counting the number of times a particular sequence of two siblings occurred in the RHS's of the common rule set, and normalizing by counting all pairs from the particular left-sibling to *any* right sibling. Two distinguished nodes, a START node and an END node, are included among the children of every grammar node. A subset of the grammar nodes are terminal nodes whose children are a list of vocabulary words.

A functional block diagram of the control strategy is given in Figure 2. At any given time, a distinguished subset of "active" *parse nodes* are arranged on a priority queue. Each parse node contains a pointer to a grammar node of the same name, and has access to all the information needed to pursue its partial theory. The top node is popped from the queue, and it then creates a number of new nodes (either children or right siblings depending on its state), and inserts them into the queue according to their probabilities. If the node is an END node, it collects up all subparses from its sequence of left siblings, back to the START node, and passes the information up to the parent node, giving that node a completed subparse. The process can terminate on the first successful completion of a sentence, or the Nth successful completion if more than one hypothesis is desired.

A parse in TINA is begun by creating a single parse node linked to the grammar node SENTENCE, and entering it on the queue with probability 1.0. This node creates new parse nodes with categories like STATEMENT, QUESTION, and REQUEST, and

Three Context-free Rewrite Rules:

NP ==> ARTICLE NOUN
NP ==> ARTICLE ADJECTIVE NOUN
NP ==> ARTICLE ADJECTIVE ADJECTIVE NOUN

The Resulting Probabilistic Network:

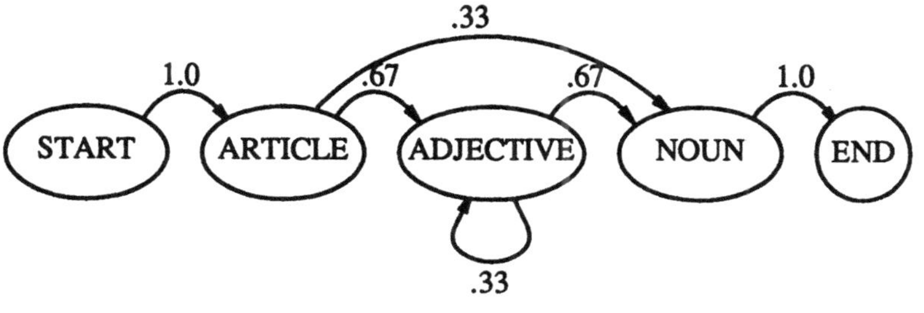

Figure 1. Illustration of Process to Convert Context-free Rules to probabilistic Network Form

places them on the queue, prioritized. If STATEMENT is the most likely child, it gets popped from the queue, and returns nodes indicating SUBJECT, IT, etc., to the queue. When SUBJECT reaches the top of the queue, it activates units such as NOUN-GROUP (for noun phrases and associated post-modifiers), GERUND, and NOUN-CLAUSE. Each node, after instantiating first-children, becomes inactive, pending the return of a successful subparse from a sequence of children. Eventually, the cascade of first-children reaches the terminal-node ARTICLE, which proposes the words "the," "a," and "an," testing these hypotheses against the input stream. If a match with "the" is found, then the ARTICLE node fills its subparse slot with the entry (ARTICLE "the"), and activates all of its possible right-siblings.

Whenever a terminal node has successfully matched an input word, the path probability is reset to 1.0. Thus the probabilities that are used to prioritize the queue represent not the *total* path probability but rather the probability *given* the partial word sequence. Each path climbs up from a terminal node and back down to a next terminal node, with each new node adjusting the path probability by multiplying by a new conditional probability. The resulting conditional path probability for a next word represents the probability of that word in its syntactic role given all preceding words in *their* syntactic roles. With this strategy, a partial sentence does not become increasingly improbable as more and more words are added.[3].

[3]Some modification of this scheme will be necessary when the input stream is not deterministic. See [10] for a discussion of these very important issues regarding scoring in a best-first search.

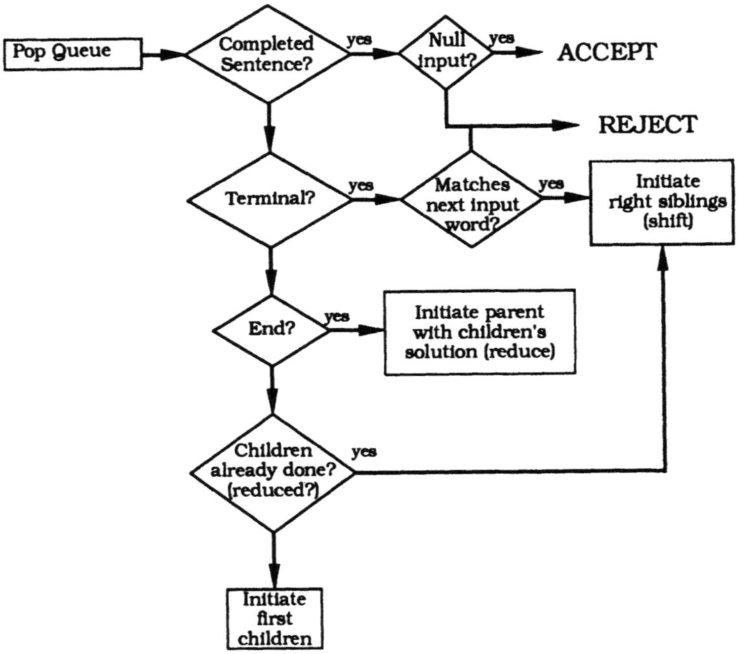

Figure 2. Functional Block Diagram of Control Strategy

Natural Language Issues

This section describes how TINA handles agreement constraints and long distance movement, issues that are usually considered to be part of the task of a syntactic parser. Movement concerns a phenomenon of displacing a unit from its natural position in a phrase, usually to a preceding position. Such "gaps" occur commonly, for instance, in questions and passive voice, as in "(Which article)$_i$ do you think I should read (t_i)?".[4] TINA is particulary effective in handling gaps. Complex cases of nested or chained gaps are handled correctly, and most ill-formed gaps are rejected. The mechanism resembles the "hold" register idea of ATN's [12] and the treatment of bounded domination metavariables in LFG's ([2], p. 235 ff), but seems to be more straightforward than both of these.

Each parse node comes equipped with a number of slots for holding information that is relevant to the parse. Included are person and number, verb-form (*root*, *finite*, etc.) and two special slots, the current-focus and the float-object, that are concerned with long-distance movement. This information is passed along from node to node: from parent to child, child to parent, and left-sibling to right-sibling. Certain nodes have the power to adjust the values of these features. The adjustment may take the form of an unconditional override, or it may be a constraint that must have a non-null union with the value for that feature passed to the node from its relative, as will become clear in the next section.

[4] *Which article*, the object of the verb "read," has been pulled out of place and moved to the front of the question.

Verb-form and Agreement

Certain nodes have special powers to set the verb-form either for their children or for their right-siblings. Thus, for example, HAVE as an auxilliary verb sets verb-form to *past-participle* for its *right-siblings*. The category GERUND sets the verb-form to *present-participle* for its *children*. Whenever a PREDICATE node is invoked, the verb-form has always been set by a predecessor.

Certain nodes specify person/number restrictions which then propagate up to higher levels and back down to later terminal nodes. Thus, for example, A NOUN-PL node sets the number to [*PL*], but only if the left sibling passes to it a description for number that includes [*PL*] as a possibility (otherwise it dies, as in "each boats"). This value then propagates up to the SUBJECT node, across to the PREDICATE node, and down to the verb, which then must agree with [*PL*], unless its verb-form is marked as non-finite. A more complex example is a compound noun phrase, as in "Both John and Mary have decided to go." Here, each individual noun is singular, but the subject expects a plural verb (*have* rather than *has*). TINA deals with this by making use of a node category AND-NOUN-PHRASE, which sets the number constraint to [*PL*] *for its parents*, and blocks the transfer of number information *to its children*. The OBJECT node blocks the transfer of any predecessor person/number information to its children, reflecting the fact that verbs agree in person/number with their subject but not their object.

Gaps

The mechanism to deal with gaps involves four special types of grammar nodes, identified as *generators*, *activators*, *blockers*, and *absorbers*. *Generators* are parse nodes whose grammatical category allows them to fill the current-focus slot with the subparse returned to them by their children. The current-focus is passed on to right-siblings and their descendents, but not to parents, and thus effectively reaches nodes that are *c-commanded* by the generator and its descendents [4,9]. *Activators* are nodes that move the current-focus into the float-object slot. They also require that the float-object be absorbed somewhere among their descendants. *Blockers* (such as SUBJECT) are nodes that block the transmission of the float-object to their children. Finally, *absorbers* are allowed to use the float-object as their subparse.

A simple example will help explain how this works. For the sentence "(How many pies)$_i$ did Mike buy (t_i)?" as illustrated by the parse tree in Figure 3, the Q-SUBJECT "how many pies" is a generator, so it fills the current-focus with its subparse. The DO-QUESTION is an activator; it moves the current-focus into the float-object position. Finally, the object of "buy," an absorber, takes the Q-SUBJECT, as its subparse. The DO-QUESTION refuses to accept any solutions from its children if the float-object has not been absorbed. Thus, the sentence "How many pies did Mike buy the pies?" would be rejected. Furthermore, the same DO-QUESTION node deals with the sentence "Did Mike buy the pies?," except in this case there is no current-focus and hence no gap.

More complicated sentences involving nested or chained traces, are handled staightforwardly by this scheme. For instance, the sentence "(Which hospital)$_j$ was (Jane)$_i$ taken (t_i) to (t_j)?" can be parsed correctly by TINA, identifying "Jane" as the object of "taken" and "which hospital" as the object of "to." This works because the VERB-PHRASE-P-O, an activator, writes over the float-object "Which hospital" with the new entry "Jane," but only for its children. The original float-object is still available to fill the OBJECT slot in the following prepositional phrase.

The example used to illustrate the power of ATN's [12], "John was believed to have been shot," also parses correctly, because the OBJECT node following the verb "believed" acts as both an absorber and a (re)generator. Cases of crossed traces, which are blocked by the Strict Cycle Condition and the Subjacency Condition in the Government/Binding rule system [3] are automatically blocked here because the second current-focus gets moved into the float-object position at the time of the second activator, overriding the preexisting float-object set up by the earlier activator. The wrong float-object is available at the position of the first trace, and the parse dies, as in the following agrammatical sentence:

*(Which books)$_i$ did you ask John (where)$_j$ Bill bought (t$_i$) (t$_j$)?

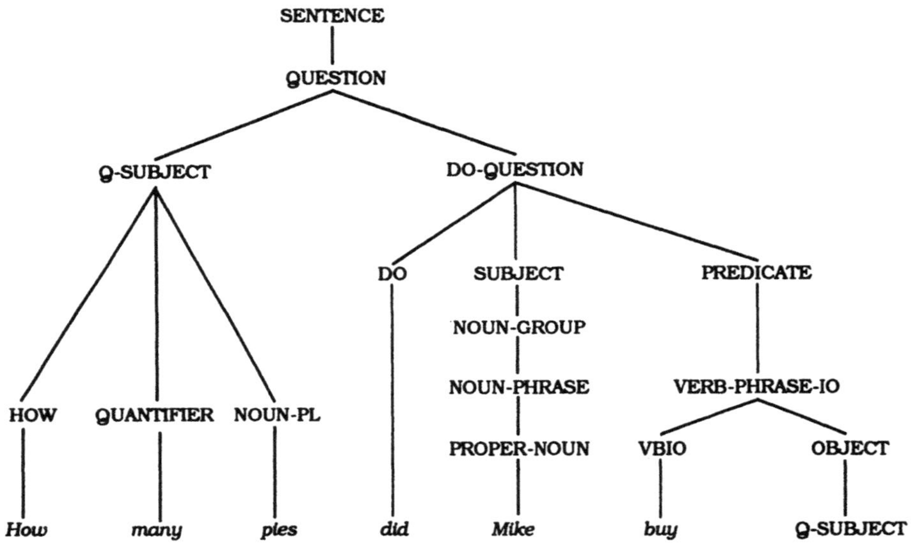

Figure 3. Example of a Parse Tree Illustrating a Gap

The current-focus slot is not restricted to nodes that represent nouns. Some of the generators are adverbial or adjectival parts-of-speech (POS). An absorber checks for agreement in POS before it can accept the float-object as its subparse. As an example, the question, "(How oily)$_i$ do you like your salad dressing (t$_i$)?" contains a Q-SUBJECT "how oily" that is an adjective. The absorber PRED-ADJECTIVE accepts the available float-object as its subparse, but only after confirming that POS is adjective, as shown in the parse tree in Figure 4.

The current-focus has a number of other uses besides its role in movement. It plays an important part in identifying the subject of verbs and in establishing the references for pronouns. For a complete description of these and other topics, the interested reader is referred to [10].

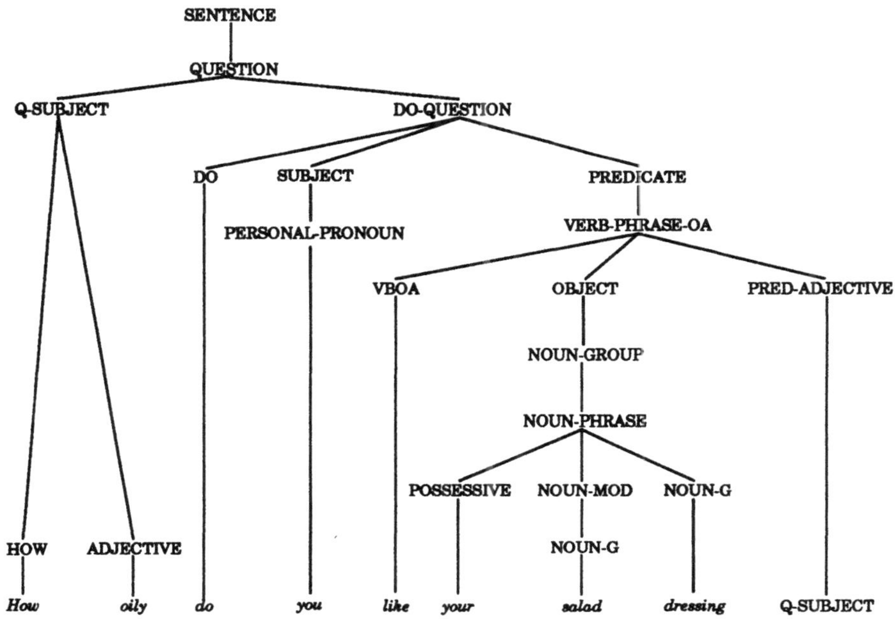

Figure 4. Parse Tree for the sentence, "How oily do you like your salad dressing?"

Evaluation Measures

This section addresses several distinct performance measures for a grammar, including coverage, overgeneration, portability, perplexity and trainability. Coverage/overgeneration are concerned with the degree to which the grammar is able to capture appropriate generalities while rejecting ill-formed sentences. Perplexity, roughly defined as the geometric mean of the number of alternative word hypotheses that may follow each word in the sentence, is of particular concern in spoken language tasks. Portability and trainability concern the ease with which an existing grammar can be ported to a new task, as well as the amount of training data necessary before the grammar is able to generalize well to unseen data.

To address these issues, we used two sets of sentences. The first set is the 450 *compact acoustic-phonetic* sentences of the TIMIT database [7]. These sentences represent a fairly complex syntax, including questions, passive voice, compound and complex sentences, relative clauses, subjunctive form, comparatives, etc. They represent a semantically unrestricted space, which makes it hard to use them for tests of constraint reduction due to semantic filtering. The second set of sentences has become popular in the DARPA speech research community for both speech recognition and natural language processing applications. They concern a naval Resource Management (RM) task, and are fairly restrictive semantically. A particular subset of 791 designated training sentences and 200 designated test sentences from this task have been selected by researchers at Bolt Beranek and Newman, Inc. for studies in natural language. We have used these two sets for testing portability, perplexity and coverage.

One of the unique aspects of TINA is that a grammar can be acquired automatically from a set of parsed sentences. A typical procedure is to gradually build up the rule system by parsing new sentences one by one, introducing new arcs as needed. Once a full set of sentences has been parsed in this fashion, the parse trees from the sentences are automatically converted to the set of rules used to generate each sentence. The training of both the rule set and the probability assignments is established directly from the provided set of parsed sentences; i.e. the parsed sentences *are* the grammar.

We took advantage of this feature to test the system's capability of generalizing to unseen data from a small set of sentence examples. Since there were only 450 sentences in the TIMIT task, we were unwilling to set aside a portion of these as designated test sentences. Instead, we built a grammar that could parse all of the sentences, and then generated a subset grammar from 449 of the sentences, testing this grammar for coverage on the remaining one. We cycled through all 450 sentences in this fashion.

Our experiments were conducted as follows. We first built a grammar from the 450 TIMIT sentences. We tested coverage on these sentences using the above jackknifing strategy. We assessed overgeneration by generating sentences at random from the TIMIT grammar, checking whether these sentences were well-formed syntactically. We then tested portability by beginning with the grammar established from the TIMIT task and then deriving a grammar for the 791 designated training sentences of the RM task. Within this task it was possible to make use of semantic categories, particularly within noun phrases, in order to reduce perplexity. We measured both perplexity and coverage on the remaining 200 test sentences of this task, using a grammar built automatically from the 791 parsed training sentences.

Coverage within TIMIT

The result of the jackknifing experiment was that 75% of the unseen sentences were successfully parsed based on structures seen in the remaining 449 sentences. In most cases where the system failed, a single unique form occurred somewhere in the unseen sentence that had not appeared in any of the other sentences, as illustrated in Table 1. We do not mean to suggest that a grammar should not be able to handle such forms; however, we are encouraged that three quarters of the sentences could parse based on such a small amount of training data. It suggests that the system can learn generalizations fairly quickly.

Table 1. Some examples of sentences that failed in the jackknife experiment

as-as construction:	Withdraw only *as* much money *as* you need.
rather x than y:	I'd *rather* not buy these shoes *than* be overcharged.
why predicate:	*Why* buy oil when you always use mine?
triple verb:	*Laugh, dance, and sing,* if fortune smiles on you.
"are both":	The patient and the surgeon *are both* recuperating from the lengthy operation.
adjunct as subject:	*Right now* may not be the best time for business mergers.

An example of a sentence that succeeded is given in Table 2, along with a list of sentences that could be used to build the portions of the parse tree necessary for the successful parse. The parse of this sentence is given in Figure 4. Because of the sharing of structures among rules with the same LHS, the system is capable of synthesizing new rules from pieces of other rules, which allows it to learn more quickly.

Table 2. Example of generalization capability of parser

New Parsable Sentence:
"How oily do you like your salad dresssing?"

Useful sentences in training set:

Q-subject as Adjective:	"*How permanent* are their records?"
Q-subject Do-question:	"*How do oysters make pearls*?"
Verb Object Predicate-adjective:	"Calcium *makes bones and teeth strong.*"
Possessive Noun-mod Noun:	Gwen planted green beans in *her vegetable garden.*

Table 3. Some sample sentences generated by the parser

Wash, of course, but puree the high hats and article colleges straight ahead.
How did the income of gold open the algebraic hit?
A child of execution stole the previous attitude.
Which tunafish would lots of the muscles smash?
Make a scholastic marriage; then only get ski fangs under a medical film.
Whenever a simple perfume must diminish near stew, enter.
It is fun to pledge to break down.
Hyenas might be used to eat every coach.
The screen is surely blistered occasionally.

Overgeneration

The issue of overgeneration is extremely important for spoken language tasks, because of the need to keep perplexity as low as possible. TINA can be run in generation mode, where, instead of proposing all alternatives at a decision point, a random number generator is used to select a particular decision. Generation mode is an extremely useful tool for discovering errors in the grammar. Randomly generated sentences are typically syntactically correct but semantically anomalous. Occasionally an ill-formed sentence is generated, due to inappropriate generalities in the rules. Usually the situation can be corrected through rule modification.

Since all of the arcs have assigned probabilities, the parse tree is traversed by generating a random number at each node and deciding which arc to take based on the outcome, using the arc probabilities to weight the alternatives. Some examples of sentences generated in this way are given in Table 3. While many of these sentences are clearly nonsense, due to the complete absence of semantic constraint, they all appear to be syntactically representative of the language. It is clear that proper semantic constraint would greatly decrease the perplexity, although, given the rich semantic base of the 450 sentences, it is not reasonable to expect to build a suitable semantic component for them. Applying semantic constraint appropriate in a sublanguage that a natural language system can interpret, however, is a much more feasible undertaking.

Portability

We tested ease of portability for TINA by beginning with a grammar bulit from the 450 TIMIT sentences and then deriving a grammar for the RM task. These two tasks represent very different sentence types. For instance, the overwhelming majority of the TIMIT sentences are statements, whereas there are no statements in the RM task, which is made up exclusively of questions and requests. The process of conversion to a new grammar involves parsing the new sentences one by one, and adding context-free rules whenever a parse fails. The person entering the rules must be very familiar with the grammar structure, but for the most part it is straightforward to identify and incrementally add missing arcs. The parser identifies where in the sentence it fails, and also maintains a record of the successful partial parses. These pieces of information usually are adequate to pinpoint the missing arcs. It required less than one person-month to convert the grammar from TIMIT to the RM task.

Perplexity and Coverage within RM Task

We built a subset grammar from the 791 parsed RM training sentences, and then used this grammar to test coverage and perplexity on the unseen 200 test sentences. The grammar could parse all of the training sentences and 78.5% of the test sentences. We are unwilling to examine the test sentences, as we may be using them for further evaluations in the future. Therefore, we cannot yet assess why a particular sentence failed, or whether the parse found by the grammar was actually the correct parse.

A formula for the test set perplexity is [8]:

$$Perplexity = 2^{-\frac{1}{N}\sum_{i=1}^{N} log_2 P(w_i|w_{i-1},...w_1)}$$

where the w_i are the sequence of all words in all sentences, N is the total number of words, including an "end" word after each sentence, and $P(w_i|w_{i-1},...w_1)$ is the probability of the ith word given all preceding words.[5] If all words are assumed equally likely, then $P(w_i|w_{i-1},...w_1)$ can be determined by counting all the words that could follow each word in the sentence, along all workable partial theories. If the grammar contains probability estimates, then these can be used in place of the equally-likely assumption. If the grammar's estimates reflect reality, the estimated probabilities will result in a reduction in the total perplexity.

An average perplexity for the 157 test sentences that were parsable was computed for the two conditions, without (Case 1) and with (Case 2) the estimated probabilities. The result was a perplexity of 374 for Case 1, but only 41.7 for Case 2. This is with a total vocabulary size of 985 words, and with a grammar that included several semantically

[5]in the case of TINA, all words up to the current word within each sentence are relevant.

restricted classes such as SHIP-NAME and READINESS-CATEGORY. The incorporation of arc probabilities reduced the perplexity by a factor of nine, a clear indicator that a proper mechanism for utilizing probabilities in a grammar can help significantly.

Discussion

This paper describes a new grammar formalism that addresses issues of concern in building a fully integrated speech understanding system. The grammar includes arc probabilities reflecting the frequency of occurrence of the syntactic structures within the language. These probabilities are used to control the order in which hypotheses are considered, and are trained automatically from a set of parsed sentences, which makes it straightforward to tailor the grammar to a particular need. Ultimately, one could imagine the existence of a very large grammar that could parse almost anything, which would be subsetted for a particular task by simply providing it with a set of example sentences within that task.

The issue of semantic analysis has not yet been dealt with properly within TINA. We were able to achieve significant perplexity reduction by defining some semantically restricted classes, particularly within noun phrases, but this is probably not the appropriate way to represent semantic constraint. I would rather see semantic filters introduced through co-occurrence data on syntactic pairs like Adjective-Noun and Subject-Verb, adopting methods similar to those proposed in [5,6].

At the time a set of word candidates is proposed to the acoustic matcher of a recognizer, all of the constraint available from the restrictive influence of syntax, semantics, and phonology should have already been applied. The syntactic parse tree of TINA can be used to express other constraints ranging from acoustic-phonetic to semantic and pragmatic. The syntactic node would contain slots for various kinds of constraint information – syntactic filters such as person/number and verb-form, semantic filters such as the permissible semantic categories for the subject/object of the hypothesized verb, and acoustic-phonetic filters (for instance, restricting the word to begin with a vowel if the preceding word ended in a flap). As the parse tree advances, it accumulates additional constraint filters that further restrict the number of possible next-word candidates. Thus the task of the predictive component is formulated as follows: given a sequence of words that has been interpreted to the fullest capability of the syntactic/semantic/phonological components, wnat are the likely words to follow, and what are their associated a priori probabilities?

While TINA's terminal nodes are lexical words, I believe that the nodes should continue down below the word level. Prefixes and suffixes alter the meaning/part-of-speech in predictable ways, and therefore should be represented as separate subword grammar units that can take certain specified actions. Below this level would be syllabic units, whose children are subsyllabic units such as onset and rhyme, finally terminating in phoneme-like atomic constituents. Acoustic evidence would enter at several stages. Important acoustic matches would take place at the terminal constituents, but duration and intonation patterns would contribute to scores in nodes at many higher levels of the hierarchy.

One issue that has not been addressed here but which I feel is very important is the notion of a *fully* automatic training capability. TINA trains automatically from a set of parsable sentences, but when a sentence fails to parse there is no recourse except human intervention to provide new rules. One possibility is to generate some new rules automatically by applying meta-rules similar to those used in the Unification framework, as in the generation of passive voice forms for verbs. Berwick's research in the area of automatic language acquisition [1] represents an important effort along these lines as well, and it is likely that some of his ideas could be extended to apply in TINA's framework.

We plan to integrate TINA with the SUMMIT speech recognition system [13] shortly. Two important issues are 1) how to combine the scores for the recognition component and the predictive component of the grammar, and 2) how to take advantage of appropriate pruning strategies to prevent an explosive search problem. In the case of the first problem, we would need to modify the mechanism for using path probabilities in TINA, in order to deal with the nondeterministic nature of the acoustic evidence. With regard to the second problem, each parse node in the tree can prune all but the best-scoring equivalent path by maintaining a record of all solutions returned to it by its children, and checking against this list whenever a new solution comes in. With this chart in place, there would not be a duplication of effort for paths beyond that point.

Acknowledgements

This research has benefitted from interactions with Lynette Hirschman, Mark Johnston, and Victor Zue. Christine Pao implemented a preliminary version of this parser, and Jim Glass developed the user interface. I am also grateful to Jim Glass, Mike Phillips, and Victor Zue for critically reading an early draft and offering perceptive comments.

References

1. Berwick, R.: *The Acquisition of Syntactic Knowledge,* The MIT Press, Cambridge, MA 1985.
2. Bresnan, J., ed.: *The Mental Representation of Grammatical Relations*, MIT Press, 1982.
3. Chomsky, N.: "On wh-movement" in P. Culicover, T. Wasow, and A. Akmajian, eds. *Formal Syntax* Academic Press, New York, 1977.
4. Chomsky, N.: *Lectures on Government and Binding*, Second Revised Edition, Foris Publications, Dordrecht, the Netherlands, 1982.
5. Grishman, R., Hirschman,L. and Nhan, N.T.: "Discovery Procedures for Sublanguage Selectional Patterns: Initial Experiments," Computational Linguistics, Vol. 12, No. 3, pp. 205-215, 1986.
6. Hirschman, L., Grishman, R., and Sage, N: "Grammatically-based Automatic Word Class Formation," Information Processing and Management, Vol. 11, pp. 39-57, 1975
7. Lamel, L., Kassel, R.H., and Seneff, S.: "Speech Database Development: Design and Analysis of the Acoustic-Phonetic Corpus," DARPA Speech Recognition Workshop Proceedings, Palo Alto, CA, Feb 19-20, 1986.
8. Lee, K.F.: *Automatic Speech Recognition: The Developemnt of the Sphinx System,* Appendix I, Kluwer Academic Publishers, Boston, 1989.
9. Reinhart, T., *Anaphora and Semantic Interpretation,*" The University of Chicago Press, Chicago, Il, 1983.
10. Seneff, S.: "TINA: A Probabilistic Syntactic Parser for Speech Understanding Systems," LCS Technical Report, Laboratory for Computer Science, MIT, Cambridge, MA, forthcoming.
11. Tomita, M.: *Efficient Parsing for Natural Language,* Kluwer Academic Publishers, Boston, MA, 1986.
12. Woods, W.A.: "Transition Network Grammars for Natural Language Analysis," Commun. of the ACM 13, 591-606, 1970.
13. Zue, V., Glass, J., Phillips, M., and Seneff, S: "The MIT SUMMIT Speech Recognition System: A Progress Report," *These Proceedings,* Feb., 1989.

The Voyager Speech Understanding System: A Progress Report [1,2]

Victor Zue, James Glass, David Goodine, Hong Leung, Michael Phillips, Joseph Polifroni, and Stephanie Seneff

Spoken Language Systems Group, Laboratory for Computer Science, Massachusetts Institute of Technology, Cambridge, MA 02139

Abstract: As part of the DARPA Spoken Language System program, we recently initiated an effort in spoken language understanding. A spoken language system addresses applications in which speech is used for interactive problem solving between a person and a computer. In these applications, not only must the system convert the speech signal into text, it must also understand the linguistic structure of a sentence in order to generate the correct response. This paper describes our early experience with the development of the MIT VOYAGER spoken language system.

Keywords: Speech Recognition, Natural Language Processing, Dialogue Systems, Knowledge Representation.

Introduction

Recently, we have been directing our research effort towards spoken language understanding as part of the DARPA Spoken Language System program. The project is motivated by the belief that many of the applications suitable for human/machine interaction using speech typically involve interactive problem solving. That is, in addition to converting the speech signal to text, the computer must also *understand* the linguistic structure of a sentence in order to generate the correct response. We have focused our attention on three main issues. First, the system must integrate speech recognition with natural language in order to achieve speech understanding. Second, the system must have a realistic application domain, and be able to translate spoken input into appropriate actions. Finally, the system must begin to deal with spontaneous speech, since people do not always utter grammatically well-formed sentences during a spoken dialogue.

Over the past six months, we have constructed the skeleton of a spoken language system. The purpose of this paper is to describe the various components of this system.

[1]This paper first appeared in the proceedings of the second DARPA Speech and Natural Language Workshop, held at Harwichport, MA, 15-18 October, 1989.

[2]This research was supported by DARPA under Contract N00014-89-J-1332, monitored through the Office of Naval Research.

NATO ASI Series, Vol. F 75
Speech Recognition and Understanding. Recent Advances
Edited by P. Laface and R. De Mori

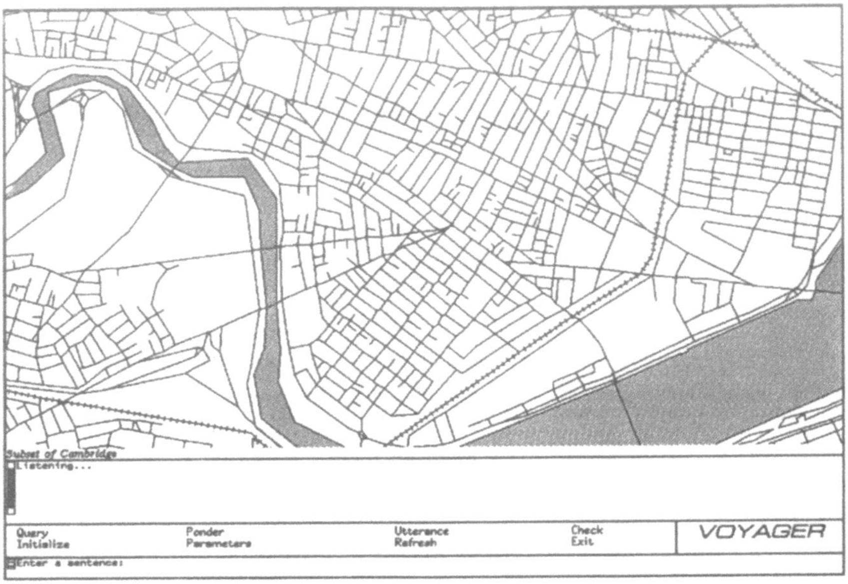

Figure 1. A display showing the geographical region known to the VOYAGER system

In related activities, we have collected a sizeable spontaneous speech database, and have used the data for analyses, system training and evaluation. The collection and analysis of the spontaneous speech database, and the preliminary evaluation of our spoken language system are described in two companion papers that appear elsewhere in these proceedings [6,7].

Task Description

In order to explore issues related to a fully-interactive spoken language system, we have selected a task in which the system knows about the physical environment of a specific geographical area as well as certain objects inside this area, and can provide assistance on how to get from one location to another within this area. The system, which we call VOYAGER, currently focuses on the the city of Cambridge, Massachusetts, between MIT and Harvard University, as shown in Figure 1. It can answer a number of different types of questions about certain hotels, restaurants, hospitals, and other objects within this region. At the moment, VOYAGER has a vocabulary of 324 words. Within this limited domain of knowledge, it is our hope that VOYAGER will eventually be able to handle any reasonable query that a native speaker is likely to initiate. As time progresses, VOYAGER's knowledge base will undoubtedly grow.

System Description

VOYAGER is made up of three components. The speech recognition component converts the speech signal into a set of word hypotheses. The natural language component then provides a linguistic interpretation of the set of words. The parse generated by the natural language component is subsequently transformed into a set of query functions, which are passed to the back-end for response generation. The back-end is an enhanced version of a direction assistance program developed by Jim Davis of MIT's Media Laboratory [1]. We will describe each component in sequence, paying particular attention to those parts that have not been previously reported.

Speech Recognition Component

The first component of VOYAGER uses the SUMMIT speech recognition system developed in our group. SUMMIT places heavy emphasis on the extraction of phonetic information from the speech signal. It achieves speech recognition by explicitly detecting acoustic landmarks in the signal in order to facilitate acoustic-phonetic feature extraction. The system can be trained automatically, since it does not rely on extensive knowledge engineering. The design philosophy, implementation, and evaluation of the SUMMIT system have been described in detail previously [8]. As a result, we will only report in this paper modifications to the system since the last workshop. These include the development of a new module for lexical expansion via phonological rules, and a new corrective training procedure.

Lexical Expansion

The original SUMMIT system used a phonological expansion capability provided to us by SRI [5]. Within the last year, however, we have decided to rewrite this part of the system in order to establish increased flexibility and speed. The new version, named MARBLE, offers several new properties. A canonic set of phonemes is represented by a set of default values for distinctive features. Specified allophonic information due to context dependencies can be represented in the particular instance of the phoneme generated in a word lattice. Thus, for instance, when a word-final /s/ and a word-initial /s/ merge, the resulting /s/ can be marked as [+geminate]. This information can then be incorporated into the scoring for the particular allophone. The allophonic slot can also be used to indicate place of articulation of adjacent consonants, for example, to facilitate the decoding of context-dependent models. The rule-writing process is also straightforward, and it is simple to keep track of the rule ordering. Finally, the time it takes to expand a lexicon has been reduced. We believe this new rule system will be a powerful tool for effectively representing context dependencies.

Corrective Training

The training of SUMMIT is performed iteratively, after being initialized on the TIMIT database [3,8]. For each iteration, the recognizer computes the *best* alignment between a path in the acoustic phonetic network and a path in the lexical network, i.e., the recognized output. The recognizer also computes a *forced* alignment using only the correct string of words. The system then trains the next iteration of phonetic models based on the matches between lexical arcs and phonetic segments in the forced alignments. The recognizer also adjusts lexical arc weights based on a comparison of the number of times the arc was used incorrectly (present in the best alignment but not in the forced alignment) to the number of times the arc was missed (present in the forced alignment but not in the best alignment). The goal of this corrective training procedure is to equalize the number of times an arc is missed and the number of times the arc is used in the wrong way. If sufficient training data is not available for a particular arc, then the weights are derived by collapsing it with other phonetically-similar arcs.

Presently, lexical decoding is accomplished by using the Viterbi algorithm to find the best path that matches the acoustic-phonetic network with the lexical network. Since the speech recognition and natural language components are not as yet fully integrated, we currently use a word-pair language model with a perplexity of 22 to constrain the search space.

Natural Language Component

In the context of a spoken language system, the natural language component should perform two critical functions: 1) to provide constraint for the recognizer component, and 2) to provide an interpretation of the meaning of the sentence to the back end. Our natural language system, TINA, was specifically designed to meet these two needs. The basic design of TINA has been described elsewhere [4], and therefore will only be briefly mentioned here. Instead, we would like to focus on the issue of how to incorporate semantics into the parses. We have found that an enrichment of the parse tree with semantically loaded categories at the lower levels leads to both improved word predictions and a relatively straightforward interface with the back end.

General Description

The grammar is entered as a set of simple context-free rules which are automatically converted to a shared network structure. The *nodes* in the network are augmented with constraint filters (both syntactic and semantic) that operate only on locally available parameters. Typically, several independent hypotheses are simultaneously active, and parameter modifications include protection of shared information, such that a parallel implementation would be possible. Efficient memory use is achieved through a recycling mechanism for the node structures, such that they become available in a resource pool whenever their current assignment is completed. These issues will become important as we move towards a fully integrated system.

One of the key features of TINA is that all arcs in the network are associated with probabilities, acquired automatically from a set of example sentences. It is important to note that the probabilities are established *not* on the rule productions but rather on arcs connecting sibling pairs in a shared structure for a number of linked rules. For instance, all occurrences of SUBJECT are pooled together for probability assignments on their children, regardless of the structural positions of these occurrences within a clause. The effect of such pooling is essentially a hierarchical bigram model. We believe this mechanism offers the capability of generating probabilities in a reasonable way by sharing counts on syntactically/semantically identical units in differing structural environments.

Semantic Filtering

For VOYAGER, we were interested in designing a parser that could handle all reasonable ways a person might request information within the domain, but that would also reject any ill-formed sentences, on the grounds of both semantic and syntactic anomalies. Building such a tight grammar not only leads to a very low perplexity for the recognizer, but also virtually eliminates the problem of multiple parses. This is because all parses that are syntactically legitimate but semantically anomalous are weeded out. It has the added benefit of improving computation time, if the semantic constraints are integrated early in the parsing process, more or less at the first chance of resolution.

We also wanted to maintain our criterion that a node should only have access to information locally available to it by default. That is, it should not be allowed to hunt back through the parse tree looking for a resolution of, for example, number agreement. By default, all nodes pass along the parameters passed to them by near relatives. The hard part is to come up with a compact representation that contains all information necessary to carry out the constraints. In terms of syntax, there are patterns describing properties such as person, number, case, and determiner. Semantics are represented by patterns that include an automatic hierarchical inheritance of broader properties from more specific ones. Thus for example the semantic category *Restaurant* automatically acquires *Building* and *Place* as additional semantic features. In addition to the syntactic features, nodes also pass along semantic features that are automatically reset by designated nodes, such as terminal vocabulary entries.

The two slots, `Current-Focus` and `Float-Object` that are used for dealing with gaps, turned out to also be very useful for providing semantic constraint. In fact, we decided to take the approach of *only* using these two parameters for semantic filtering, to see whether in fact that would be adequate. Their use in the gap mechanism is described elsewhere [4], but for clarification we will briefly review it here. *Generators* are nodes that enter their subparse into the `Current-Focus` slot. *Activators* move the `Current-Focus` into the `Float-Object` position, for their children. *Absorbers* can accept the `Float-Object` as their subparse, in place of something from the input stream. The net result of this mechanism, aside from its intended use in gap resolution, is that it provides a second-order memory system for identifying the semantic categories of certain key content words in the history.

As an example, consider the sentence, "(What street)$_i$ is the Hyatt on (t_i)?". The *Q-Subject* places "what street" into the `Current-Focus` slot, but this unit is activated to `Float-Object` status by the following *Be-Question*. The *Subject* node refills the now empty `Current-Focus` with "the Hyatt." The node *A-Street*, an absorber, can accept the `Float-Object` as a solution, but only if there is tight agreement in semantics, i.e., it requires the identifier *Street*. Thus a sentence such as "(What restaurant)$_i$ is the Hyatt on (t_i)" would fail on semantic grounds. Furthermore, the node *On-Street* imposes semantic restrictions on the `Current-Focus`. Thus the sentence "(What street)$_i$ is Cambridge on (t_i)?" would fail because *On-Street* does not permit *Region* as the semantic category for the `Current-Focus`, "Cambridge."

1.	Do you know the most direct route to Broadway Avenue from here?
2.	Can I get Chinese cuisine at Legal's?
3.	I would like to walk to the subway stop from any hospital.
4.	Locate a T-stop in Inman Square.
5.	What kind of restaurant is located around Mount Auburn in Kendall Square of East Cambridge?

Table 1. Sample sentences generated consecutively by the VOYAGER version of TINA

The `Current-Focus` always contains the subject whenever a verb is proposed, and therefore it is easy to specify filtering constraints for subject-verb relationships. Thus for example, the verb "intersect" demands that its subject be *Street* and the verb "eat" demands *person*. We have not yet incorporated probabilities into the semantic predictions, mainly because our domain is simple enough that they don't seem necessary. However, in principle probabilities could be added. Furthermore, these probabilities could be acquired automatically by parsing a collection of sentences and counting semantic co-occurrence patterns.

An indicator of how well our semantic restrictions are doing can be obtained by running the sentence generator with the semantic filters in place. Table 1 gives a list of five consecutively generated sentences from the Voyager domain. For the most part, generated sentences are now well-formed both semantically and syntactically.

Application Back-End

After an utterance has been processed by TINA, it is passed to an interface component which constructs a command function from the natural language parse. This function is subsequently passed to the back-end where a response is generated. In this section, we will describe VOYAGER's current command framework, the interface between TINA and the back-end, and some of the discourse capabilities of the back-end.

Command Framework

We will illustrate the current command framework of VOYAGER by way of the simple example shown below:

Query:	Where is the nearest bank to MIT?
Function:	`(LOCATE (NEAREST (BANK nil) (SCHOOL "MIT")))`

`LOCATE` is an example of a major function that determines the primary action to be performed by the command. It shows the physical location of an object or set of objects on the map. Table 2 lists some of the major functions currently implemented in VOYAGER.

Functions such as `BANK` and `SCHOOL` in the above example access the database to return an object or a set of objects. Such functions search for all entries that match the string pattern provided. When null arguments are provided, all possible candidates are returned

LOCATE	locate a set of objects
DESCRIPTION	describe a set of objects
PROPERTY	identify a property of a set of objects
DISTANCE	compute distance between two objects
DIRECTIONS	compute directions between two objects

Table 2. Examples of some major functions in the VOYAGER back-end

STREET	return a set of streets
ADDRESS	return a set of addresses
INTERSECTION	return a set of intersections of two streets
SQUARE	return a set of squares

Table 3. Some examples of data functions in the VOYAGER back-end

from the database. Thus, for example, (SCHOOL "MIT") and (BANK nil) will return the objects MIT and all known banks, respectively. Table 3 contains some examples of current data functions.

Finally, there are a number of functions in VOYAGER that act as filters, whereby the subset that fulfills some requirements are returned. The function (NEAREST X y), for example, returns the object in the set X that is closest to the object y. Table 4 contains several examples of filter functions.

Note that these functions can be nested, so that they can quite easily construct a complicated object. For example, "the Chinese restaurant on Main Street nearest to the hotel in Harvard Square that is closest to City Hall" would be represented by,

```
(NEAREST
  (ON-STREET
    (SERVE (RESTAURANT nil) "Chinese")
    (STREET "Main" "Street"))
  (NEAREST
    (IN-REGION (HOTEL nil) (SQUARE "Harvard"))
    (PUBLIC-BUILDING "City Hall")))
```

AT	the subset of objects that are at a location
ON-STREET	the subset of objects that are on a street
SERVE	the subset of objects that serve a particular kind of food
NEAREST	the single object of a set that is nearest to a location

Table 4. Examples of filter functions in the VOYAGER back-end

Interface between TINA and Back-End

Our natural language component does not produce a logical form that is a separate entity from the parse itself. Instead, structural roles such as *Subject* and *Dir-Object* are an integral part of the parse tree. Furthermore, prepositional phrases are given case-frame-like identities such as *From-Loc* and *In-Region* [2]. Because of the availability of such semantic labels within the parse tree, the nested command sequence required by the back-end can be generated by a recursive procedure operating directly on the parse tree.

The parses are transformed to a set of commands in a two-stage process. The first stage establishes the major function of the sentence, and the second stage fills in any arguments required by the major function. Each stage uses a list of entries that contain a parse pattern, a back-end function specification, and one or more argument specifications. In the first stage, each parse pattern corresponds to a sequence of one or more nodes in the parse tree and can specify a hierarchical constraint between certain nodes. Each argument specification corresponds to one or more entries in the second-stage list. In the second stage, a parse pattern can only be a single node, and each argument specification may be either one or more entries in the second-stage list, a terminal node, or a null value. Terminal nodes return a string from the parse tree, such as "MIT". In most cases, each function specification corresponds to a single back-end function or a null value. When present, a function will be called with its associated arguments, such as `(SCHOOL "MIT")`. A function specification can also designate that the arguments are wrapped around each other. This mechanism is useful for generating nested filtering operations.

When a parse is passed to the interface component, the patterns in the first-stage list are compared to the parse tree. Whenever a match is found, any argument specifications are passed one at a time to the second stage for resolution. If the argument specification is a single node, only the portion of the parse tree found below this node is processed, thus restricting the domain of the second stage analysis. If there are multiple entries in an argument specification, the first one found is processed in the same way as a single node.

In our previous example, the presence of the word "where" in a trace would result in the specification of `LOCATE` as the major function. In this case, the first node found in the associated argument specification is a *Subject*. The portion of the parse tree found below *Subject* is thus passed to the second stage. In the second stage analysis the *Subject* node would find an *A-Place* node in the subparse. Evaluation of this node would subsequently generate two arguments `(BANK nil)` and `(NEAREST (SCHOOL "MIT"))` which are wrapped to produce the desired result.

Discourse Capabilities

The discourse capabilities of the current VOYAGER system are simplistic but nonetheless effective in handling the majority of the interactions within the designated task. Currently, anaphora resolution is dealt with in the back-end as opposed to the natural language component. We will describe briefly here how a discourse history is maintained, and how the system keeps track of incomplete requests, querying the user for more information as needed to fill in ambiguous material.

User:	Can you tell me how to get to the Chinese restaurant?
Voyager:	Where are you?
User:	I am near a bank.
Voyager:	I know of many banks. Can you specify which one you mean?
User:	The one on Main Street.
Voyager	The only bank on Main Street that I know of is the Baybank at 226 Main Street in Cambridge. The Chinese restaurants that I know of are Hong Kong and Royal East. Can you specify which one you mean?
User:	The one nearest me.
Voyager:	The nearest Chinese restaurant to the Baybank at 226 Main Street that I know of is the Royal East. I will try to find the best way to go from the Baybank at 226 Main Street to the Royal East. (– Directions –)

Table 5. Example dialogue illustrating clarification capabilities

Two slots are reserved for discourse history. The first slot refers to the location of the user, which can be set or referred to. The second slot refers to the most recently referenced set of objects. This slot can be a single object, a set of objects, or two separate objects in the case where the previous command involved a calculation involving both a source and a destination. Because of the location slots, user queries can include pronominal reference, such as "What is their address?" or "How far is it from here?"

VOYAGER can also handle ambiguous queries, in which a function argument has either no value or multiple values, when a single value is required. Examples of ambiguous queries would be "How far is a bank?" since there are several banks, or "How far is MIT?" when no default location has been specified. VOYAGER points out such ambiguity to the user, by asking for specific clarification. The ambiguous command is also pushed onto a stack of incompletely specified commands. When the user provides additional information that is evaluated successfully, the top command in the stack is popped for reevaluation. If the additional information is not sufficient to resolve the original command, the command is again pushed onto the stack, with the new information incorporated. In the case where the clarification is also ambiguous, it is pushed onto the stack itself, until it can be clarified. A protection mechanism automatically clears the history stack whenever the user decides to abandon that line of discussion before all ambiguous queries are clarified. An example dialogue illustrating clarification capabilities is given in Table 5.

Discussion

This paper summarizes the present status of our spoken language system development effort. In particular, we focused on those aspects of VOYAGER that have not been reported previously. The system is clearly in its infancy, since it has been in existence for only a few months. In the coming months, however, we hope to improve its performance and expand its capabilities.

Our strategy for system development was to bring up a complete working system as soon as possible, so that we may study component behavior from a system's perspective. As a result, we devoted most of our resources to constructing missing components, as opposed to refining existing ones. Nevertheless, we have begun to improve phonetic recognition accuracy of SUMMIT by incorporating context-dependent phoneme models. Preliminary evaluation indicates that phonetic recognition accuracy is improved by about 5% with these models on test utterances drawn from TIMIT. However, we have not incorporated this refinement into VOYAGER. We view the improvement of phonetic recognition accuracy as one of the most critical areas to pursue in the near future.

In order to provide an early indication of baseline performance, we have recently started to evaluate VOYAGER using the spontaneous speech database that we have collected. Performance evaluation for spoken language systems is an important issue with which the entire research community is beginning to grapple. In a companion paper, we report our own experience with spoken language system evaluation.

Finally, we should again point out that currently the speech recognition and natural language components are connected in a serial manner, in which the recognizer proposes a string of words and passes it along for linguistic analysis. Such a loose coupling is again a reflection of our desire to bring up a working system at the sacrifice of flexibility and overall performance. In the long run, we firmly believe that speech recognition and natural language must be fully integrated. Experiments in a fully integrated system have been under way, and encouraging results have been obtained. We hope to be able to report on this and other improvements in the near future.

Acknowledgments

We gratefully acknowledge Jim Davis of MIT's Media Laboratory for providing us with the map drawing, path finding, and response generation parts of the VOYAGER back end.

References

1. Davis, J.R. and Trobaugh, T.F.: "Directional Assistance," Technical Report 1, MIT Media Laboratory Speech Group, December 1987.
2. Fillmore, C.: The Case for Case. In *Universals in Linguistic Theory*, E. Bach and R.T. Harms, Eds., Holt, Rinehart, and Winston, Chicago, pp. 1–90, 1968.
3. Fisher, W., Doddington, G., and Goudie-Marshall, K.:"The DARPA Speech Recognition Research Database: Specification and Status," *Proceedings of the DARPA Speech Recognition Workshop*, Report No. SAIC-86/1546, February, 1986.
4. Seneff, S.: "TINA: A Probabilistic Syntactic Parser for Speech Understanding Systems," *Proceedings of the First DARPA Speech and Natural Language Workshop*, pp. 168-178, February, 1989.
5. Weintraub, M., and J. Bernstein: "RULE: A System for Constructing Recognition Lexicons," *Proc. DARPA Speech Recognition Workshop*, Report No. SAIC-87/1644, pp. 44–48, February 1987.
6. Zue, V., Daly, N., Glass, J., Goodine, D., Leung, H., Phillips, M., Polifroni, J., Seneff, S., and Soclof, M.: "The Collection and Preliminary Analysis of a Spontaneous Speech Database," These Proceedings.
7. Zue, V., Glass, J., Goodine, D., Leung, H., Phillips, M., Polifroni, J., and Seneff, S.: "Preliminary Evaluation of the VOYAGER Spoken Language System," These Proceedings.
8. Zue, V., Glass, J., Phillips, M., and Seneff, S.: "The MIT SUMMIT Speech Recognition System: A Progress Report," *Proceedings of the First DARPA Speech and Natural Language Workshop*, pp. 178-189, February, 1989.

The Interaction of Word Recognition and Linguistic Processing in Speech Understanding[1]

H. Niemann, G. Sagerer[2], U. Ehrlich, E.G. Schukat-Talamazzini, F. Kummert

Lehrstuhl für Informatik 5 (Mustererkennung), Universität Erlangen-Nürnberg
8520 Erlangen, F.R. of Germany

Abstract: This contribution describes an approach to integrate a speech understanding and dialog system into a homogeneous architecture based on semantic networks. The definition of the network as well as its use in speech understanding is described briefly. A scoring function for word hypotheses meeting the requirements of a graph search algorithm is presented. The main steps of the linguistic analysis, i.e. syntax, semantics, and pragmatics, are described and their realization in the semantic network is shown. The processing steps alternating between data- and model-driven phases are outlined using an example sentence which demonstrates a tight interaction between word recognition and linguistic processing.

Keywords: speech understanding, semantic networks, scoring, linguistic analysis

1 Introduction

The main effort of automatic *speech recognition* is directed towards reliable, speaker-independent, and fast recognition of spoken sentences and words drawn from a sufficiently large vocabulary. Problems of linguistic analysis, understanding of the meaning, or derivation of an answer are of less interest. However, in many systems a *language model* is used to constrain the allowed word sequences. Examples of work in this direction are [4,5,6,13,20,25,26,28,32,38,42]. In *natural language* processing the main effort is towards linguistic analysis of printed texts, including text and story understanding as well as generation of answers to questions. Problems of falsely recognized or unrecognized words, incomplete sentences, or conversion of techniques developed for text understanding to speech understanding are of less interest. Examples of work in this direction are

[1] *Acknowledgment:* The work reported here has been supported by the German Ministry of Research and Technology (BMFT) in a 'Joint Research Project on Speech Understanding' and by the German Research Foundation (DFG) in the special program 'Models and Structural Analysis for the Evaluation of Image and Speech Signals'. This support is gratefully acknowledged; only the authors are responsible for the content of this paper.

[2] now AG Angewandte Informatik, Universität Bielefeld

NATO ASI Series, Vol. F 75
Speech Recognition and Understanding. Recent Advances
Edited by P. Laface and R. De Mori

[3,16,49,17]. The interest in speech is motivated (besides the pure scientific interest) in the facts that it provides an information channel which can be used independent of and in parallel to hand and eye, that it is a natural means of communication, that it can be used via standard telephone links, and that it allows a higher data rate in comparison to a keyboard. This allows interesting potential and actual applications [2,19,22,29], for example, dialog and language translation systems using speech for input and output.

A tacit assumption often is or has been that speech recognition sooner or later will achieve close to 100% recognition rate and that then the text input of a natural language system can just be replaced by the spoken words obtained from a speech recognition system. *Speech understanding* then consists of the two decoupled steps of word recognition and understanding of the meaning of an utterance. The decoupling implies that no interaction between the two steps is possible.

It has been argued frequently and supported experimentally that human speech understanding is a process incorporating *all* sources of evidence simultaneously. For example, the human recognition rate in phonetic transcription of meaningless words increases if the phonological rules of English are met. This supports the view that knowledge about speech on all levels of processing should be used [14], and in fact this is a standard approach in the automatic systems given in the references. It is supported from neurophysiolgy that certain subtasks are realized by independent modules [15]. This suggests a modular approach also to automatic systems. An early example of an automatic system having distinct modules (or knowledge sources) is HEARSAY II [11], and an early example of a system compiling all available knowledge into one network of states is HARPY [28]. Some systems for speech understanding are described in [18,27,23,31,47,52,51,53].

Of course, a *modular* system architecture *not* necessarily implies a *hierarchical* system. In such a system the i-th module accepts an input from the $(i-1)$-th module and passes an output to the $(i+1)$-th module. No information can be passed from module i to module $(i-1), (i-j)$, or $(i+j)$. A more flexible processing strategy should allow the system to focus on promising or important parts of the input at first, and then to inspect other parts in an order and amount of detail infered from intermediate results. This type of focusing or planning has been employed, for example, in [50]. In speech understanding the system might at first concentrate on semantically and/or pragmatically important words. An approach to implementing this type of flexibility is to use a modular system architecture with a distinct control module that determines the processing strategy. This means that the control module determines which processing module should be activated at which time using which data. In principle, it is well possible to activate p modules on p processors. This type of approach has been introduced and used, for example, in [11,34,33,36].

The purpose of this paper is to describe an approach to implement a speech understanding and dialog system which

- provides a well structured representation of general linguistic and special task specific knowledge such that
  - a homogeneous framework for knowledge representation on all levels of processing results,
  - the means for incorporating procedural knowledge of an arbitrary type are given;
- and simultaneously allows a flexible control strategy alternating between data-driven and model-driven phases of processing which includes
  - a theoretically based approach to achieve a global optimum of system behaviour,
  - the possibility to include local control heuristics if needed.

In particular, we consider the interaction between word recognition and linguistic processing which is enabled by the flexible control strategy. In this paper linguistic processing means syntactic, semantic, and pragmatic analysis. The goals of the system are described briefly in the next paragraph. The basis for system implementation is a homogeneous framework for knowledge-based speech understanding described in the next section. Sect. 3 concentrates on the judgment of word hypotheses and only gives a very short overview of word recognition, Sect. 4 describes linguistic processing up to the level of pragmatic interpretation of an utterance. The integration of these processing steps in a homogeneous system architecture is shown in Sect. 5, where also the processing strategy is presented. Results and a conclusion are given in Sect. 6 and 7, respectively.

The main goal of the system is to answer questions concerning a certain task domain using (German) speech for input and output. It has to perform the subtasks of *recognition* of words, *understanding* of the meaning of an utterance, generation of an *answer*, and if necessary generation of a *further inquiry* of the system. Derived from the four subtasks is the acronym EVAR. The vocabulary is between 1000 and 4000 words, the syntax accepts a reasonably large subset of German, word recognition is speaker-independent (but dialects are excluded), speech quality is limited to telephone bandwidth, the task domain is inquiries about German intercity trains, and the dialog module is in development [30]. Descriptions of the individual modules are available from [37,38,35]. This paper will provide recent approaches and results concerning the integration of modules.

2 An Approach to Knowledge Based Speech Understanding

Speech understanding is viewed as a sequence of operations transforming the speech input via different levels of abstraction to a desired output or also to an internal representation. The data structures for representation of the relevant knowledge and the results of operations are introduced, the task independent rules for using knowledge are defined, and an outline of a control algorithm which determines a processing strategy is given.

2.1 Data Structures

Representation and use of knowledge in a system for speech understanding is a primary problem. The knowledge consists of syntactic, semantic, pragmatic, and dialog knowledge, where the latter type is meant to allow the understanding of a sequence of utterances. In addition the relevant knowledge for the generation of an answer must be represented; as mentioned above, in the EVAR system this knowledge concerns intercity train connections. The required knowledge is represented in a *model* $\mathcal{M}$ which is a *semantic network*. The elements used here in the definition of a semantic network are illustrated in Fig.2.1. The network consists of

- three types of nodes:
  - the *concept*, which is a computer representation of a general definition or an *intensional definition* of a conception (e.g. of an object, event, fact, or meaning),
  - the *modified concept*, whose attribute values are more restricted (due to available results) than in the corresponding concept,
  - the *instance*, which corresponds to an actual occurrence of the corresponding concept in the sensor data,
- six types of links:
  - the *part link* relating a concept to one of its parts,
  - the *concrete link* relating the concept to one on a lower level of abstraction,
  - the *specialization link* relating a concept to a more specialized one,
  - the *reference link* relating a concept (e.g. the noun 'man') to one referencing it (e.g. the pronoun 'he'),
  - the *instance link* relating a concept and an instance of it,

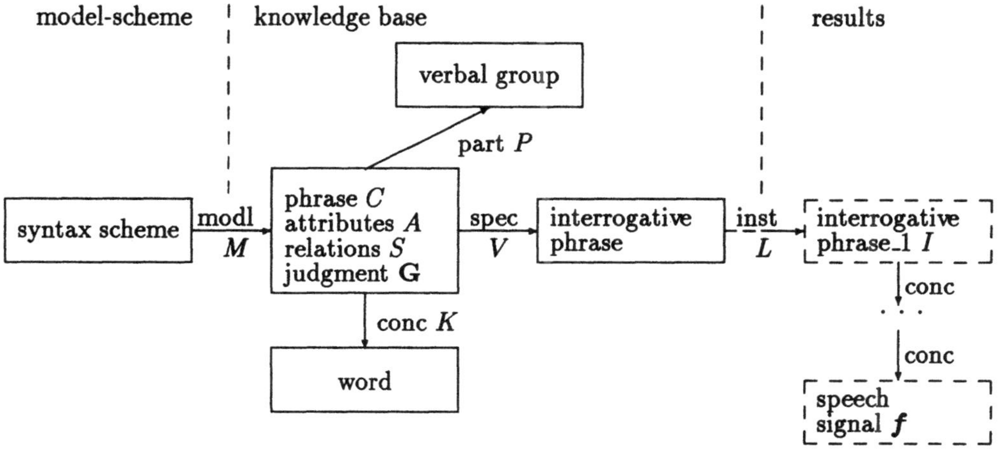

Figure 2.1. An illustration of a concept C and its links P, K, V, M, L to a part, concrete, specialization, scheme-concept, and instance, respectively

  - the *model link*, relating a scheme-concept containing a priori knowledge to a concept derived from it (this is used for automatic knowledge acquisition and is not considered here),

- nine types of substructures which are used to define, for example, links, attributes, relations, or functions needed by the concept.

Hence, the only data type for knowledge representation is a concept

$$\begin{aligned} C &= \Big(D : T_C, (A : (T_A \mapsto F))^*, [H_{OBL}, H_{OPT}, H_{INH}]^*, (V : C)^*, \\ &\quad (R : C^+)^*, (M : C)^*, (L : I)^*, (S(A_C, A_P, A_K) \mapsto F)^*, (\mathbf{G} \mapsto F)\Big) \\ H &= \Big((P_{ci} : C^+)^*, (P_{cd} : C^+)^*, (K : C^+)^*\Big) \end{aligned} \tag{2.1}$$

An example is given in Fig.2.1. It has a name D (e.g. 'phrase'), a type T_C , attributes A (e.g. 'case, number, gender'), parts P (e.g. 'verbal group'), concretes K (e.g. 'words'), specializations V (e.g. 'interrogative phrase'), structural relations S (e.g. 'a noun must follow an adjective in time'), and a judgment vector $\mathbf{G}$. Each attribute, structural relation, and judgment references a function F for the computation of a value.

A concept may have an arbitrary number of *sets of modality*. Each set is sufficient to define an instance of the concept and it consists in turn of an obligatory, optional, and inherent element denoted by H_{OBL}, H_{OPT}, and H_{INH}, respectively. Each element H may consist of an arbitrary number of context-independent and context-dependent parts as well as of concretes, denoted by P_{ci}, P_{cd}, K, respectively. As the notation implies, all parts and concretes of an obligatory element must be available in order to instantiate the concept, and any subset of parts and concretes (including the empty subset) of an optional element is sufficient. An inherent element is one which is not observed in the sensory input, but which is implied by the meaning of the concept. For example, in EVAR a time table of train connections is inherent to a concept defining train connections. A context-independent part is one which can be instantiated without reference to an instance of the superior concept, a context-dependent part needs for its instantiation an instance of the superior concept.

A concept represents general knowledge which is used to analyse an utterance. Depending on the speech signal *instances* I of concepts C are computed. An instance ultimately relates a concept to an interval of the signal, possibly via several other instances. It has the same data structure as a concept, but references to functions F are replaced by computed *values.*

At a certain state of analysis of a speech signal some concepts will already have instances and for some concepts C_i it will still be impossible to compute instances because some prerequisites are missing. Nevertheless, the data available from the instances may allow one to impose additional constraints on the yet uninstantiated concepts. These more constrained concepts are called 'modified concepts'. For example, the available instances may request that the 'number' of an attribute may only be 'plural' or that the start-time of a noun group may only be 'behind' an already instantiated verbal group. The additional constraints are represented as restrictions on the set of values which can be computed by the functions F in a concept. These restrictions are defined in the above mentioned substructures.

2.2 Rules for Using Knowledge

In this subsection general rules are introduced to obtain an instance $I(C)$ of a concept C and to compute a modified concept $Q(C)$. These rules are *task-independent* in the sense that they can also be used, for example, in image understanding; task-dependent procedural knowledge is introduced by the functions F referenced by a concept.

There are three rules for computing an instance of a concept and three rules for computing a modified concept. Only the basic ideas are given here, a detailed definition of the rules is given in [34,40,44]. The six rules perform the following actions:

RULE_1 : create a partial instance ignoring context-dependent parts as well as optional parts and concretes; consider referential links during instantiation

RULE_2 : create an instance out of a partial instance taking into account context-dependent parts

RULE_3 : extend an instance by considering optional parts and concretes

RULE_4 : create a modified concept by bottom-up propagation of constraints

RULE_5 : create a modified concept by top-down propagation of constraints

RULE_6 : create a modified concept from attribute values or initial goal concepts

2.3 A Control Algorithm

Usually, at any state of analysis several out of the above defined six RULES will be applicable to several subsets of concepts, modified concepts, and instances. In order to achieve an efficient analysis focused to the goal of analysis it is mandatory to select the one or the few most promising alternatives. This makes up a processing strategy which is determined by a *control algorithm*. The general idea of such an algorithm is described in the following. It is based on an adaptation of the well-known A^*-graph search algorithm to the peculiarities of semantic networks [41,34]. The *advantages* of using this algorithm are that

- it provides a means for finding an optimal solution, hence enforces a clear definition of what is a 'good (intermediate) result',
- provides global control of the whole analysis process,
- but nevertheless allows one to introduce local and heuristic criteria via the procedures for the computation of the judgment attached to a concept.

It has been discussed in detail, for example, in [33,39,34], that the control problem may be viewed as the problem of finding an optimal path, the *solution path*, in an implicitly defined search graph $\mathcal{G}$ which is restricted here to a *search tree*. A node in this tree represents a state of analysis. It is important to distinguish the model $\mathcal{M}$ from the search tree $\mathcal{G}$. In an abstract sense both consist of nodes and links, but the *nodes* in $\mathcal{M}$ are *concepts*, whereas the *nodes* in $\mathcal{G}$ represent *states* of analysis. Hence, a goal concept in the model has to be distinguished from a goal node in the search tree. The goal concept may be known in advance, for example, it may be stated in advance that an answer to an uttered question is desired, that is the instantiation of an *ANSWER*-concept, or it may be stated that a pragmatic analysis of an utterance is desired, that is the instantiation of a *PRAGMATIC*-concept. The goal node in the search tree is *not known* in advance, rather its content will be the result of a successful analysis. A criterion for a goal node may be, for example, that a good scoring instance of the goal concept has been computed and that this instance covers at least $p\%$ of the utterance.

The main steps of a control algorithm are summarized in Fig.2.2. In the context of semantic networks there are two types of transformations for generating successors of a node, the *expansion* and the *instantiation* of a model concept. Expansion of a model concept means that its successor concepts along the part and concrete links are generated and attached to the successor node; this corresponds to top-down or model-driven processing. Instantiation, of course, means the computation of an instance of a concept; this corresponds to bottom-up or data-driven processing. Hence, a flexible processing strategy *alternating* between model-driven and data-driven phases of processing is possible. Due to incomplete, noisy, and ambiguous data several alternative instances of one concept may be generated, for example, several instances of a noun group may result due to competing word hypotheses or ambiguous syntax. Every alternative instance is attached to a separate successor node in the search graph. Possible modifications of concepts are carried out during these steps.

The main assumption for determining a processing strategy is that a *judgment* ϕ can be computed for every node v in the search tree. The judgment ϕ of a search tree node should be distinguished from the judgment $\mathbf{G}$ of an instance of a concept. This point will be discussed below in more detail. According to the judgment it is possible to select from among the unexpanded nodes in the search tree the best scoring node v_k to be processed next. If a concept or modified concept in v_k can be instantiated, this is done first. If no instantiation is possible, it is tried to expand the node v_k along concrete and obligatory part links. If this is not possible, expansion along optional part and specialization links is tried.

2.4 Discussion

In the preceding subsections we presented a general framework for implementation of a knowledge-based speech understanding system. It is designed such that an *interaction* between the recognition and the understanding phases is possible. The main ideas are briefly summarized.

The system is structured into two main phases. The first one is a segmentation phase containing those processing steps which are executed mainly data-driven, in a fixed order, and without using task-specific linguistic knowledge. This is also called the speech front-end. The result is a set of word hypotheses which also may be checked syntactically by a grammar or a language model.

<table>
<tr><td colspan="5">Input: application function to provide a list of goal concepts C_{gi}, $i = 1, \ldots, n$</td></tr>
<tr><td colspan="5">Initialize: application function and shell function to provide list OPEN with nodes v_{gi}, $i = 1, \ldots, n$ and their judgments $\phi(v_{gi})$</td></tr>
<tr><td colspan="5">WHILE OPEN is not empty DO:</td></tr>
<tr><td rowspan="10"></td><td colspan="4">select from the list OPEN the best scoring node v_k by application function select(OPEN) , remove v_k from OPEN</td></tr>
<tr><td>IF</td><td colspan="3">application function end_analysis(v_k) decides that an analysis goal has been achieved</td></tr>
<tr><td>THEN</td><td colspan="3">STOP - successful end of search or end of resource</td></tr>
<tr><td>IF</td><td colspan="3">application function goal_conc(v_k) defines nonempty set S of new goal concepts</td></tr>
<tr><td>THEN</td><td colspan="3">shell function gen_goal(v_k, S) generates new goals and corresponding nodes on OPEN</td></tr>
<tr><td rowspan="5">ELSE</td><td>IF</td><td colspan="2">one object in v_k can be instantiated by one of the RULES_1,2</td></tr>
<tr><td>THEN</td><td colspan="2">shell function instant(v_k) to instantiate v_k and to perform modifications by RULES_4,5</td></tr>
<tr><td rowspan="3">ELSE</td><td>IF</td><td>there is one object in v_k with an unfulfilled premise</td></tr>
<tr><td>THEN</td><td>shell function expand(v_k) to expand v_k, to perform modifications by RULE_5, and to consider referential links</td></tr>
<tr><td>ELSE</td><td>shell function opt_spec(v_k) to consider optional parts and specializations</td></tr>
<tr><td colspan="5">STOP - unsuccessful end of search</td></tr>
</table>

Figure 2.2. An outline of the main steps of a general control algorithm; it distinguishes actions which are task-independent and can be supplied in a system shell (**shell** functions) and actions which are task-dependent and have to be specified by the system designer (**application** functions)

The second phase is knowledge-based processing where a processing strategy is determined for every input by a control algorithm. The declarative knowledge is represented by a model $\mathcal{M}$ which is a semantic network of concepts C. Problem-specific procedural knowledge is referenced by functions attached to concepts. Speech understanding and dialog amounts to the computation of a description which is consistent with the model and the dialog context and optimally fits to the input speech signal. Instantiation of concepts is defined in a problem-independent manner by six RULES. A strategy for the instantiation of concepts is defined by a control algorithm which is a version of the A^*-algorithm adapted to semantic networks. Since it can alternate between data-driven and model-driven phases of processing it allows an interaction between word recognition and linguistic analysis. The task-independent part of this approach is implemented in the system shell ERNEST [40].

3 Scoring of Results from the Acoustic Front End

The main purpose of this section is to introduce a method for judging the quality of word hypotheses or also of chains of words generated by an acoustic front end. The judgment function should be in accordance with the requirements of subsequent knowledge-based processing. It is not intended here to describe the acoustic front end in detail; details are given in [24,43,46].

3.1 Overview

In the acoustic front end the speech signal $\boldsymbol{f}$ is processed to obtain a lattice of word hypotheses. The first processing step is a parametric representation of the speech signal which in EVAR is based on smoothed delta Bark spectrum Lem cepstrum coefficients. In addition the parametric representation includes suprasegmental features. A segmentation into phonemic subword units is carried out next. Based on a lexicon, the structure of which is discussed in detail in Subsect.4.1, a set of word hypotheses is determined. A word is represented by a hidden Markov model (HMM). In addition to the lexicon syntactic constraints may be used to reduce the number of word hypotheses. These processing steps make up the *acoustic front end.* The result is a set of words or of syntactically correct word chains.

3.2 Judgment of Acoustic Quality

The *judgment of a hypothesis* for a word or a word chain should meet the conditions of the A*-algorithm for graph search (see for example [41,34]), since it is used to control linguistic processing. The search must be *admissible* in the sense that the best sequence of words is found even if not all sequences are evaluated. This can be achieved if the *judgment of a node* v_i in the search space is estimated by

$$\hat{\phi}_q = \hat{\psi} + \hat{\chi}, \tag{3.1}$$

where

- $\hat{\psi}$ is an estimate of the cost of a path from the start node to v_i
- $\hat{\chi}$ is an estimate of the cost of a path from v_i to a goal node

  (this path is not yet known if the search proceeded only to v_i)

In order to achieve an *admissible algorithm*, the estimate $\hat{\chi}$ must be *optimistic*, that is it must be *smaller* than the true costs χ.

For the word hypotheses this means that the judgment should evaluate the cost of that part of the utterance *covered* by the hypothesis ($\hat{\psi}$) plus the cost of the part *not covered* by the hypothesis ($\hat{\chi}$). The term $\hat{\psi}$ is based on the score obtained from word recognition. For $\hat{\chi}$ we use an estimate of the rest which is optimistic in most cases.

The term $\hat{\psi}$ is based on (but not necessarily identical to) the score G_q obtained from *word recognition*, that is in DTW the sum of local distances and in HMM the product of pairs of probabilities for state transition and output, or the sum of their (negative) logarithms, that is $G_q = -\log[p(o_b \ldots o_e \mid HMM)]$. Hence, a simple *'linear model'* of the quality is

$$G_q = \sum_{\tau=1}^{T} G_\tau \,, \tag{3.2}$$

where T is the length (in frames of speech) of the word hypothesis.

Let us assume *independent identically distributed* random variables G_τ with mean μ_c and variance σ_c^2 for a *correct hypothesis* and mean μ_f and variance σ_f^2 for a *false hypothesis.* In this case the cost of a hypothesis of length T has a distribution with

- $(T\mu_c, T\sigma_c^2)$ for a correct hypothesis,
- $(T\mu_f, T\sigma_f^2)$ for a false hypothesis.

It has been *verified experimentally* that mean and variance of the quality of correct and false hypotheses depend linearly on T as predicted by the above 'linear model' for the judgment of hypotheses [45].

During *word hypothesization* the set of N best scoring hypotheses is selected. The *first step of linguistic processing* is to select from the set of word hypotheses the N' best scoring pragmatically relevant hypotheses. The *subsequent steps of linguistic processing* generate linguistically meaningful sequences of words, for example, syntactic constituents, case frames of verbs, instances of pragmatic concepts (e.g. an instance of the concept *TRAIN_CONNECTION*), or instances of dialog concepts. Each such word sequence

- causes the generation of a node v in the search tree,
- is judged by a vector **G**, one of whose components measures the *acoustic quality* G_q of n words in a sequence,
- is used to estimate the quality $\hat{\phi}_q(v)$ of the corresponding search tree node.

Let $w1, w2, \ldots, wn$ be the words in the word sequence. The acoustic quality is defined by

$$G_q = \sum_{i=1}^{n} G_{qi} = \sum_{i=1}^{n} -\log[p(o_{bi} \dots o_{ei} \mid HMM_{wi}])$$ (3.3)

This is also a good estimate for non-adjacent words.

The *quality of the search tree node* is in accordance with the requirements of the A*-algorithm defined to be

$$\begin{aligned} \hat{\phi}_q &= G_q + G_r \\ &= \hat{\psi} + \hat{\chi} \end{aligned}$$ (3.4)

An approach to determine G_r is from the 'linear model' of scoring. With some constant γ we define

$$G_r = T_u \mu_c - \gamma \sigma_c \sqrt{T_u},$$ (3.5)

where T_u is the duration of that part of the utterance *not covered* by the word sequence. This ensures an optimistic estimate $\hat{\chi}$ in a certain percentage of cases determined by the constant γ. For example, if the distribution of quality judgments G_q of a word of length T were normal (which is at best an idealization), $\gamma = 2$ would ensure that at most 3% of hypotheses of length T would have a better quality score.

This estimate of χ is better than the 'optimistic' estimate $\hat{\chi} = 0$ and according to our experiments it is also better than the 'shortfall score' [50].

4 Linguistic Constraints for Speech Understanding

This section describes the linguistic constraints used in EVAR. The constraints range from recognized words or word chains through the levels of syntactic, semantic, and pragmatic analysis up to carrying out a man–machine dialog. The so–called 'raw linguistic knowledge base' is the lexicon which is used by various levels in the system; for each level a preprocessor extracts the relevant information from the lexicon. Syntactic processing as understood here determines the syntactic structure of an utterance taking into account only rules for the combination of (syntactic) word classes without considering their meaning. Semantic processing determines the meaning of words, but only considers general task–independent semantic properties of words. Pragmatic processing interprets *one* utterance with respect to a model of the task domain. Finally, dialog processing — among others — interprets one utterance in the *context* of preceding utterances. This allows a layered or stratified approach to the representation of declarative knowledge and facilitates changes in the linguistic competence of one level without affecting others. Since according to Subsect.2.3

a distinct control module is used, the stratified approach nevertheless allows a flexible processing strategy.

4.1 The Lexicon

Since detailed descriptions of the lexicon and justifications of the design choices are available from [9,10,35], only a short overview of the main points is given in the following.

The unit of the lexicon, i.e. an individual entry in the lexicon, is defined by the spelling of a word or by the *graphematic word*. A unit of the lexicon is simply called a *word*. Every word has associated with it a unique integer number for its identification. The information attached to a word consists of the tuple [spelling, word number, pronunciation, syntactic information (may be multiple per word), semantic information (may be multiple per syntactic definition) and possibly dialog attributes, pragmatic information (may be multiple per semantic definition)]. A word may have only a part of this information. The detailed format of a lexicon entry is specified in [10,35]. Each alternative for the syntactic, semantic, or pragmatic definition is called a syntactic, semantic, or pragmatic word, respectively. In order to facilitate word recognition it was decided to include all inflections of a word as separate entries.

Presently the lexicon has 4427 words whose entries at least contain [spelling, word number, pronunciation]. They are used for word recognition experiments with vocabularies of different sizes. The lexicon contains 3953 syntactic words, out of which 841 are base forms, 1124 semantic words, or lexemes, and 246 pragmatic words in the above defined sense. The syntactic, semantic, and pragmatic information is described in the corresponding subsections. A set of preprocessors has been implemented which extracts the information necessary for a certain processing algorithm and transforms it to a useful format. The result is called a sub–lexicon. Since all sub–lexica are derived from the same lexicon, consistency of sub–lexica is guaranteed. For example, syntactic processing only needs word number and syntactic information. This approach separates acquisition and maintenance of the lexicon from the design and implementation of processing modules.

The lexicon is implemented in LISP as a database which in particular has an editor and search functions. The editor supports the addition of new words and the modification of existing words; it also checks the formal consistency of entries.

4.2 Syntactic Constraints

Syntactic analysis is based on constraints defined by an augmented transition network (ATN) grammar of a subset of German language. Only *simple syntactic constituents* are determined, but not complete sentences. By a 'simple' syntactic constituent we mean a group of words containing only one nucleus, for example, 'the next train', but not 'the next train to Hamburg'; the latter group would be treated as two constituents. This approach has the following reasons. In a spoken dialog it often occurs that not a complete sentence but only a fragment is uttered which usually is fully understandable in the context of a dialog. Since word recognition is not perfect, some words in a sentence may be missing, but many of the constituents may still be recognized correctly. Due to missing correct words and wrongly recognized words a large number of wrong sentences which are accepted by the ATN may be generated. The separation of syntax, semantics, and pragmatics is useful for a modular representation of knowledge. As mentioned already, this does not prevent a flexible processing strategy. The syntactic constituents are checked immediately for semantic consistency as described in the next subsection. This is done because syntax alone imposes only weak constraints on a constituent, in particular if a grammar for a sufficiently large subset of German is used. Finally, if a syntactic constituent is pragmatically relevant, it can be tried at a very early stage of processing to verify the corresponding pragmatic concept top–down or model–based. For these reasons the so–called 'constituent grammar' does not specify sentences.

Usually, a syntactic constituent is coherent in the sense that the words belonging to it follow each other in time. An exception are verbal groups VG which in German may be non-coherent. Therefore, verbal groups are not checked initially.

The grammar only contains constructions occurring in inquiry dialogs. Therefore, it does not contain passive forms. Meta–communicative forms, for example, 'thank you' or 'good morning', are treated by the dialog module. Ideally, a grammar should be task–independent. However, since it accepts only a subset of German, this subset is chosen to meet the requirements of a task–domain. For example, the syntactic constituent 'DATUM' (date) is important for inquiries about intercity trains, but it may be unimportant and omitted for inquiries about traffic rules.

Syntactic analysis using the full ATN is performed under the control of the algorithm described in Subsect.2.3. Depending on the judgment of search tree nodes syntactic analysis may proceed data–driven or model–driven, and it may switch between those two phases. Hence, there is no distinct parser. But it is emphasized that a special parsing algorithm could be integrated into the semantic network approach if desired. This can be done, for example, in one of the following two ways: first, by placing the interface between the semantic network and the other processing steps not at the level of words but at the level of syntactically parsed word strings; second, by leaving the interface at the level of words and attaching the parser as a procedure to one concept *SYNTAX*.

4.3 Semantic Constraints

The general task independent meaning of words and utterances is represented on the level of *semantics*. It is based on case and valency theory as developed in [1,12,48], on a system of semantic classes of words, and on constraints between semantic classes of words within a syntactic constituent. Semantic analysis consists of three steps. First, the semantic consistency of simple syntactic constituents generated according to the ATN grammar is checked. Second, simple syntactic constituents are combined to complex syntactic constituents. Examples are constituents like 'the next train to Hamburg' or 'on Monday morning at about 8 o'clock'. Third, the constituents are summarized to a sentence hypothesis including also verbal groups.

Valency theory is outlined shortly using the verb valency as an example. In EVAR a noun, adjective, and adverbial valency is defined in a similar way as well. A verb used in a particular meaning requires a certain set of obligatory and optional elements having well–defined syntactic and semantic properties. These elements define the 'valency frame' of a verb in a particular meaning. If the same verb (the same graphematic word in the sense of Subsec.4.1) is used in another meaning, it will require a different set of elements thus defining a different valency frame. The above mentioned obligatory elements are those necessary to obtain a grammatically correct sentence. The optional elements are those which in addition to the obligatory ones are necessary to fully define the verb meaning. In German it is possible to add an almost arbitrary number of so–called 'free elements' which can be used fairly independent of a certain verb meaning. Therefore, free elements are not checked in order to determine the meaning of a verb, but they may have to be considered in order to determine an interpretation spanning a complete utterance. The emphasis of valency theory is on the syntactic structure of a verb and on selectional semantic restrictions between words.

Case theory concentrates on the 'functional role' or the 'deep case' of a syntactic constituent in a sentence. For example, in the sentence 'Tom washes the car' and 'the car is washed by Tom', the deep case of 'Tom' as an AGENT performing a certain action is the same although the syntactic properties are different. Thus the logical relations of a word with respect to a verb are considered in case theory. In EVAR 29 deep cases are distinguished which can be found in [35]. Examples are the GOAL, PATH, SOURCE, and TIME of a journey.

According to case and valency theory the *case frame* of a verb (or a noun, adjective, adverb) is determined. The relevant raw semantic knowledge is contained in the lexicon and extracted by a preprocessor. The elements of this knowledge depend on the word class and are given in Fig.4.1.

	word class	semantic knowledge	
ADJ	attributive adjective	selection class	case frame
ADV	adverb	semantic class	relation
DET	determiner	negation or not	reference
N	noun	semantic class	case frame
PRAEP	preposition	semantic class	selection class
VERB	verb	semantic class	case frame

Figure 4.1. Types of semantic knowledge associated with a certain syntactic word class

Semantic classes of words are selected from a set of 107 general task-independent meanings used in EVAR. Possible meanings of a class of words are specified in a tree structure of semantic classes. For example, the syntactic word class VERB may have the semantic classes Act (action, e.g. drive a car), Chg (change of state, e.g. drive to Hamburg), Pro (process, e.g. to fall down), Stt (state, e.g. to sit), and Tru (truth value, e.g. could be). A semantic class may have subclasses. For example, the class 'Chg' mentioned above has the subclasses Mov (movement, e.g. to drive), Per (perception, e.g. to see), and Cmm (communication, e.g. to speak).

According to Fig.4.1 a noun has a 'semantic class' and a 'case frame', a preposition has a 'semantic class' and a 'selection class'. For the combination of a preposition (or an adjective) with a noun the selection class and the semantic class are used to represent restrictions on the combination of words used in a particular meaning. The restriction is defined by the rule

IF (a preposition or an adjective is used in the meaning defined by its semantic class),

THEN (the associated noun must have a semantic class corresponding to the 'selection class' of the preposition). (4.1)

An example is given in Fig.4.2. It shows a noun and a preposition which both have two possible meanings resulting in $2 \times 2 = 4$ possible combinations for the prepositional noun group 'in the coach'. But only the selection class Location of the preposition occurs as semantic class of one meaning of the noun. Hence, only one combination of meanings is selected in this case.

In summary, semantic analysis consists of the three steps:

1. Checking of simple syntactic constituents for semantic consistency. This is the *filtering of syntactic constituents*. It is mentioned that in order to increase the efficiency of this step the pragmatic consistency of constituents is checked, too. There is a test for consistency of adjectives and prepositions with the corresponding noun (thus

noun: coach	
meaning_1: railway carriage	meaning_2: trainer in athletics
semantic class: Transport, Location	semantic class: Acting_person

preposition: in	
meaning_1: in the evening	meaning_2: in the room
semantic class: Duration selection class: Time	semantic class: Place selection class: Location

Figure 4.2. Example of a noun and a preposition both having two possible meanings; note that the 'meaning_i'-entry ($i = 1, 2$) is only for readability of the lexicon

eliminating combinations like 'the fast tree' or 'inside the morning'). It is tested for nouns used in the singular number and having no article (thus eliminating combinations like 'how much is ticket'). The contradiction of semantic features is tested (thus eliminating cases like 'take a next train'). Finally, there is a test for constituents in the genitive and depending not on a noun but on a verb or an adjective, since the latter two are rarely used in inquiry type dialogs; however, they are reconsiderd when searching for complex constituents.

2. The *construction of complex constituents.* It consists of the construction of genitive constituents (for example, 'the departure of the train'), the construction of denominal actants (for example, 'the train to Heidelberg' or 'ten kilometers away'), the identification of anaphoric references requiring a dialog memory, the generation of sentence hypotheses for one sentence with a verb (for example, 'I want to go to Hamburg') using valency and deep case information from the lexicon, but not using all available restrictions in this phase, and finally the construction of temporal constituents (for example, 'tomorrow in the morning at about 8 o'clock').

3. The instantiation of a sentence hypothesis in a semantic network representing the full semantic knowledge. It consists of the two phases of the structural interpretation of a sentence hypothesis in the semantic network including *all* available restrictions, and the inclusion of free elements if there are any.

It is emphasized that the above steps *need not be executed sequentially* since the semantic network environment provides a very flexible control algorithm. Pragmatic interpretation is discussed briefly in the next Section, in particular Fig.5.3.

5 System Integration

In the preceding Sects.2-4 we presented a general approach to knowledge-based speech understanding, the scoring of results from the acoustic front end, and the main steps of linguistic analysis. This section shows the ***integration in a homogeneous system*** based on semantic networks. An important point is that a semantic network allows a ***well structured representation of knowledge*** maintaining the idea of a stratified system and simultaneously facilitates a ***flexible use of knowledge*** thus enabling an alternation between data-driven and model-driven processing steps as well as immediate transmission of results and requests via several levels of representation. The network implementation of linguistic conceptions is discussed, the handling of model-based requests passed from the linguistic analysis to the acoustic front end, the processing strategy, and the judgment of instances and search tree nodes.

5.1 General Ideas

The basic ideas of integrating a complete speech understanding system into a semantic network, for example, into the ERNEST environment are:

- map declarative knowledge to concepts, that is in particular knowledge about syntax, semantics, pragmatics, dialog, task domain (Intercity train connections in EVAR), and answer generation;
- attach procedural knowledge to concepts (e.g. for computation of attributes, relations, judgment);
- provide a judgment function meeting the requirements of the A*-algorithm for graph search;
- provide the application-functions for the control algorithm (or otherwise be content with the default functions provided in ERNEST);
- provide a (software) interface between acoustic front end and linguistic processing.

An overview of the network is given in Fig.5.1.

5.2 Representation of Linguistic Knowledge

In this subsection we consider the representation of syntax with the example of a noun group and the representation of a section of the task domain in ERNEST.

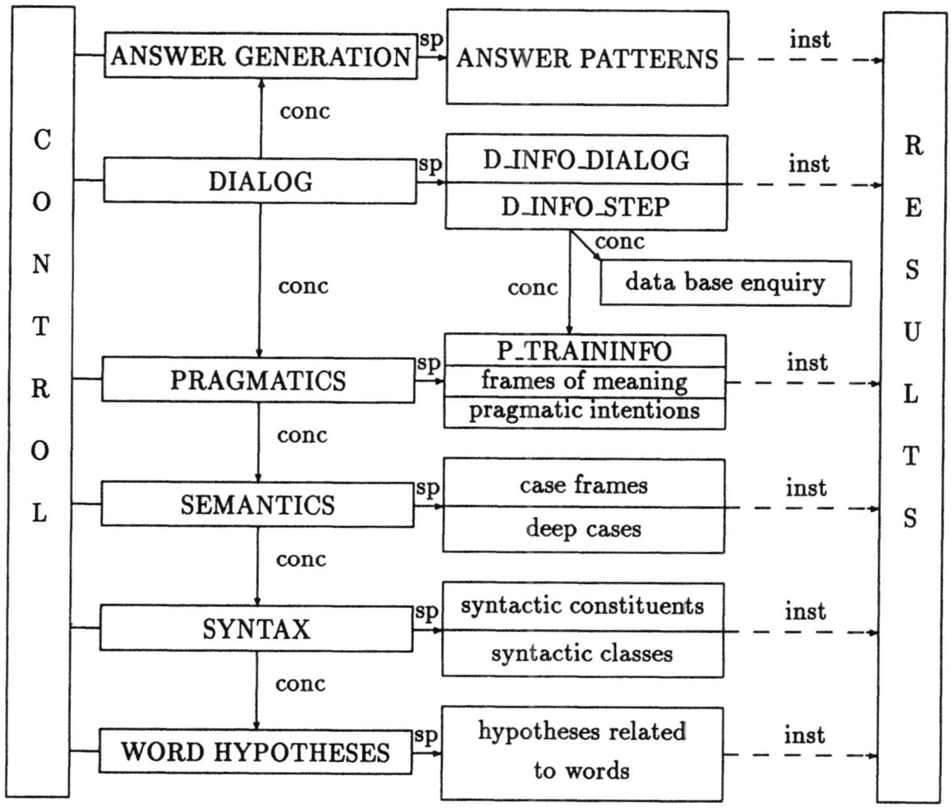

Figure 5.1. An overview of the semantic network representing the speech understanding system

Fig.5.2 shows a noun group in the common RTN representation and as concepts in a semantic network. The other syntactic constituents of the constituent grammar are represented in a similar way. The elements occurring on the edges of the RTN are represented by concepts (rectangles in the figure), the alternative definitions of a noun group correspond to different sets of modality (H_{OBL} and H_{OPT} in the figure), and the time order of elements is represented by an adjacency matrix (ADJAC in the figure).

Next we consider a section of the task domain as shown in Fig.5.3, where concepts are denoted by P_(name) on the pragmatics level, S_(name) on the semantics level, and SY_(name) on the syntax level. It defines on the pragmatics level the obligatory and optional elements of an inquiry for a train connection (*P_TIME_TABLE*). Since this model defines the elements necessary for such an inquiry, it is possible for the dialog module to generate a request for further particulars if one of the obligatory elements is missing

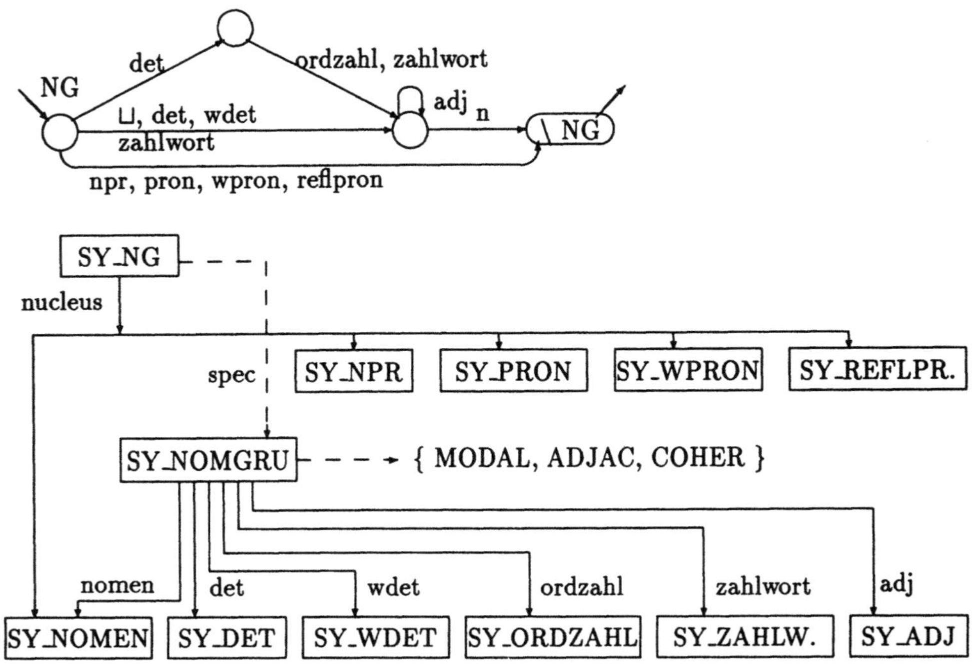

MODAL : H_{OBL} : { nomen }
H_{OPT} : { det, wdet, ordzahl, zahlwort, adj }

ADJAC :

$$\begin{array}{c} \left[\begin{array}{cccccc} 0 & 1 & 1 & 1 & 1 & 1 \\ 0 & 0 & 0 & 0 & 0 & 0 \\ 0 & 0 & 0 & 0 & 0 & 0 \\ 0 & 1 & 0 & 0 & 0 & 0 \\ 0 & 1 & 0 & 0 & 0 & 0 \\ 0 & 1 & 1 & 1 & 1 & 2 \end{array}\right] \begin{array}{l} \text{nomen} \\ \text{det} \\ \text{wdet} \\ \text{ordzahl} \\ \text{zahlwort} \\ \text{adj} \end{array} \\ \text{nomen det wdet ordz zahl adj} \end{array}$$

⇒ a word class in column may precede a word class in row

COHER : YES

⇒ constituents must be adjacent in time

Figure 5.2. An RTN of a noun group and its representation by a semantic network

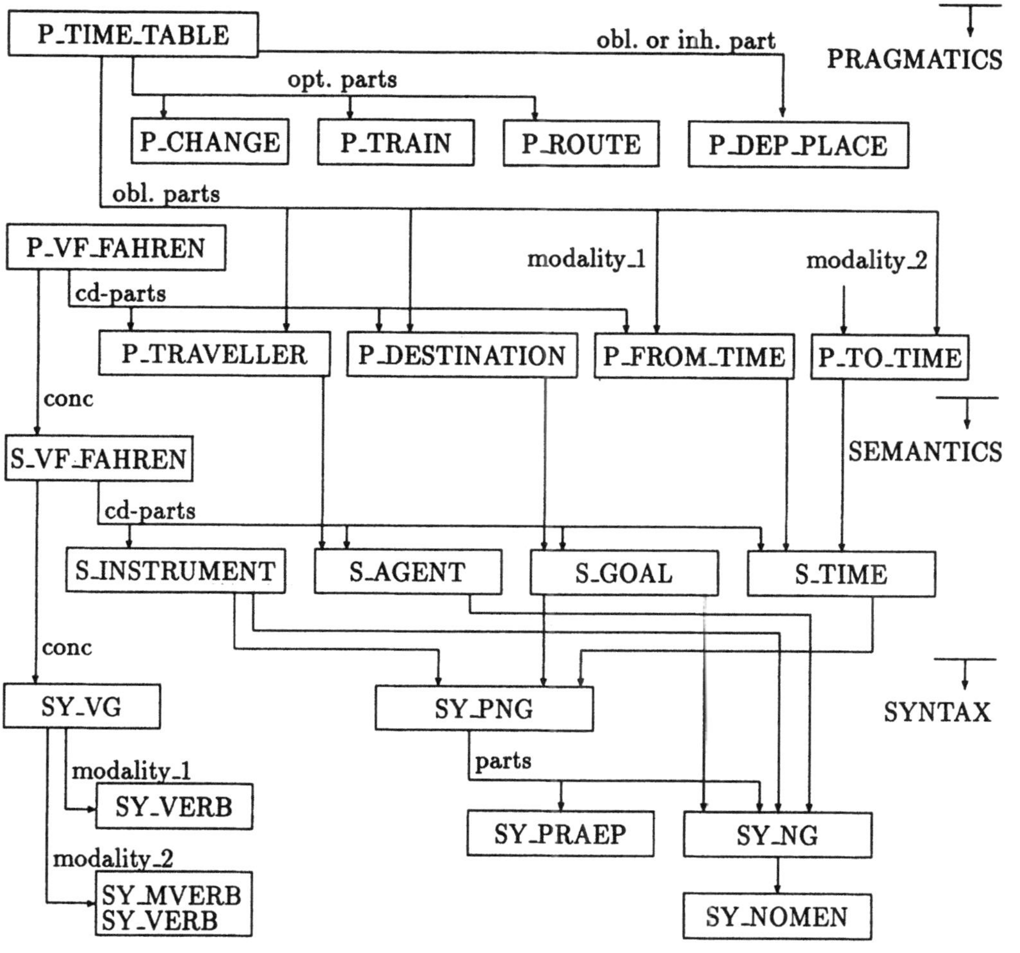

Figure 5.3. An overview of the concepts in a *section* of the model of the task domain; not all concepts and not all links are shown

(e.g. because it was not recognized or not uttered). Among the parts is, for example, a concept *P_DESTINATION* which in turn is referenced as a context-dependent part by the verb frame *P_VF_FAHREN*. Hence, *P_DESTINATION* can only be instantiated if a partial instance of *P_VF_FAHREN* is available. Of course, there are other verb frames (and also noun frames) referencing *P_DESTINATION* such that different alternatives for its instantiation are possible. A concrete of *P_VF_FAHREN* is on the semantics level *S_VF_FAHREN* and on the syntax level *SY_VG* for a verbal group. It is seen that, for example, *P_VF_FAHREN* has a part *S_TIME* which in the pragmatic context is specialized to either *P_FROM_TIME* (departure time) or *P_TO_TIME* (arrival time).

5.3 Judgment

Two types of judgment have to be distinguished:

1. The judgment $\mathbf{G}$ of modified concepts $Q(C)$ and instances $I(C)$ of a concept C.

2. The judgment ϕ of a node v in the search tree.

   The node contains (among others)

   - a goal concept C_g from the model which has been modified to $Q(C_g)$ or instantiated to $I(C_g)$,
   - the modified concepts and instances generated so far in order to modify and instantiate C_g,
   - potentially the whole model and all intermediate results.

The judgment of modified concepts and instances is defined by functions referenced by the corresponding concept.

The estimate $\hat{\phi}$ of the judgment of a search tree node is defined to be the current judgment of the modified or instantiated *goal concept* C_g, subject to the requirements of the A*-algorithm.

The judgment $\mathbf{G}$ of an instance I or a modified concept Q of a concept C is the vector $\mathbf{G} = (G_c, G_q)_t$. The components are

- G_c: the compatibility of a hypothesis with the linguistic knowledge (a binary number),

- G_q: the quality of a word sequence (that is, not necessarily adjacent words) making up the hypothesis with respect to the speech signal — in the present implementation *without* the estimate of the remainder according to the A*-algorithm.

Since G_c is a binary number we do not use a stochastic language model presently. The reason is that a stochastic model needs a large sample for the estimation of its parameters and that it is adapted to the average statistics of this sample, but not to the statistics occurring in a particular dialog situation. Meanwhile significant progress has been made in estimating the statistics of speech and in statistical modeling of dialog situations [21,8]. If an appropriate statistical model is available, it can be used in the above approach by modifying G_c, G_q appropriately.

The judgment of a search tree node is based on the idea that

- the computed result should be compatible with the linguistic knowledge and the dialog context,

- and it should have maximal 'similarity' to the speech signal.

The judgment $\hat{\phi}$ of a search tree node v having the associated goal concept C_g with current modification $Q(C_g)$ is the vector

$$\hat{\phi} = (\phi_c, \phi_q, \phi_r, \phi_t, \phi_p)_t. \tag{5.1}$$

It is evaluated in lexicographical order using intervals for some components.

The components are

- ϕ_c: the compatibility of a hypothesis with the linguistic knowledge

  $$\phi_c = G_c(Q(C_g)) \in \{0, 1\},$$

  since first of all, a hypothesis must be compatible with the linguistic knowledge;

- ϕ_q: the quality of the word chain making up the hypothesis with respect to the speech signal according to (3.4)

  $$\phi_q = G_q(Q(C_g)) + G_r(Q(C_g)),$$

  since among the compatible hypotheses we prefer the one having best quality;

  the above two components ensure the computation of compatible hypotheses having optimal acoustic quality; the following components are used to further reduce the amount of search:

- ϕ_r: the reliability of the hypothesis

  $$\phi_r = [\text{number of speech frames in the longest word chain (adjacent words)}],$$

  since among those having high quality we prefer the most reliable one (long words are recognized more reliably than short ones);

- ϕ_t: the total coverage of the utterance

  $$\phi_t = [\text{number of speech frames covered by word hypotheses in } Q(C_g)],$$

  since among the reliable hypotheses we prefer those best covering the utterance (i.e. yielding a final result more quickly);

- ϕ_p: the pragmatic relevance of the hypothesis

  $$\phi_p = G_p(Q(C_g)),$$

  since finally, we prefer the pragmatically most relevant one. The definition of pragmatic relevance uses fuzzy functions [7] and details are given in [9].

6 Processing Steps and Results

In the following we consider as an example some processing steps of the sentence

Wir möchten am Wochenende nach Mainz fahren

(We would like to go to Mainz at the weekend)

After initial generation of 100 word hypotheses processing on a RISC work station (about 14 mips, 6MB) took 101 sec CPU time until the instantiation of the (correct) instance of the pragmatic concept *P_TIME_TABLE*. During the analysis the control algorithm generated 1053 search tree nodes, 1233 modified concepts, and 777 instances.

The following are the test conditions:

1. Sentences in continuous speech sampled at 10 kHz using telephone bandwidth.
2. Word recognition and word verification are speaker independent.
3. The lexicon for recognition and verification has 1250 entries.
4. The judgment vector is as described previously.
5. Tests were performed with 100 word hypotheses including *all* words actually spoken.

So far only 12 sentences were analysed, but a larger sample is in preparation.

Linguistic processing is initialized using as initial goal concepts C_{gi} the following concepts from the syntax level:

- *SY_ADJ*, *SY_ADJU*, *SY_ADV*, *SY_NOMEN*, *SY_NPR*, *SY_ORDZAHL*, *SY_PRON*, *SY_ZAHLWORT*, *SY_WADV*

The control algorithm initializes a search tree node v_{gi} for each goal concept. From the 100 word hypotheses the 10 best scoring pragmatically relevant ones are used to instantiate the above goal concepts.

Next the control algorithm provides the option to generate a list S of new goal concepts as evident from Fig.2.2. This option is implemented presently in a 'look-up table' providing as the list S the following concepts on the pragmatics level:

$$\begin{aligned} S &= \{P\_TRAVELLER, P\_DEP\_PLACE, P\_ROUTE, P\_DESTIN \\ &\quad P\_TRAIN, P\_CHANGE, P\_FROM\_TIME, P\_TO\_TIME \,\} \end{aligned} \tag{6.1}$$

In the model — a section of which is shown in Fig.5.3 — it is now tried to find a path along 'concrete-of' or 'part-of' links from an instantiated concept (e.g. *SY_NOMEN*) to one of the new goal concepts (e.g. *P_FROM_TIME*). In order to increase the efficiency of this step

- only concepts on the path are considered having a pragmatic class compatible to that of the instance of the starting concept,
- modifications and instantiations possible on this path are carried out.

After this bottom-up step the verification of the pragmatic concept *P_FROM_TIME* follows as a top-down step.

Next the case frame of a verb is tried since a verb (or a noun) references pragmatic concepts as context-dependent parts in its case frame. In general, in a pragmatic concept (like *P_FROM_TIME*) *all* possible contexts (verbs, nouns) are referenced, but only one example is shown in Fig.5.3. For every context a modified concept is created by RULE_4 and a new search tree node is generated. By iterative application of RULE_5 the model is expanded, in this case for the verb frame.

The control algorithm allows in principle an arbitrary number of steps for the generation of new goal concepts. If the verb frame is instantiated, new goal concepts on the level of inquiries are generated. For the concept *P_FROM_TIME* this gives

- *P_CONNECT_INFO*, *P_TIME_TABLE*

As an example the content of a search tree node representing this intermediate state of analysis is shown in Fig.6.1. It is seen that it contains the complete structure of the utterance analysed so far.

Finally the goal concept on the pragmatics level is instantiated. The instance has to meet the conditions for a successful analysis:

1. instantiation of a top-level concept,
2. coverage of at least 80% of the utterance.

This way control alters between expansion of the model and modification of expanded concepts (by RULE_5) and instantiation of the model (by RULES_1,2,3) and thus achieves an interaction between word recognition and linguistic processing.

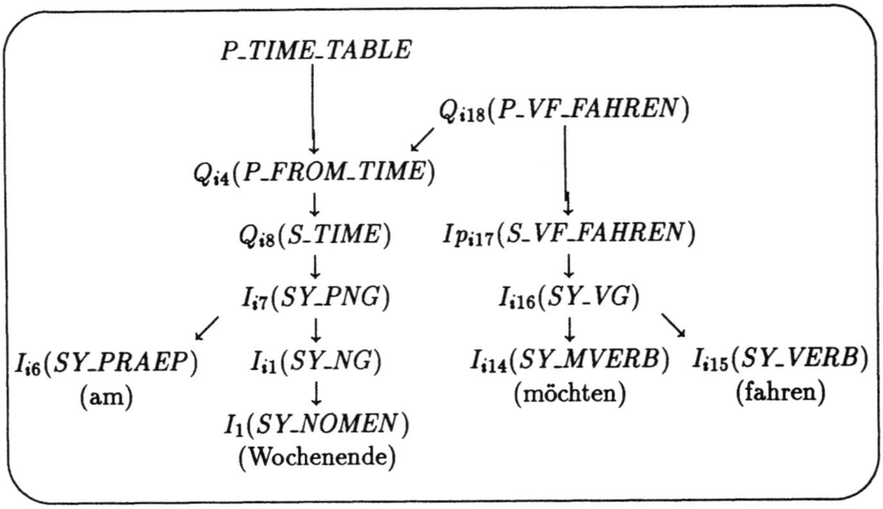

$$\phi = (1, -4818, 77, 97, 2)_t$$

Figure 6.1. A search tree node showing an intermediate state of analysis of an utterance and its judgment vector

7 Conclusions

The main points of this contribution are:

1. *Knowledge representation and use* in a general system shell based on semantic networks and a flexible control algorithm allowing well-structured knowledge representation as well as alternating phases of data-driven and model-driven processing.

2. *A system integration* of a speech understanding system from the level of word hypotheses to the pragmatics level interfacing the acoustic front end to the linguistic knowledge base.

3. *Results* for the understanding of several sample sentences, illustrated by one particular sentence, and implemented on a standard hardware using standard software.

Our present and future work is directed towards improved word recognition, extended linguistic competence including a larger vocabulary, integration of dialog and answer generation, improving the efficiency of control and hence of processing, coping with utterances containing more than one sentence, and integrating prosodic cues to word recognition and linguistic analysis.

References

[1] W. Abraham, editor. *Valence, Semantic Case, and Grammatical Relations, Vol. 1.* John Benjamins, Amsterdam, 1978.

[2] W.A. Ainsworth. *Speech Recognition by Machine.* Volume 12 of *IEE Computing Series*, Peter Peregrinus Ltd., London, 1988.

[3] J. Allgayer, K. Harbusch, A. Kobsa, C. Reddig, N. Reithinger, and D. Schmauks. XTRA: a natural language access system to expert systems. *Int. Journ. of Man Machine Studies*, 31:161–195, 1989.

[4] L.R. Bahl, F. Jelinek, and L.R. Mercer. A maximum likelihood approach to speech recognition. *IEEE Trans. Pattern Analysis and Machine Intelligence*, 5:179–190, 1983.

[5] J.M. Baker. Dragon dictate TM-30K: A natural language speech recognition with 30,000 words. In *Proc. European Conf. on Speech Communication and Technology*, pages 161–163, Paris, 1989.

[6] J.S. Bridle. Alphanets: a recurrent 'neural' network architecture with a hidden Markov model interpretation. *Speech Communication; special issue on Neurospeech*, 6, 1989.

[7] R. DeMori. *Computer Models of Speech Using Fuzzy Algorithms.* Plenum Press, New York, 1983.

[8] R. DeMori, J. Bourdeau, and R. Kuhn. A probabilistic approach to person-robot dialogue. In P. Laface and R. De Mori, editors, *Recent Advances on Speech and Language Modeling, NATO ASI Series F*, Springer, Berlin, Heidelberg, 1990.

[9] U. Ehrlich. *Bedeutungsanalyse in einem sprachverstehenden System unter Berücksichtigung pragmatischer Faktoren. Sprache und Information Bd. 22*, Max Niemeyer, Tübingen, 1990.

[10] U. Ehrlich. *Ein Lexikon für das natürlich-sprachliche Dialogsystem EVAR.* Volume 19 of *Arbeitsberichte des Inst. für Mathematische Maschinen und Datenverarbeitung*, Universität Erlangen-Nürnberg, Erlangen, F. R. of Germany, 1986.

[11] R.D. Fennell and V.R. Lesser. Parallelism in artificial intelligence problem solving, a case study of HEARSAY II. *IEEE Trans. Computers*, 26:98–111, 1977.

[12] C. Fillmore. The case for case. In E. Bach and R. Harms, editors, *Universals in Linguisitic Theory*, pages 1–90, Holt, Rinehardt, and Winston, New York, 1968.

[13] L. Fissore, P. Laface, G. Micca, and R. Pieraccini. A word hypothesizer for a large vocabulary continuous speech understanding system. In *Proc. Int. Conf. on Acoustics, Speech, and Signal Processing*, pages 453–456, 1989.

[14] H. Fujisaki, K. Hirose, H. Udegawa, and N. Kanedera. A new approach to continuous speech recognition based on considerations on human process of speech perception. In *Proc Int. Conf. on Acoustics, Speech, and Signal Processing*, pages 1995–1962, Tokyo, 1986.

[15] N. Geschwind. Specializations of the human brain. *Scient. American*, 241:158–168, No. 3 1979.

[16] B.J. Grosz, D.E. Appelt, P.A. Martin, and F.C.N. Pereira. TEAM: An experiment in the design of transportable natural language interfaces. *Artificial Intelligence*, 32:173–244, 1987.

[17] L. Hirschman, F.-M. Lang, J. Dowding, and C. Weir. Porting PUNDIT to the resource management domain. In *Speech and Natural Language Workshop*, pages 277–282, Philadelphia, 1989.

[18] H. Hoege and H. Ney. Das Projekt SPICOS: Organisation und Systemarchitektur. *Kleinheubacher Berichte*, 29:29–36, 1986.

[19] J. Hollingum and G. Cassford. *Speech Technology at Work.* Springer, Berlin, Heidelberg, 1988.

[20] F. Jelinek. The development of an experimental discrete dictation recognizer. *Proc. IEEE*, 73:1616–1624, 1985.

[21] F. Jelinek. Stochastic methods for context free grammars. In *Proc. NATO ASI Speech Understanding*, Cetraro, Italy, 1990.

[22] H. Kitano, H. Tomabechi, T. Mitamura, and H. Iida. A massively parallel model of speech-to-speech dialog translation. In *Proc. European Conf. on Speech Communication and Technology*, pages 198–201, 1989.

[23] D.H. Klatt. Review of the ARPA speech understanding project. *J. Acoust. Soc. America*, 62:1345–1366, 1977.

[24] S. Kunzmann. *Einsatz von Suchstrategien für Worthypothesen in Spracherkennungssystemen*. PhD thesis, Technische Fakultät, Universität Erlangen-Nürnberg, 1989.

[25] K. Lari and S.J. Young. The estimation of stochastic context-free grammars using the inside-outside algorithm. *Computer Speech & Language*, 4(1):35–56, 1990.

[26] K.-F. Lee, H.-W. Hon, and R. Reddy. An overview of the SPHINX speech recognition system. *TransASSP*, 38:35–45, 1990.

[27] S.E. Levinson and L.R. Rabiner. A task-oriented conversational mode speech understanding system. *Bibliotheca Phonetica*, 12:149–196, 1985.

[28] B.T. Lowerre. *The HARPY Speech Recognition System*. PhD thesis, Dept. Comput. Sci., Carnegie-Mellon University, Pittsburgh, PA, 1976.

[29] T.B. Martin. One was to talk to computers. *IEEE Spectrum*, 14(5):35–39, 1977.

[30] M. Mast. *Entwicklung und Realisierung eines Dialogmoduls für ein System zum Verstehen kontinuierlich gesprochener Sprache*. Technical Report, Lehrstuhl für Informatik 5 (Mustererkennung), Universität Erlangen-Nürnberg, Erlangen, 1989.

[31] R. Moore, F. Pereira, and H. Murveit. Integrating speech and natural-language processing. In *Speech and Natural Language Workshop*, pages 243–247, Philadelphia, 1989.

[32] H. Ney. The use of a one-stage dynamic programming algorithm for connected word recognition. *IEEE Trans. Acoustics, Speech, and Signal Processing*, 32:263–266, 1984.

[33] H. Niemann. Control strategies in image and speech understanding. In *Proc. GWAI*, pages 31–49, Springer, Berlin, 1983.

[34] H. Niemann. *Pattern Analysis and Understanding, 2. ed. Springer Series in Information Sciences 4*. Springer, Berlin, Heidelberg, 1990.

[35] H. Niemann, A. Brietzmann, U. Ehrlich, S. Posch, P. Regel, G. Sagerer, R. Salzbrunn, and G.Schukat-Talamazzini. A knowledge based speech understanding system. *Int. J. on Pattern Recognition and Artificial Intelligence*, 2:321–350, 1988.

[36] H. Niemann, A. Brietzmann, U. Ehrlich, and G. Sagerer. Representation of a continuous speech understanding and dialog system in a homogeneous semantic net architecture. In *Proc. Int. Conf. on Acoustics, Speech, and Signal Processing*, pages 30.6.1–30.6.4, Tokyo, Japan, 1986.

[37] H. Niemann, A. Brietzmann, R. Mühlfeld, P. Regel, and G. Schukat. The speech understanding and dialog system EVAR. In R. DeMori and C. Y. Suen, editors, *New Systems and Architectures for Automatic Speech Recognition and Synthesis*, Springer, Berlin, Heidelberg, New York, Tokyo, 1985.

[38] H. Niemann, M. Lang, and G. Sagerer, editors. *Recent Advances in Speech Understanding and Dialog Systems*. Volume 46 of *NATO ASI Series F*, Springer, Berlin, 1988.

[39] H. Niemann, G. Sagerer, and W. Eichhorn. Control strategies in a hierarchical knowledge structure. *Int. J. on Pattern Recognition and Artificial Intelligence*, 2:557–572, 1988.

[40] H. Niemann, G. Sagerer, S. Schröder, and F. Kummert. ERNEST: A semantic network system for pattern understanding. *IEEE Trans. Pattern Analysis and Machine Intelligence*, 9:883–905, 1990.

[41] N.J. Nilsson. *Principles of Artificial Intelligence*. Springer, Berlin, Heidelberg, New York, 1982.

[42] L.R. Rabiner. Mathematical foundations of hidden markov models. In H. Niemann, M. Lang, and G. Sagerer, editors, *Recent Advances in Speech Understanding and Dialog Systems,*, pages 183–205, Springer, Berlin, 1988.

[43] P. Regel. *Akustisch-Phonetische Transkription für die automatische Spracherkennung. Fortschrittberichte VDI Reihe 10 Nr. 83*, VDI Verlag, Düsseldorf, 1988.

[44] G. Sagerer. *Automatisches Verstehen gesprochener Sprache.* Volume 74 of *Reihe Informatik*, BI Wissenschaftsverlag, Mannheim, 1990.

[45] E.G. Schukat-Talamazzini. *Generierung von Worthypothesen in kontinuierlicher Sprache.* Volume 141 of *Informatik Fachberichte*, Springer, Berlin, 1987.

[46] G. Schukat-Talamazzini and H. Niemann. Generating word hypotheses in continuous speech. In *Proc. Int. Conf. on Acoustics, Speech, and Signal Processing*, pages 1565–1568, Tokyo, 1986.

[47] S. Seneff. TINA: A probabilistic syntactic parser for speech understanding systems. In *Speech and Natural Language Workshop*, pages 168–178, Philadelphia, 1989.

[48] L. Tesniere. *Elementes des Syntaxe Structurale.* Klincksieck, Paris, 1966.

[49] W. von Hahn, W. Hoeppner, W. Jameson, and W. Wahlster. The anatomy of the natural language dialog system HAM-RPM. In L. Bolc, editor, *Natural Language Based Computer Systems*, chapter , pages 119–153, Carl Hanser, München, 1980.

[50] W.A. Woods. Optimal search strategies for speech understanding control. *Artificial Intelligence*, 18:295–326, 1982.

[51] S.J. Young and C.E. Proctor. The design and implementation of dialogue control in voice operated database inquiry systems. *Computer Speech & Language*, 3(4):329–353, 1989.

[52] S.R. Young and W.H. Ward. Towards habitable systems: use of world knowledge to dynamically constrain speech recognition. In *Proc. 2. Symp. Advanced Man-Machine Interface*, pages 30.1–30.12, Hawaii, 1988.

[53] V. Zue, J. Glass, D. Goodine, H. Leung, M. Phillips, J. Polifroni, and S. Seneff. The Voyager speech understanding system: a progress report. In *Proc. Second DARPA Speech and Natural Language Workshop*, Harwichport MA, USA, 1989.

Linguistic Processing in a Speech Understanding System

Egidio P. Giachin Claudio Rullent

CSELT - Centro Studi e Laboratori Telecomunicazioni
Via G. Reiss Romoli 274 - 10148 Torino, Italy

Abstract

The goal of a speech understanding system is to correctly identify the action to be taken as a response to a user's voiced request. To this purpose, the system has to rely on some type of linguistic knowledge beside merely recognise words. Several approaches have been proposed to employ language modeling in speech understanding. They include unified architectures integrating modular knowledge sources that account for every level of knowledge from acoustics to linguistics, and two-level architectures in which the separation between recognition and linguistic processing is well defined. Within this approach, two main methods may be conceived: linguistic constraints are integrated into the recognizer, which decodes one string of words that is treated by a natural language interface; or the recogniser produces a scored word lattice that is subsequently processed by a suitable linguistic module. For the present study, this latter approach was considered the most promising one, provided a satisfactory solution to efficient *word lattice parsing* could be found.

Parsing a word lattice is a search activity whose space is extremely large. It may be performed in two basic modes, namely the left-to-right mode and the score-driven middle-out mode. Optimal algorithms based on the left-to-right mode induce a computation that grows polynomially with the lattice length, while those based on the middle-out mode work exponentially with length. However, it is possible to devise score-driven middle-out methods so that the amount of computation they induce depends on the average likelihood score of the word sequence they are expected to output. Hence, if these words are recognised with a good score, computation may get lower than with left-to-right methods.

This paper describes in detail an algorithm that was experimentally proven to exhibit high parsing efficiency in the task it was designed for (1000-word continuous speech understanding, restricted semantic domain, and high syntactic freedom). Improved efficiency is reached through the use of heuristics which, exploiting the redundancy of the middle-out parsing approach, permit to cut down search without sensibly invalidate the optimality of the method. Problems like imperfect determination of start and ending points of words and the absence of short function words from the lattice are also kept into account.

Experimental results, evaluated on lattices produced on the speaker-dependent version of the recogniser available in 1988, show that high-speed speech understanding is feasible compatibly with habitable language models (for a specific application) and reasonable accuracy of comprehension. Parsing time is about 1.8 seconds on a Sun 4 workstation and correct sentence understanding is 82% for a language model of perplexity 25.

1 Introduction

The ultimate goal of a speech understanding system is to correctly identify the action to be taken as a response to a user's voiced request. To this purpose, the system has to rely on some type of linguistic knowledge beside merely recognize words. Generally speaking, we distinguish in the language the *semantic* component and the *syntactic* component. The former component permits to formally define meanings of words and sentences. Both components, moreover, act as constraints on the type and the order of words that may appear in a user's request, and hence may contribute to make word recognition easier.

This research has been partially supported by the EEC within the Esprit 26 project.

NATO ASI Series, Vol. F 75
Speech Recognition and Understanding. Recent Advances
Edited by P. Laface and R. De Mori

There are several approaches for integrating linguistic processing with acoustic-phonetic speech recognition. The choice between one approach or the other is influenced, among other things, by the state-of-the-art of the technologies employed in the system and by the requirements posed by the application for which the system is developed. This paper describes the linguistic processing developed at CSELT within a speech understanding system project started in the mid 80's. Our research group is interested in developing automated voice-based information retrieval services over the telephone network. This task poses some precise implications. First, we have to accept continuous-speech sentences expressed in relatively free syntax, otherwise the interaction with the service would be too unnatural. Second, the system should be speaker independent and able to process noisy and distorted speech. Third, the response time of the system should be short (ideally, it should work in real time), hence system efficiency is crucial.

The linguistic processor operates on a lattice of scored word hypotheses. Language is modeled through a grammar-like formalism related to unification grammars. The language model is used by a fast parsing algorithm able to select the best-scored sequence of word hypotheses from the lattice that is correct according to the model itself. This approach was deemed to provide the best trade-off between accuracy of understanding and efficiency of operation at the time the project was started. However, ongoing research has provided new results for high accuracy speech recognition, which suggest that the system architecture may be modified in the future.

The most relevant component of the linguistic processing system is the fast parsing algorithm. The algorithm is able to take advantage of the fact that the mean score of the sentence to be extracted from the lattice should be good, even though some of its word hypotheses may be bad. Hence the computational activity is very sensitive to such mean score and is on the average low. The description of the parsing method will make the core of this paper. The method allowed to reach about 90% correct sentence understanding, in a speaker-dependent mode, with a 1003-word lexicon, a language model of perplexity 25, and about 3 seconds of processing time per sentence on a Sun 4 workstation.

2 Analyzing the task

2.1 Speech recognition and language processing

As we have said, there are several ways to employ language modeling in speech understanding. Four different approaches may be outlined:

1. A single architecture is used for recognition and for linguistic processing. The different pieces of knowledge involved (acoustic, phonetic, phonological, syntactic, semantic, etc.) are distributed among knowledge sources that individually contribute to the whole work. Under this general paradigm several systems have been developed [7,14]. This approach is interesting from many points of view. A weakness, in our view, is that it is not easy to cope with problems of different nature through specific, focused techniques that may rely on detailed procedural knowledge. Also, this approach poses nontrivial problems of control of the knowledge source activities. The paradigms outlined below, on the contrary, are characterized by more "algorithmic" approaches to the problem of speech understanding.

2. All the linguistic constraints are exploited at the recognition level, after being translated into a formalism suitable for the recognizer itself. The advantage of this approach is that the recognizer need not change its basic architecture and need only decode a single (best-scored) word string. The best grammar formalism for a Viterbi-based recognizer is a Finite State Network (FSN) grammar, possibly reduced to an n-gram grammar. There are two ways of using such a formalism:

   - The FSN is precompiled, or
   - The FSN is dynamically created, on request by the recognizer, by a linguistic module working in parallel.

   These approaches are not risk free, however. The first one, which has been used in several systems [18,2,25], is suitable when the language model is small, but leads to an impractically large FSN if a reasonably free syntax has to be accounted for. The second one, which received interest in very recent times [20], is more promising but posits a frequent dialogue between the recognizer and the linguistic module. Hence, it requires an efficient communication strategy between these two modules, which may be difficult to obtain.

3. The recognizer works with reduced or no language constraints, and produces a set of n word strings ordered by score, which are passed to a written language analyzer. The analyzer selects the best-scored string that satisfies the language constraints. This approach eliminates the need for space-consuming networks at the recognition level and for potentially time-consuming communications with the linguistic processor. However, in order to include the correct string in the n decoded strings, n may grow extremely large. This approach may receive refreshed interest now, after more accurate recognizers and more efficient n-best decoding algorithms have been provided [26].

4. The recognizer produces a lattice of word hypotheses and the linguistic processor extracts from the lattice the best-scored word sequence that satisfies the language constraints. This approach preserves the advantage that the communication between the recognizer and the linguistic processor only happens once, and, with respect to point 3, adds the advantage of a reduced load for the recognizer. In fact, it is true that producing a word lattice is heavier than decoding one single sentence, but a word lattice accounts for such a large number of sentences that an n-best decoding recognizer producing the same number of solutions would be computationally more expensive. The word-lattice approach is the basis of the systems described in [3,4,22,15,5,29]. Using lattices however involves some problems: the lattice may be inaccurate (missing words, ill-defined boundaries of word hypotheses), and the linguistic processor must be able to deal with uncertain input rather than a written sentence, which greatly complicates the search.

Approach no. 4 appeared to be the most promising one, provided a satisfactory solution to efficient parsing of a word lattice could be devised. Also, lattice inaccuracy has to be tackled; however, this problem, though important, is the lesser one and will be treated later below. We will suppose that each word hypothesis (or simply word for brevity) has one definite initial point, final point, and score. ("Full" lattices with many instances per word hypothesis, each having a slightly different location and score, will be mentioned below.) In the following we discuss our approach to fast lattice parsing.

2.2 Parsing word lattices

A word lattice poses many more search problems to a parser than a written sentence. Consequently, limiting search suggests to avoid to allow "gaps" of unparsed input during analysis, that is, every phrase hypothesis generated by combining words should be an "island" of adjacent words. Under the assumption that this principle is followed, there are two basic modes to perform parsing:

1. Left-to-right mode. Parsing begins with the leftmost items and continues to the right, preserving only the best-scored of the subparses that, at a certain time frame, have exactly the same successors. Under this class several parsing methods may be included. If the grammar is expressed by a FSN, then the method outlined above reduces itself to the Dynamic Programming (DP) procedure. If the grammar is represented by a context-free formalism, then this class includes many of the classical parsing algorithms for such a formalism (the CKY and the Earley parsers [1], and their modified versions, in particular those adapted to treat uncertain input and easily amenable to deal with word lattices [17,21,23,19,27]).
2. Middle-out mode. Parsing begins with the best-scored item, which may lie anywhere in the lattice, and expands it with adjacent words. The original item is deleted from the list of active items; the produced items are assigned a score and are introduced into such list. The list is then reordered according to the items' score. In the subsequent cycles, word groups may be expanded by other word groups beside simple words. The procedure is optimal, that is it ends when it finds the first word sequence covering the whole utterance, and that sequence is insured to be the best-scored one. This method is reminiscent of the classical A* search algorithm and has been used in the first DARPA speech understanding project [28].

Methods of the first class are attractive for several reasons. First, with most grammars the parsing time depends *polynomially* on the input length. For instance, if the grammar is a FSN and a DP procedure is employed, time grows linearly with length; with a context-free grammar, the CKY algorithm grows at most as the cube of length and the Earley algorithm may grow as low as its square. Second, linguistic processing may begin before the whole utterance has been processed by the recognizer, thus reducing the total processing time. However, these methods do not exploit the fact that often (according to the accuracy of the recognizer) the word sequence to be decoded has a good overall score; the amount of computation is the same whatever the overall score of the solution may be. The computational load may be high if the grammar imposes constraints between words that are temporally apart from one another (which happens e.g. when morphological agreements have to be taken into account), because in that case it is very unlikely that any two subparses will have identical successors, and hence many paths have to be kept alive.

The methods of the second class, on the other hand, are much more sensitive to the sentence length, but do exploit the value of the solution score. Such statement is justified by the following discussion. (Throughout the rest of the paper, we interpret the word scores produced by the recognizer as log-likelihood scores obtained through a Viterbi-like decoding, i.e. "good" scores are low, and "bad" scores are high.) The goal of parsing is to reach the best-scored *solution*, or word sequence that covers the whole utterance. If word scores are intended as the log-likelihood scores, then for coherence the score of a word sequence should be computed as the sum of the scores of its words.

However, to evaluate the best-scored solution, we may use their ***density score***, defined as the score divided by the time span, because all solutions span the same time interval. Then we may generalize this idea and use the density score as a sorting criterion for *all* phrase hypotheses, which involves the additional benefit that long phrases are not penalized with respect to shorter phrases or single words.

Because optimality is insured, the analysis terminates as soon as a phrase hypothesis is found that covers the whole utterance. Given the way analysis is done, the total amount of computation will depend on the number of different phrase hypotheses that are generated before reaching the solution. Let us compute this number under some simplifying assumptions, for the no grammar case. First, suppose that the score of the word hypotheses is uniformly distributed between a minimum and a maximum value s_1 and s_2. Second, let us suppose the words are grouped in n columns each of which has a time width of t/n frames, where t is the total duration of the utterance, and contains w/n words, where w is the total number of words in the lattice. The first assumption is acceptable for real lattice; the second is not, but we shall see below how we can overcome it. Let s be the density score of the solution to be found. Before finding it, all and only the phrase hypotheses having density score lower than s will be examined. The number of such phrases is

$$G(s,n) = \sum_{i=1}^{n} G_i Pr(s_i < s) \tag{1}$$

where G_i is the total number of phrases having i words, and $Pr(.)$ the probability that the score s_i of such a phrase is less than s. Phrases with i words may be constructed by choosing a word from every possible set of i contiguous columns; hence G_i is expressed by

$$G_i = (n+1-i)(w/n)^i$$

$Pr(.)$ may be estimated by considering that, the hypotheses having equal width, the score s_i is simply the arithmetic average of the scores of the component words. Hence the probability density function $f_\eta(x)$ of s_i is

$$f_\eta(x) = i f_\xi^{*i}(ix)$$

where $f_\xi(x)$ is the pdf of the score s in each column, that we have supposed uniform between s_1 and s_2, and $f_\xi^{*i}(x)$ is obtained by convolving $f_\xi(x)$ with itself i times. If s is small, as we suppose it to be, we can approximate $P(s_i < s)$ by considering only the highest-degree component of $f_\eta(x)$ and integrating up to s. In this way we get

$$Pr(s_i < s) \approx \frac{1}{i!}\left(i\frac{s-s_1}{s_2-s_1}\right)^n$$

Combining the expressions for G_i and $Pr(s_i < s)$ into Eq. 1, and using the upper bound $i^i/i! \leq e^{i-1}$, we finally get the approximate expression

$$G(s,n) \leq \left(\frac{s-s_1}{s_2-s_1}\frac{w}{n}\right)^n n^2 e^{n-1} \tag{2}$$

In all practical situation, this upper bound is highly conservative. It shows, anyway, that $G(s,n)$ grows ***exponentially*** with the input length n and ***polynomially*** with s. This is confirmed by a lower bound obtained considering that $G(s,n)$ must be greater than the number of phrases consisting of words whose score is lower than s. This leads to

$$G(s,n) \geq \left(\frac{s - s_1}{s_2 - s_1} \frac{w}{n}\right)^n$$

Thus the middle-out methods become rapidly intractable as the length of the utterance increases; however, they have the great advantage that the amount of computation decreases enormously when the score of the solution is good. Notice that only the density score matters, so that the presence of a single bad-scored word in the solution does not hamper the analysis.

The situation is not different if we realistically allow the words of the lattice to be uniformly distributed over time. What is changed is the expression of the number of words that will be adjacent to a given word or phrase. This depends on the tolerance we admit for "adjacency". Strict adjacency is very rare because of inaccuracies in the determination of word boundaries. Assuming that such inaccuracies never exceed a value of d frames, the number of adjacent words is given by wd/t. Finally, if we have a grammar of perplexity q, such number is furtherly reduced by a factor of q/l, l being the size of the lexicon. We get the final expression

$$G(s,n) \leq \left(\frac{s - s_1}{s_2 - s_1} \frac{dwq}{tl}\right)^n n^2 e^{n-1} \tag{3}$$

which is similar to Eq. 2.

We must warn that the above analysis is intended to give a qualitative idea of the behavior of middle-out methods and does not take into account other aspects that contribute to the determination of an algorithm complexity; on the other hand, a thorough analysis would require a more detailed description of the algorithm itself. First, we did not consider the amount of time devoted to sorting the phrases at each cycle, an operation that may become critical in the A* or related algorithms. The sorting time however was experimentally proven to be small, provided a moderate amount of memory is devoted to build optimized data structures.
Second, the analysis of the grammar case did not account for the necessity of checking the language constraints before generating a phrase. The load of this operation is critically dependent on how language constraints are practically implemented.
Third, again in the grammar case, a phrase has generally more than one grammatical representation (or parse); these versions must be treated separately because they may join with different words. This causes a computation overload.
Fourth, the above discussion does not evaluate storage requirements. Storage is less crucial than time, though it may grow exceedingly large for certain grammar structures and for big lattices. We do discuss storage problems briefly in the next section.
Finally, and more importantly, the analysis is heavily *redundant*: since parsing "jumps" from one place of the utterance to another, a phrase may be generated in several ways starting from different subphrases or words. One way will correspond to the optimal path (generation of a phrase after all and only the phrases having a worse score have been examined); the other ways will appear later and will correspond to suboptimal paths. Since there is no way of blocking such suboptimal paths, duplication of a phrase may be eliminated only through checks after the phrase has been generated. This is a time-consuming operation; however, we shall see below that analysis redundancy is also a key factor to devising efficient parsing heuristics.

In summary, middle-out strategies have several potentially dangerous bottlenecks and pitfalls. However, there is a powerful advantage intrinsic to them, namely the dependency of computation on the score of the solution. This consideration suggested to devote an effort to find a satisfactory set of heuristics that permit to avoid most of its shortcomings without destroying its optimality.

3 Representing the language model

The linguistic knowledge employed in the system is defined by hand. Consequently, the formalism that should be adopted is difficult to choose because of the conflicting requirements posed by the human who defines the model and by the parser that will use it. The most appropriate formalism to be used by the parser would be a FSN, or, since that would be intractably large, a kind of context-free-like grammar. On the other hand, at the human level it is necessary in the first place that the model naturally suits the structure of language. Also, for the sake of maintenance and development, it would be useful to keep separate representations for syntax and for semantics. This requirement stems from two main reasons. First, the same sets of syntactic constructs may be applied to form sentences expressing different semantic concepts, and it would be impractical to explicitly declare all the combinations syntactic construct - semantic concept. Second, a separate syntactic representation might be reused to a great extent when the semantic component has to be changed because, say, the system has to be adapted to a new application domain; if syntax and semantic were kept together in one single model, the whole model would have to be rewritten.

The above considerations suggest to adopt *two* representations, one for human declaration, the other to be used during parsing. A *compiler* translates the first representation into the second one. During this process, the compiler also brings together constraints of syntactic and of semantic nature. To perform this function, the compiler relies on a further source of information, that is called the *mapping* knowledge, expressing with a high degree of generality how syntactic constituents may be functionally associated with semantic concepts. A more detailed account on the "human-level" representations and the work of the compiler is reported elsewhere [24,11], thus we only recall here the essential ideas.

The compiler would not be able to perform its task, whatever the mapping information, if the formalisms for syntax and semantics were not similar to some extent. This consideration weighed on the choice of language representations together with suggestions deriving from natural language processing expertise. Semantics is represented via the Caseframe formalism [8]; syntax via the Dependency Grammar formalism [16]. Caseframes permits to represent the semantics of a language through "main" concepts, or *headers*, generally verbs or nouns, connected through relations to "secondary" concepts, the *fillers*. A Dependency Grammar expresses the syntactic structure of sentences through rules whose symbols represent morphological categories. The right-hand side of the rule contains one distinguished terminal symbol, that represents in a sense the "main" symbol of the rule. It is called *governor*; the other symbols are called *dependents*. The similarity in structure (both caseframes and dependency rules have a distinguished terminal symbol) is essential for the compiler to perform its work.

3.1 The language model as used during parsing

It was mentioned above that a suitable formalism to be used during parsing would be a context-free grammar (CFG) embedding syntax and semantics. However, using a "pure" CFG poses a problem. Often in a language there are relationships between words so that the presence of one word constrains the other word to have certain characteristics. One typical example is the case of morphologic agreement, in which the gender or number of a word affects the gender or number of a related word; the two words are not necessarily contiguous. A similar case happens when the semantic meaning of a word constrains the semantic type of another word. To build a CFG able to cover all and only the correct constructs, then, it would be necessary to explode all the combinations allowed by the morphologic agreement rules and by semantic compatibility, and hence to write a very large number of rules. This would cause two problems. First, memorizing the rules would simply require a large amount of storage space. This would be particularly hard for Italian, which is very rich in morphological variability. The second problem may be briefly outlined as follows. In a grammar in which all the possible morphologic agreements and semantic compatibility are explicited into different rules, there would be groups of many "similar" rules, having exactly the same structure and symbols, except maybe for one or two nonterminal symbols. Consider a phrase consisting of words pertaining to the "common" part of such rules. This phrase has as many different parses as there are rules. Since all these parses have to be preserved, many memory-consuming operations are required.

In order to avoid the above problems, the compiler generates rules augmented with variables attached to their symbols. The variables refer to morphologic or semantic features. A set of data structures called *conditions* defines the values that may be assigned to the variables and the constraints existing between variables of different symbols. At the beginning, the variables may assume any of the values expressed by the conditions. When a symbol of the rule is filled with a word or a phrase, the values of the corresponding variables are reduced according to the morphologic and semantic features of the word or phrase. Then the values of the other symbols are reduced according to the relationships expressed by the conditions. This will constrain the feature of words or phrases that may be attached to the symbols. Since similar constructs are merged into one single rule, both problems outline above disappear.

The idea of making a grammar rule more general by adding variables to its symbols and defining relationships between variables of different symbols dates back to the formalism of unification grammars. There are several modified versions of this approach, and some have been recently been studied for application to speech recognition [5]. Checking and propagating constraints introduces a computational overload, but an accurate implementation can keep it small as compared to the total analysis time.

The rules employed by the parser depart from the rules of a classical CFG in another respect. We always impose that each rule has at least one terminal symbol, that is called *header* for analogy with the Caseframe header. This insures that, whenever a rule is invoked to parse a fragment of input, it is always possible to check whether the rule may parse another piece of input, simply by looking into the lattice and see whether there are words that may fill the header symbol. Whether header check is performed or not depends on the algorithm that will be explained below, but in this way we definitely avoid long chains of rule-application that may happen if there are rules consisting of nonterminal symbols only. This subject will be discussed again during the description of the parsing algorithm. The constraint of having one terminal symbol per rule is present in some classical CFG formats, such as the Greibach normal form [1]. In the present case it does not imply a loss of generality neither an increase of redundancy (such as a higher number of rules or of symbols) because it naturally mirrors a structural property of the human-level representations (the presence of a header and of a governor), and even facilitates the task of the compiler.

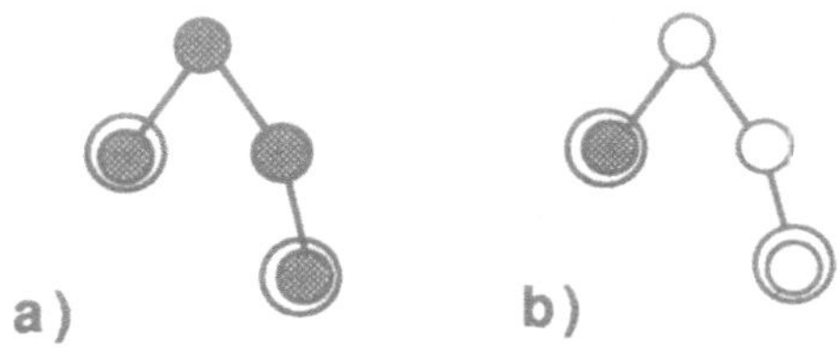

Figure 1. Example of complete and incomplete parse trees

4 Designing an efficient parsing method

4.1 The strict middle-out principle

Throughout this section it may be assumed that the language model is a CFG with a terminal symbol for each rule. The notion of variables added to the rule symbols and of constraint propagation and checking is not relevant here.
The middle-out, best-first parsing principle outlined in Section 2 guarantees optimality but is not practical if one has to follow it rigorously. This principle is shown in Table 1, independently of any implementation issues. (We remind that strict time adjacency is assumed here.) There, by "phrase" we mean any group consisting of one or more words that is consistent according to the language model. A phrase may reduce to a single word, or may be a group of adjacent words contained in a structure called *parse tree* which is constructed applying the grammar rules. A parse tree may be *incomplete* if there are terminal or nonterminal symbols that do not account for any word, or *complete* otherwise. Two examples are shown in Fig. 1; simple circles represent nonterminal nodes; double circles represent terminal nodes; circles are shaded if they account for a piece of parsed input, and are empty if the input corresponding to that node has not yet been parsed. In the following, the terms "phrase" and "parse tree" will be used interchangeably if no ambiguity arises. Different operations may be involved in the generation of a tree; operations will be briefly reviewed below.
Note, finally, that when a new phrase is inserted into the list of generated phrases L_2, it is masked during the cycle in which it was generated; thus, when L_2 is scanned, only phrases generated in the previous cycles are visible.

The generation of phrases must be accompanied by the storage of their corresponding parse tree, since they define the syntactic structure of the phrase itself. In general, a group of adjacent words has more than one parse tree, and all of them have to be memorized. Hence, the parsing method shown in Table 1, when applied with a CFG-like grammar, would create a large quantity of complex parse trees. However, as was noticed earlier, the middle-out parsing method is redundant, i.e. a single parse tree may be obtained through different paths. This assertion will be justified in detail in the next subsection. Hence, several pruning heuristics exist, allowing a broad range of trade-offs between speed and optimality. The main result reported in this paper consists of the definition of a group of heuristics that have been experimentally proven to have a great pruning power with little non-optimality effects. The resulting parsing algorithm, though not complex in itself, is wordy to describe because it involves a large number of elementary steps (such as solving a terminal symbol, or expanding a nonterminal one) that have lengthy description. To make the presentation more concise, let us describe

1. Define a list L_1 of active phrases, ordered by score.
2. Define a list L_2 of generated phrases.
3. Initialize L_1 and L_2 with the words of the lattice.
4. Select the best-score phrase $e(L_1)$ in L_1, and delete it from L_1.
   (a) If L_1 is empty declare failure and stop the process.
   (b) If the element $e(L_1)$ extracted from L_1 covers the whole utterance and is a sentential form according to the language model, output it as the "solution" and stop the process.
   (c) Otherwise, scan L_2 and consider each phrase element $e(L_2)$ in it. Check if it is possible, using the language model, to construct a phrase that include both $e(L_1)$ and $e(L_2)$. Check if temporal adjacency between $e(L_1)$ and $e(L_2)$ is satisfied. Check that the phrase to generate is not already present in L_2. If the checks are positive,
      i. Compute the phrase score.
      ii. Insert the phrase into L_1.
      iii. Insert the phrase into L_2.
5. Loop to point 4.

Table 1. Score guided, middle-out parsing method, with grammar

such operations once for all, and associate conventional names to each of them.

4.2 A list of elementary operations on parse trees

The number of elementary operations that may be performed with a grammar such as the one we have described above is very small. They are shown in Fig. 2, where arbitrary names have been assigned to them. Simple circles represent nonterminal nodes. Double circles represent terminal nodes. Let us briefly describe them in the following. These names will be used below to discuss in more detail the bottlenecks of the parsing method and the heuristics that may be used to cope with them.

1. ACTIVATE - A word from the lattice creates a "phrase" if it matches the terminal symbol of a rule. The parse tree of such phrase may contain some empty slots, which correspond to the nonterminal symbols of the rule. In this way new sentence constituents are hypothesized.
2. FILL - The header of a phrase whose parse tree has already some adjacent constituents is filled by a word from the lattice.
3. EXPAND - A phrase having an empty nonterminal slot expands that slot according to a rule. New constituents are thus hypothesized.
4. PREDICT - A phrase whose root symbol matches a nonterminal symbol of a rule creates a new phrase according to that rule. No new words are added but, like in ACTIVATE, new constituents are hypothesized.
5. JOIN - A phrase whose root symbol matches a nonterminal node of another phrase fills that node. This is similar to FILL, the difference being that in JOIN a nonterminal slot is involved rather than a terminal one.

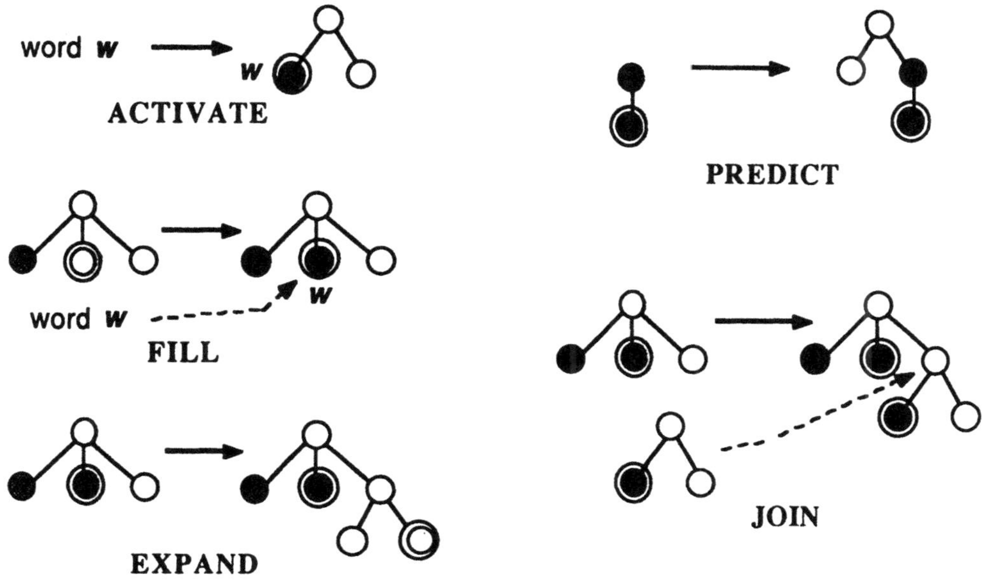

Figure 2. Elementary operations on parse trees

4.3 Optimality and efficiency

The rigorous middle-out, score-driven algorithm of Table 1, though optimal, would create a large number of parse trees. Parse trees would also have complex structures. Consider for example the case shown in Fig. 3. If the word A starts a cycle with the ACTIVATE operation, then in order to complete all the possible ways of joining adjacent hypotheses it may repeatedly go through a series of PREDICT and EXPAND steps until it finds a terminal slot that can be filled with word B through a FILL step. The resulting parse tree is very large, and a lot of time is spent in trying all the possible rules during PREDICT and EXPAND.

The redundancy of the method, however, can be exploited to prune many operations while maintaining a satisfactory level of suboptimality. To illustrate this concept, let

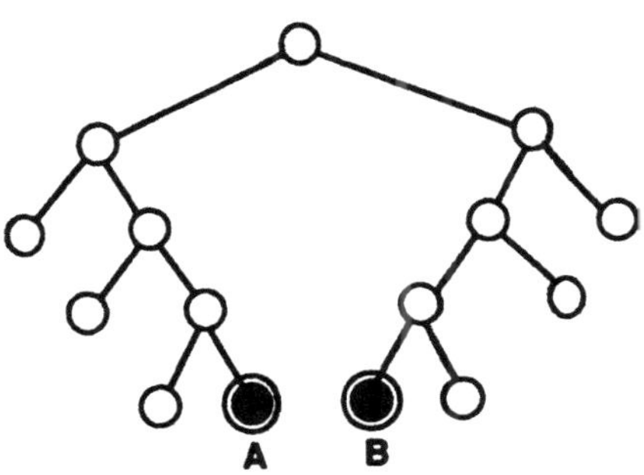

Figure 3. A possible parse tree

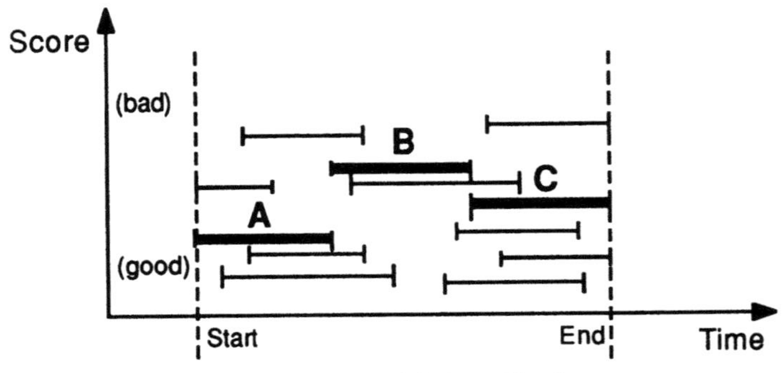

Figure 4. A simplified word lattice

us consider the lattice shown in Fig. 4. (We suppose there is no grammar in this example.) Among other words, three of them, A, B, and C have scores such that $s(A) < s(C) < s(B)$. They may form the three phrases AB, BC, and ABC, with scores such that $s(AB) > s(C)$. Let us suppose that ABC is the solution to be found, i.e. the lowest-score phrase that covers the whole time span. The optimal sequence of operations through which ABC should be formed would be

- 1.1 A is selected, connects with B, creates AB.
- 1.2 C is selected (its score is better than AB), connects with AB, creates ABC.

Suppose now that the activation of A is blocked, thus preventing the generation of AB. Then ABC will be generated as follows:

- 2.1 C is selected, connects with B, creates BC.
- 2.2 BC is selected, connects with A, creates ABC.

Thus the creation of ABC is not prevented by the blocking of AB, because another path is possible that passes through the "alternative" phrase BC instead of AB. This path is suboptimal because, since $s(BC) > s(C)$, the step 2.2 will be performed later than the step 1.2. That is, the output of the solution ABC will be delayed. This in principle destroys optimality, because if there are solutions with score included between $s(BC)$ and $s(ABC)$, these solutions will appear before ABC. However, B and C are a subset of the *correct* solution, thus their mean score $s(BC)$ is likely to be good; this means that the probability that an incorrect solution is generated with score better than $s(BC)$ should not be very high. The above discussion is more and more realistic if the alternative partial phrases, unlike BC which has only two words, have many words.

Redundancy may be exploited in several heuristics. For instance, consider again the phrase with the parse tree shown in Fig. 3. We may avoid the generation of such parse trees simply by preventing the word A from going through series of PREDICT and EXPAND; then the word B will be considered only after the whole left subtree is filled, but there are many words in such subtree, and hence it is unlikely that this kind of pruning will allow incorrect solution to come up first.

The great advantage of heuristics of this kind is that they prune *incorrect* phrases (i.e., phrases that are not part of the best-scored solution) beside correct phrases. Incorrect phrases are many more than correct phrases, so it may be expected that the speed of analysis should be greatly increased. Unfortunately, it is virtually impossible to analytically predict the effect of a heuristic on optimality and on efficiency. Thus experimentation is crucial to assess their effectiveness. The next section discusses the heuristics that have been studied and presents the resulting parsing algorithm as is implemented at the end of the project.

4.4 Finding useful heuristics

Here follows a series of heuristics for pruning search that have been experimentally proven to have limited or no effect on optimality. The increase of sentence error rate due to the non-optimality induced by the heuristics is about 1.5%, i.e. about one order of magnitude less than the error rate due to every other possible cause. (Performances figures are discussed in more detail below.) The decrease of computational load with respect to a completely optimal system, estimated to be more than one order of magnitude, has been evaluated through partial experimentation on the individual heuristics. Heuristics are applied at each cycle. We remind that, according to the basic method, any combination of operations may take place; the only limitation is that the selected element has to be expanded with other words, hence any series of operations must be concluded with a FILL or JOIN operation.

All these heuristics, in principle, invalidate optimality, under the strong assumption that every phrase we construct consists of adjacent words. This assumption was taken as early as in Section 2.2 on the grounds that, if you allow the language model to combine words from the lattice leaving unparsed segments between them, a tremendous amount of incomplete parse trees will be generated. There are some heuristics, however, such as no. 4 (related to filling terminal nodes with words of the lattice), that are very helpful for pruning search, but decrease optimality too much under the above assumption. In this case, we may permit a word to fill a terminal node of a phrase even if there is an unparsed segment between this word and the rest of the phrase. This subject will be better described below.

1) Limit "sparse" parse trees - Given a parse tree, the PREDICT and EXPAND operations create bigger parse trees having a higher number of empty (unsolved) terminal or nonterminal nodes. Clearly, applying repeatedly such operations creates a large number of "sparse" trees such as the one shown in Fig. 3. This work is time- and memory-consuming, though in principle it is essential for optimality. However, as was said in the previous section, optimality is likely to take place in practice even if such trees are avoided. Therefore, in a cycle it is imposed that *only one* PREDICT *or* EXPAND *operation per rule may be applied to the selected element.* In other words, given the selected element, we may apply one PREDICT or EXPAND on it for every rule, but the next operation on the new generated trees must be a FILL or JOIN.

2) Select one node to expand - In general, the selected parse tree has two nonterminal nodes to which EXPAND or JOIN may be applied (they are the nodes at the left and at the right of the nodes that are already filled with words). In this case, it is imposed that *only one is selected*, and the other is discarded. Several criteria exist to operate the selection.

3) Limit the tree branching - Suppose we have the phrase shown in Fig. 5. Two possible initial operations on it are to apply EXPAND to node A or to node B, using all the suitable rules. Let us suppose that, in both cases, the next operation is a FILL, in agreement with the first heuristic. According to the second heuristic we have to choose either node A or B. While choosing B is not necessarily better than choosing A for the sake of parsing efficiency, we would like to apply EXPAND to B for the following reason. If we apply EXPAND, followed by FILL, to A, we get the structure in Fig. 6a which has two incomplete subtrees at the same level (at nodes A and B). This structure has to be

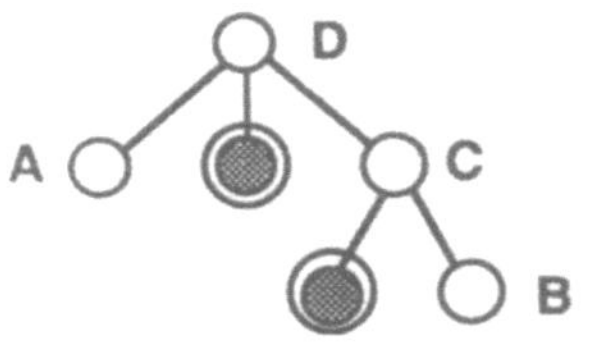

Figure 5. A parse tree to treat

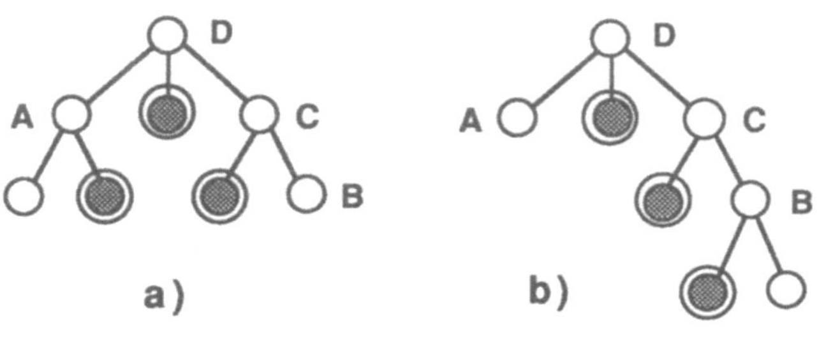

Figure 6. Two possible expanded parse trees

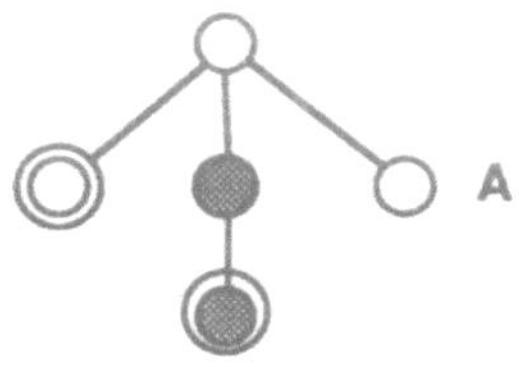

Figure 7. A parse tree with an empty header node

memorized; the parent structure is deleted from the list of active parses L_1 but must still be kept in L_2 because it might be used by some other parse to expand itself. The common part to the trees in Figs. 5 and 6a might be shared using suitable structures. However, we must remember that the grammar rules are not exactly those of a CFG, because there are *conditions* between symbols, as was said in Section 3. This means that mutual constraints may exist between the empty nodes of the subtrees rooted in A and C respectively. The need to represent such constraints makes it virtually useless the sharing of structures. Suppose, instead, that node B is selected. The structure in Fig. 6b is then obtained, which has a single (longer) incomplete subtree rooted in C. Of course, even in this case there are mutual constraints between the node A and the empty node of the subtree rooted in B. However, if the present heuristic is regularly applied during parsing, in no occasion will the node A be selected for expansion as long as there is an incomplete subtree in C. Therefore, constraints need only to be memorized for the subtree in B. Constraints to the node A will be set only when the whole subtree in C will be completed with words. This permits an easy sharing of common parts and a substantial saving of memory. The following heuristic is then imposed: When the EXPAND operation is performed, ***the node to be expanded is always chosen among the highest-depth ones.***

When the tree of Fig. 6b is generated, a single new subtree is created in practice, namely the one rooted in B; to indicate that there are constraints on it, a pointer to its *father* structure shown in Fig. 5 is added to this subtree. For brevity, we say that this subtree is a *context tree*, the context being its father structure. Notice that, in its turn, the father tree is actually composed of a context tree (rooted in C) and a regular (no context) tree, rooted in D.

4) Check terminal symbols first - The above heuristics still allow a large amount of work to do in order to "grow" the selected phrase with other words or phrases during a cycle. Consider a case in which a phrase may be grown in two ways, one way being the solving of a terminal symbol, or header, through the FILL operation. An example is shown in Fig. 7, where the other operations are an EXPAND and a JOIN on node A. If node A is expanded according to all the possible rules, looking for phrases to be joined, many new phrases will be created, all of which have their header unresolved. This work may prove useless if no word from the lattice will be able to fill the headers. Hence we may expect that a large quantity of pruning can be achieved if we introduce the following rule: ***When a phrase has its header unresolved, just apply*** FILL ***to it; no other***

operation should be performed. This heuristic proved itself so useful that an extension of it has been experimented: The check for words to fill the header is done even if there are no adjacent words to it. As was said in the introduction of this section, relaxing the time adjacency constraint decreases efficiency because more incomplete parse trees may be generated. However, experiments proved that the pruning power of this heuristics prevails, and the shift from optimality is sensibly reduced, at least for some particular type of rules.

5) **Freeze bad-scored headers** - When a header is solved, in general many words will be found to be suitable for a FILL operation. These words will have different scores and hence many new phrases with different scores will be generated. All of them have to be visible by other phrases willing to use them to expand themselves, and hence they are inserted into the list L_2 at the end of the cycle. However, phrases with bad-scored headers are unlikely to be essential for building the correct solution. Then we impose the following rule: *Only the best-scored new phrase is inserted into* L_2. All of them are regularly inserted into L_1. When one of them is popped from L_1, it is also inserted into L_2.
This heuristic makes some unlikely phrases invisible to the parser for some time. In the actual implementation, it is not even necessary to actually generate these phrases before popping them from L_1, thus saving also memory occupation.

6) **Limit expansion** - It has experimentally been observed that, even applying the previous heuristics, many incomplete parse trees were still created, that were never successful in growing to a solution. To decrease the number of incomplete parses, we *limit the* JOIN *operation to join only complete phrases.*

7) **Limit prediction** - A further limitation on the generation of incomplete phrases is implemented by *constraining the* PREDICT *operation to be applied only on complete phrases.*

8) **Avoid duplication** - It has been noted before that, owing to the algorithm redundancy, it is impossible to avoid phrase duplications except by checking it a posteriori. This operation is time-consuming. The above heuristics reduce but do not eliminate redundancy, thus a certain amount of duplication checking still remains. However, we may take advantage of it by using it to reject phrases that, although different, are very similar, for instance because they differ only for short function words having the same meaning.

4.5 The parsing algorithm

The algorithm of Table 1, once the heuristics are applied, is transformed into the one shown in Table 2. This description leaves out some implementation issues. For example, the lists L_1 and L_2 are compacted into one single structure, that is optimized in order to permit a fast access to its elements when operations like JOIN are involved. Similarly, to facilitate the FILL operation, the lattice of word hypotheses is organized into a suitable structure.

The algorithm should be read as a main task with some routines. For each routine the type of argument that is passed to it is given. A remark must be made. Consider again the tree shown in Fig. 6b. When the empty node of the subtree rooted in B is

filled, two complete trees are generated as a by-product of this operation, namely those rooted in C and in B. (This may happen in different situations from the one shown here.) These trees are then duplicated as separate entities and placed individually in the lists. This operation is referred to as Decontext in the table.
Note, finally, that when a new element is inserted into L_2, it is assumed that it is also masked during that particular cycle. By "terminal rule" we mean a rule that has no nonterminal symbols in its right-hand side.

---

1. Define a list L_1 of active elements (words from the lattice or generated parse trees), ordered by score.
2. Define a list L_2 of generated parse trees.
3. Initialize L_1 and L_2 with the words of the lattice.
4. Select the best-score element $e(L_1)$ in L_1, and delete it from L_1. Do different operations according to its characteristics:
   (a) If L_1 is empty declare failure and stop the process.
   (b) If the element $e(L_1)$ extracted from L_1 covers the whole utterance and is a sentential form according to the language model, output it as the "solution" and stop the process.
   (c) If $e(L_1)$ is a single word, do **Activation** on it.
   (d) If $e(L_1)$ is a complete parse, do **Completion** on it.
   (e) If $e(L_1)$ is an incomplete parse with a filled header, do **Resolution** on it.
   (f) If $e(L_1)$ is an incomplete parse with an empty header, do **Verify** on it.
   (g) If $e(L_1)$ is marked as header-frozen, do **Reinsert** on it.
5. Loop to point 4.

Activation (word w):

1. Find all the terminal rules that may use the word w as a header. For each of them:
   (a) Apply FILL and create a complete tree C containing only the word w.
   (b) Insert C into L_2 only.
   (c) Do **Completion** on C.
2. Find all the nonterminal rules that may use the word w as a header. For each of them:
   (a) Apply FILL and create an incomplete tree I containing only the word w.
   (b) Insert I into L_2 only.
   (c) Do **Resolution** on I.

Completion (complete parse C):

1. Find all the rules that may use the complete tree C to fill a nonterminal symbol. For each of them:
   (a) Apply PREDICT and hypothesize an incomplete tree I containing only the C as a subtree. (I will have an empty header node.)
   (b) Do **Verify** on I, creating new trees with a filled header.
   (c) Compute the scores of the new trees.
   (d) Insert the new trees into L_1, marking them as header-frozen, except the best-scored one.
   (e) Insert the best-scored tree into L_2.
2. Scan L_2 looking for incomplete trees I that may use C to fill a nonterminal node, satisfying all the temporal and language constraints.

3. Apply JOIN on I and C, creating new trees.
4. Compute the scores of the new trees.
5. Insert the new trees into L_1.
6. Insert the new trees into L_2.

Resolution (incomplete parse I, with a filled header):

1. Select an empty nonterminal node to expand.
2. Scan L_2 looking for complete trees C that may fill the selected node, satisfying all the temporal and language constraints.
3. Apply JOIN on I and C, creating new trees.
4. Compute the scores of the new trees.
5. Insert the new trees into L_1.
6. Insert the new trees into L_2.
7. If the new trees are complete trees, and if I is a context-tree, do **Decontext** on them.
8. For each rule that may expand the selected node,
   (a) Apply EXPAND, hypothesizing a tree I_1 with new empty nodes (among which a header node).
   (b) Do **Verify**, creating new trees.
   (c) Compute the scores of the new trees.
   (d) Insert the new trees into L_1, marking them as header-frozen, except the best-scored one.
   (e) Insert the best-scored tree into L_2.
   (f) If the new trees are complete trees (which happens only if the rule is terminal), do **Decontext** on them (the new trees are context-tree by construction).

Verify (incomplete parse I, with an empty header node):

1. Scan the lattice looking for words that may fill the header node of I, satisfying the temporal and language constraints.
2. Apply FILL, creating new trees having such words as headers of I.

Reinsert (generic tree T marked as header-frozen):

1. Insert the tree T into L_2.
2. If T covers the whole utterance and is a sentential form according to the language model, output it as the "solution" and stop the process.
3. If T is a complete parse, do **Completion** on T.
4. If T is an incomplete parse with a filled header, do **Resolution** on T.

Decontext (complete parse C):

1. Trace back the father I of the complete context-tree C.
2. Apply JOIN to I and C, creating a new tree.
3. Compute the score of the new tree.
4. Insert the new tree into L_1.
5. Insert the new tree into L_2.
6. If the new tree is a complete tree, and if I was in its turn a context-tree, do **Decontext**.

Table 2. Score guided, heuristic middle-out parsing algorithm

5 Dealing with inaccuracies of word lattices

Up to this point we have assumed that a lattice is *perfect*, i.e. that there is always a solution (or a sequence of words satisfying the language constraints) that spans exactly the whole utterance and consists of words that are perfectly aligned, so that their begin and end times match precisely with one another. Unfortunately, this is not a realistic situation. There are several reasons why this does not happen:

1. Word begin and end times are approximate. This is an intrinsic limitation of a recognition system. The transition from a phone to another one is not generally abrupt; consequently, when the recognizer tries to decode a word, it will find many different versions of that word, each having different begin and end times. All these different versions may get into the lattice as independent words; this permits to find a perfect temporal match between two adjacent words, simply by choosing the most appropriate versions, but greatly complicates search. The solution that has been chosen here consists in putting into the lattice only the best-scored version for each word. Now, if we consider all the versions of a word that have, say, the same begin time, we will notice that in general the score remains good for a wide range of different ending times, and falls down abruptly outside this interval. The best-scored version may have its end point anywhere in this range, whose average duration is of the order of 0.1 s. Consequently, the estimated begin and end times for words are largely arbitrary within this range. The result is that there are usually gaps or overlaps between words that had been uttered in sequence.
2. Short words are not reliable. On the one hand, in fact, short words (and especially function words, such as articles, some prepositions, auxiliary verbs, etc.) tend to be uttered less clearly and are affected by coarticulation with neighboring words. Consequently, they are difficult to recognize and their scores may be worse than the average scores for "normal" words. On the other hand, the phonetic description of short words is often very similar to a portion of the phonetic description of a longer word. Hence, short words may match a portion of the input even if they have not been uttered. These "false" words, which may have good scores, mislead the parser and increment the search activity.

A simple way of overcoming the problem of gaps and overlaps is to declare "adjacent" two words even if their terminal points do not match perfectly. Two thresholds are used, one for the maximum value of a gap, the other for an overlap. Low thresholds may prevent the correct solution from being assembled; high thresholds increase the number of possible word groupings, among which some incorrect solutions may exist; the search activity is also increased. In practice, the threshold values are experimentally determined in order to optimize performance.

This method proved satisfactory in the majority of cases, but is not able to account for regularities such as a big gap before a word beginning with a plosive, or a big overlap between two words that have a common phone at their boundary. Hence a more detailed model of coarticulation has been experimented [6]. The models represents the probability of having a gap or an overlap of given amplitude between various classes of phones. Whenever two words are connected into a phrase during parsing, these probabilities are used as a penalty associated to the boundary between the words. The penalty is added to the score of the generated phrase. Frequent cases of gaps or overlaps, such as the ones cited above, receive a very low penalty and do not prevent words to be considered "adjacent" and hence grouped into a phrase. The converse is true for unlikely coarticulations.

The case of the short words is more complex. Some of them are "content" words, that is they are necessary to determine the meaning of the sentence. In that case, if a word of such kind is not present in the lattice, the analysis will fail. On the other hand, many short words are function words that act as syntactic markers, but are not semantically meaningful. In this case, it would be desirable to skip them when parsing a lattice. Notice, however, that not all function words are short and unreliably scored, and there is no reason to skip such kind of words.

The parsing algorithm has been modified so that every terminal symbol pertaining to a function word may be marked as "short", "long", or "unknown" according to its morphologic features. If it is marked as "long" it cannot be skipped. If it is marked as "short" it is always skipped, and a temporal "hole" is set in the place where the word should be. If it is marked as "unknown" both options are followed. Details of this method are given in [11]. Results show that about 75% of lattices in which at least one function word is missing may be successfully parsed with this technique at the expense of a moderate increase of computational activity.

A more refined method has been tested to keep into account regularities in the misrecognition of function words [6]. As in the case of coarticulation, penalties are added to a phrase whenever a function word is skipped. The penalties reflect the probability that such a word is not recognized correctly. This method yielded a further increase of understanding accuracy.

6 Experimental results

Two series of results are presented here. The first one shows the performance of the algorithm described above, used in combination with a language model of the type described in Section 3. The second series assesses the robustness of the system against lattice inaccuracies. Unless differently stated, the analyses are performed on word lattices produced by a speaker-dependent recognizer as described in [10]. The main features of the recognizer are summarized as follows. The signal is sampled at a 12 kHz rate and 18 Mel-based cepstrum coefficients are computed. The samples are vector-quantized using a size 144 codebook. Words are phonetically modeled via 123 subword units represented by discrete HMMs. Words are decoded using a modified Forward-backward procedure; this algorithm has been experimentally proven to provide more precise likelihood scores and word boundaries than a Viterbi decoder, though the computational load is substantially higher [9].

6.1 The algorithm performance

The parser has been evaluated on 210 lattices obtained with the recognizer outlined above. Preliminary experimentation with a refined, speaker independent recognizer based on continuous density HMMs and context-dependent subword units is still in course. The overall performance of the parser are given in Table 3. The perplexity of a language model can be computed directly if the language is represented as a regular or context-free grammar, but if, as in the present case, the language representation is more complex, perplexity can be evaluated only indirectly through an experimental measure of the average branching factor. The value resulted to be about 25. The

Perplexity	Ave. lattice size	Correct sent. understanding	Correct sent. recognition	Word accuracy	No. of phrases	Parsing time
25	632	82%	65%	92%	268	1.8 s

Table 3. Experimental results

System	Correct identification rate %	Failures %						
		Resources	Missing words	Gap	Overlap	Better-scored sentence	Non optimality	Total
Understanding	82	1	4.5	1.5	2.5	7.5	1.0	18
Recognition	65	1.5	8	2.5	4	16.5	2.5	35

Table 4. A breakdown of analysis failures

sentence correct understanding rate is the most significant accuracy measure from our viewpoint. However, since the goal of the understanding level also includes the completion of recognition, we have added two typical measures used in speech recognition, namely the *sentence correct recognition rate* and the *word accuracy*, defined as

$$a = 100 * (n - s - d - i)/n$$

where s, d, i, and n are the number of word substitutions, deletions, insertions and the total number of words. All errors are equally treated. The parser has been implemented on a Sun 4 workstation and is written in C. The analysis time has been measured on such an environment. Such a figure makes the system performance promising even when it will treat speaker-independent lattices, which are supposed to contain more incorrect words able to lead astray the analysis.

Table 4 gives a breakdown of the type of failures encountered by the system. Failures have been partitioned into six classes, according to their causes: *resources*, i.e. the time required for parsing exceeded the allowed timeout (10 seconds); *missing* correct *words* in the lattice; exceeding *gaps* and *overlaps* between words; *better scored* consistent wrong *sentences*; and finally errors due to the residual *non optimality* of the strategy. The number of recognition or understanding errors due to the nonoptimality of the heuristics is indeed quite low, about one order of magnitude lower than the total number of errors. A significant number of errors is due to the absence of some significant word in the lattice. This type of inaccuracy, however, should become less frequent once the new acoustic-phonetic models are used in the recognizer. In the same way the failures due to excessive gaps or overlaps between correct words should decrease in number.

6.2 The effect of lattice inaccuracies

The results presented in the previous section were obtained with a modified parsing strategy able to occasionally skip short function words. The details of the modified strategy are described in [11]. In this way, the parser has been able to successfully analyze lattices in which some of these words were not recognized and hence would have led the baseline version of the parser to failure. Notice that it is possible to skip even two or three function words in sequence. This is a novel improvement over systems that, to our knowledge, only admit one single skippable word [27] or recognize any configuration of missing words but do not distinguish cases in which the information content of an absent word can be ignored [13]. To better show the advantage of using

missing jolly words per sentence	1	2	3
n. of sentences	40	18	4
successfully parsed	35	15	3
average n. of generated phrases	318	440	563

Table 5. Successful parsing

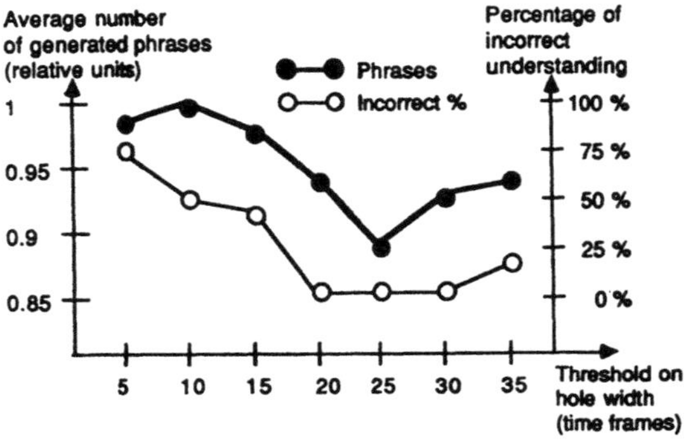

Figure 8. Performance vs. width threshold

the modified parsing strategy, Table 5 displays the number of lattices, corresponding to the sentences containing at least one function word, in which some of such words are missing. It is seen that about 75% of them have been successfully understood. This figure does not change substantially as the number of missing function words per sentence increases, and hence indicates robustness. The computational load, given by the number of generated phrases, is somewhat affected by the number of missing jollies. However, this is mainly due to the fact that sentences with many jollies are also longer and syntactically complex. The actual efficiency can be better estimated from Fig. 8, where the average number of generated phrases is plot as a function of the threshold on the maximum width of the "hole" that goes in the place of the missing word. The figure also displays the amount of parsing failures related to function word problems (failures due to other reasons have been ignored for simplicity). The curve indicates that raising the threshold does not change much the number generated phrases (the relative oscillations of the values are small). This means that the relaxation of constraints during the application of the modified parsing method is not a source of inefficiency. Moreover, there is a large range of values for which the parsing failure remains low.
The curve also shows that relaxing constraints may even speed up the parsing. This can be easily explained. When the threshold is low, no function word is skipped, and failure occurs when any function word is missing from the lattice. When the threshold is raised, skipping begins to work: good-scored function words that were not actually uttered are no more a source of disturbance, and correct but bad-scored words are skipped thus avoiding to delay the parsing; as a consequence the overall number of

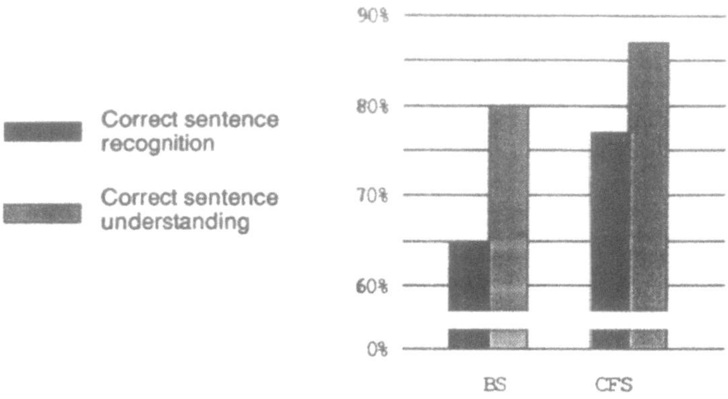

Figure 9. Sentence correct recognition and understanding rates

phrases decreases. Further enlarging the threshold reverts this tendency, since the too-much-relaxed constraints allow the aggregation of words that would have been discarded with stricter constraints.

All the previous data were obtained using fixed thresholds for maximum gaps and overlaps between supposedly adjacent words. Some improvements have been obtained using a simple probabilistic model for gaps and overlaps depending on the type of coarticulation between the phones of adjacent words. This method, mentioned in Section 5, is described in [6]. A similar method is used to account for function word losses in the lattices. Fig. 9 gives the sentence correct recognition and understanding rates for the Baseline System (BS) and a modified version (CFS) incorporating the probabilistic model for coarticulation and for function word loss. Sentence understanding is considered successful if its semantic representation is correct. Sentence recognition is successful if word-by-word recognition is correct. Considerable improvements are obtained over the BS. However, it must be kept present that, while the baseline system could deliver the first solution that was found, the modified system had to wait until the maximum timeout is reached or a fixed maximum number of phrases had been generated.

7 Further developments

A major current goal is to increase the word and sentence accuracy of the linguistic processor. On one side, this means to take advantage of the recent advancements in speech recognition, notably more powerful statistical models of speech, such as continuous-density HMMs, and more refined sets of sub-word units. On the other side, the overall architecture of the system is supposed to change. Up to this point, it has be assumed that the information flows only in one direction, from the recognizer to the linguistic processor, and that such information consists of a lattice of word hypotheses. This approach has been experimentally shown to provide satisfactory results, especially in terms of efficiency, however it involves a lack of precision in determining the word location and score that may only partially be recovered with the techniques described in the previous section. Since the speed of the linguistic processor is high, we are now evaluating the following approach. The linguistic processor, instead of stopping at the first solution found, proceeds until a fixed amount of time is exhausted. By this time, n different solutions will be produced. All these sentences are then passed back to the

recognizer in order to be aligned against the acoustic data and to receive a likelihood score. Finally, the most likely sentence is selected.
The advantage of this method is that the recognizer, when called by the linguistic processor, works on a limited set of words and hence its operation is very fast. This permits to use more detailed sources of acoustic/phonetic information, such as models of inter-word coarticulation, that have been proven to yield significant accuracy improvements, at least for other languages [12]. Another advantage is that it is possible to skip also semantic meaningful short words during parsing, as the discrimination will be performed during this last phase and the resulting parse tree consequently modified.

A second major objective is to add to the system a dialogue capability. The importance of dialogue is manyfold. It may permit a more natural interaction between the user and the machine in cases where the requests would be too cumbersome to express in one single sentence. This happens in several situations, for example when a reference to concepts mentioned in earlier sentences is necessary. On the other hand, a dialogue component should enhance understanding accuracy because, in the middle of an interaction, the machine may have some expectations on what the user will utter in his or her next sentence, and consequently may use more restricted linguistic knowledge and/or recognition dictionary. To achieve this result, however, it is essential that the dialogue be driven mostly by the machine, rather than the human. Finally, dialogue permits to recover information that is lost owing to recognition or understanding failures, for example by querying the user on an essential word that has not been recognized, or asking him or her to use simpler language constructs if there had been a parsing failure.

References

[1] A.V.Aho, J.D.Ullman, *The Theory of Parsing, Translation, and Compiling*, Englewood Cliffs, NJ: Prentice-Hall, 1972.

[2] L.R.Bahl, F.Jelinek, and R.L.Mercer, "A maximum likelihood approach to continuous speech recognition", *IEEE Trans. on PAMI*, vol. PAMI-5, pp. 179-190, March 1983.

[3] D.Bigorgne, A.Cozannet, M.Guyomard, G.Mercier, L.Miclet, M.Querre, and J.Siroux, "A versatile speaker-dependent continuous speech understanding system", *Proc. ICASSP 88*, New York, April 1988.

[4] A.Brietzmann and U.Ehrlich, "The role of semantic processing in an automatic speech understanding system", *Proc COLING 86*, Bonn, August 1986.

[5] Y.-L.Chow, S.Roukos, "Speech understanding using a unification grammar", *Proc ICASSP 89*, Glasgow, Scotland, May, 1989.

[6] M.DeMattia and E.P.Giachin, "Experimental results on large vocabulary continuous speech understanding", *Proc ICASSP 89*, Glasgow, Scotland, May, 1989.

[7] L.D.Erman, F.Hayes-Roth, V.R.Lesser, D.Raj Reddy, "The Hearsay-II speech understanding system: integrating knowledge to resolve uncertainty", *ACM Computing Survey*, vol. 12, 1980.

[8] C.J.Fillmore, "The case for case", in Bach, Harris (Eds.), *Universals in Linguistic Theory*, Holt, Rinehart, and Winston, New York, 1968.

[9] L.Fissore, E.P.Giachin, P.Laface, G.Micca, R.Pieraccini, and C.Rullent, "Experimental results on large vocabulary continuous speech recognition and understanding", *Proc ICASSP 88*, New York, April 1988.

[10] L.Fissore, P.Laface, G.Micca, and R.Pieraccini, "Interaction between fast lexical access and word verification in large vocabulary continuous speech recognition",*Proc ICASSP 88*, New York, April 1988.

[11] E.P.Giachin and C.Rullent, "Robust parsing of severely corrupted spoken utterances", *Proc. COLING-88*, Budapest, August 1988.

[12] E.P.Giachin, A.E.Rosenberg, and C.-H.Lee, "Word juncture modeling using phonological rules for HMM-based continuous speech recognition", *Proc ICASSP 90*, Albuquerque, NM, April, 1990.

[13] G.Goerz and C.Beckstein, "How to parse gaps in spoken utterances", *Proc. 1st Conf. Europ. Chapt. ACL*, 1983.

[14] Y.F.Gong, J.P.Haton, "A specialist society for continuous speech understanding", *Proc ICASSP 88*, New York, April 1988.

[15] P.J.Hayes, A.G.Hauptmann, J.G.Carbonell, and M.Tomita, "Parsing spoken language: a semantic caseframe approach", *Proc. COLING 86*, Bonn, WG, August 1986.

[16] D.G.Hays, "Dependency theory: a formalism and some observations", Memorandum RM4087 P.R., The Rand Corporation, 1964.

[17] S.E.Levinson, "Structural methods in automatic speech recognition", *Proc of the IEEE*, vol. 73, no. 11, pp.1625-1650, Nov.1985.

[18] B.T.Lowerre and D.R.Reddy, "The Harpy speech understanding system", in *Trends in Speech recognition*, W.Lea, Ed., Englewood Cliffs, NJ: Prentice-Hall, 1980, pp.340-360.

[19] L.G.Miller and S.E.Levinson, "Syntactic analysis for large vocabulary speech recognition using a context-free covering grammar", *Proc ICASSP 88*, New York, April 1988.

[20] H.Murveit and R.Moore, "Integrating natural language constraints into HMM-based speech recognition", *Proc ICASSP 90*, Albuquerque, NM, April, 1990.

[21] H.Ney, "Dynamic programming speech recognition using a context-free grammar", *Proc ICASSP 87*, Dallas, April 1987.

[22] G.T.Niedermair, "Divided and valency-oriented parsing in speech understanding", *Proc. COLING 86*, Bonn, August 1986.

[23] A.Paeseler, "Modification of Earley's algorithm for speech recognition", in Niemann, Lang, and Sagerer (Eds.), *Recent advances in speech understanding and dialogue systems*, NATO ASI series, Springer Verlag, 1987.

[24] M.Poesio and C.Rullent, "Modified caseframe parsing for speech understanding systems", *Proc. IJCAI 87*, Milano, August 1987.

[25] L.R.Rabiner and S.E.Levinson, "A speaker-independent, syntax-directed, connected word recognition system based on HMMs and level building", *IEEE Trans. on ASSP*, vol. ASSP-33, pp.561-573, June 1985.

[26] R.Schwartz and Y.-L.Chow, "The n-best algorithm: an efficient and exact procedure for finding the n most likely sentence hypotheses", *Proc ICASSP 90*, Albuquerque, NM, April 1990.

[27] M.Tomita, "An efficient word lattice parsing algorithm for continuous speech recognition", *Proc ICASSP 86*, Tokio, April 1986.

[28] W.A.Woods, "Optimal search strategies for speech understanding control", *Artificial Intelligence*, vol. 18, 1982.

[29] V.Zue, J.Glass, D.Goodine, H.Leung, M.Phillips, J.Polifroni, and S.Seneff, "The VOYAGER speech understanding system: preliminary development and evaluation", *Proc ICASSP 90*, Albuquerque, NM, April, 1990.

LINGUISTIC TOOLS FOR SPEECH RECOGNITION AND UNDERSTANDING

Rodolfo Delmonte

Dipartimento di Linguistica e Glottodidattica - Ca' Garzoni-Moro - San Marco 3417 - Università degli Studi 30124 VENEZIA (Italy)

Abstract
A realistic model of speech recognition and understanding should be heavily based both on linguistic and acoustic knowledge. If this fact seems to be acknowledged by most reasearch people in the field it is not yet clear how to thread that knowledge into acoustic evidence. We present a proposal that tries to solve some of the problems involved in this task. In particular, acoustic patterns made up of word models are substitued by a model which makes use of features and syllables. The phone (the phoneme is too abstract!) and the word are regarded as abstract objects which are built up the former from feature matrixes, and the latter from syllables and morphological parsing into morphemes. Thus the lexicon is not a list of word forms but is composed of root morphemes and affixes in graph structure, and is traversed by a parser which makes use of rules for the composition of legal words of the language from subword units. Hypotheses are fired out both at phone and at syllable level on the basis of feature extraction.

Keywords
Speech recognition & understanding, Linguistic Tools, Syllable Structure, Morphological Parsing, Feature extraction .

Introduction
In order to cope with continuous speech recognition and understanding(SRU), a number of systems have been created which rely on mathematical or statistical techniques to account for contextual variability of the speech signal. In order to produce the desired output, the system must perform pattern matching on a set of adequately smoothed sample dataset, which is produced by recording a large enough amount of data. The model is usually based on word units - and sometimes subword units like triphones - and is trained on the recorded data. The model will subsequently be used to generate predictions on the incoming observed data, and to guide the recognition process accordingly. Fitting procedures of the observed data against the model database are also mathematically guided.

These models are introduced with the explicit aim to cover a domain limited vocabulary, and are rigidly restrained in that they are bottom up or data driven. Even though their performance attains remarkable results, their portability to other domains is equal to null.

The same drawback are suffered by pragmatically guided SRU systems. In this case, parsing of the input stream is performed strictly top-down, so that poorly recognized or neglected function words are hypothesized and directly inserted via a set of task domain linguistic restrictions. The efficiency of the system can be computed as an inverse function of the dimension of the domain guided linguistic encoding: by increasing the vocabulary size and dialogue functions, the overall efficiency is bound to be reduced since the pragmatically guided process of word hypothesization is bound to lose its predictive power, having to cope with too many alternatives available at a certain point.

Our Proposal
We propose a SRU system which is linguistically based and can efficiently take advantage of both bottom-up decoding of acoustic events via data-driven adequately cast mathematical models, and top down rule driven linguistic structuring grammars.

The system we propose is organized as follows:

1. It uses an augmented syllable model to cope with coarticulation phenomena both at interword and intraword level: this constitutes the bottom-up procedure;
2. It classifies acoustic events into place and manner distinctive features - a richer, more acoustically based, set of features could be used as well, in order to produce partial segmental recognition as long as they are hierarchically organized;
3. It produces phones hypothesis on the basis of partially filled distinctive feature matrixes
4. It discriminates between stressed and unstressed syllables in order to guide word composition at a morpho-phonological level;
5. It produces word hypotheses on the basis of a lexicon of root morphemes and affixes plus word nuclei for invariable function words;

NATO ASI Series, Vol. F 75
Speech Recognition and Understanding. Recent Advances
Edited by P. Laface and R. De Mori

6. It restricts the number of function word hypotheses by the use of a top-down parser based on a LL(1) grammar, which incorporates statistical measures as to the most likely local sequence of grammatical categories in a certain task domain;
7. It uses prosodic cues like pauses, amplitude and F_0 contour variations to build up phonological phrases (PhPh) and help disambiguate attachment of PP adjuncts and other modifers as well as other more complex phenomena, like parentheticals, vocatives, afterthoughts or tails, topicalizations, left dislocations, extraposition adverb and locative preposition, etc.

Features and Syllables

Why using syllables? They constitute the most natural correlate to an acoustic template which encompasses both low level coarticulatory variations, in terms of transitions from one sound segment to the following one; and high level prosodic suprasegmental traits. Training syllable models is quite simple and requires a low amount of memory and computation time as compared to word models. Moreover, since all the possible syllable of a certain language are present both in the oral and written texts and discourses, we propose to build a stochastic model from data derived from written texts, and independent from the application domain.

This could be modified in case the task is related to a microlanguage like the one of chemistry or any other purely scientific domain. However, data on the syllable structures regarded as legal sequences both at interword and intraword level are clearly very easily portable to any domain whatsoever. As to the relevance of a stochastic model for predicting legal syllable sequences in a certain language, it should serve to indicate two things:

a. whether a syllable can be used in word initial position;
b. its eventual restricted use in word internal position.

Consider in this respect, the role played by consonant clusters, whose initial sound is a sonorant: this would simply indicate that parsing is positioned at intraword segmental level, or interword level, in case of function words like "il"(the) "non"(not) and a few others. We decided to compute frequency of occurrence of syllables structured according to the following general rule:

$$^{3}C_{0}\ (C_{[glide]})\ ^{2}V_{1}$$

In this way, syllable onsets are triggered by the acoustic event characterizing the end of a vowel and the beginning of a non-vowel. This is different from the approach outlined in [5] based on a loudness model in neural nets. In their model, syllable boundaries are computed by localizing syllable nuclei first and then by approximating the segments intervening between two consecutive syllable nuclei with the use of demisyllables or diphones: the first demisyllable contains a part of the vowel and the transitions towards the following consonant cluster and the second demisyllable contains the initial consonant cluster as well as transitions towards the central vowel. However we believe this method could be useful, as well.

We include a brief exerpt from the Table of the most frequent syllables computed over a corpus of approximately half a million word occurrences. We use ":" for geminates, "ts" for the unvoiced dental affricate, and "c" for the unvoiced palatal affricate, and indicate absolute frequency up to the ninth position within a word. No provision is given for open and closed "o" and "e". But we surmise that this problem might be solved by indicating stressed syllables.

<u>List 1. Exerpt of the most frequent syllables</u>

l:a	0	6121	290	211	19	3	0	0	0
tsjo	3	438	1765	2351	1190	1140	97	17	4
ni	81	1723	1919	1713	669	250	233	11	9
mi	2706	1222	1045	692	390	267	83	7	2
li	1220	1493	1454	904	632	259	78	51	38
mo	3047	1400	779	470	152	117	6	1	5
ce	1130	2852	1017	598	91	21	14	0	1
vo	1459	1305	1669	754	377	97	22	5	1
pa	4031	362	554	350	112	2	0	0	0
lo	1357	1834	1014	629	271	85	35	9	1
sa	1911	1449	822	462	272	24	6	0	0
ro	366	1832	1596	711	178	48	13	0	0
vi	1627	1537	772	367	125	90	29	2	0
pro	4162	58	236	2	37	2	2	0	0
sta	1684	1295	596	725	54	17	1	0	0
ci	925	1221	1084	564	273	57	30	4	2
kue	3477	91	185	130	8	3	1	0	0

As can be seen from the list, most syllables are unevenly distributed, their informative value being thus highly predictive. Consider for example, "tsjo": its presence in the corpus is high in the fourth and third position within a word, but is low in first and second position. Here below we give the summary of syllable counts relatively to their position within the word.

Table 1. Summary of syllable counts relatively to their position within the word

Position / Samples	Total Corpus	Most frequent syllables	Percent of the Total C.	Present in two positions	Syll. not TCorp.	Pres. in one pos.	Present in more than 2 positions
Position 1.	312799	271635	86,84	2702		980	302366
Position 2.	254716	174735	68,6	9995	923	42827	200325
Position 3.	149615	110016	73,53	4987		2480	140168
Position 4.	79742	68033	85,31	108		34	78500
Position 5.	37156	33982	91,45	7		0	36791
Position 6.	16401	15380	93,77	0		0	16323
Position 7.	5698	5318	93,33	0		0	5624
Position 8.	1307	1228	93,95	0		0	1301
Position 9.	330	330	100	0		0	330
Total no. samples	740	66		77	355	300	367

The total number of syllables computed is 1905, but the 66 most frequent syllables make up the 87.5 per cent of the total number of occurrences.

The Hypothesizer and Pruning

In a SRU system using word models or other non syllabic unit, pruning is justified by the task to be performed: matching observed data onto the previously stored data leads to the creation of sets of candidates more or less adequately fitting the data. Pruning is performed on the basis of some likelihood measurements and/or consideration of some threshold scoring apriori assumed to represent a confident enough approximation. However, in a syllable model, no such task can be performed: syllables constituting the stored database on which training for HMMS or discriminations for classifications to be evinced from neural networks are all and only the legal syllables of the language. A syllable can thus either be completely recognized - according to some likelihood measurement or confidence threshold - or only "partially" recognized: in this case a partial or incomplete feature matrix will result.

In both cases no pruning seems allowed. Syllables recognition rate will vary according to the global intelligibility of the speech signal. The procedure remains strictly bottom-up at this stage. In case of low intelligibility and partially recognized syllables the choice of the best fitting candidate will not be left entirely to mathematical measurements.

Consider now a simple feature model which discriminates sonorants on the basis of seven features:

Table 2. Feature matrix for the discrimination of sonorants

features/phones	l	λ	r	n	m	ɲ	ɱ	ŋ	j	w
VOWEL	-	-	-	-	-	-	-	-	+	+
LABIAL	-	-	-	-	+	-	+	-	-	-
PALATAL	-	+	-	-	-	+	-	-	+	-
VELAR	-	-	-	-	-	-	-	+	-	+
LATERAL	+	+	-	-	-	-	-	-	-	-
NASAL	-	-	-	+	+	+	+	+	-	-
GLIDE	-	-	-	-	-	-	-	-	+	+
INTERRUPT	-	-	-	-	-	-	-	-	-	-
FRICATIVE	-	-	-	-	-	-	-	-	-	-

We use [+interrupt, +stop] for stops, [+interrupt, -stop] for affricates, and [-interrupt, +fricative] for fricatives.

All completely recognized syllable candidates and all partially recognized ones will have to be passed up to a Hypothesizer. Consider how it works on a simple example. Let's take a Clitic Group(CG) like "lariproduzione"(the reproduction) with the following possible output from the acoustic level:

Table 3. Output of feature/syllable based recognizer

σ1	σ2	σ3	σ4	σ5	σ6
////////// A	////////// I	P ////////// O	D | U	Z | J | O	N | E
[-vowel]	[-vowel]	[-vowel]	M | O	S	
[-velar]	[-lateral]	[-nasal]	B | U		
[-nasal]	[-glide]				
[-interrupt]	[-interrupt]				

Possible candidates hypothesized by the system for partially recognized syllables 1, 2, 3 are respectively:
| la, ra, ia | for syllable 1; | ri, ni, mi, gni | for syllable 2; finally | pio, puo, pro, plo | for syllable 3.
Some of the syllables hypothesized make legal morphemes, but most of them do not. In particular, |ia| does not fit into any word initial position of Italian given the possible continuation already computed: it can thus be discarded from the list. As to syllable 2, "rani" does not constitute a legal word of Italian, thus "ni" can be discarded from the list. Coming now to syllable 3, "plo" is not contextually a legal syllable ("mplo" is a legal syllable). However only some of these syllable will be accepted by the morphemic parser which has the aim to build up legal sequences of syllables into legal morphemes and words of Italian. Consider some of these hypothesized root morphemes: "produ" and "promo" are legal ones, but *"probu", *"piodu", *"puodu", *"piobu", *"puobu",*"piomo",*"puomo", are all illegal root morphemes and illegal suffixes of Italian. At this point only certain words are filtered out by the parsing process: "la", "rari, rami, ragni", "mi", "produzione, promozione", "riproduzione, ripromozione". Hypotheses at this level are taken care for by the syntactic parser which checks whether a certain sequence of words constitutes a grammatical phrase. For instance, "la mi produzione/promozione" does not constitute a grammatical sequence and "mi" is discarded as possible intervening clitic word. "Rari produzione" is also discarded owing to the lack of agreement between the adjective and the noun. Compound words like "rami produzione" or still worse "ragni produzione" are also absent from the Italian dictionary. The two remaining possible legal sequence on a syntactic basis is "la riproduzione" and "la ripromozione": however the latter will be discarded by a morphological filter which disallows prefixes like "ri" indicating iterative events, to combine with stems like "promuovere"/promote, pass for semantic reasons.

Morphological analysis

The Hypothesizer should fire possible syllable hypotheses on the basis of the available partial or incomplete feature matrix. All syllable candidates will be used in turn by the morpheme scanner or creator, and subsequently passed onto the morphological parser which attempts to create words out of morpheme sequences. The Hypothesizer should also account for prosodic effects at word boundaries such as syllable deletion or consonant doubling at word initial position.

Morphemes could be organized in a graph structure: stressed syllable information is relevant in that it can guide the search into the graph. Most inflectional morphemes in Italian are stress attractors; however, a lot of function words like clitics are unstressed and can sometimes become a cause of confusion in the attempt to find the word ending. Stress information would help the hypothesizer to choose the most adequate syllable sequence to be fed onto the morphological analyzer. Here below we show a brief exerpt from the list of inflections of Italian verbs which carry stress on the first syllable - the number at the end is the morphological code which allows it to be selected by the appropriate root morpheme:
ESSA579; ESSE520; ESSERO520; ESSI520; ESSIMO520; ESSO579; ESTE520 ESTI520; ETE530; ETTE566; ETTERO520; ETTI566; EVA523; E'VAMO520; EVANO520; E'VATE520;EVI520; EVO520; E'520; IA570; IAMO511;IAMO530; IAMO550; IATE501; IATE520; IATE540; IBILE435; IMMO540; I'RAI540; IRA'540.

When morphemes have been recognized the morphological parser will try to compose words out of morpheme sequences. This is done using a lexicon of the language where all roots, affixes - i.e. prefixes, derivational and inflectional suffixes - and invariable words are organized into continuation dictionaries and the parser is used to determine possible concatenation of the basic units into words.We include here below a diagram showing the data flow from the acoustic level onto the morphological analysis and the syntactic parser.

Diagram 1. Main Components of a Linguistic Model for SRU

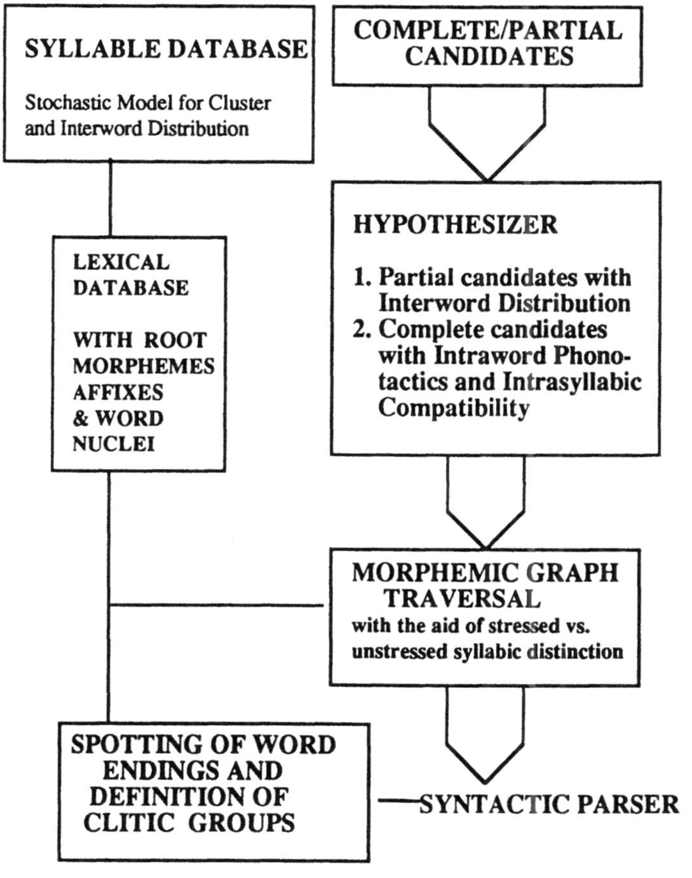

The main advantage we see in using this model is the possibility to overcome the major drawback caused by uncertainty that may result from subparts of a word by introducing legal hypotheses on a higher abstract level. Uncertainty will never result in rejection of a word unit because words are the abstract processing of morphological and syntactic parsing. Uncertainty will only cause the hypothesizer to produce a certain number of legal candidates which may or may not be discarded by the system. In this case, what is required is a system which allows semantic and pragmatic knowledge to be used to filter out possible legal word sequences.

Bibliography and References

1. Delmonte, R.:Parsing Difficulties & Phonological Processing in Italian, Proceedings of the 2nd Conference of the European Chapter of ACL, Geneva, 136-145, 1985.
2. Delmonte R., G. Mian, G.Tisato: A Grammatical Component for a Text-to-Speech System, Proceedings of the ICASSP'86, IEEE, Tokyo,2407-2410, 1986a.
3. Delmonte R.: A computational model for a text-to-speech translator in Italian, REVUE-Informatique et Statistique dans les Sciences humaines,XXII,1-4, 23-65, 1986b.
4. Delmonte R.:The Realization of Semantic Focus and Language Modeling, in Proceedings of the International Congress of Phonetic Sciences ,Tallinn (URSS),100-104,1987.
5. Ruske, G. Weigel W.: Syllable-based stochastic models for continuous speech recognition, this volume.

Evidential reasoning and the combination of knowledge and statistical techniques in syllable based speech recognition

Luke Boucher, Tony Simons, Phil Green
SPLASH, Speech Laboratory,
Sheffield University, UK

INTRODUCTION

In what follows we describe SYLK: a project which aims to deploy expert phonetic knowledge within an admissible statistical framework for syllable based speech recognition - hence 'SYLK', Statistical sYLlabic Knowledge. The project has been running at the Universities of Sheffield and Leeds, in the UK, since mid 1989 and has both HMM and Syllabic IKBS constituents. See the overview in Fig.1. It is funded by the UK IED programme and follows on from previous Alvey-funded work on the Speech Sketch (Green et al [6]) and HMM-based broad phonetic classification (P.Roach[10]). In this paper an overview of the system is presented, with particular attention paid to its statistical and hypothesis refinement aspects.

The next section describes the syllable model, within which classification takes place. This is followed by a description of the system's various components. We then, finally, consider the scheme of evidential reasoning used for hypothesis refinement.

THE SYLLABLE MODEL

The syllable has often been proposed as a useful unit within which to account for phone context dependency (eg Church [3] Allerhand [1]) and in SYLK, recognition is centred around a syllable model developed by Simons (Green et al [7]) from the work of Church [3] and Allerhand [1]. The number of different syllables makes it impractical to build recognisers with one model per syllable, except in some cases (eg Lin-shan Lee et al [9]). The model used in SYLK reduces this combinatorial problem by defining a network structure of syllable constituents, which act to constrain the co-ocurrance of their different possible refinements. This leaves us with the easier job of producing statistical models for a reduced number of syllable constituents and the new problem of finding a method of hypothesis refinement which satisfies the constituent's constraints.

In brief, the model has two separate planes; the Constituent Plane and the Refinement Plane. In the Constituent Plane arcs define what constituents any node is allowed to have. In Figure 2, then, we see that the syllable is made up of constituents ordered in time; Optional Onset then Rhyme, and that the Rhyme has constituents Peak followed by Optional Coda. The Rhyme serves to account for co-articulatory effects which exist between the peak and coda, as will be shown in later figures.

In the Refinement Plane, arcs indicate the alternative refinements that can be made at any node. During classification these nodes serve as hypotheses, so In Figure 3 the Rhyme can be seen as a Hypothesis set which is refined in terms of its members, the hypotheses open rhyme and closed rhyme. There is more space for structure in the lower reaches of the network and new nodes and arcs are expected to be filled in - from gathered statistics - as the project proceeds. The lowest point of refinement, the "SYLK symbol", is at a mid-class acoustic-phonetic level. This makes 22 SYLK symbols for onset types and 62 for coda types. Work already under way, however, is developing the refinement power of the system to a finer level for some phone classes.

Known co-articulation is modelled explicitly, and the rhyme refinements are used to constrain the sorts of peak and coda which occur together. Thus we see, in figure 4, that in the Constituent Plane, Closed Lax Rhyme contains only a Shorter Peak followed by a Coda.

NATO ASI Series, Vol. F 75
Speech Recognition and Understanding. Recent Advances
Edited by P. Laface and R. De Mori

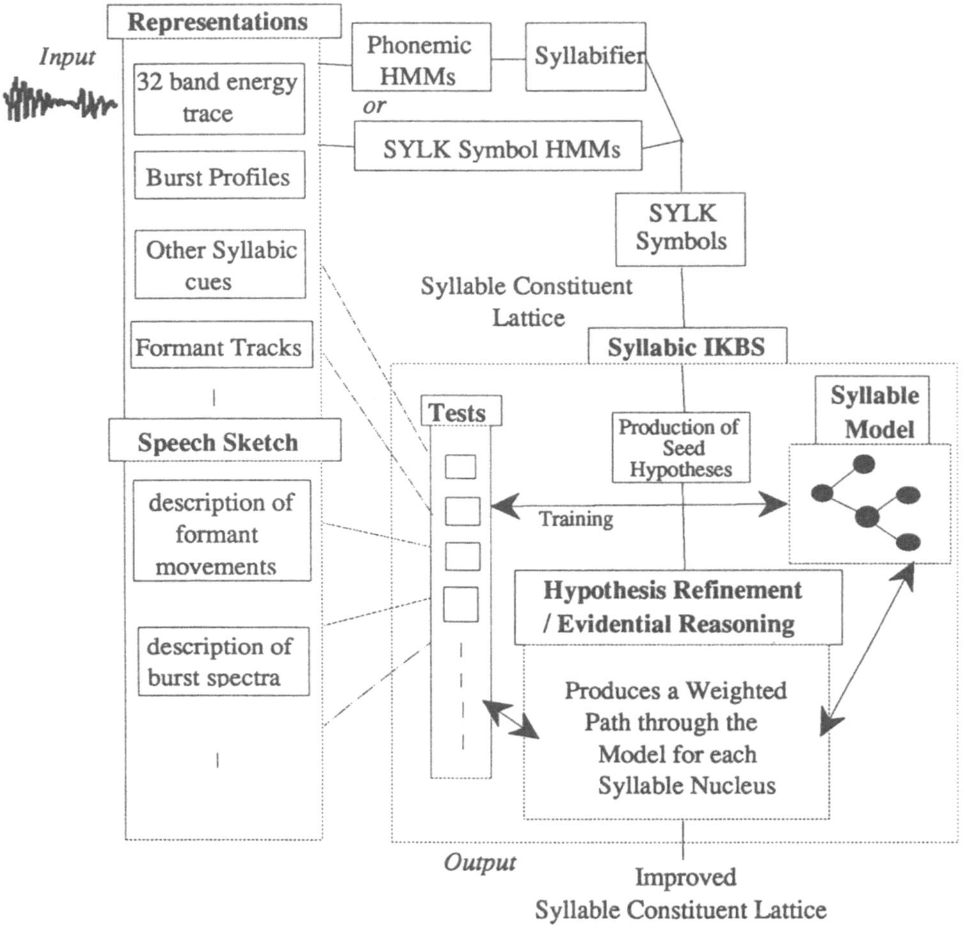

Figure 1. The SYLK Project

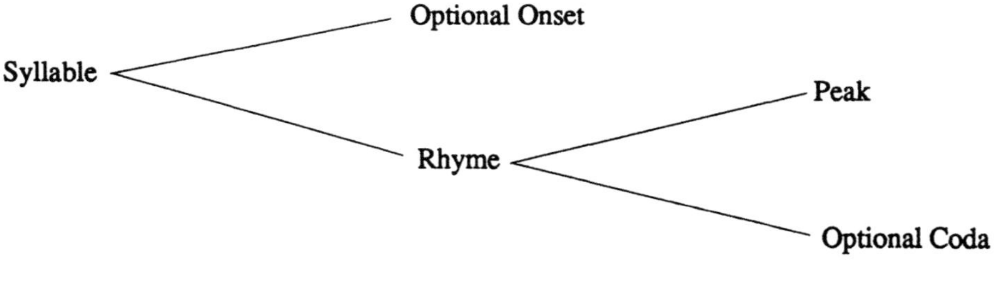

Figure 2. Constituents of the Syllable

The knowledge-driven constraints in the model are written explicitly into the implementation, through the use of object oriented data structure definitions, and, in fact CLOS, the Common Lisp Object System, has been used to develop a whole environment for creating and running the SYLK system.

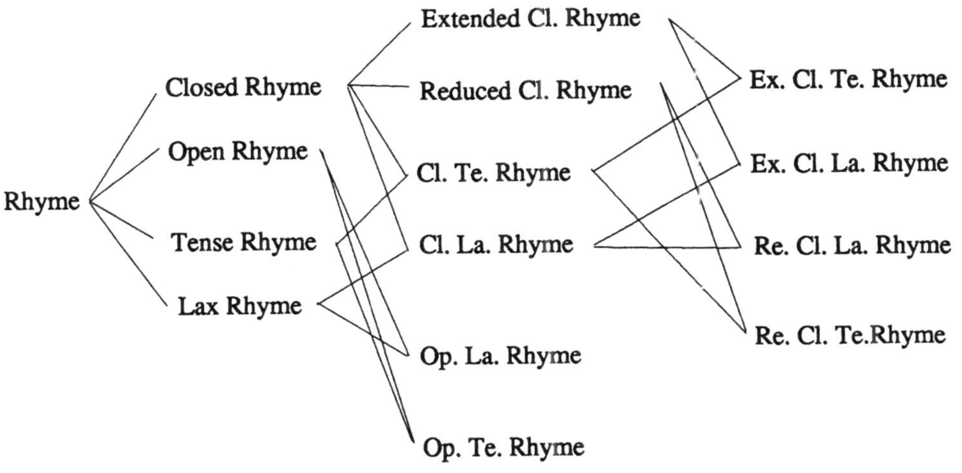

FIgure 3. Refinements of the Rhyme

Figure 4. Constituents of a Closed Lax Rhyme

COMPONENTS OF THE SYSTEM (figure 1)

Initially, HMM-based processing, developed at Leeds, is used to produce a syllable-constituent lattice. An HMM is trained for each of the SYLK symbols. The job of SYLK's Evidential Reasoning component is then to improve on this transcription, both correcting errors and in some places making it more specific.

Each Peak in the lattice is used to instantiate a Syllable Hypothesis. The Onset and Coda transcriptions are then used to produce initial "syllable structure hypotheses" as instantiations of the Syllable hypothesis's constituents and refinements, with the explicit constraints of the model affecting both the Rhyme and Peak hypotheses. Initial measures of probability, derived from a confusion matrix of the HMMs' performance, are assigned to these refinements.

Each "syllable structure hypothesis" is then refined independently. For this, refinement "tests" are used to modify the probabilities (and original segmentation) using the evidential reasoning scheme described in the next section and based on evidence from a variety of representations.

EVIDENTIAL REASONING

As mentioned in the last section, the idea of the evidential reasoning part of SYLK is to use the evidence from knowledge based processing to update the probabilities associated with various syllable constituents in a gradual process of hypotheses refinement, down through the network of our Syllable model. The main stages of this refinement process involve the use of refinement Tests.

REFINEMENT TESTS

Each Test is associated with one or more refinement nodes in the model and is used to improve the discrimination between alternative permissible syllable structures. For this, the system exploits the statistical relationship that exists between some evidential representation and the alternatives in the model. No restriction is placed on the form of processing that a test uses or on the region of time it considers. It is simply required to return a vector.

The emphasis for test development is for the system to provide the best sort of discrimination a test can make, given the developer's particular choice of parameters. This differs from previous approaches which chose the distinction first and then tried to find the best set of parameters which could make that discrimination, (eg the FEATURE (Cole et al [4]) system, and to some extent SUMMIT (Zue [14])). It can be claimed, then, that the model is developed to fit the 'natural' distributions of classes that are found in our training data.

At any refinement node of the Syllable model, a test can be trained to discriminate between that node's successors. This is done, at present, by producing a simple Bayesean classifier. However, the system is more flexible than that, in that tests can be trained over any set of basic classes to distinguish between mutually exclusive subsets of them. This means that at run time, a test can be associated with any number of refinement nodes, while a scheduling system orders the tests to construct a path through the network down to, and perhaps below, the SYLK Symbol level.

From the point of view of classification, when "fired", a test will simply return a vector, of numbers and/or symbols. Given this output, a training sample can be used to produce the test's Bayesean classifier, by estimating the probability densities which govern the variability of the various classes over that feature. In training, however, classes with only a small distance between them (for that measured feature), or for which there exists only a small sample, will be clustered - using a distance measure that compares between and within sample variability (Hand [8]) - and deemed inseparable. In this way, new refinement nodes in the syllable model become defined implicitly in the tests statistics.

For estimating the class's probability densities an Adaptive Kernel classifier (Silverman [13]) - which generalises both the nearest neighbour and Kernel classifiers - was initially considered. However, given the prohibitive sizes of training data needed for these estimates in higher dimensions, and the scarcity of some components of the syllable model, a combination of these and Gaussian (exponential etc) estimates has been adopted.

Unlike the discovery of new refinement structure, no provision is yet made for the automatic inclusion of new structure in the constituent planes, ie new co-articulation effects that the statistical analysis might find. This feature could possibly be the subject of future work.

HYPOTHESIS REFINEMENT

Two methods of combining the probabilistic evidence from different tests have been considered for SYLK. The received Bayesean updating scheme was feared to be too centered around singleton hypotheses to make use of the implicit structuring over the power set of singleton hypotheses given us by the syllable model. The more maverick Belief Functions (Shafer[11]), whilst appearing to cater for this structure, however, seemed to lack a strong semantics and thus a firm justification, in the context that it was envisaged they be used. These issues, and others relating to claims made of Belief Functions, were investigated in Boucher [2], and based upon the findings a form of Dempster's rule - a claimed generalisation of Bayes also used in Belief Functions- has finally been adopted. It is constrained, however, by the Syllable model, in that prior probabilities are chosen at a level in the network which is only as specific as the information returned by the new evidence. This is illustrated by an example;

Consider the test (see figure 3),

T1:rhyme -> {open-rhyme closed-rhyme} that distinguishes between an open and closed rhyme

and, given some trial syllable say,

T1 returns (0.6 open-rhyme) (0.4 closed-rhyme)

The system must start from a position of no Knowledge. This is represented by

(1 Rhyme)

Now consider updating this with the information from T1. We must choose prior probabilities, and taking this simply as an equal partitioning of probability amongst these hypotheses, we must choose between;

(1/2 Closed-rhyme)(1/2 open-rhyme) In future CR and OR

or

(1/6 open-lax-rhyme)	OLR
(1/6 open-tense-rhyme)	OTR
(1/6 extended-closed-lax-rhyme)	ECLR
(1/6 reduced-closed-lax-rhyme)	RCLR
(1/6 extended-closed-tense-rhyme)	ECTR
(1/6 reduced-closed-tense-rhyme)	RCTR

Using the second of these alternatives, Bayes gives us;

(0.21 OLR) (0.21 OTR) (0.41 ECTR) (0.41 RCLR) (0.41 ECLR) (0.41 RCTR)

Which can also be re-written, with loss of information, as;

(0.42 OR) (0.58 CR)

Both of which we see as undesirable in that they obscure the actual discrimination suggested by the test. The desired result is in fact produced by choosing the first alternative for prior probabilities, which is;

(0.6 OR) (0.4 CR)

an assumption that also preserves the specificity of our knowledge, which is also desirable.

Finally, then, the hypothesis refinement stops when either, a path has been traced to some basic class, at the end of the syllable network, with enough confidence for further tests not to lower its probability significantly in favour of any other class or, when all the available tests have been applied. When this is complete for each of the refinement graphs contained in the syllable (ie the Onset Rhyme Peak and Coda refinement graphs) an output is given in the form of a weighted Syllable constituent Lattice.

SUMMARY

Following the encouraging results produced by the SUMMIT system at MIT (Zue [14] and Seneff's paper in this volume) it is believed that statistical classification systems can be produced which successfully support, and are augmented by, an increasing amount of domain dependant and expert knowledge. The main condition seems only to be that sufficient time, and training data, is at hand. With this fact it would appear that the SYLK system with its adaptable framework for the introduction of new representations and model structure, has good potential for development as a speech recognition system.

Its advantages include; a syllabic model of co-articulation, the scope for continual addition of new knowledge and representations and the ability to classify at various levels of specificity, automatically choosing that most appropriate to the systems current ability.

There are also areas for possible addition within the statistical framework, and these include; the ability to augment the co-articulatory model automatically from the distribution of applied training data.

REFERENCES

[1] M H Allerhand (1986) 'A knowledge-based approach to speech pattern recognition', Phd dissertation, Darwin college, University of Cambridge.
[2] L.Boucher and P D Green (1990) 'Syllable based hypothesis refinement in SYLK', IOA meeting. Windermere.
[3] K W Church (1983) 'Phrase structure parsing: a method for taking advantage of allophonic constraints', Phd dissertation MIT, Cambridge MA.
[4] R A Cole, Stern, Larry (1983) 'Performing Fine Phonetic distinctions: templates vs features', Conf on invariability in speech, MIT.
[5] Dempster (1968) 'A generalisation of Bayesian inference' J.Royal statistical society, Series B 30, 205-247.
[6] P D Green, M P Cooke, M D Crawford and A J H Simons (1988), 'Acoustic-phonetic reasoning with the Speech Sketch: a progress report', Proc. 7th Symposium of Fed. Acoust. Soc. Europe (SPEECH-88), eds. W A Ainsworth and J N Holmes, Institute of Acoustics, 353-360.
[7] P D Green et al (1990) 'The SYLK project: Foundation and Overview', IOA meeting Windermere.
[8] D.J.Hand (1981) 'Discrimination and Classification' J.Wiley & sons.
[9] Lin-shan Lee et al (1990) 'A real-time Mandarin dictation machine for Chinese language with unlimited texts and very large vocabulary' proc ICASSP 90 pp65
[10] P Roach (1989) 'Phonetic feature extraction by automatic segmentation and labelling' final report, Science and Engineering Research Council. UK
[11] G Shafer (1981) 'A mathematical theory of evidence' Princeton University Press.
[12] G Shafer (1981), 'Constructive probability', Synthese 48, 1-60.
[13] B W Silverman (1986) 'Density estimation for statistics and data analysis' Chapman and Hall.
[14] V W Zue, J Glass, M Phillips and S Seneff (1989), 'Acoustic segmentation and phonetic classification in the SUMMIT system', Proc. IEE ICASSP, Glasgow, Vol. 1, S8.1, 389-392.

Part 6

Speech Analysis, Coding and Segmentation

Data Base Management for Use with Acoustic-Phonetic Speech Data Bases

Jan P. Hendriks[1], Sandra H. Swagten[2], and Louis Boves[3]

[1]PTT Research, P.O.Box 421, 2260 AK Leidschendam, The Netherlands
[2]Speech Processing Expertise Centre, P.O.Box 421, 2260 AK Leidschendam, The Netherlands
[3]Nijmegen University, P.O.Box 9103, 6500 HD Nijmegen, The Netherlands and PTT Research, P.O.Box 421, 2260 AK Leidschendam, The Netherlands

Abstract: In this paper, the requirements for an acoustic-phonetic data base are outlined and three different managements systems (flat files, relational DBMS's, and Object-oriented DBMS's) are compared with respect to these demands.

Keywords: Acoustic-phonetic data base, Management Systems.

Why an acoustic-phonetic data base?

Although research into Knowledge Based (KB) speech recognition has lost most of its appeal to development of Integrated Search techniques based on Hidden Markov Models, we do not think that it will ever disappear completely. Now that the facilities for processing huge amounts of speech data are becoming available, we expect a renewed interest in KB approaches. Even if it is not the only cause -and perhaps not even the most important one- of the less than satisfactory results of KB techniques in the past, the limited amount of data that was available to base rules on (or to extract rules from) certainly has played a nonnegligible role.

The future KB systems that we have in mind will not rely on deterministic rules. If sufficient speech material is available, it will be possible to get to grips with the essentially probabilistic relation between the acoustic description of speech signals and the underlying linguistic representations that one is trying to find in doing speech recognition. Given large pools of data, it will become possible to obtain realistic and reliable estimates of the relevant probability distributions.

KB speech recognition research needs more than just large amounts of speech from many different speakers, produced under many different circumstances. It needs all that speech together with reliable transcriptions and a large set of reliably extracted acoustic parameters. The collection of this kind of data involves an enormous amount of work, not only during the recording and verification of the speech material itself, but mostly for the detailed annotation and for the interactive correction of errors made by presently available feature extractors.

A sufficiently large pool of annotated and hand corrected speech sata can only be formed if many researchers contribute during long periods of time. And in order for the informa-

NATO ASI Series, Vol. F 75
Speech Recognition and Understanding. Recent Advances
Edited by P. Laface and R. De Mori

tion to be accessible to a large research community, that may ask many diverse questions, most of which cannot be foreseen before the research has actually started, a powerful and flexible system for storing and retrieving that data is indispensable.

What is an acoustic-phonetic data base?

An Acoustic-Phonetic data base (APDB) is a collection of speech recordings with related data, such as meta-information, and linguistic, phonetic and acoustic data. The meta-data give, among other things, information about speaker characteristics and recording conditions.
The linguistic descriptions are derived from the written representation of the text to be spoken. Since they derive from the text, they seem to be essentially unambiguous; still, it may become necessary to account for generally accepted variations in pronunciation by including alternative phonemic representations of words. Similar problems will arise when one tries to include the "theoretically optimal" prosodic structure of sentence material in the linguistic part of the data base. Of course, there is the well know class of sentences that can have several different syntactic parses, corresponding with different meanings.
Whereas it may not be absolutely necessary to tune the linguistic descriptions to a specific rendering of the textual material, the phonetic description can only be obtained by means of a transcription of the actual realisations of a text. This is probably the most time consuming (and the most error-prone) part of the data. Little is known about the reliability of phonetic transcriptions, especially if they encompass diacritics.
The acoustic description (a large set of parameters, like LPC coefficients, cepstrum coefficients, and filter bank outputs, but also formant values, and pitch) at e.g. 10 ms intervals can for the most part be obtained by means of automated acoustic analyses. However, parameters like formants and pitch, that require an interpretation of raw analysis results will require at least some hand correction for a long time to come.
It is one of the basic issues in speech research to determine the relationships between the linguistic, the phonetic and the acoustic level. The mapping of phonetic and acoustic data poses the problem of having to map a discrete phonetic description onto a continuous acoustic signal. That mapping should allow for at least some amount of uncertainty and that poses considerable problems in a data base management system.

Basic Structure

The APDB we are working on is described by three structures: Objects, Relations between the objects, and Features of the objects; Figure 1 depicts a possible structure for an APDB as it is actually implemented in DEMSI([HEND88]) in a relational DBMS (VAX Rdb/VMS).
Objects are depicted by the ovals, whereas the arrows represent relations. Features of the objects are not shown explicitly in Figure 1, but one can think of e.g. the acoustic parameters mentioned above as the features of the *Frame* object. Another example of features is part-of-speech information attached to the words.
The objects and relations distinguished in a data base reflect the way in which the developer thinks the real world data is structured and what parts of that structure are sufficiently important to be included in the data base design. Different researchers may

have different beliefs about the optimal data structure. It is quite conceivable, for instance, that researchers might want to include additional objects on the level of prosody, whereas others might feel that objects like the *Morpheme* and *Grapheme* are not necessary.
The structure of an APDB derives from relations between objects. Sometimes these relations are hierarchical (as between a sentence and a word), sometimes not quite (word and syllable), and sometimes not at all (syllable and morpheme).

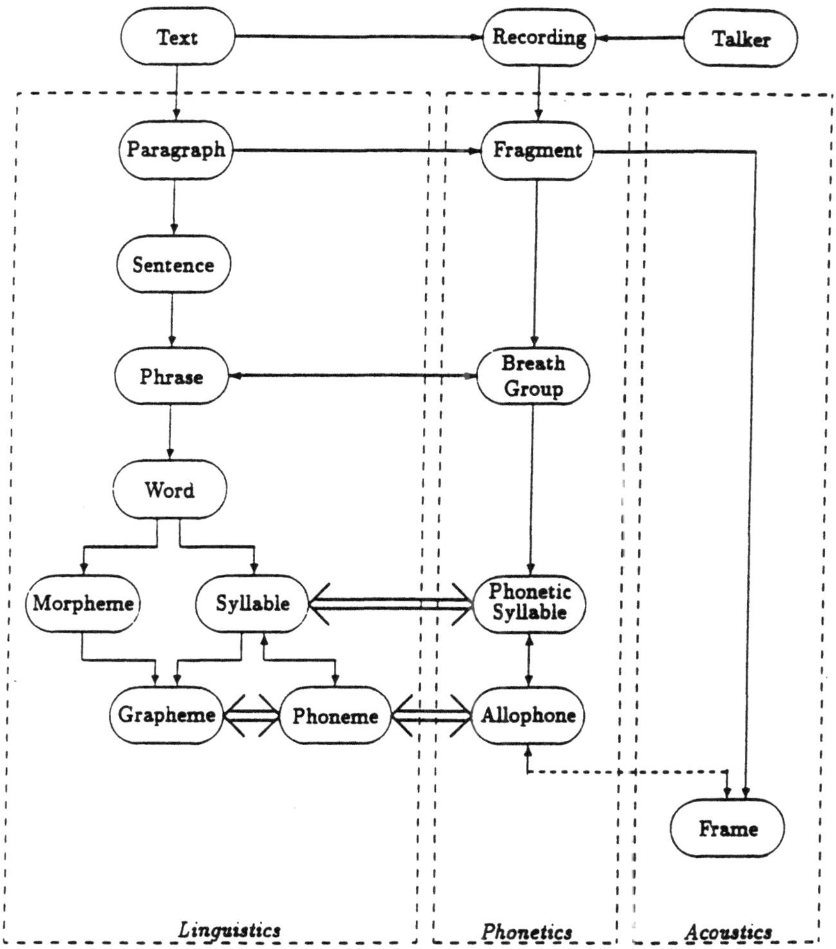

Figure 1. The structure of the database (DEMSI-approach)

The four types of relations existing within the DEMSI structure and depicted in Figure 1 are ([TERL89]):

1. UNIQUE PARENT ($\longrightarrow$) This is a one-to-many relation, e.g. a phrase may consist of one or more words.
2. NON UNIQUE PARENT ($\longleftrightarrow$) This is a many-to-many (N:M) relation; as an example: a syllable can consist of one ore more phonemes, but so-called ambisyllabic phonemes may belong to two syllables.
3. BIRTH AND DEATH ($\Longleftrightarrow$) This is also a many-to-many (N:M) relation, only now either N or M may be zero.
4. FUZZY ($\leftarrow - \rightarrow$) This is a many-to-many (N:M) relation that expresses the (un)certainty that a record of one object relates to a record of another object. An example is the relation between Allophone and Frame.

Requirements for a Data Base Management System

What demands do we have to make on the definition of objects, their relations, their features, their storage, and their retrieval when an APDB should be used as a tool in developing a probabilistic KB speech recognition system?

Flexibility From what has been said above, it is clear that in an APDB it should be possible to redefine the objects, relations, and features in an easy way. Only then can queries addressing new and unforeseen relations be handled efficiently. Flexibility is also required in order to allow the integration of data coming from various and probably uncoordinated sources.

Abstraction The definition of the data base structure must be independent of the underlying physical structure. The way in which the data are physically organised and stored is determined by technical factors, like the size and type of storage media available. These technical factors are irrelevant for the conceptual meaning of the data, and therefore should not affect the definition of the conceptual structure of the data base.

Consistency The abstract model of the data that determines the design of the data base should coincide as closely as possible with the structure of the data in the real world. This tenet must be maintained even if the structure of the real world is not unequivocal; if necessary, the data model must be adapted so as to reflect the real world structure.

Conceptual Queries This requirement is closely related to the two previous ones. In order for a data base to be useful for a wide variety of researchers who cannot (and should not) afford the time to study the structure of the data base in great detail, it must be possible to formulate queries at the conceptual level and in a familiar formalism ([CARL86]).

Performance Performance should be high, the researcher must be able to work interactively with the data base. Quasi-instantaneous response must be guaranteed at least for subsets of the data stored in special purpose tailored data bases of a realistic size.

Integrity The data should be (kept) complete. This is just a technical requirement. In the case of an APDB the practical importance of this topic is probably limited because most interactive use such a data bases will be limited to retrieval actions, that cannot compromise data integrity.

Possible Management Systems

When developing an APDB one has a choice of four classes of Data Base Management Systems, i.e. a flat File System, a Relational DBMS (RDBMS), a Semantic DBMS (SDBMS), or an Object-Oriented DBMS (OODBMS). The only advantage of a flat File System is that is potentially very fast. The major disadvantage is that all burden is placed on the programmer/user. Consequently, File Systems are only suited for repetitive applications, provided that the user is an accomplished programmer who knows exactly how the data are organised. Also, File Systems are insecure and inflexible. Finally, data stored in File Systems cannot be combined to a growing pool of data that can be readily made available to other researchers for their own purposes. Due to the importance of the disadvantages, flat File systems are obviously not adequate for our purposes.

Unfortunately, we were not able to obtain and test a Semantic DBMS. Thus, although this type of DBMS is potentially useful for implementing an APDB, we will not consider it in this paper. Thus, we are left with Relational and Object-Oriented DBMS's as possible viable candidates.

Relational Data Base Management Systems

Modern commercial RDBMS's offer so much support for the conversion of a Data Base Schema to a Data Base System that they fulfill the requirements of flexibility, integrity and consistency, at least as far as data base construction is concerned. The major limitation of RDBMS's is their lack of support of abstraction on the level of relations and features. Relations that are implicitly present in the data model cannot be exploited, unless they are explicitly addressed. E.g. the fact that *Phrases* consist of *Words* that consist of *Syllables* can only be used to obtain all syllables in a phrase by explicitly asking for all syllables in all words in that phrase. This complicates the formulation of unusual queries, thus somewhat detracting from the flexibility on the level of construction. Another possible disadvantage of RDBMS's is their slow response, especially when answering a query involves the combination of data in several large tables.

Object Oriented Data Bases

The commercially available OODBMS ONTOS has been tested by emulating a part of the DEMSI data base system that was previously built using an RDBMS ([SWAG90]). ONTOS offers little support to the data base implementor: the complete array of consistency checks that one has come to expect from an RDBMS is absent. Consequently, the implementation of especially the first Data Base System may be quite time consuming and costly, if only because all consistency checks must be programmed explicitly. However, the strong point of OODBMS's is their support of abstraction, both at the levels of relations and features, that can easily be defined and implemented in terms of a function of available data. Finally, and quite importantly, an OODBMS seems to be much faster than a conventional RDBMS, because the relations are implemented by means of direct pointer mechanisms instead of tables.

Conceptual Queries
DEMSI has an interface that allows a user to formulate queries in an *Acoustic Phonetic Formalism*, or *APF*, reminiscent of the context sensitive rewrite rules known from transformational phonology. The design and implementation of this interface was not especially supported by tools coming with the RDBMS in which DEMSI was first implemented. In the meantime it has appeared that the lack of abstraction for relations and features makes it very difficult to adapt the APF to changes in the data base structure or to unexpected requirements of the users.
Also when using an OODBMS, the burden of relieving an occasional user from complex query formulations, remains on the shoulders of the data base developer. However, the improved level of abstraction helps to facilitate the design and implementation, as well as the adaptation and maintenance of the query interface quite considerably.

Conclusions

No available DBMS meets all requirements. RDBMS's are strong on Flexibility, Consistency and Integrity during the Data Base construction phase, whereas OODBMS's are strong in the fields of Abstraction and Performance. Both kinds of DBMS's require the construction of a *shell* to offer sufficient abstraction for the end-user.
In our view, Abstraction and Performance are more important than Flexibility and Integrity checks. Due to the inherent software re-usability when working in an Object Oriented environment, we expect that one will need to write only very little new code when changes -even quite drastic ones- are made to an existing APDB built on an OODBMS. Thus, the major disadvantage of an OODBMS when compared to an RDBMS seems to be limited to the first implementation. The most difficult part in building a useful APDB appears to be the shell or query interface; moreover, it seems to be this shell where changes are required most often. This, combined with the fact that many queries in an RDBMS are inherently slow, makes that we would prefer an OODBMS over an RDBMS when all technical and financial conditions were equal.

References

[HEND88] Hendriks, J. & L. Boves: Definition of relations in an acoustic phonetic database. Proceedings ICASSP-'88, page 643-646, 1988.

[SWAG90] Swagten, A.: The ONTOS Test Document. SPEX Report, September 1990.

[TERL89] Terlingen, J.: A Comparison between two possible Structures for a speech-database. M.A. Thesis 304 RNL/89, May 1989.

[CARL86] Carlson, R & B. Granström: A search for durational rules in a real-speech data base. Phonetica, Vol. 43, 140-154, 1986.

BPF Outputs Compared with Formant Frequencies and LPC's for the Recognition of Vowels

Atalay Barkana, Atila Barkana

Electrical Engineering Department, Anadolu University, Eskisehir,TURKEY

Abstract: The purpose of this paper is to compare the existing different measurement methods for the recognition of Turkish vowels. These methods use the band-pass-filter (BPF) outputs, the formant frequencies, the linear prediction coefficients (LPC's). The recognition rates obtained by using the minimum distance principle show that the BPF outputs seem to be the best.

Keywords: Band-pass-filter output, formant frequency, linear predictive coefficient, vowel recognition.

Introduction

Limited number of words are taken from zero to nine in Turkish to be able to distinguish the vowels within the words. Only male voices are recorded in personal computers. Each voice is sampled at 10 kHz with 8 -bit A/D converters. There are only eight vowels in Turkish. Four of them are front vowels i,ü,e,ö and the other four are back vowels ı,u,a,o. All of them are used in the observed limited number of words. Only steady-state wave shapes are taken for each vowel, that is, transitions between CV's are omitted. Three methods are used to parameterize the wave-shape of each vowel. These are BPF (band-pass-filter) outputs, formant frequencies, and LPC's (linear predictive coefficients).

NATO ASI Series, Vol. F 75
Speech Recognition and Understanding. Recent Advances
Edited by P. Laface and R. De Mori

Recognition Using the BPF Outputs

The sampled signal from each vowel is passed through eight digital filters which are written to cover the whole frequency band of the voice spectrum [1]. The transfer function of each digital filter is, second-order band-pass, that is in z-domain

$$H(z) = K\frac{z^2 - 1}{(z - \alpha e^{j\theta})(z - \alpha e^{-j\theta})}$$

The output of each BPF is changed to its absolute value and summed every 20 ms shown in Figure 1.

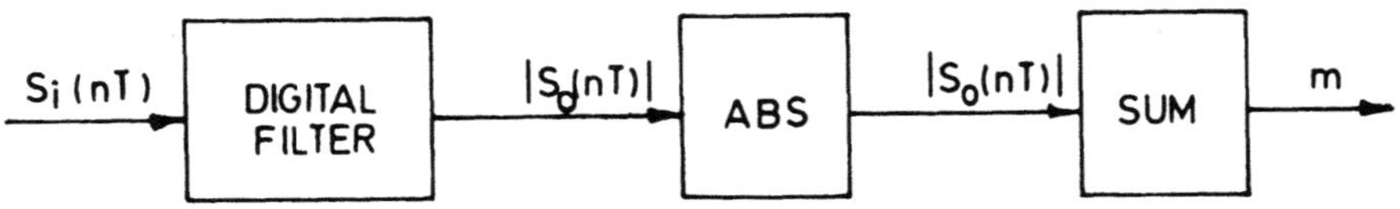

Figure 1. The block diagram of BPF outputs

The frequency band of each filter is listed below. The first filter has the frequency band from 400 to 600 Hz. The second is from 600 to 1000 Hz, the third is from 1000 to 1400 Hz, the fourth is from 1400 to 1800 Hz, the fifth is from 1800 to 2400 Hz, the sixth is from 2400 to 3000 Hz, the seventh is from 3000 to 3800 Hz, and the eighth is from 3800 to 4800 Hz.

The summed outputs of BPFs for each vowel are drawn with respect to the output of the first filter, m_1, for different speakers. Linear relations can easily be seen between the outputs of filters as shown in Figure 2.

Since these linear relations do not coincide for the different vowels, a hyperplane is fitted for the summed outputs of each vowel in an eight-dimensional space.

$$m_1 = k_2 m_2 + k_3 m_3 + \ldots + k_8 m_8$$

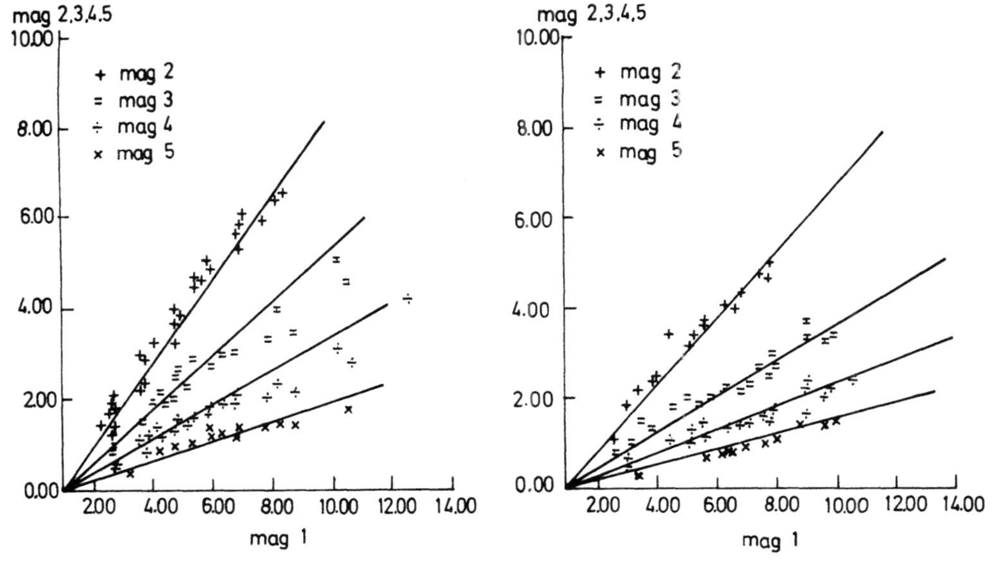

Figure 2. The summed outputs for "a" and "ì"

The k_i coefficients are determined for each vowel from the measurements. In matrix notation

$$B = MK$$

and

$$K = (M^T M)^{-1} M^T B$$

where B ìs a column vector whose elements are the summed outputs of the first digital filter, K ìs a column vector of the k_i coefficients.

Cross-tables are prepared [1] toward building a recognition algorithm which uses some type of distance idea from the hyperplanes, which can be expressed as

$$\text{Distance} = \left| m_{1,\text{spoken}} - m_{1,\text{obtained from hyperplane}} \right|$$

About 85 % recognition rate is obtained for each wowel.

Recognition Using the Formant Frequencies

The first four formant frequencies are obtained by using digital narrow band-pass-filters for each vowel FFT is also used to recheck the location of the formant frequencies. Then, the formant frequencies are drawn with respect to each other as shown in Figure 3.

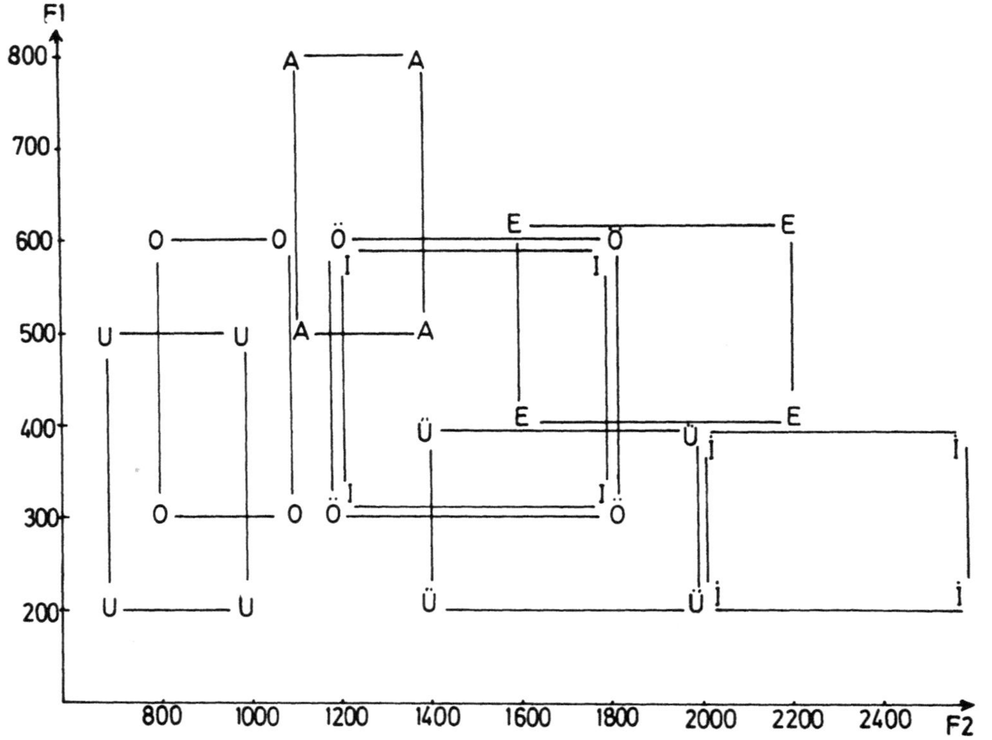

Figure 3. The second formant frequency versus the first one.

The scaling of formant frequencies is also tried in [2] and [3], but this did not help much toward recognition. Similar to the hyperplane equations of the summed outputs of BPFs, the plane equations are obtained in the three-dimensional formant frequency space for each vowel, that is,

$$f_1k_1 + f_2k_2 + f_3k_3 = 1$$

This time, the distance idea is modified as below:

$$\text{Distance} = f_1k_1 + f_2k_2 + f_3k_3 - 1$$

The recognition algorithm based on the minimum distance has a very low recognition rate, but after the tables of the distances for each vowel are prepared, it is seen that some vowels are too far away from the plane equations of certain other vowels. Thus, by using the smallest and the largest distance measures, a recognition algorithm with 80 % recognition rate for each vowel (except e,i) is obtained [3]. The vowels e and i are confused in all cases.

Recognition Using the Linear Prediction Coeffients

In recognition of the wovels, the linear predictive coding (LPC) is also employed [4]. The linear prediction coefficients are calculated for a 10 th order predictor for each pitchperiod. Then, the average value vectors of the coefficients are found for each vowel. For the recognition process, minimum distance square is used in decision making. The success rate was between 35 to 80 % generally. The reason for the low recognition rate was the large spread observed in some of the coefficients. A single coefficient was able to cause a wrong recognition, even though the other nine coefficients tended toward the recognition of the correct vowel.

As another alternative, log area ratios (LAR) are used for the recognition, since these were expected to have better distributions. Hence, a 10 section model is assumed and LARs with respect to the first cross-section are calculated. Again, the average value vectors of LARs for each vowel are found and minimum distance square is used for recognition. The recognition rate has been 93 % and 95 % for a and o vowels, respectively, but was quite poor (around 40 %) for ı,e,ö, and u. The recognition rate was about 70 % for i and u, respectively. In some cases, i was confused with ü and vice versa, ı was confused with o and u, e with o, ö with o and ı, and u with ü and vice versa.

As yet another method, the roots of the all pole predictor are examined [5]. These roots are obtained from the LPC coefficients used earlier. The five pairs of complex conjugate roots are marked on a polar graph paper. The spread of the roots were extremely wide which practically made it impossible to devise a successful recognition algorithm. The recognition rates were around 30 to 50 %.

Conclusion

The BPF bank with absolute value summers seem to yield higher recognition rates compared to LPCs and formant frequencies. The deviations from the mean values of LPCs and related log-area functions are rather large, and this causes low recognition rates. The same is true for formant frequencies. Also, sometimes a formant frequency may not be seen in the frequency spectrum at all. The low recognition rates may also be due to the phonetic structure of Turkish which is thought of as a phonetically-spoken and phonetically-written language.

References

1. Özek, B.; Barkana, Atalay: Recognition of Turkish vowels, Diploma Thesis, Electrical Eng., Anadolu University, July 1989 .
2. Yalçın, A.; Barkana, Atalay: Analysìs of Turkìsh vowels for recognition, Diploma Thesis, Electrical Eng., Anadolu University. February 1989
3. Uluturk, F.; Barkana, Atalay: An algorithm that recognizes Turkish vowels, Diploma Thesìs, Electrical Eng., Anadolu University, July 1989.
4. Tafan, H.; Barkana, Atila: LPC analysìs of Turkish vowels, Master's Thesis, Electrical Eng., Anadolu University, February 1990.
5. Cengiz, D. Barkana, Atila: Vowel analysis with complex roots, Diploma Thesis, Electrical Eng. Anadolu University, September 1989.

A CODIFICATION OF ERROR SIGNAL BY SPLINES FUNCTIONS

M.C.Benitez, J.A.Galvez, A.Rubio, J.Diaz

Departamento de Electronica y Tecnologia de Computadores. Universidad de Granada.

Abstract: One possible way of increasing the low-bit rate speech-coders performance is to consider a parametric glottal model as source for voiced sounds and to approach the glottal waveform using trigonometric functions. The main problem of this procedure is how to determine the opening and closure times. This work proposes an alternative method based on an approximation of the glottal signal by spline functions (particularly cubic splines).

Another objective is to remove the influence from the non-white glottal source in the estimation of the parameters of the production filter. In the proposed algorithm, this is obtained using conventional predictive analysis.

1. INTRODUCTION

It is well known that an increase in the speech quality is obtained using a glottal model as source for voiced sounds.

A model of the glottal source for one pitch period can be formed of three phases, one closed, one opening and one closing phases. Both, the opening and the closing phase, can be modeled with sine functions [1,2]; but the main problem of this procedure is how to determine the opening and closure times. In this work we propose an alternative method, that consists of approximating the glottal signal by splines functions. In order to get that, we must solve two problems:

First, we use a glottal waveform train as a model for the input, so the non-whiteness of the input signal must be compensated by the vocal tract parameters.

Second, it is necessary to extract the most representative points of the glottal waveform.

2. THE LPC ANALYSIS

The residual signal, after inverse filtering with a standard LPC filter A(z), is maximally white [3]; in other words, the filter A(z) cancels the vocal tract formants as well as the effects of lip-radiation and the glottis.

The speech production system can be approximated by a linear model. Within this approximations, it is assumed that the glottal source and the vocal tract systems are linearly separable.

The speech production model for voiced sounds is:

$$S(z) = E(z)L(z)G(z)\frac{1}{A(z)} \tag{1}$$

in which E(z) is a train of impulses; L(z) is the lip radiation filter; G(z) is the filter for simulating the glottal waveform and 1/A(z) is the vocal tract filter.

NATO ASI Series, Vol. F 75
Speech Recognition and Understanding. Recent Advances
Edited by P. Laface and R. De Mori

These filters are supposed to be time-invariant for the analysis frame. As a first approximation in order to find the estimated glottal signal, we adopt [4]:

$$G(z) = \frac{1}{(1-\mu z^{-1})^2} \qquad L(z) = (1-\mu z^{-1}) \tag{2}$$

where μ is close to 1. If we define

$$\frac{1}{D(z)} = G(z)L(z) = \frac{1}{(1-\mu z^{-1})} \tag{3}$$

then

$$S(z) = \frac{E(z)}{D(z)} \frac{1}{A(z)} \tag{4}$$

and

$$S(z)D(z) = \frac{E(z)}{A(z)} = S_1(z) \tag{5}$$

The signal $S_1(z)$ will not depend on the filters G(z) and L(z), so we can calculate its LPC parameters, (we will call them LPC_1), and build the filter $1/A_1(z)$. Now we make an inverse-filtering of S(z) with the filter $A_1(z)$ to get the signal Silpc(z); that is:

$$S(z)A_1(z) = \frac{E(z)}{D(z)} = Silpc(z) \qquad Silpc(z) = E(z)G(z)L(z) \tag{6}$$

The glottal signal to be estimated must have strong similarities with Silpc(z), then this signal must have some information about the opening and closure times of the glottal waveform.

Now, in order to get suitable LPC parameters for an input without a flat spectrum, we perform a LPC analysis on the signal Silpc(z), (which we will called LPC_2) and another inverse-filtering of S(z) with the filter $A_2(z)$ to get $S_2(z)$. This signal is cleaned of the influence of the glottal source and the lip radiation in a greater degree than $S_1(z)$. Another LPC analysis performed on the $S_2(z)$ allows to get good parameters for the vocal tract filter (which we will called LPC_3).

This whole procedure can be reiterated to improve the estimation of A(z) and Silpc(z).

Combining these equations we have the next outline:

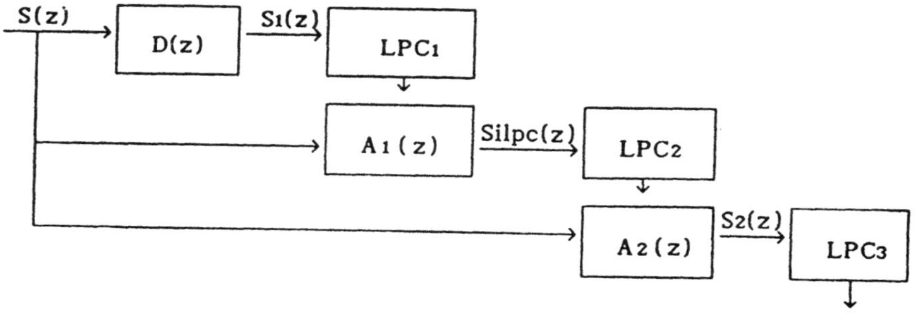

fig[1]

3. STUDY OF THE Sglpc SIGNAL

If we continue the study of the Silpc(z) signal, the opening and closure time can not be easily evaluated; but we have noticed that if we eliminate the influence of the lip radiation in Silpc(z), a new signal, called Sglpc(z), is obtained:

$$Sglpc(z) = \frac{Silpc(z)}{L(z)} \tag{7}$$

For differents sounds, voiced, unvoiced and mixed, this signal has the following shape:

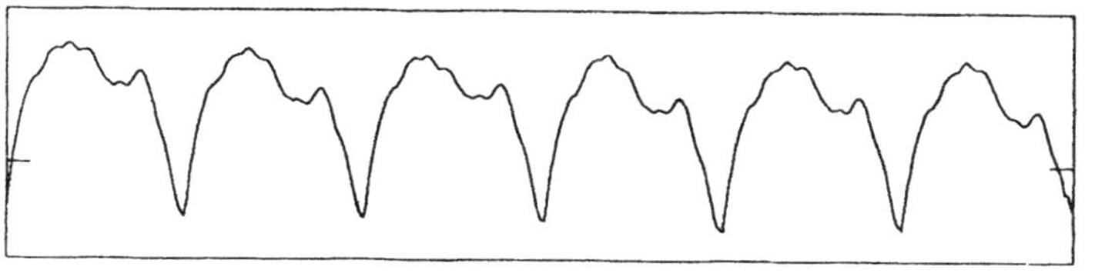

fig[2]

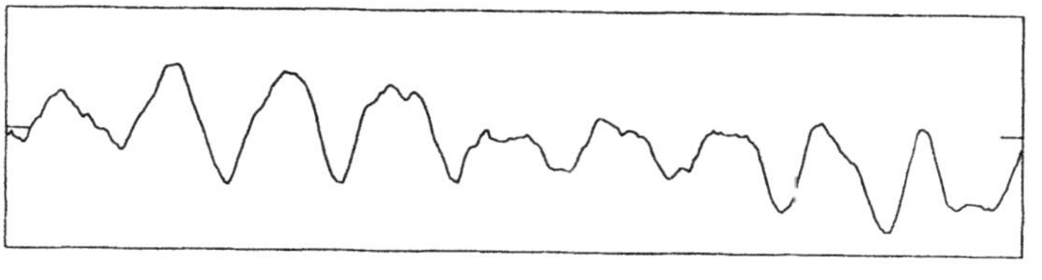

fig[3]

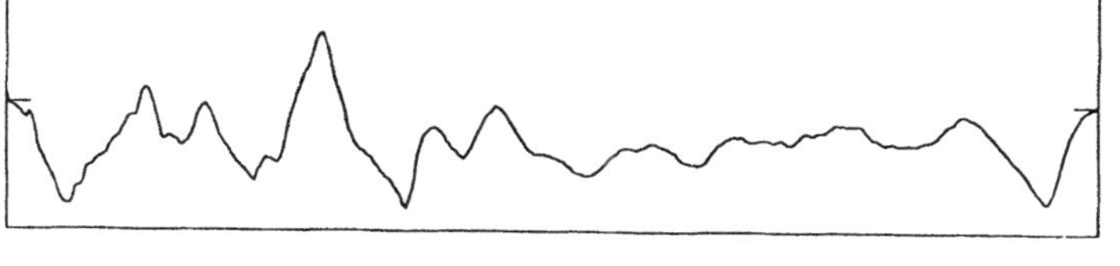

fig[4]

Where fig[2] correspond to a voiced sound, fig [3] to a mixed sound and fig[4] to a unvoiced sound.

4. ALGORITHM FOR THE EXTRACTION OF Sglpc SIGNAL PARAMETERS

Given the waveform Sglpc, we can obtain a set of parameters which will allow us to rebuild the Sglpc signal. For example, we can make a new LPC analysis of this signal, or we can take out the more representative points and use any mathematical procedure of interpolation.

The most significant characteristics of this signal are:

1.- For voiced sounds, this signal is periodic, with the period equal to the fundamental frequency or pitch.

2.- For one period, this signal has only two zero crossing, two minima separating the pitch period, and one or two maxima.

The interpolation of the Sglpc signal is done for one period, and we shall use it as a pattern which will be shifted along the frame. In order to choose the values of the parameters, we shall consider two axes coordinated system, the time will be represented in x-axis, and the amplitude of the signal in the y-axis. The parameters chosen as the most representative in the signal within a period are:

- The coordinates of the deep peaks which are separated by one pitch period.
- If there is only one maximum, its coordinates and the coordinates of the middle between each one of two minima and the maximum.
- If there are two maxima, we shall take their coordinates and the coordinates of the absolute minimum between the two maxima.

The algorithm used to extract the period of the signal is the following:

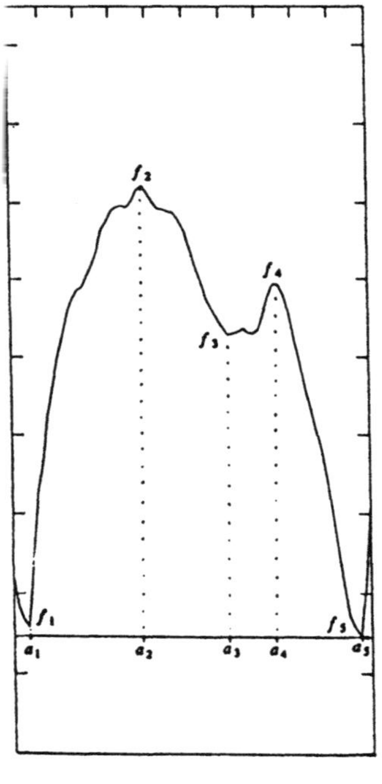

fig[5]

1. We take the absolute minimum over three frames of the Sglpc signal and we perform a clipping of about the 40% of the minimum.
2. We extract the minima.
3. We calculate the differences between one given minimum and the next one.
4. We sort the differences from the largest (Md), to the smallest (md), and they are compared in the following way:
   - -If Md - md < u (a threshold value of the number of samples), then we shall take the average of all the differences as fundamental period.
   - -If Md - md > u and if we call cd to the central differences, then:
     - - If Md - cd < u (or cd - md < u), we shall take the average of the differences between cd and Md (md and cd) as the pitch period.
     - - If Md - cd > u and cd - md > u, if we call cMd to the central differences between Md and cd (cmd , cd and md), then:

       If cMd - cd < u (or cd - cmd < u) we shall take as the pitch period the average of the differences between cMd and cd (cd and cmd).

       If cMd - cd > u and cd - cmd > u, we shall consider this frame as unvoiced.

5. FUNCTION INTERPOLATION USING CUBIC SPLINES

This interpolation procedure consists of the approximation of the Sglpc function using a third order polynomial, (cubic splines) within discrete intervals. From a mathematical point of view, a spline function is defined in the following way:

Let's $a = a_1, < a_2 < ...... < a_n = b$ a partition of the interval [a,b]. We shall call spline functions of order m with nodes $a_1, \cdots ,a_n$ to a function S(x) defined on [a,b] that verifies the next conditions:

1. The restriction of S(x) within every subinterval $[\ a_i ,a_{i+1}\ ]\ i = 1, \cdots ,n-1$ is a polynomial of order not greater than m, for cubics splines $m = 3$.
2. S(x) and its derivates until order m-1 are continuous in [a,b].

If we have five pairs of points $(\ a_i ,f_i)$ to rebuild a period of Sglpc, there will be four subintervals of length $h_i = a_{i+1} - a_i$; in every subinterval we shall approximate the function by a cubic spline with the following expression :

$$S_i(x) = \frac{Z_{i+1}}{6h_i}(x - a_i)^3 + \frac{Z_i}{6h_i}(a_{i+1} - x)^3 + \left[\frac{f_{i+1}}{h_i} - \frac{Z_{i+1}h_i}{6}\right](x - a_i) + \left[\frac{f_i}{h_i} - \frac{Z_i h_i}{6}\right](a_{i+1} - x) \quad (8)$$

Where Z_i and Z_{i+1} are the values of $\ddot{S}_i(x)$ in a_i and a_{i+1} respectively.

It can be shown [5] that this expression satisfies all the conditions above mentioned, and that S(x) takes the values f_i in the nodes a_i.

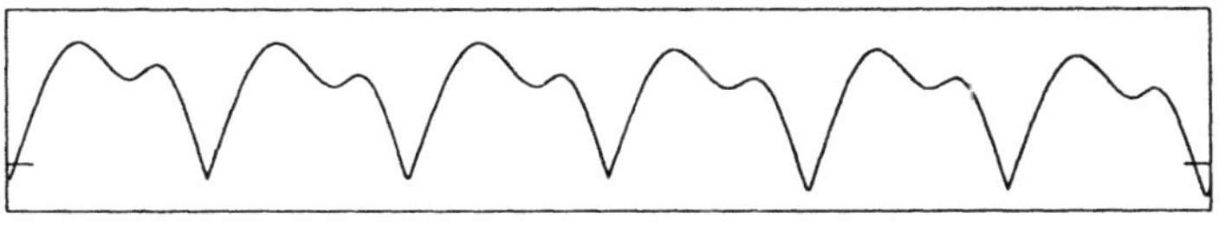

fig[6]

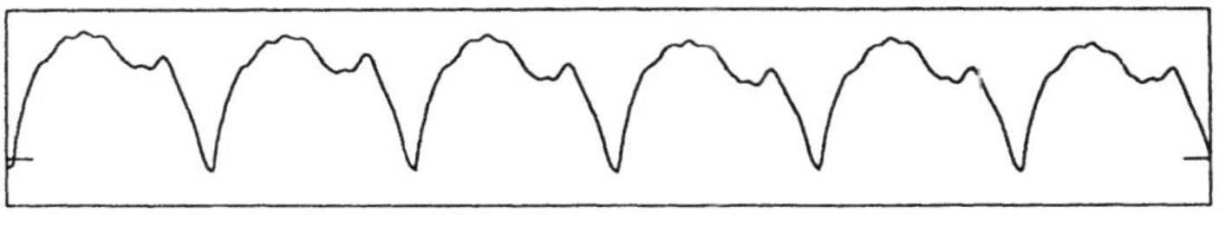

fig[7]

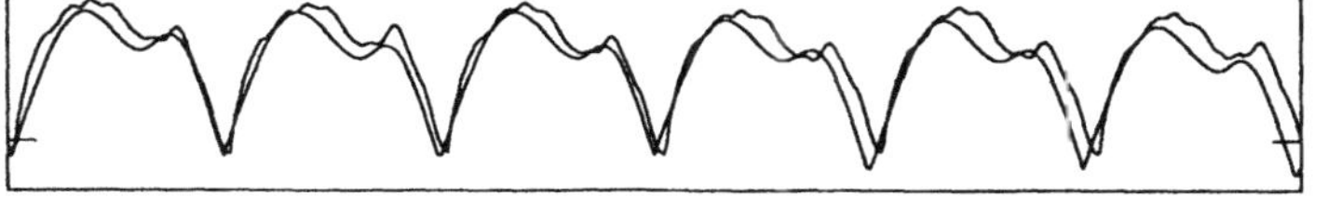

fig[8]

So, in fig. [6], we can see the shape of the interpolated Sglpc pulse for a voiced frame, in fig[7] the same frame for the original Sglpc pulse and in fig[8] the overplot of the two signals.

6. RESULTS AND CONCLUSIONS

First of all, some remarks about working environment. Speech signal was sampled at 8 kHz and windowed every 128 samples. Then an LPC analysis was made using autocorrelation methods. So, filter stability is guaranteed. In order to obtain LPC coefficients we use three frames: the previous, the current and the following frames. This three frames have been weighted with a Hamming window. The remaining procedures are developed in this work.

Given the observed characteristics of Sglpc signal we can conclude:

1. It is useful for the classification of different sounds in voiced, unvoiced and mixed classes. This can be obtained from the shape of Sglpc for this three types of sounds.
2. Sglpc signal can be used to obtain fundamental frequency or "pitch" of a speech frame.
3. In view of the easy parametrization of this signal, it can be used for low bit rate coding and speech synthesis purposes, obtaining improved quality, as we can see in the following figures.

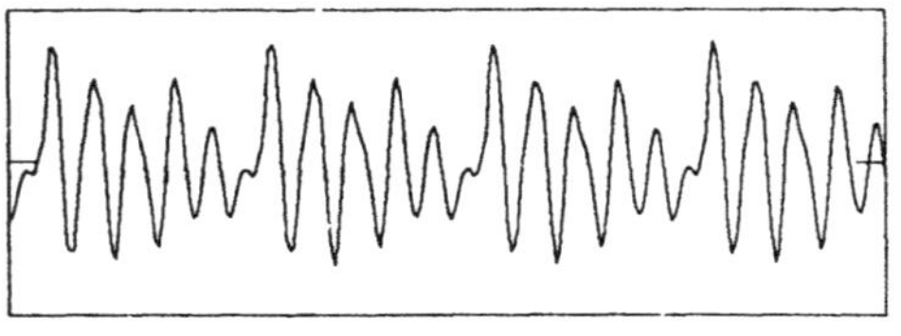

fig[9]

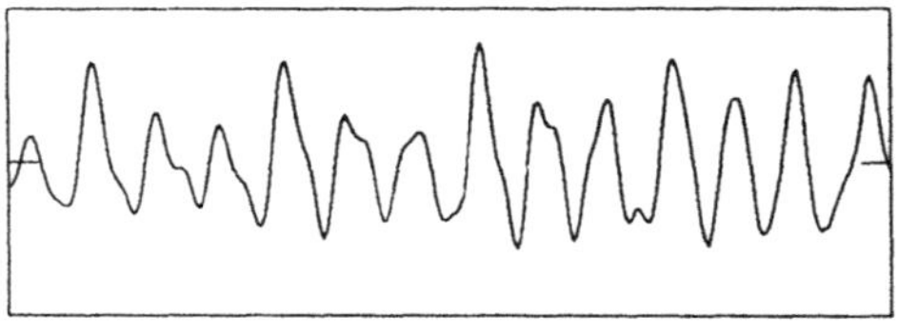

fig[10]

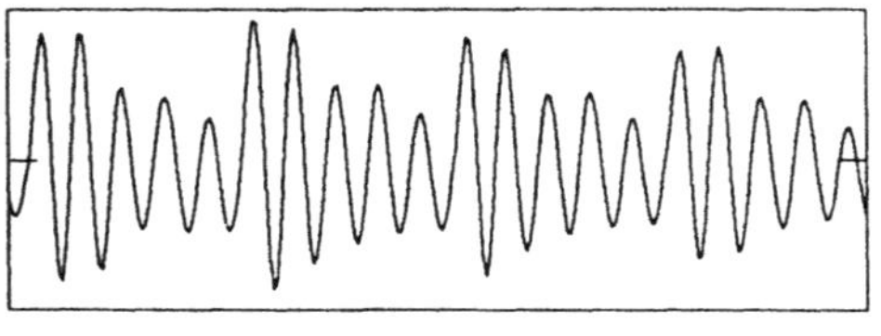

fig[11]

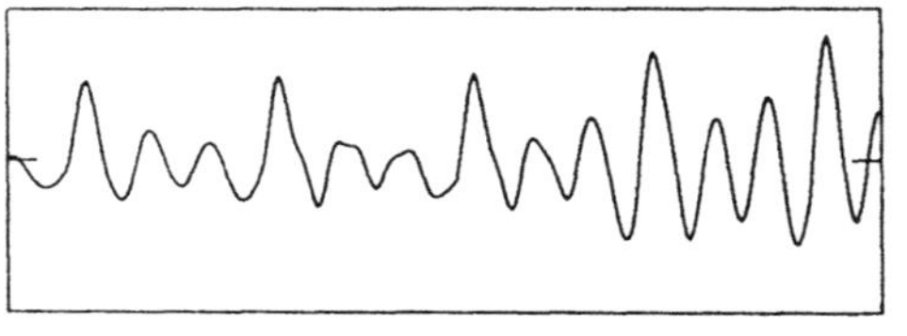

fig[12]

Where the fig [9],[10] are the original signal and the fig [11],[12] are the synthesied signals.

REFERENCES

1. Hedelin: "A Glottal LPC-vocoder", Proc.ICASSP-84, pp 161-164.
2. Hedelin: "High Quality Glottal LPC-vocoder", Proc.ICASSP-86, pp 991-994.
3. Markel-Gray: "Linear Prediction of speech", Springer Verlag, 1976
4. Bergstrom: "Glottal Analysis Using LPC", Proc.IASTED-87, pp 22-26.
5. Gasca Gonzalez: "Calculo Numerico I", UNED, 1988.

Specific Distance for Feature Selection in Speech Recognition

E. Lleida, C. Nadeu, J.B. Mariño, E. Monte, A. Moreno

Dpto. of Signal Theory and Communications,U.P.C., Apdo. 30.002, 08080 Barcelona, Spain

Abstract: In this paper, the use of a specific metric as a feature selection step is investigated. The feature selection step tries to model the correlation among adjacent feature vectors and the variability of the speech. We propose a new procedure which performs the feature selection in two steps. The first step takes into account the temporal correlation among the N feature vectors of a template in order to obtain a new set of feature vectors which are uncorrelated. This step gives a new template of M feature vectors, with M << N. The second step defines a specific distance among feature vectors to take into account the frequency discrimination features which discriminate each word of the vocabulary from the others or a set of them. Thus, the new feature vectors are uncorrelated in time and discriminant in frequency.

Keywords: Feature selection, Specific distance, Data Compression, Karhunen-Loève transform, Principal component analysis, Discriminant analysis.

I. INTRODUCTION

The problem of feature selection in speech recognition can be studied by different ways. One way is to study the feature selection as a problem of *data compression* to reduce the computational time and memory requirements. From this point of view, a lot of similar techniques were implemented in the last years where a new template is obtained removing those feature vectors which are similars. Trace segmentation and variable frame coding are two classical techniques [3,4,6]. Another point of view, proposed in this paper, is to assume that there is an underlying set of "real" uncorrelated features, and the features we are working on are "impure" in the sense that they are a linear combination of those "real" features. Then, the objective is to find a transformation which recovers the "real" features [5]. These two points of view select the feature in the time dimension, that is, it is a *temporal selection*. Thus the *temporal selection* obtain a new template where the feature vectors are uncorrelated or without temporal redundancy. However, the *temporal selection* doesn´t take into account the variability and separability among words. Thus, a *frequency selection* step which reduces the within-class variability and increases the separability among words is needed. This is the second step proposed in our feature selection procedure which is done by defining a *specific distance* for each feature vector. Thus, we propose a feature selection procedure which makes use of a representation criteria for *temporal selection* and discriminant criteria for *frequency selection*. Figure 1 shows the process of *temporal* and *frequency selection*.

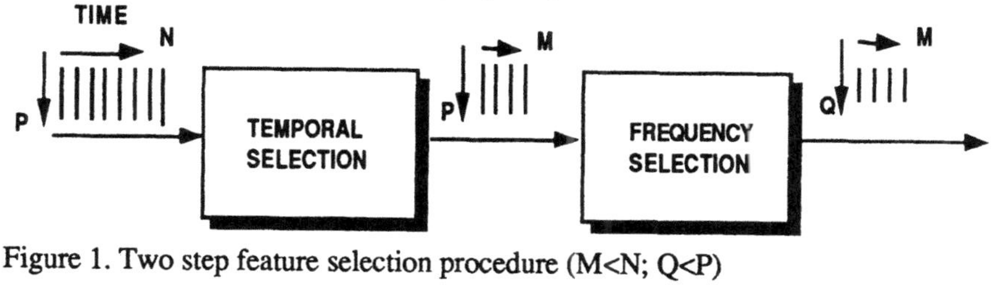

Figure 1. Two step feature selection procedure (M<N; Q<P)

This work was supported by the PRONTIC grant nº 105/88

NATO ASI Series, Vol. F 75
Speech Recognition and Understanding. Recent Advances
Edited by P. Laface and R. De Mori

II. TEMPORAL SELECTION

Temporal selection is the first step in our feature selection process. Its purpose is to obtain a time compression by removing the correlation of the temporal evolution of the spectrum. Basically, the problem is to represent the sequence of spectra by a superposition of the members of an orthogonal family of functions where the input template is representated with less coefficients.

Given a NxP matrix Y of spectral parameters $\{y_i(n)\}$ representing N frames of P "impures" features, the transformation obey the following formulation

$$y_i(n) = \sum_{m=1}^{N} \alpha_m(i)\phi_m(n) \qquad 1\leq i\leq P\ ;\ 1\leq n\leq N \tag{1}$$

where $\alpha_m(i)$ is the mth "real" feature vector and $\phi_m(n)$ is the mth transformation function of the transformation matrix $\{T\}$.

There are two definitions of transformation functions:

1- *Data independent transformation functions*. The transformation functions are members of a deterministic family of orthogonal functions as the Discret Cosinus Transform (DCT).

2- *Data dependent transformation functions*. The transformation functions are found from the data using a criteria of minimum square error. The error of representing $y_i(n)$ with M "real" feature vectors is defining as follow

$$\varepsilon = E\{(y_i-\hat{y}_i)(y_i-\hat{y}_i)^t\} \tag{2}$$

where $\hat{y}_i = \sum_{m=1}^{M} \alpha_m(i)\phi_m(n)$ is the estimation of "impure" feature vectors with M "real" feature vectors. With the constraint that the transformation functions are orthogonal, the transformation functions are found solving the eigensystem

$$C_{yy}\phi_m = \lambda_m\phi_m \tag{3}$$

where C_{yy} is the covariance matrix defined as

$$C_{yy} = \frac{1}{P-1}\sum_{i=1}^{P} (\underline{y}_i - \bar{\underline{y}})(\underline{y}_i - \bar{\underline{y}})^t \tag{4}$$

with

$$\bar{\underline{y}} = \frac{1}{P}\sum_{i=1}^{P} \underline{y}_i \tag{5}$$

$$\underline{y}_i = \{y_i(1),y_i(2),\ldots\ldots,y_i(N)\}$$

From this eigensystem, N eigenvalues and their corresponding eigenvectors are obtained. However, only the M eigenvectors with the *largest* eigenvalues are retained. Thus, the transformation matrix $\{T\}$ is composed by the M eigenvectors with the M largest eigenvalues, ranking them from the largest to the smallest one. Because of the orthogonal property of the transformation functions, the new "real" feature vectors are computed as a linear combination of the "impure" feature vectors as follows [2]

$$\alpha_m(i) = \sum_{n=1}^{N} y_i(n)\phi_m(n) \qquad 1\leq i\leq P\ ;\ 1\leq m\leq M \tag{6.a}$$

which is known as the Karhunen-Loève transform (KLT) or

$$\alpha_m(i) = \sum_{n=1}^{N} (y_i(n)-\bar{y}_i(n))\phi_m(n) \qquad 1\leq i\leq P\ ;\ 1\leq m\leq M \tag{6.b}$$

which is Known as Principal Component Analysis (PCA).

The principal properties of the new representation are:

-Coefficients with the largest variances are the "real" features.
-The new "real" features are uncorrelated.
-Feature vectors are arranged in variance order, thus, *no time-alignment* is needed in the comparison step.

The transformation matrix is computed in the training process. We distinguish two cases:

1- *General matrix.* A transformation matrix T_g for all the words of the vocabulary. In this case, the covariance matrix C_{yy} is obtained averaging the covariance matrix of each training word.

2- *Specific matrix.* A transformation matrix T^w for each word of the vocabulary. Then, the covariance matricx C_{yy} is obtained using several repetitions of the word 'w'.

Figure 2 shows the evolution of the first three transformation functions for the *General matrix* and for the *Specific matrix* of the word /sɛt/.

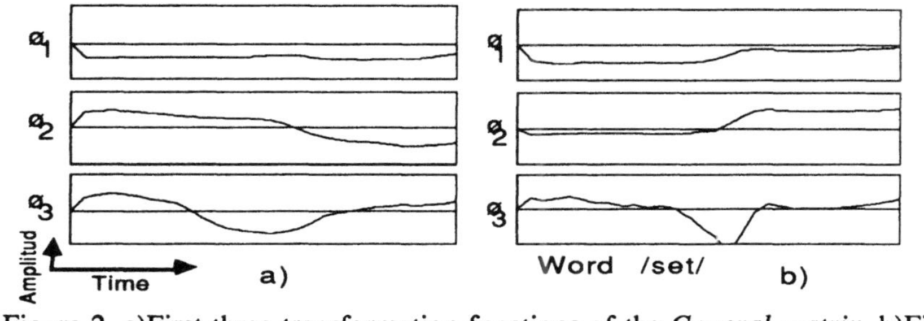

Figure 2. a)First three transformation functions of the *General matrix*. b)First three transformation functions of the word /sɛt/.

III. FREQUENCY SELECTION

The second step of the feature selection process is to compute a transformation matrix for each new uncorrelated feature vector obtained in the *temporal selection* in order to discriminate between words. This step can be seen as a method for finding a *specific distance measure* for each reference vector. This *specific distance* takes into account the discriminant properties of the feature vectors which reduces the within-class variability and increases the separability among feature vectors. Thus, the *frequency selection* step is related with the comparison step. Defining the weighted Euclidean distance between the test vector α_i and the reference vector α_j as

$$d(i,j) = \| F_j (\alpha_i-\alpha_j)\|^2 \qquad (7)$$

a *specific distance* matrix $F_j=\{fd_1,fd_2, ...,fd_Q\}$ of Q weighting vectors for each reference vector of each word has to be computed. Figure 3 shows the relation between *specific distance* and *frequency selection.*

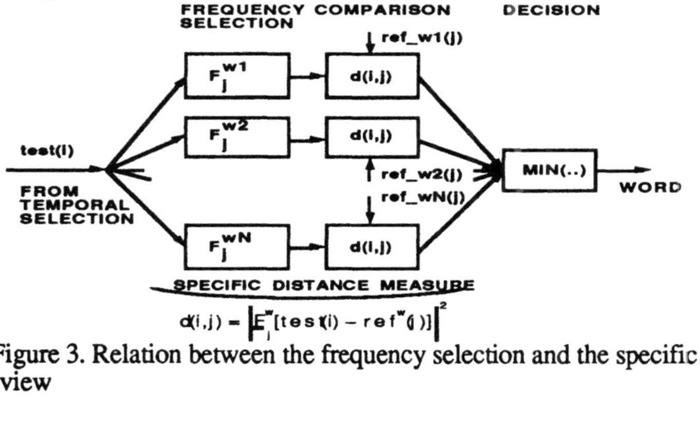

Figure 3. Relation between the frequency selection and the specific distance measure point of view

In order to find the discriminant matrix F_j, two classes of vectors are defined. For a word 'w', the mth feature vector of any utterance of it forms the *correct class* (α_{ce}) and the mth feature vector of the other words forms the *incorrect class* (α_i). Thus, defining the *mean interclass distance* as the mean distance between the incorrect class vectors and the mean correct class (α_c) feature vector

$$E(D_{inter}) = E(\|F_c(\alpha_i - \alpha_c)\|^2) \tag{8}$$

where α_c is mean vector of the correct class, and taking into account the matrix relation

$$\|a\|^2 = tr(a\, a^t) \tag{9}$$

we can rewrite (.) as

$$E(D_{inter}) = tr(F_c\, E((\alpha_i-\alpha_c)(\alpha_i-\alpha_c)^t)F_c^t)) \tag{10}$$

defining the *between-class mean distance* matrix as

$$B = E((\alpha_i-\alpha_c)(\alpha_i-\alpha_c)^t) \tag{11}$$

the mean interclass distance is as follow

$$E(D_{inter}) = tr\,(F_c\, B\, F_c^t) \tag{12}$$

In order to take into account the within-class variability a *mean intraclass distance* is defined as

$$E(D_{intra}) = tr\,(F_c\, W\, F_c^t) \tag{13}$$

where

$$W = E((\alpha_{ce} - \alpha_c)(\alpha_{ce} - \alpha_c)^t) \tag{14}$$

In this way, the criterion function to be maximized is [1,5]

$$J = tr(fd_k\, B\, fd_k^t) - \lambda(tr(fd_k\, W\, fd_k^t) - 1) \tag{15}$$

The solution of this optimization problem is the eigensystem

$$(W^{-1}B)fd_k = \lambda_k fd_k \tag{16}$$

therefore, the specific distance matrix is formed by the Q eigenvectors with the Q *largest* eigenvalues of $W^{-1}B$, whenever

$$E(D_{inter}) = \sum_{k=1}^{Q} \lambda_k \; >> \; Q = E(D_{intra}) \tag{17}$$

The discriminant properties can be found in the matrix $W^{-1}B$ because

$$tr(W^{-1}B) = \sum_{k=1}^{P} \lambda_k \tag{18}$$

This process is made for each reference vector of each word of the vocabulary. In the training process, a mean vector (α_c) for each feature vector is computed and used as reference to compute de within-class and between-class mean distance matrices as well as reference vector for the recognition process. Figure 4 shows of the projection of the first feature vector of several words using the first two discriminant eigenvectors of the word /dos/.

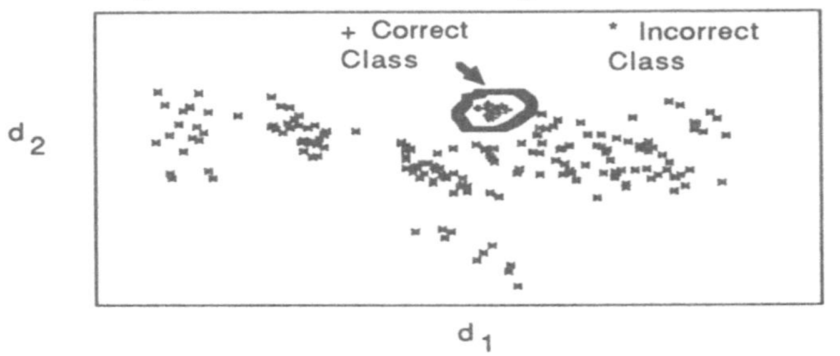

Figure 4. Projection of the first feature vector of several words using the first two discriminant eignevectors of the word /dos/ (correct class).

IV. RESULTS AND CONCLUSIONS

The recognition experiments were made using two data base. A data base consists of ten repetitions of the Catalan digits {u,dos,trɛs,kuatra,sink,sis,sɛt,vuit,nou,zɛru} uttered by six male and three female speakers (900 words) and recorded in a quiet room. The second data base is the Spanish E-set {b,c,d,e,g,p,t} uttered by two male and one female speakers, seven repetitions, recorded in a laboratory enviroment.

For feature extraction, the speech signal was sampled at 8 KHz, preemphasized ($H(z)=1-0.95z^{-1}$) and 8 Log-Area ratios were computed each 15 ms for the digit data base and 10 ms for the E-set data base using the LPC analysis of 30 ms. A typical Hamming smoothing window was applied to the data. After the LPC analysis, template were normalized to a fixed number N of frames, with N equal to 30, to apply the temporal selection step.

A classical pattern recognition system which compares an input template with a set of reference templates by means of a linear frame to frame comparison was used. The references, obtained in the trainning process, are constituted by the feature vectors obtained in the feature selection process and two transformation matrices. One is used to temporal selection and the other is to frequency selection which is specific for each frame.

We present two experiments. The first experiment is a multispeaker experiment. Six repetitions of the nine speakers digit data base are used as training. In each recognition experiment, an evidence measure was computed as Ev=(D2-D1)100/D1, where D2 is the distance to the second candidate and D1 is the distance to the first candidate. Figure 5 shows the recognition results obtained for different values of M and Q using both temporal transformation matrices T_g (General matrix) and T^w (Specific matrix). For the best result, M=3 and Q=2 with T_g, the mean evidence is 85,4 % showing the good discrimination properties of this feature selection process.

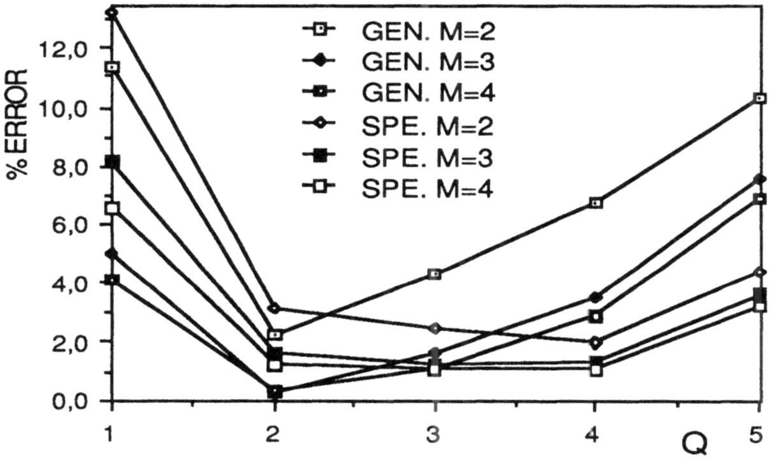

Figure 5. Error rate for several values of M and Q

With the E-set data base, the results are quite different. In this case, the optimal number M of temporal features is 9 (General matrix) and the discriminant features is 4. With these values of M and Q, the recognition rate is 12.53 % averaging seven experiments where six different repetitions were used as training in each experiment. Figure 6 shows the confusion matrix.

	B	C	D	E	G	P	T
B	129	1	7				
C	1	137		8			1
D	7	1	141				
E				136		4	7
G		3			144		
P	1	9		15		109	13
T				32		9	106

Figure 6. Confusion matrix for the E-set experiment

The second experiment is a speaker independent experiment. In this case, the training set is made up by ten repetitions of six speakers of the digit data base, using three speakers are as test. Figure 7 shows the results for three different versions: a) classical independent system with a clustering process, b) only temporal selection step and c) temporal and frequency selection steps. This figure shows also the computational load in number of multiplications per recognized word. It can be noted that the best result is obtained using the two steps feature selection process which increase the recognition rate decreasing the number of features and the computational load.

		% error	evidence	# references per word	# features per word	computational load
a)	Classical system	2.00	45 %	2	480	48000
b)	Temporal selection	3.00	54 %	1	24	1000
c)	Two steps selection	1.66	77 %	1	9	1500

Figure 7. Results for the speaker independent experiments

As conclusion, a two step feature selection procedure has been introduced. This feature selection procedure transforms an input template of features to a new representation where the correlation among vectors is removed by using the idea of *Principal Component Analysis* or *Karhunen-Loéve transform* and the variability and separability among words is taking into account using the idea of *Discriminant Analysis* and *Specific distance measure*. As a result, the new template has the feature vectors arranged in variance order, therefore, no time-alignment is needed in the comparison step and each reference vector has associated a specific distance measure which select the discriminant features. The result is an important improvement in the error rate and a very small computational load.

V. REFERENCES

1. Fukunaga K.:Introduction to statistical pattern recognition. Academic Press, 1972.
2. Gerbrands Jan J.: On the relationships between SVD,KLT and PCA. Pattern Recognition, vol 14, pp. 375-381, 1981.
3. Kuhn M.H., Tomaschewski H.: Improvements in IWR. IEEE. trans. on ASSP, vol-31, pp 157-167, Feb. 1983.
4. Lleida E., Nadeu C., Mariño J.B: Speech parametrization and recognition using block and recursive linear prediction with data compresion. European Conference on Speech Technology, pp. 300-303, Edinburgh-1987.
5. Lleida E.:Feature compression and selection in speech recognition. Ph.D. thesis (in spanish), Universidad Politecnica de Cataluña, 1990.
6. Pieraccini R., Billi R.: Experimental comparison among data compression techniques in IWR. ICASSP-83, Boston.

Multiple Template Modeling of Sublexical Units*

P. Aibar[1], M.J. Castro[2], F. Casacuberta, E. Vidal
Departamento de Sistemas Informáticos y Computación.Universidad Politécnica de Valencia, Spain

Abstract: Automatic training procedures are developed to obtain the model or models for a certain type of linguistic unit, under the framework of a Distance-based approach. The chosen units are phonetic-units and the models are templates. In a first approach, one prototype (centroid) per phonetic-unit is obtained through an iterative process and by using Dynamic Time Warping techniques. A refinement is performed through a Clustering procedure that obtains several prototypes per phonetic-unit. Another refinement, which is based on Multiedit Condensing techniques, is also proposed. Some preliminary experimental results for single-speaker and multi-speaker tasks are reported.

Keywords: Acoustic-Phonetic Decoding, DTW, Clustering, Multiedit Condensing.

1. Introduction

Within the general framework of Pattern Recognition (Non-Parametric) (K-)Nearest Neighbours techniques often prove more powerful in practice than other Parametric approaches [3]. However, while successful Automatic Speech Recognition systems have been achieved both through Nearest Neighbours (DTW) and Parametric (HMM) approaches, it seems that the possibilities of the interesting Distance-based techniques have not sufficiently been explored in Acoustic-Phonetic Modeling [2].

This work focuses on the Acoustic-Phonetic Decoding of the speech signals in terms of subword units. When an Acoustic-Phonetic Decoder is designed, there are two main problems to solve, the election of the concrete *subword units* and, secondly, the technique to be used in the decoding process and, consequently, the *type of models*. Nowadays, the most widely used subword unit is the phoneme (or phoneme-like), since the number of units is limited, and some coarticulation effects can be efficiently represented through extensions of the models (context-dependent phonemes) and/or by rules [6] [11] [8]. The technique we have used is based on Dynamic Time Warping, under the framework of a Distance-based approach, and the models are (sets of) templates. The aim of this work is to develop automatic learning procedures to obtain the model or models for each subword unit from training utterances.

---

* Supported in part by the Spanish CICYT, under grant TIC 89/0448.
[1] Supported by a grant from the "Ministerio de Educación y Ciencia".
[2] Supported by a grant from the "Conselleria de Cultura, Educació i Ciencia de la Generalitat Valenciana".

NATO ASI Series, Vol. F 75
Speech Recognition and Understanding. Recent Advances
Edited by P. Laface and R. De Mori

Two ways for using DTW techniques to automatically extract phonetic-units from a set of training utterances are proposed: Connected Units and Unit Spotting. The first approach is reminiscent of Connected Words and Forced Recognition [2] [5] [10], and the second one is based on Word Spotting techniques [4] [7] [1] [2].

2. Two Techniques to Segment Speech Signals into Subword Units

Given a set of training samples and their corresponding phonetic transcriptions, we use two techniques that segment a fluently spoken sentence into individual subword units by DTW comparing the sentence with the prototypes of the phonetic-units which are given by the phonetic transcription of the sentence. Both techniques, Connected Units and Unit Spotting, build a set of templates for each phonetic-unit. The main differences between the proposed approaches are the distinct use of DTW techniques.

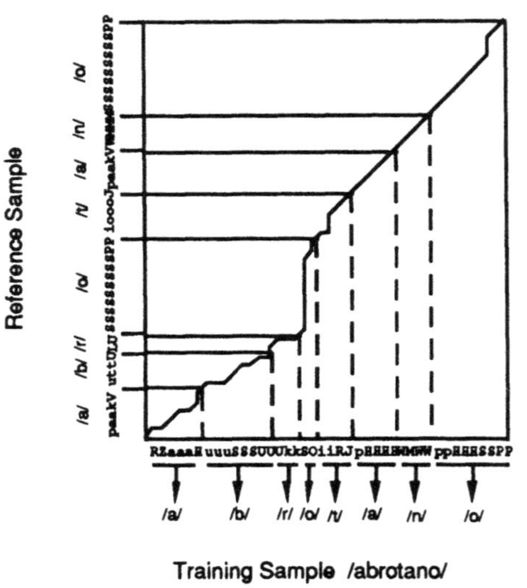

Figure 1. An example of segmentation of the training sample /abrotano/ by using the Connected Units approach.

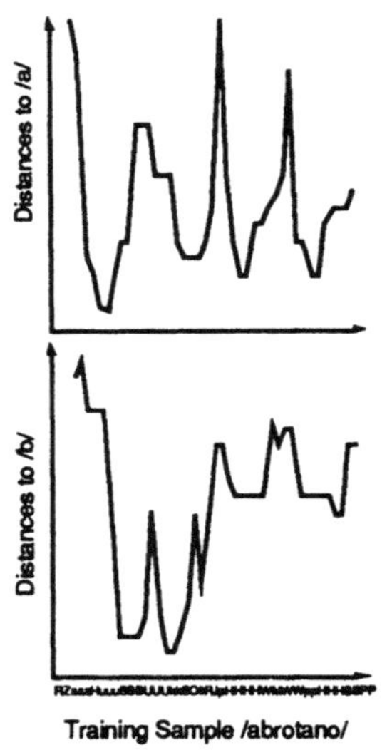

Figure 2. Distances of all possible optimal paths between the training sample /abrotano/ and the phonetic-units /a/ and /b/ (Unit Spotting approach).

In the first approach, the templates are obtained by using a conventional DTW algorithm. Each training sample is DTW compared with a reference sample that is formed by the concatenation of the prototypes of each phonetic-unit which is given by the phonetic transcription of the word. The result of the comparison is a time warping path which relates each frame of the training sample with the reference sample that is obtained by concatenation. From this path and the known boundaries of the concatenated prototypes, a segmentation and labelling of the training sample can be performed (figure 1).

The second approach, Unit Spotting, is similar to the first one, except in the application of DTW. In this case, the boundary conditions of the conventional DTW procedure are relaxed, in the sense that a warping path can start at any frame of the training sample and can reach any successive frame of this sample. Each training sample is thus DTW compared with all the prototypes of the phonetic-units which are given by the phonetic transcription of the sample (figure 2). As a result each frame of the sample has attached the prototype and the distance that has produced the minimum distance up to this frame. Then, by using some heuristics, a segmentation of the training sample can be performed according to the order given by the phonetic transcription of the training sample. The same as before, the new segments are incorporated to the set of templates of each phonetic-unit.

3. Iteration Optimization Techniques

Both techniques which are described above are used within an iterative process. In each iteration, the training samples are segmented and labelled by using Connected Units or Unit Spotting techniques, and the prototype or prototypes of each phonetic-unit from each set of templates can be obtained according to one of the following approaches. First, only one prototype per phonetic-unit is obtained by calculating the Centroid of each set of templates corresponding to one phonetic-unit. This technique has been improved by obtaining a *variable number of prototypes* per phonetic-unit, using Clustering and Multiedit Condensing techniques. With these improvements, the algorithms should be able to capture most of the contextual variability of each phonetic-unit.

3.1. Centroid

This technique has been implemented for both approaches: Connected Units and Unit Spotting. The initialization is based on bootstrapping from a small initial set of prototypes which is built from a subset of training samples that is segmented manually. Once all the training samples are segmented and labelled, the Centroid of each set of templates is obtained and is considered as the new prototype for the phonetic-unit. This procedure is iterated until the prototypes do not change from one iteration to another. The prototypes (centroids) that are obtained in this way can be directly used for the Acoustic Phonetic task or they can be used for bootstrapping the other techniques. Both for Connected Units and Unit Spotting approach, the computational cost of the algorithm is mainly due to the computation of the centroid.

3.2. Clustering

The Clustering technique [3] has been developed for both approaches: Connected Units and Unit Spotting, although for Unit Spotting no iteration was done because no improvement was observed. With this technique, the representation of each phonetic-unit consists of a *set of prototypes* which is obtained by clustering the set of templates of each phonetic-unit. The number of prototypes per phonetic-unit depends on the distortions supplied by the Clustering procedure for the different units. Depending on the required number of classes, the computational cost is due to the number of times that the DTW procedure is performed or to the computation of the distances for the clustering algorithm.

3.3. Multiedit Condensing

Finally, the Pattern Recognition technique which is known as Multiedit Condensing [12] has been applied (only for Connected Units) in an attempt to obtain a variable number of *'discriminative'* prototypes per phonetic-unit. The main scheme of the procedure is the same as for Clustering, but applying Multiedit Condensing over all sets of samples of phonetic-units.

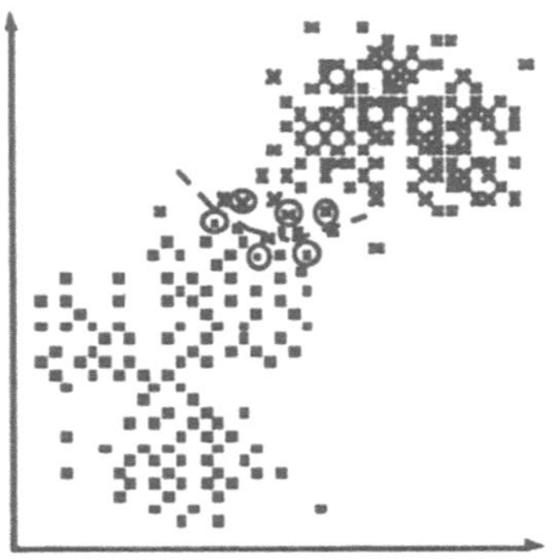

Figure 3. The samples inside the circles are the prototypes obtained by applying Multiedit Condensing to a two-class problem.

This technique selects a small subset of prototypes from the initial set of templates, such that the decision frontiers which are associated to the Nearest-Neighbour classification rule with these prototypes approach the Bayes frontiers for minimum classification error. Therefore, this technique can be seen as a method to obtain *discriminant subsets of prototypes*. A hypothetical example of the application of this technique is shown in figure 3. The computational cost of this procedure comes almost completely from the Multiedit algorithm.

4. Experimental Results

Two speech corpora have been used to test the feasibility of the described approaches: *'Herbs'* and *'Spanish Sentences'*. The corpus of Herbs consisted of 4 repetitions from a single male speaker of 209 names of aromatic and medicinal herbs with a total of 25 phonetic-units. The corpus of Spanish Sentences was used to perform some preliminary experiments. This corpus consisted of 76 phonetically balanced sentences, each encompassing an average of 30 phonemes. Each of these sentences was (continuously) uttered by 10 different speakers, 5 males and 5 females (760 utterances in total). The number of different phonetic-units was 25.

The data representation is based on a Vector Quantization of 32 codewords. The acquisition and parametrization of both corpora resulted in sequences of 10-dimensional parameter vectors of Cepstrum coefficients, which were obtained at a rate of 66.66 vectors per second for the corpus of Herbs and a rate of 100 for the Spanish Sentences. Each frame was then given a label from a set of 32 microphonetic categories for both corpora but with different codebooks. The actual Acoustic-Phonetic Decoder which is used in the recognition phase is based on the Dynamic Programming One-Stage algorithm [9].

The results of the experiments are reported using the following assessment parameters:

$$\text{Percent total} = 100\,\frac{c}{c+s+i+d} \qquad \text{Insertion rate} = 100\,\frac{i}{c+s+i+d}$$

$$\text{Substitution rate} = 100\,\frac{s}{c+s+i+d} \qquad \text{Deletion rate} = 100\,\frac{d}{c+s+i+d}$$

where c is the number of correctly recognized phonetic-units, and i, s and d are the number of insertions, substitutions and deletions, respectively. These parameters were obtained by dynamic warping the output of the Acoustic-Phonetic Decoder with the correct phonetic transcription of each word.

With the corpus of Herbs experiment H was made using all the repetitions of 100 words for training, and the rest (different words) for testing, with a total of about 3,000 phones. With the Spanish Sentences in experiment S, 35 sentences of five speakers were used for training, about 5,000 phones, and the rest of the sentences by the other five different speakers for testing. The results of all the performed experiments with the different techniques are reported as follows:

	CONNECTED UNITS																			
	Centroid				Clustering no iteration				Clustering iteration				Multiedit Condensing no iteration				Multiedit Condensing iteration			
	pt	pi	ps	pd	pt	pi	ps	pd	pt	pi	ps	pd	pt	pi	ps	pd	pt	pi	ps	pd
H	57	3	23	15	61	4	23	11	61	4	23	10	61	7	19	11	62	7	17	13
S	39	3	30	26	45	5	31	18	44	7	31	17	-	-	-	-	-	-	-	-

Table 1. Experimental results for the Connected Units approach. H and S are the experiments which are described in the text.

	UNIT SPOTTING							
	Centroid				Clustering no iteration			
	pt	pi	ps	pd	pt	pi	ps	pd
H	57	9	23	9	59	9	21	8
S	35	9	38	16	37	5	36	20

Table 2. Experimental results for the the Unit Spotting approach.

5. Conclusions and Future Work

In this work several methods to automatically learn template-based models of phonetic-units from a training set of utterances are presented. The necessity for several prototypes per phonetic-unit to represent the contextual variability of each phonetic-unit is evident. So far, the best results are obtained by using Multiedit Condensing techniques for the single-speaker task.

The obtained results are quite satisfactory, if we take into account that the results are given at phonetic-unit level. Although the results of the experiments with the corpus of Spanish Sentences are not as good as could be expected, it must be pointed out that this is a very new application and the experiments are very preliminary. We think that the use of a parametric representation of the speech data, instead of a string of labels from a codebook, will improve the results significantly. We are now adapting all existing techniques to use a parametric representation of the speech signal. In order to compare results with other models of the literature, we plan to implement Isolated Word Recognition based on DTW techniques, by using word models built by concatenation of the template-based models of the phonetic-units we have obtained. Finally, we will perform more experiments with the entire corpus of Spanish Sentences, which consists of about 2,000 phonetically balanced sentences.

REFERENCES

[1] F. Casacuberta, E. Vidal: "Reconocimiento Automático del Habla", Marcombo, 1987.

[2] A. M. Colla: "On training a large vocabulary speech recognition system", Proc. International Conference on Speech Technologies, Verba 90, pp 107-116, 1990.

[3] R. O. Duda, P. E. Hart: "Pattern Classification and Scene Analysis", ed. John Wiley & Sons, Inc., 1973.

[4] A. Komatsu, A. Ichikawa, K. Nakata, Y. Asakawa, H. Matsuzaka: "Phoneme recognition in continuous speech", Proc. ICASSP 82, pp 883-886, 1982.

[5] J. B. Kruskal, D. Sankoff: "An Anthology of algorithms and concepts for sequence comparison", D. Sankoff and J. B. Kruskal (eds.), Addison Wesley, chp. 10, 1983.

[6] K. F. Lee: "Large-Vocabulary Speaker Independent Continuous Speech Recognition: the SPHINX System", Ph. Thesis, Tec. Rep. CMU-CS 88-148. Carnegie Mellon Univ., 1988.

[7] S. Nakagawa, A. Hauptmann:"On Quick Word Spotting Techniques", Proc. ICASSP 86, pp 2311-2314,86.

[8] S. Nakagawa: "Speaker-independent continuous-speech recognition by phoneme-based word spotting and time-synchronous context-free parsing", Computer Speech and Language, vol. 3, pp 277-299, 1989.

[9] H. Ney: "The Use of a One-Stage Dynamic Programming Algorithm for Connected Word Recognition", IEEE Trans. ASSP, vol. 32, no. 2, pp 263-271, 1984.

[10] H. Sakoe, S. Chiba: "Dynamic Programming Algorithm Optimization for Spoken Words Recognition". IEEE Trans. ASSP, vol. 26, pp 43-49, 1978.

[11] R. M. Schwartz et al.: "Acoustic-Phonetic Decoding of Speech" in "Recent Advances in Speech Understanding and Dialog Systems", H. Nieman, M. Lang, G. Saguer, Springer-Verlag, pp 25-50, 1988.

[12] J. Voisin, P. A. Devijver: "An Application of the Multiedit-Condensing Technique to the Reference Selection Problem in a Print Recognition System", Pattern Recognition, vol. 20, pp 465-474, 1987.

Learning Structural Models of Sublexical Units†

E. Sanchis, F.Casacuberta, S.Carpi

Depto. Sistemas Informáticos y Computación. Universidad Politécnica de Valencia. SPAIN

Abstract: The use of structural-stochastic models has been proved to be adequate for the Acoustic-Phonetic Decoding problem. Whereas there exist good algorithms to estimate the parameters of the models, the structures are chosen according to the a-priori knowledge about the speech objects which are to be represented. This structural information is represented through a finite state network and therefore the use of a Grammatical Inference algorithm is a natural way to infer the structure of the models. In this paper, we propose the use of a specific grammatical inference algorithm (the Error Correcting Grammatical Inference algorithm) to learn structural models of the phonetic units.

Keywords: Acoustic Phonetic Decoding, Structural Models, Phonetic Units.

1. Introduction

The lexicon of every automatic speech recognition system is defined in terms of units. These units can be whole words, but when the size of the lexicon is large, the training of the corresponding word models becomes a difficult or impossible task. The current approach to avoid this problem is the use of a reduced set of subword units such as phonemes, generalized context dependent phonemes, etc. In this case, the word models are usually obtained by concatenating subword models according to a known transcription of the word in terms of subword units.

The subword units can be represented under different formalisms: rules, templates, hidden Markov models, neural nets, etc, [8]. The most popular formalism is, without any doubt, hidden Markov modeling. One of the main reasons is their capability of adequately modeling acoustic events. This capability is achieved through the possibility of accurately estimating the statistic parameters of the structured model from a set of training utterances [2]. The structural component of the model is a finite-state network (a finite set of states and interstate transitions), and is always supplied through a priori knowledge of the designer and/or by experimentation.

† Supported in part by the Spanish CICYT, under grant TIC 89/0448.

NATO ASI Series, Vol. F 75
Speech Recognition and Understanding. Recent Advances
Edited by P. Laface and R. De Mori

Most of the usually adopted structures or topologies are so simple that many important spectral and/or temporal features of the actual speech data tend to be poorly represented. A consequence of this is the fact that no canonical structure is yet commonly accepted.

Hidden Markov models can be considered as particular cases of stochastic finite-state automata, and some types of hidden Markov models are closely related to the stochastic regular grammars [3]. Consequently, a possible way to obtain the structure of the subword units is the use of the *Grammatical Inference techniques* [5].

In [6], an algorithm (ECGI algorithm) has been proposed to automatically infer very accurate finite-state automata or grammars from samples of the corresponding language. Further developments and studies of this algorithm have been presented in [11] and [7]. The application of this algorithm to obtain the structure of a hidden Markov model was presented in [4]. For other approaches which are more or less related with ECGI, see [9] and [10].

In this work, the *subword units are represented through stochastic-finite state automata.*. These automata are obtained through the ECGI algorithm and training samples of speech. In the next section, we present a brief description of this algorithm.

In section 3, the application of this algorithm to the model subword unit is discussed. Sections 4 and 5 deal with experiments and conclusions, respectively.

2. The Error Correcting Grammatical Inference Algorithm

The type of stochastic finite state automata that we will use in this work is of the form $A = (\Sigma, Q, T, E, q_i, q_f)$, where Σ is a finite alphabet, Q is a finite set of states, $T \subset QxQ$ is a set of transitions, such that $\forall (q,q') \in T$ $q'<q$, $E:Q \dashrightarrow \Sigma$ represents an association of symbols to states, and $q_i, q_f \in Q$ are the initial and final states whose associated symbols are the start and end markers E(qi)='#' and E(qf)='$' $(\#,\$ \in \Sigma)$.

The ECGI algorithm builds a stochastic finite state automata through the following incremental procedure: Initially, a trivial network is built from the first training string. Then for every new string which cannot be exactly recognized by the current network, the network is updated by adding those states and transitions which are required for the new string to be accepted. In order to determine such states and transitions, a Dynamic Programming procedure is used to find the best alignment of the input string with the current network. In fact, this is an error correcting parsing. The current network is updated so as to take advantage of as much of this network as possible; that is, only the sequences of error transitions (substitutions, insertions and deletions) in the error correcting parsing lead to the addition of new states and/or transitions.

The recognition process with the inferred finite state automata is based on a stochastic error correcting parser. To obtain the transition probabilities of the automaton and the corresponding error probabilities, some counters are used. An array is used for the counters of substitution errors ($|\Sigma| \times |\Sigma|$), two vectors for the counters of insertions and deletions ($|\Sigma|$), and a counter for each transition for the number of times that these transitions have been used. In the process of inference of the finite state automata, the error counters are updated according to the error rules that are used for parsing each training sample. Moreover, when a transition is used, the corresponding counter is also updated. At the end of the inference process, the counters are adequately normalized. For more details on the ECGI see [6], [11], and [7].

3. Learning Structural Models of Phonetic Units

The application of the ECGI algorithm to model phonetic units is shown in the following procedure.

```
ALGORITHM: Learning the structural model of phonetic units.
INPUT: F = a set of phonetic units.
    S = { (x,t) | x is a string of codeword symbols; t ∈ F* is the corresponding phonetic transcription}
    Ss ⊂ S; A subset of segmented and labelled samples by hand.
OUTPUT: A stochastic finite automaton Af per phonetic unit f ∈ F.
METHOD:
    For all f ∈ F do    / INITIALIZATION /
            let sf be the set of segments in Ss that are labelled by f
            Af := ECGI(sf)
    endfor
    Repeat                    / ITERATION /
            For all f ∈ F do sf := empty set endfor
            For all (x,t) ∈ S do
                    if t = t1t2...tn then A := At1t2...tn endif
                    segment x according to VITERBI(A,x)
                    For all f in t do
                            add the corresponding segments labelled by f to sf
                    endfor
            endfor
            For all f ∈ F do Af := ECGI(sf) endfor
    until convergence
```

The input data is a set of training samples (words or sentences), which are represented by strings of codeword symbols, and the phonetic transcription. A subset is segmented by hand.

The initialization of this procedure consists of the inference of an initial stochastic finite automaton for each phonetic unit by using the hand-segmented training samples.

In each iteration, the entire training data are used. The automata that are obtained in an iteration are used to segment the training data in the following iteration. This segmentation process is performed by using a forced recognition of each utterance x by a connected automaton A. This automaton is obtained by concatenation of the automata of each unit according to the corresponding phonetic transcription of the training sample t. Once the entire training set has been segmented, new stochastic automata are obtained by using the ECGI algorithm with the new segments.

In figure 1, the finite state automata of the Spanish phonetic units /m/, /p/ and /s/ are shown. These figures are organized in such a way that all states at the same level have the same symbol.

4. Experiments and Results

In this section we describe the experiments for acoustic- phonetic decoding that were performed to prove the adequacy of the proposed method. In every experiment the decoding consists of the application of the Viterbi algorithm to a whole automaton. In this automaton, all automata of phonetic units are placed in parallel with a common initial state and a common final state. These states are connected through a feedback nil transition.

The experiments that were performed correspond to a single-speaker and isolated work task. For this type of experiment, the used corpus consisted of four repetitions of 209 medical and aromatic herbs (HER). In this task, the number of phonetic units was 24. Two different experiments were made with this corpus. In the first experiment, two repetitions of each word were used as the training set, and the rest of repetitions as the test set (appoximately 2500 phones). In the second experiment, four repetitions of 100 words were used as the training set, and the rest of the words as the test set.

For each experiment, the following values were computed: 1) the number of phonetic units that were recognized correctly (c); 2) the number of phonetic units that were inserted (i), deleted (d) and substituted (s). These values were obtained by warping the output of the acoustic-phonetic decoding and the correct phonetic transcription of each utterance.

From this values, the following parameters were obtained:

$P_c= c / (i+s+d+c)$ (correct percent)

$P_s= s / (i+s+d+c)$ (sustitution rate)

$P_d= d / (i+s+d+c)$; (deletion rate)

$P_i= i / (i+s+d+c)$ (insertion rate)

The experimental results are shown in table 1.

	P_c	P_s	P_d	P_i
Experiment 1	65	5	20	9
Experiment 2	63	2	22	12

Table 1 Experimental results for single-speaker and isolated word corpus (HER).

The low value of insertion errors is due to the use of a penalty weight when the feedback nil transition was used.

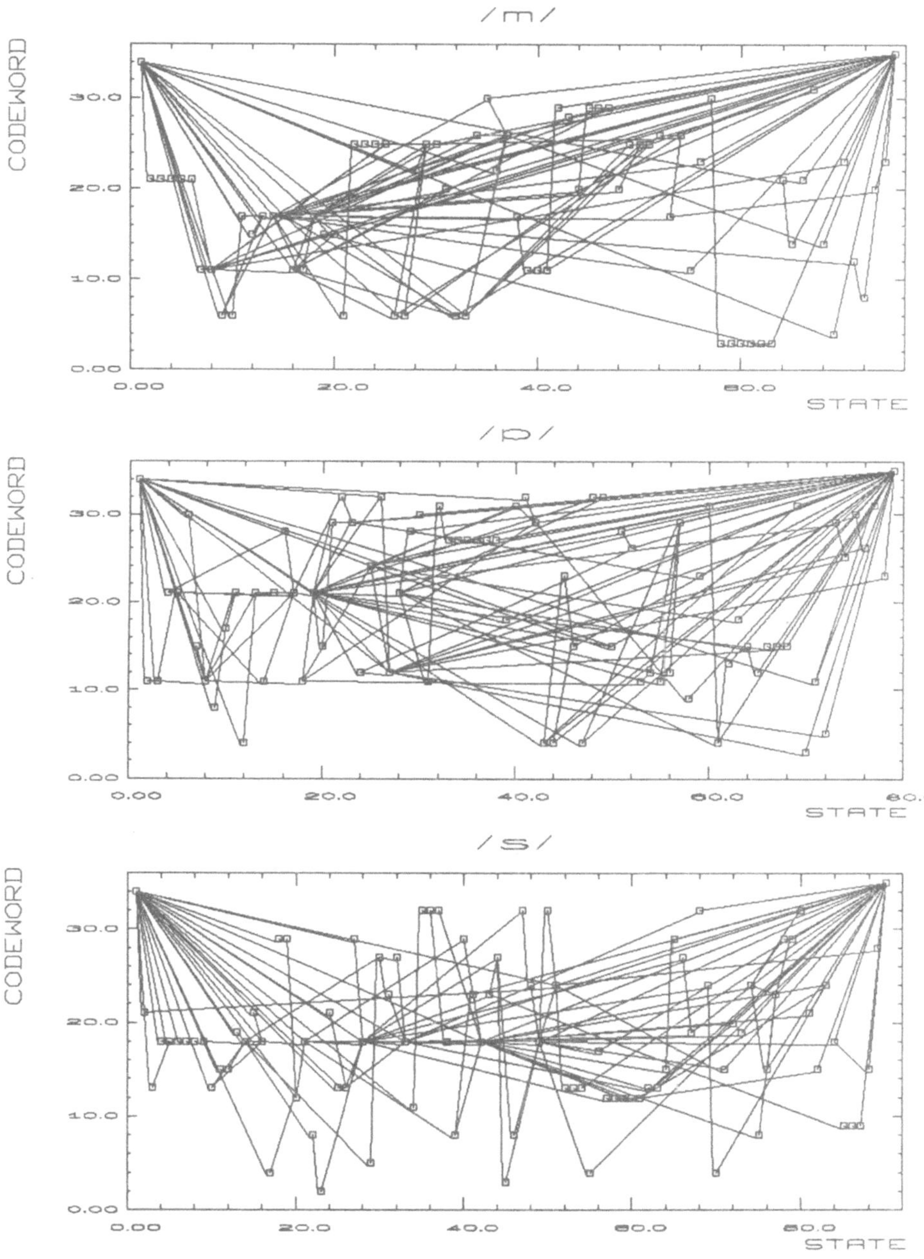

FIGURE 1. The finite state automata of the Spanish phonetic units /m/, /p/ and /s/ that are obtained by using the ECGI algorithm

5. Conclusions and Future Work

The proposed method actually seems to be capable of learning meaningful structural information of each phonetic unit from the training samples. However, the automata that are obtained by this method are large. To reduce the size of this automata, some pruning procedures can be used [3].

On the other hand, this technique can be applied to other kind of units, such as diphones, demisyllables, etc. In this case, the differences are the bootstrap step and the transcription of the training utterances.

The approach presented in this paper opens a door to other interesting grammatical inference algorithms for finite state automata [5], [11], [12].

The results presented in this paper can be improved by adding other known techniques. First, the estimation procedure of the probabilities of the stochastic automata is very close to the Viterbi re-estimation algorithm. Therefore, an adequate version of the Baum-Welch procedure should be necessary to learn the transition probabilities and the error probability distributions. Second, the use of a semicontinuous or continuous approach is necessary, instead of the use of strings of codeword symbols. Third, there is no doubt about the necessity for context-dependent models. With this approach the context can be represented explicitely in a single model. Finaly, two different types of experiments are being performed. The first one corresponds to a multispeaker and continuous speech task, and in the second one, for the purpose of comparison, classical Hidden Markov Models have been used.

6. References

1. H.Nieman, M.Lang, G.Sager: "Recent Advances in Speech Understanding and Dialog Systems". Springer-Verlag. 1988.
2. L.R.Rabiner: "Mathematical Foundations of Hidden Markov Models", in (1), pp. 183-206. 1988.
3. F.Casacuberta: "Some Relations among Stochastic Finite State Networks used in Automatic Speech recognition". IEEE Trans. Pattern Analysis and Machine Intelligence. Vol. 12. No. 7. pp. 691-695.1990.
4. F.Casacuberta, E.Vidal, B.Mas, H.Rulot: "Learning the Structure of HMM's through Grammatical Inference Techniques". Proc. ICASSP90. pp 717-720. 1990.
5. K.S.Fu and T.L.Booth: "Grammatical Inference: Introduction and Survey". Parts 1 and 2. IEEE Trans. on System, Man and Cybern. Vol. 5. 1975.
6. H.Rulot, E.Vidal: "Modeling (sub)string-length Based Constraints through a Grammatical InferenceMethod". In "Pattern Recognition Theory and Applications". Devijver and Kitler (eds). Springer Verlag. 1987.
7. H.Rulot, E.Vidal, N.Prieto: "Learning Accurate Finite State Structural Models of Words through the ECGI algorithm" Proc. ICASSP89. pp. 643-646. 1989.
8. Schwartz et al.: "Acoustic-Phonetic Decoding of Speech". in (1) Springer-Verlag. 1988.
9. K.F.Lee: "Incremental Network Generation in Word Recognition". Proc. ICASSP-86. 1986.
10. M.Thomason et al: "Experiments in Dynamic Programming Inference of Markov Networks with String representating Speech Data". Pattern Recognition. Vol 19. No. 5. 1986.
11. E.Vidal et al.: "Application of the Error Correcting Grammatical Inference Method (ECGI) to Multispeaker Isolated Word Recognition". in (1) pp' 317-321. 1988.
12. P.Garcia,E.Vidal,F.Casacuberta: "Local Languages, the succesor method, and a step towards a general methodology for the inference of Regular Grammars". IEEE Trans on PAMI, Vol 9, n.6, pp.841-845, 1987.

On the use of Negative Samples in the MGGI Methodology and its application for Difficult Vocabulary Recognition Tasks†

Encarna Segarra, Pedro García, Jose M. Oncina, Armando Suarez

Dept. de Sistemas Informáticos y Computación. Universidad Politécnica de Valencia. Spain.

Abstract: The inference methods which are proposed in Syntactic Pattern Recognition in practice only make use of positive data and generate a heuristic generalization of strings in the data. However, the use of positive data becomes insufficient when very discriminatory models are needed. This is the case of Difficult Vocabularies in Isolated Word Recognition tasks. This paper is a first attempt at using positive and negative data that presents two main characteristics: it respects the computational efficiency with moderate-sized training sets, and it is suitable for tasks in Syntactic Pattern Recognition, specifically in Automatic Speech Recognition.

Keywords: Grammatical Inference, Syntactic Pattern Recognition, Inference Method, Difficult Vocabulary Isolated Word Recognition Tasks, Positive and Negative Data, Clustering of States, Prefix Tree Acceptor, MGGI methodology.

Introduction

While in the Statistical Theoretic approach to Pattern Recognition the training problem usually entails feature selection and parameter estimation, the Structural or Syntactical approach asks for primitive selection and *Grammatical Inference* [3]. Therefore, it is accepted that *Grammatical Inference* constitutes the natural framework for learning in Syntatic Pattern Recognition.

An *Inference Method* is an algorithm that, given a data set (or strings) of an unknown language, obtains a description (such as a grammar, automaton, etc.) for such a language. To be useful an *Inference Method* must be computationally efficient and use moderate-sized training sets. Consequently, in practice, the inference methods which are proposed in Syntactic Pattern Recognition only make use of positive data [2] and generate a heuristic generalization of strings in the data. Although the class of regular languages remains unidentifiable in the limit from only positive data, the practical methods generate sufficient approximations by using careful probability estimations of parameters in the inferred models.

† Supported in part by the Spanish CICYT, under grant TIC 89/0448.

NATO ASI Series, Vol. F 75
Speech Recognition and Understanding. Recent Advances
Edited by P. Laface and R. De Mori

However, the use of positive data becomes insufficient when very discriminatory models are needed. This is the case of difficult vocabularies in Isolated Word Recognition tasks. This paper is a first attempt at using positive and negative data that presents two main characteristics: it respects the computational efficiency with moderate-sized training sets and it is suitable for practical tasks of Syntactic Pattern Recognition.

Inference by Clustering of States

One of the general schemes of regular language inference from positive data is the *Clustering of States*. Any inference method can be seen as an instance of the following scheme:

Let R+ be a positive sample of a regular language L.

Let $A_0 = (Q,\Sigma,\delta,q_0,F)$ be the Prefix Tree Acceptor, i.e.: $Q = Pr(R+) = \{u\in\Sigma^* \mid uv\in R+, v\in\Sigma^* \}$, $q_0=\lambda$, $\delta(u,a)=ua \quad u,ua\in Pr(R+)$. (The finite-state automaton with tree form that generates all and only the strings in R+).

If R+ is a subset of L with certain representative conditions, there is a partition π of the states of A_0 such that: the quotient automaton A_0/π (which results by clustering all the states of the A_0 which are in the same block of the partition in one state) is isomorphic to the minimun deterministic finite-state automaton which recognizes L.

For an arbitrary partition π, the following condition holds: $L(A_0/\pi)\supset R+$, that is, the positive sample is included in the language which is generated by the quotient automaton of π. Therefore, we obtain a generalization of the positive sample by clustering the states of A_0.

For all inference methods that generate a non-ambiguous finite-state automaton $A'=(Q',\Sigma',\delta',q_0',F')$, we can recover the corresponding partition of states of A_0 as follows:

if $B(u,\pi)$ denotes the block of the state *u* of A_0 in π then $\forall u,v\in Pr(R+)$ $B(u,\pi)=B(v,\pi)$ iff $\delta'(q_0',u)=\delta'(q_0',v)$.

Inference refinement which uses Negative Data

We are going to see how, from the partition of states in the Prefix Tree Acceptor, this Acceptor is refined by using negative data.

Let $R+\cup R-$ be the training data. Let π be the partition of states of A_0 (Prefix Tree Acceptor), which is produced by an algorithm from positive data. The incorporation of negative data is done as follows: we refine the partition π and obtain a partition π' such that $L(A_0/\pi')\cap R- = \emptyset$. That is, if *u* and *v* belong to the same block of π, then *u* and *v* will belong to the same block in π', only when the resulting automaton of clustering *u* and *v* do not accept any strings of the negative sample.

This refinement can be done in several ways. In this paper, we have chosen an algorithm to make this refinement. In order to be clear, we are going to illustrate it with an example.

Example

Let R+={ aa,aba,abbba} and let R-={a}. Let A_0 be the Prefix Tree Acceptor of R+, which is shown in fig.1. Let A_2(R+) be the automaton that results from the application of an inference algorithm from R+, which is shown in fig.2. In this case, for example, we have applied the Two-Testable in Strict-Sense Inference (2-TSI) algorithm [5].

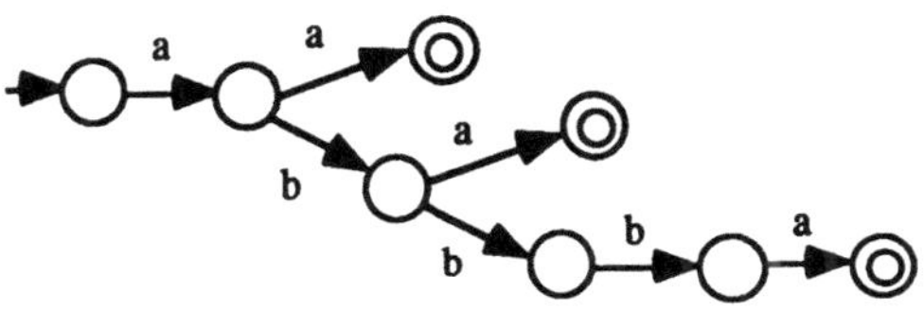

Fig.1.

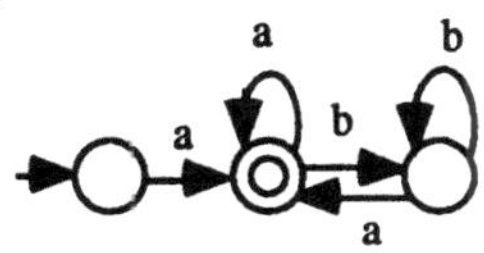

Fig.2.

The partition of states of the A_0 which is induced by this inference algorithm is: π = < B_1={λ}, B_2={a,aa,aba,abbba}, B_3={ab,abb,abbb} >.

The blocks in π are ordered by spreading the lexicographical order < in Σ^* to π as follows: $B_i<B_j$ if $\exists u\in B_i \mid \forall v\in B_j \ u<v$. Following this order with the blocks, we try to cluster the states into each block. We go left to right into a block, and attempt to cluster each state with the preceeding ones.

Block B_1 contains only one state, then we can not cluster.

Block B_2: the states *aa* and *a* are NOT CLUSTERED because the resulting automaton accepts the negative sample. The same is true for the states *aba* and *a*. But the states *aba* and *aa* ARE CLUSTERED, and the result is shown in fig.3. This follows for the rest of block B_2 and block B_3. The final automaton which is obtained is shown in fig.4.

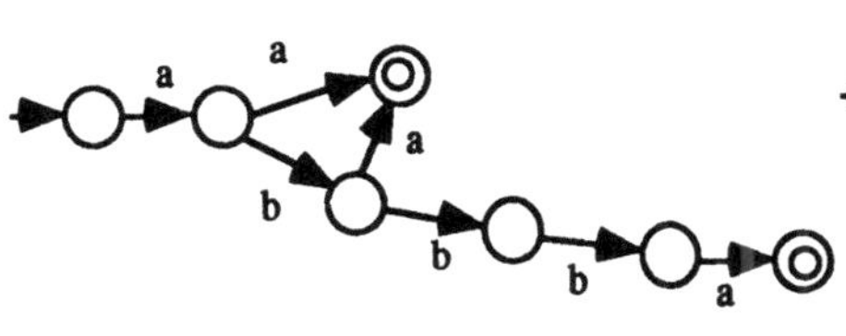

Fig.3.

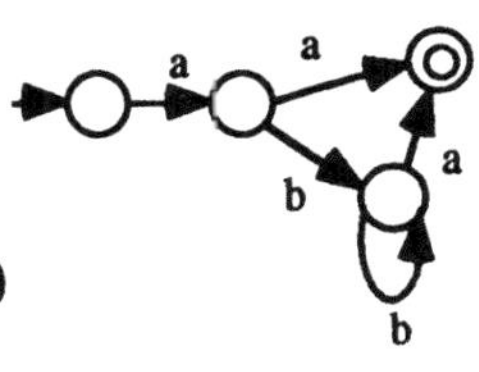

Fig.4.

In conclusion, while an inference algorithm that uses only positive data generates an automaton, fig.2, which accepts all the strings over the alphabet {a,b}, which begin and end by *a*, the process of refinement generates an automaton, fig.4, which accepts *ab**a* and obviously does not accept the negative sample.

The inference algorithm from positive data that we have used in this paper is the Morphic Generator Grammatical Inference (MGGI) methodology [4] and [7]. This algorithm is based on the fact that every regular language can be obtained from a local language by applying a morphic generator. We have chosen it because of the good results which were obtained when the MGGI methodology was applied to simple tasks in Automatic Speech Recognition.

The experimental process

In the first part of the training phase, the inference process generates a deterministic finite-state automaton for each class. Since these automata are deterministic, their transition probabilities and their continuous output symbol probabilities have been estimated by the corresponding contributions of positive samples. We have assumed, in particular, that the continuous output symbol probability that is associated with each state of automaton is given by a multivariate normal probability distribution. The assumption that this probability might be given by a mixture Gaussian distribution, as happens in the Hidden Markov Models, seems to have no sense in our case. The reason is that the obtained models are very extensive, with a large number of states, and the acoustic variability is already accounted for in the structure.

Based on the semi-continuous Hidden Markov Models [6], these continuous output symbol probabilities have been estimated by using the information which is supplied by the sequences of codewords, or what we call microphonetic label strings (derived from the vector quantization process) and by the cepstral coefficient vector strings of the training samples. For those states whose occupation index in the learning phase has been high enough (10, approx.), a particular multivariate normal probability function has been estimated. However, for the states with a low occupation index, a general, overall multivariate normal probability function has been estimated.

Due to this estimation process, the previous deterministic automata have become ambiguous stochastic models. Therefore, the Viterbi algorithm [1] has been used for the recognition process (classification following the criterium of maximum likelihood).

The speech data consist of 10 repetitions of 9-word vocabulary, which were uttered by 10 different speakers (5 male and 5 female). This small vocabulary is the most difficult subset of the Spanish alphabet, called E-letters. It consists of the following sounds: f,l,ʎ,m,n,ɲ,r,r̄,s.

The adquisition and end-point procedures were rather standard, and the signal was sampled at approximately 8.5 khz with a 12 bits A/D converter. The adquired speech data were then converted into 900 strings of 10-dimensional parameter vectors of cepstral coefficients, which were obtained from a software sixteen channel Mel frequency scale filter bank at the rate of

133.32 vectors per second. Each coefficient vector (or frame) was labeled with a name from a set of 32 microphonetic categories.

Experiments

TABLE.1

Exp.	Av. Num. of States	Branching Factor	% Classificat. Rate
1	91	5.3	76.7
2	88.8	5.2	74.4
3	88.2	5.2	58.3
4	86.3	5.0	71.1
5	88.2	5.1	61.1

AVERAGE CLASSIFICATION RATE = 68.32%

Table.1. The results with positive data experiments, threshold factor = 1.

TABLE.2

			Threshold=1	Threshold=10
Exp.	Av. Num. of States	Branching Factor	% Classificat. Rate	% Classificat. Rate
1	423.3	2.5	75.0	75.6
2	424.3	2.4	75.6	76.7
3	408.3	2.5	56.1	62.8
4	419.5	2.4	71.1	69.4
5	425.4	2.4	70.6	71.7
AVERAGE CLASSIFICATION RATE =			69.68%	71.24%

Table.2. The results with positive and negative data experiments.

A leaving-one-out experiment has been done for two cases: using only positive data, and using positive and negative data. In each leaving-one-out experiment, five test have been performed, with a training set of 720 samples, and a test set of 180 samples.

The results of the positive data experiment are summarized in table.1, and those with positive and negative data are summarized in table.2. The number of intervals in the MGGI algorithm was 5. The columns indicate the average number of states and the branching factor of the obtained automata, and the achieved recognition rate. For the last case, the experiments are repeated by changing the threshold factor, that is, the minimun occupation index for a particular estimation of continuous output symbol probability, from 1 to 10. The reason is the increase in the number of states in the second case with respect to the experiments in the first case. In the second case, using positive and negative data achieve somewhat better results.

Conclusions

The last obtained results for the E-letter task, by using other different methods, achieve recognition rates of approximately 73.8% (by using another grammatical inference algorithm) and 75% (by using Hidden Markov Models). These methods only make use of positive data.

We hoped that the use of positive and negative data for difficult vocabularies would improve the recognition rates which were obtained by using only positive data. Unfortunately, although they are within the range of rates that were obtained by other models, the results in two cases have been comparable; they present a difference of approximately 1.36% and 2.92%. Furthermore, this small improvement has proved to be computationally too expensive.

Consequently, these results, which make use of positive and negative data, do not encourage us to follow this idea. However, we must consider that the incorporation of negative data information has been done by using a heuristic approximation in particular, and other unexplored ways might exist to use positive and negative data.

References

1. Forney, G.D.: The Viterbi Algorithm. IEEE proc. N.3, pp 268-278 (1973).
2. Fu, K.S. and Booth, T.L.: Grammatical Inference Introduction and Survey, parts 1 and 2. IEEE Trans Syst. Man Cybern. SMC-5, pp 95-111 409-423 (1975).
3. Fu, K.S.: Syntactic Pattern Recognition and Applications. Prentice-Hall 1982.
4. García, P., Vidal E. and Casacuberta, F.: Local Languages, the successor method, and a step towards a general methodology for the inference of regular grammars. IEEE Trans. on PAMI, Vol Pami-9, N.6, pp 841-845 (1987).
5. García, P. and Vidal, E.: Inference of K-Testable Languages in the Strict Sense and Application to Syntactic Pattern Recognition. IEEE Trans. on PAMI, Vol PAMI-12, N.9, pp 920-925 (1990).
6. Huang, X.D. and Jack, M.A.: Semi-continuous Hidden Markov Models for Speech Recognition. Computer Speech and Language, N.3, pp 239-251 (1989).
7. Vidal, E., García, P., Segarra, E. and Galiano, I.: Multi-speaker experiments with the Morphic Generator Grammatical Inference methodology. In NATO-ASI Recents Advances in Speech Understanding and Dialog Systems. H. Nieman, M. Lang and G. Sagerer (eds.), pp323-327, Springer Verlag 1988.

A New Method for Dynamic Time Alignment of Speech Waveforms

J. Kittler and A. E. Lucas

Dept. of Electronic Engineering, University of Surrey, Guildford GU2 5XH, UK

Abstract In this paper, a new method for dynamic time alignment of speech waveforms is introduced. The method attempts to address the shortcomings of traditional time alignment approaches, commonly based on *dynamic programming* algorithms. Such methods, usually called *dynamic time warping* (DTW) algorithms, make the assumption that the samples of the speech waveform under consideration are statistically independent. The proposed method makes no such assumption. Instead, the method is based on models of speech entities with Gaussian distributions and general covariance matrices. These ideas are implemented by employing the branch and bound search algorithm [1] coupled with the Mahalanobis distance measure as the matching criterion. Hence, the new method attempts to utilise more discriminatory information than is presently incorporated. Preliminary results on a spoken letter recognition problem are reported validating the approach.

Keywords Dynamic time alignment, dynamic time warping, branch and bound algorithm, Viterbi, Mahalanobis distance

1 Introduction

It is well known that automatic speech recognition by template matching is impeded, among other things, by local and global variations along the time axis. The problem has been approached, with some success, by applying *dynamic programming* (DP) algorithms [3]. These methods attempt to remove variations in time-registration by mapping the speech waveform onto the template via some non-linear warping function.

However, traditional DP techniques use speech models based upon the first order statistics of the speech waveform (with respect to the time variable). This approach assumes that samples in the speech waveform are statistically independent.

In this paper, a new method for dynamic time alignment of speech utterances is introduced where each class of speech entity being matched is modelled by a Gaussian distribution with a *general* covariance matrix. By attempting to incorporate the dependencies of inter-frame samples on one another, full utilisation of the discriminatory information content of the data should be possible. The aim of this paper is to introduce the new method of non-linear waveform alignment.

Firstly, an introduction to the proposed matching criterion is given, including a scheme for removing speaker variation. Then, a recursive form of the criterion is presented with a companion search and optimisation algorithm. Finally, an experiment to test the effectiveness of the approach is described based on the recognition of the

NATO ASI Series, Vol. F 75
Speech Recognition and Understanding. Recent Advances
Edited by P. Laface and R. De Mori

utterances "b", "d" and "g". From this, some conclusions are drawn and some further work discussed.

2 The Matching Criterion

The problem under consideration is that of matching an unknown pattern, representing a segment of continuous speech, against a set of prototype (idealised) patterns, called *templates*. Each template represents one class of a specific speech entity, which can be a single word, phoneme or some other basic unit. The criterion described below is suitable for matching these two entities - the unknown pattern and the template - producing a criterion function value; a quantitative measure of closeness, or fit.

Considering just one class, a template consists of a *set* of J, N-dimensional vectors, $\underline{\mu}_j = [\mu_{j1}, \ldots, \mu_{jN}]^{\mathrm{T}}$, $j = 1, \ldots, J$. Each vector characterises the time-variation of a single coefficient (of J such coefficients) from a particular form of speech representation: a formant frequency or a mel-frequency cepstral coefficient (MFCC), for example.

The belief put forward here is that the traditional assumption of statistical independence in the pattern matching problem is not justifyable. While retaining the analytical tractability afforded by the normal distribution, statistical dependence and unequal variances are taken into consideration allowing the class covariance matrix $\mathbf{\Sigma}$ to have a general form.

This generalisation is achieved [2] by employing the Mahalanobis distance as the matching criterion between an unknown speech waveform, $\underline{y}_j = [y_{j1}, \ldots, y_{jN}]^{\mathrm{T}}$, and the class template, $\underline{\mu}_j$, embodying the entire set of J waveforms

$$M = \sum_{j=1}^{J} (\underline{y}_j - \underline{\mu}_j)^{\mathrm{T}} \mathbf{\Sigma}_j^{-1} (\underline{y}_j - \underline{\mu}_j). \tag{1}$$

3 Mean Level Compensation

When considering time-alignment (or warping) and matching, the primary criterion is the *shape* of the waveform. Although different speakers pronounce identical words with different frequency components, the basic shape of the waveform is similar. Consequently, this frequency variation may be reflected as a mean level shift in the resultant speech waveform. A change in the mean level does not affect the shape of the waveform, but the Mahalanobis distance is sensitive to it. Without compensation, the criterion will treat two waveforms which differ solely in their mean levels as being two distinct speech entities.

The effect of mean level differences could easily be removed if an exhaustive search procedure was employed to solve the minimisation problem. Then, the Mahalanobis distance would be computed directly for the complete waveform $\underline{y}_j$ from which its mean level had been subtracted. However, the dynamic time alignment procedure requires decisions to be made on a sample-by-sample basis in order to select the correct path. To do this, sequential mean level compensation is required.

A decision at each stage k about the constant level S_k to be subtracted from the current waveform, k samples long, needs to be made. The question arises of how to treat each of the J waveforms that constitute a segment of speech? One method would involve the removal of a constant level from *all* component waveforms. In this way, the resultant mean level compensation would be optimised over all waveforms, and yet, any inter-waveform information would be preserved.

The constant signal (mean offset) S_k removed from all of the waveforms in the speech segment generates the modified Mahalanobis distance as a function of S_k

$$M(S_k) = \sum_{j=1}^{J}[\underline{y}_{jk} - S_k\underline{e}_k - \underline{\mu}_{jk}]^{\mathrm{T}}\mathbf{\Sigma}_{jk}^{-1}[\underline{y}_{jk} - S_k\underline{e}_k - \underline{\mu}_{jk}] \tag{2}$$

where $\underline{e}_k$ is a k-dimensional unit vector and $\underline{y}_{jk}$, $\underline{\mu}_{jk}$ denote the vectors composed of the first k elements of $\underline{y}_j$ and $\underline{\mu}_j$ respectively and $\mathbf{\Sigma}_{jk}$ the corresponding matrix inverse.

Differentiating with respect to S_k we obtain the condition for a stationary point [2], and thus the optimum value of S_k. In [2] it is shown that this value corresponds to a minimum of function $M(S_k)$ as its second derivative at this point is positive.

Substituting the optimum value of S_k into (2) the corresponding optimum value of the criterion function $M(S_k)$ is found to be

$$M_k = M(S_{k\mathrm{opt}}) = \sum_{j=1}^{J}(\underline{y}_{jk} - \underline{\mu}_{jk})^{\mathrm{T}}\mathbf{\Sigma}_{jk}^{-1}(\underline{y}_{jk} - \underline{\mu}_{jk}) - \frac{\varepsilon_k^2}{\xi_k} \tag{3}$$

where

$$\xi_k = \underline{e}_k^{\mathrm{T}}\sum_{j=1}^{J}\mathbf{\Sigma}_{jk}^{-1}\underline{e}_k, \quad \varepsilon_k = \underline{e}_k^{\mathrm{T}}\sum_{j=1}^{J}\mathbf{\Sigma}_{jk}^{-1}(\underline{y}_{jk} - \underline{\mu}_{jk}). \tag{4}$$

The last term in equation (3) can be considered as a *mean level correction factor*. Thus the matching criterion (3) achieves template matching of speech waveforms with full statistical dependence and mean level compensation.

4 Recursive Form of the Criterion Function

For an understanding of the requirements of a recursive form of the criterion function, consider the following scenario. The criterion function has been evaluated at the k-th stage of the recognition process. That is, measurements of the utterance have been made up to the k-th sample. An extra sample is observed, $y_{j(k+1)}$, $j = 1, \ldots, J$, necessitating a comparison of segments $\underline{y}_{j(k+1)}$ (utterance) and $\underline{\mu}_{j(k+1)}$ (template). Now at the $(k+1)$-st stage, the criterion function value (from (3)) is given by

$$M_{k+1} = \sum_{j=1}^{J}(\underline{y}_{j(k+1)} - \underline{\mu}_{j(k+1)})^{\mathrm{T}}\mathbf{\Sigma}_{j(k+1)}^{-1}(\underline{y}_{j(k+1)} - \underline{\mu}_{j(k+1)}) - \frac{\varepsilon_{k+1}^2}{\xi_{k+1}}. \tag{5}$$

Applying the recursive formular for computing the inverse matrix given in [1], M_{k+1} can be expressed in terms of the criterion function value M_k. In particular, denoting the $(k+1)$-st column of covariance matrix $\mathbf{\Sigma}_{j(k+1)}^{-1}$ by

$$\underline{\sigma}_{j(k+1)} = \left[\sigma_{j(k+1)1}, \ldots, \sigma_{j(k+1)(k+1)}\right]^{\mathrm{T}} \tag{6}$$

and introducing the notation

$$\eta_{j(k+1)} = \underline{\sigma}_{j(k+1)}^{\mathrm{T}}(\underline{y}_{j(k+1)} - \underline{\mu}_{j(k+1)}) \tag{7}$$

$$\rho_{j(k+1)} = \underline{\sigma}^{\mathrm{T}}_{j(k+1)}\underline{e}_{k+1} \tag{8}$$

M_{k+1} in equation (5) can then be expressed recursively as

$$M_{k+1} \quad = \quad M_k + \frac{\varepsilon_k^2}{\xi_k} + \frac{\varepsilon_{k+1}^2}{\xi_{k+1}} + \sum_{j=1}^{J} \frac{\eta_{j(k+1)}^2}{\sigma_{j(k+1)(k+1)}} \tag{9}$$

where, in terms of notation (7) and (8), ξ_{k+1} and ε_{k+1} defined above respectively can be computed as

$$\xi_{k+1} = \xi_k + \sum_{j=1}^{J} \frac{\rho_{j(k+1)}^2}{\sigma_{j(k+1)(k+1)}}, \quad \varepsilon_{k+1} = \varepsilon_k + \sum_{j=1}^{J} \frac{\eta_{j(k+1)}\rho_{j(k+1)}}{\sigma_{j(k+1)(k+1)}}. \tag{10}$$

Since ξ_k and $\sigma_{j(k+1)(k+1)}$ are positive, the right hand side will always be non-negative and the criterion function is therefore *monotonically increasing.*

The main results of the paper is contained in equation (9) which enables the recursive computation of the modified Mahalanobis criterion (3) in the classification of speech waveforms utilising statistical dependence.

It should be noted that the above scheme requires only the last column of any inverse covariance for calculating the updated criterion function value (9). Storage space is thus reduced to just $N(N+1)/2$ values.

5 The Branch and Bound Algorithm

In principle, the best warping transformation can be found by exhaustive search. However, for most cases, such a search is computationally prohibitive and, instead, a more efficient method of warping is needed. Below, a search algorithm is described which achieves this and allows the optimisation of the new matching criterion function.

The search process can be represented as a tree, a portion of which is shown in Figure 1. Each branch of the tree represents a warping transformation. Descending the tree is equivalent to progressing along a speech template, sample-by-sample. At each level (or sample), a number of possible warping transformations are possible (just two in this case). The value of each node is the criterion value associated with the respective transformation.

The tree is traversed along the right-most branch calculating the criterion function value M_k at each stage k (nodes 1, 2 and 3). When the leaf node is reached (node 3), the criterion function value M_B is stored as the current best value. Then, the search backtracks up the tree to the nearest unexplored node (node 2) and then moves forward (downward) along the right-most branch.

Now, since the criterion function has been shown in Section 4 to be monotonically increasing with the depth of the tree, the search only continues forward if the current value M_k' is *less than* the current best M_B. Otherwise, if $M_k' > M_B$, M_k' will *never* decrease and hence the current path will never lead to a better match than the one currently stored. The search continues in a similar fashion until there are no more valid nodes to expand.

In order to minimise the the size of the subtree to be explored during the search process, the branches originating from any node of the tree should be searched in ascending order of the criterion function value at its successor nodes. The branch and

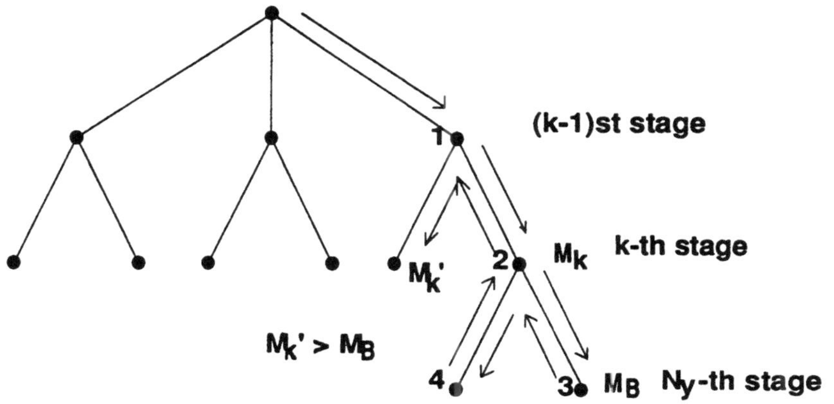

Figure 1. Example tree search using the branch and bound algorithm

bound algorithm thus eliminates the unproductive search of many candidate warping transformations without compromising the global optimality of the solution yielded by the algorithm.

6 Experiments & Results

A preliminary experiment was performed which forms the first stage of a programme to test the hypothesis that the performance of speech recognition systems can be improved by the incorporation of statistical dependence between speech samples. To validate the branch and bound search strategy, an experiment was performed which aimed to correctly classify utterances of the letters "b", "d" and "g".

We used data, kindly supplied by British Telecom Research Laboratories, Martlesham, UK, which had undergone preprocessing and was in the form of mel-frequency cepstral coefficients (MFCCs).

A training database of over 600 utterances spoken by over 100 distinct speakers was used to infer the class model statistics of mean and covariance. Inferencing was performed simply by an iterative process of normalisation (warping all training utterances to the length of the template), averaging and calculation. For the first iteration of the inferencing scheme the "bdg" template database was constructed by hand from reference speaker number 1. This utterance was not used in further iterations.

Recognition was attempted on a test set of 309 distinct utterances spoken by the same group of speakers. Template length was eight samples and the warping process could operate on the first sixteen samples of the test utterances only.

As a comparison, the whole test database was classified using the Viterbi algorithm which, of course, is a DP algorithm making the assumption of statistical independence. Results are shown in Table 1. From this simple test, it does appear as if a noticable improvement is possible by using the new method. Although the experiment reported so far is preliminary in nature, we believe this validates the method and provides a good platform from which to explore the full benefits of the idea.

Algorithm	% Classification Error
Viterbi	23.9
Branch & Bound (identity covariances)	23.0
Branch & Bound (full covariances)	21.7

Table 1. Percentage classification error for the recognition of "b", "d" and "g" using the Viterbi and branch and bound algorithms

7 Conclusions

In the preceeding sections, a new method for non-linear waveform alignment was introduced which is directly applicable to the problem of dynamic time alignment of speech waveforms for automatic speech recognition. The method incorporates the statistical dependencies of inter-frame samples within the template matching criterion, thus enhancing the full discriminatory information content of the data, by allowing the use of a general covariance matrix in the class model. A recursive form of the criterion function was presented which makes minimal demands on storage space in a very elegant solution.

In order to optimise the criterion function, a companion search scheme was described, commonly known as the branch and bound algorithm. It was shown that the branch and bound technique reduces the complexity of the search problem significantly.

Finally, an experiment was described to test the validity of the technique as applied to a real speech problem. The problem comprised recognising the letters "b", "d" and "g" in a multi-speaker environment. Class models were estimated from only a few iterations of the inferencing mechanism. From these preliminary tests, it can be concluded that the method can achieve an improvement in recognition performance when compared with a conventional method, by utilising the inherent statistical dependence in a speech signal.

There are several avenues of future work which derive from this study. Firstly, a reliable method of inferencing the class statistics needs to be investigated. So far, improvement has resulted from just a few iterations. Then, some experiments on the optimal size of template, in the sense of the parameters speed and discriminatory information content, need to be carried out. As yet, little or no research has been reported which indicates the extent of statistical dependence in the speech waveform; the ensuing studies may help to rectify this.

References

[1] P. A. Devijver and J. Kittler (1982), *Pattern Recognition: A Statistical Approach*, (Prentice Hall, Englewood Cliffs, NJ.)

[2] J. Kittler and A. E. Lucas (1990), "A New Method for Dynamic Time Alignment of Speech Waveforms. Part I: Matching Criterion", Technical Report, Dept. of Elec. Eng., University of Surrey, UK.

[3] H. Sakoe and S Chiba (1978), "Dynamic programming algorithms optimization for spoken word recognition", *IEEE Trans.*, **ASSP-26**, pp. 43–49.

A New Technique for Automatic Segmentation of Continuous Speech†

E. Vidal and A. Marzal*

Dept. de Sistemas Informáticos y Computación, Universidad Politécnica de Valencia, Valencia 46020, SPAIN

Abstract: A new procedure is proposed to automatically segment speech signals in accordance with a given linguistic-unit description of the signals. This technique is "model-free" in that no models are required or used for the units. Segmentation is obtained by (locally) minimizing an appropriate distortion function that measures the linguistic inconsistency of a segmentation through DTW acoustic dissimilarities. Experimental results show the inferiority (greater distortion) of manual segmentations with respect to the automatic segmentations obtained with the proposed technique. These automatic segmentations could be useful to improve the performance of any training procedure for whatever type of linguistic-unit models.

Keywords: automatic segmentation, linguistic constraints, bootstrapping, continuous speech.

Introduction

The *automatic segmentation of continuous speech signals* into meaningful Linguistic Units is an old problem for which no commonly accepted satisfactory solutions have been found so far. While, current state-of-the-art technology allows for Continuous Speech Recognition without resorting to any kind of segmentation, this problem is now receiving increased attention because of the growing popularity of the use of *linguistic-unit models* that are *trainable* from samples of these units. Obviously, when the units are phonemes, for instance, which cannot be uttered in isolation, the samples need to be extracted from larger utterances – through segmentation. Most current continuous Speech Recognition Systems try to avoid this problem by integrating both segmentation and model-training into a single, global procedure, which is generically known as *bootstrapping*. However, bootstrapping does not completely eliminate the need for segmented speech data, since the quality of its results significantly depends on the quality of the initial segments that are used to seed the procedure.

In this paper a new segmentation method is proposed which can be used to obtain reliable initial segmentations. Furthermore, it could eliminate the need for bootstrapping, therefore allowing a complete separation of Segmentation and Model-Training.

---

† Supported in part by the Spanish CICYT, under grant TIC 89/0448.

* Supported by a grant from the "Conselleria de Cultura, Educació i Ciencia de la Generalitat Valenciana"

NATO ASI Series, Vol. F 75
Speech Recognition and Understanding. Recent Advances
Edited by P. Laface and R. De Mori

Model-Free, Linguistically Constrained Segmentation

According to the taxonomy presented in [1], the segmentation technique introduced here can be considered as a *model-free linguistically-constrained* approach. It is called *model-free* in opposition to the name given to other techniques which attempt to segment speech data in accordance with the fitness of the acoustic segments to the corresponding models. On the other hand, it is called *linguistically-constrained* as the segmentation of each speech signal must be consistent with some linguistic description that is supplied for the signal.

Let $\mathcal{A}$ be a set of *acoustic observations* or "frames" (typically, parameter vectors), $\mathcal{L}$ a set of *linguistic categories* (e.g. phonemes) and $\mathcal{B}$ the set where *segment boundaries* are represented (usually $\mathcal{B}\equiv\mathbb{N}$). Then, a *linguistically-constrained segmentation method* is a function $\sigma:\mathcal{A}^*\times\mathcal{L}^*\rightarrow\mathcal{B}^*$ where, for any set X, X^* denotes the set of finite sequences of elements of X. For the segmentations described by σ to be proper, certain elementary constraints should be satisfied for every $a\in\mathcal{A}^*$, $l\in\mathcal{L}^*$ and $b=\sigma(a,l)\in\mathcal{B}^*$:

i) $0<|a|\geq|b|=|l|>0$

ii)† $b_1>0;\ b_{j-1}\leq b_j,\ 1\leq j\leq|b|;\ b_{|b|}=|a|$

Note that this formulation properly accounts for the segmentation of multiple (concatenated) utterances.

The chosen linguistic categories in $\mathcal{L}$ can be of a great variety of types. Typically they will be phonemes, diphones, phonemes with context, demisyllables or any kind of sublexical or lexical units that are considered appropriate for the application.

Linguistically constrained segmentation assume that the acoustic segments must be *consistent* with the linguistic information contained in the given linguistic script. This means that the segments of a specified by $b=\sigma(a,l)$ must be in sequential, one to one correspondence with the successive linguistic labels in l, and that those acoustic segments associated to the same linguistic category must be *similar* in a certain sense.

The consistency of a linguistically constrained segmentation can be measured in terms of *distortion*. For *model-tied segmentations*, distortion can be naturally defined through the fitness of the acoustic segments to the corresponding models; but for *model-free segmentations*, models are no longer assumed and distortion can only be defined in terms of *dissimilarities* between acoustic segments. Since segments are subsequences of acoustic observations, very appropriate dissimilarity measures can be obtained through Dynamic Time Warping (DTW).

Let $d:\mathcal{A}^*\times\mathcal{A}^*\rightarrow R^{>0}$ be a DTW dissimilarity measure. A model-free *linear distortion*, D, of a segmentation b of a that is constrained by l can be defined as:

$$D(a,l,b)=\sum_{\substack{\forall i,j=1\ldots|b|\\ l_i=l_j}}\frac{d(a_{b_{i-1}+1}\ldots a_{b_i},a_{b_{j-1}+1}\ldots a_{b_j})}{n_{l_i}} \tag{1}$$

† An additional, null element $b_0=0$ is assumed to exist whenever required.

where n_{ℓ_i} is the number of times that the linguistic category ℓ_i appears in ℓ. This number is introduced in order to guarantee a linear (rather than quadratic) apportioning to the overall distortion of the segments that are associated with each linguistic label of ℓ.

An optimal segmentation $\hat{b}$ of a with respect to ℓ is then defined as a b for which the *linear distortion* is the smallest: $\hat{b} = \underset{b}{\operatorname{argmin}}\, D(a,\ell,b)$.

Minimizing the Distortion through Gradient Descent

Let k be the number of segments specified by the given linguistic sequence, ℓ. Then each possible segmentation b can be seen as a point in the k-dimensional vector space $\mathbb{N}^k$. Hence, a locally optimal solution to the above minimization problem can be sought by Gradient Descent. The procedure would start with an arbitrary initial segmentation, b, and then iteratively modify the components of b in amounts proportional to $-\nabla_b(D(a,\ell,b))$. While, at first sight, the computation of the gradient of (1) can be thought to constitute a formidable task, it is shown below that the partial derivatives of D with respect to each b_i can be obtained as a subproduct of the computation of $D(a,\ell,b)$ itself. Therefore, each iteration only asks essentially for one computation of (1).

Since the dissimilarities, d, that appear in $D(a,\ell,b)$ correspond to DTW acoustic matches, the numerical computation of the derivatives required for $\nabla_b(D(a,\ell,b))$ amounts to determining the *variations* of DTW dissimilarities with variations in the end points of the acoustic subsequences. This is straightforward for variations in the positions of right boundary marks: Once we have computed the usual DTW dissimilarity for the current position, the dissimilarity value corresponding to a (small) negative increment of the position of this boundary is also available. Similarly, we can extend somewhat the DTW computation towards the right of the current boundary mark to obtain the dissimilarity value corresponding to an (also small) positive increment of the position of the mark. Then, the variation of the dissimilarity can be numerically approximated as a difference of the values obtained for both increments (Fig 1a).

Since the distortion gradient also contains derivatives with respect to the positions of left

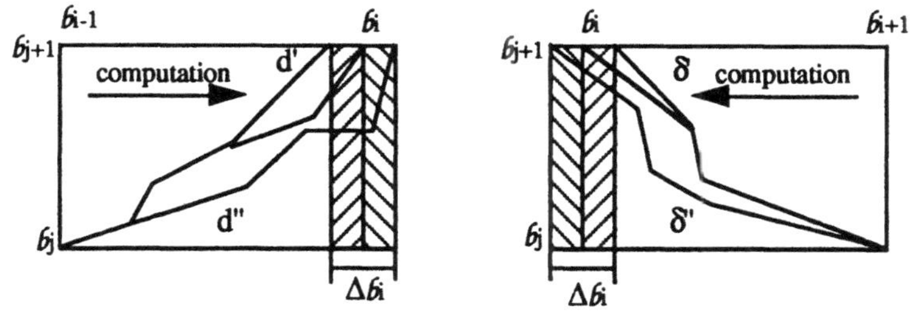

$$\frac{\partial}{\partial b_i} d(a_{b_{i-1}+1}\dots a_{b_i},\, a_{b_j+1}\dots a_{b_{j+1}}) \approx \frac{d''-d'}{\Delta b_i} \qquad \frac{\partial}{\partial b_i} d(a_{b_i+1}\dots a_{b_{i+1}},\, a_{b_j+1}\dots a_{b_{j+1}}) \approx \frac{\delta''-\delta'}{\Delta b_i}$$

forward   *backward*

Figure 1. Numerical computation of the derivatives for the ith component of the gradient

boundaries, this method should also be adapted to variations in the left end-points. Given the incremental nature of the Dynamic programming computation of DTW, this necessarily requires reversing the computing direction. That is, we should proceed *backwards* from the right boundary, which is now fixed, towards the left (now variable) end-point (Fig 1b). For this reason, a *reversible* DTW procedure is needed, such as the one proposed in [2].

Segmentation as a Greedy, Hill-Climbing Search

A direct implementation of the general procedure outlined in the last section entails certain difficulties. In fact, gradient descent does not directly apply in discrete spaces, as is our case and, what is worst, there are even stronger discontinuities which are created both by the initially fixed boundary marks (e.g. fixed end-points) and by constraints imposed on the mobility of the marks (e.g. duration constraints).Taking all these problems into account, the original descent procedure had better be transformed into a Greedy, Hill-Climbing algorithm in which a reduction in distortion can in fact be guaranteed in each step.

The proposed Hill-Climbing algorithm examines each mark in turn in order to determine which of its permitted movements yield the greatest distortion decrease. This is accomplished by using essentially the same forward or backward procedures seen before for the computation of the gradient. Consequently, each iteration either reduces the distortion or leaves things unchanged. In the latter case, the procedure is stopped with the current sequence of boundary marks corresponding to a locally optimal linguistically constrained segmentation.

```
Algorithm HCS
input     a ∈ 𝒜* ; ℓ ∈ ℒ* ; b ∈ ℕ*          // b is an initial segmentation // ;
          minseg, maxseg :ℕ                 // duration constraints //;
output    b' ∈ ℕ* ;
var  δ : array of ℕ;
repeat    b := b';
     for i:=1 to |b|-1 do
          δ := [0,...,0] ;
          w_right := min (b_i+1 - minseg, b_i-1 + maxseg) - b_i ;
          w_left := b_i - max (b_i-1 + minseg, b_i+1 - maxseg) ;
          ∀ i,j ∈ [1...|b|] : ℓ_i = ℓ_j and i ≠ j do
               δ := δ + d_forward(a_{b_{i-1}+1}...a_{b_i+w_right} , a_{b_{j-1}+1}...a_{b_j}) ;
          ∀ i,j ∈ [1...|b|] : ℓ_i+1 = ℓ_j and i+1 ≠ j do
               δ := δ + d_backward(a_{b_i-w_left+1}...a_{b_{i+1}} , a_{b_{j-1}+1}...a_{b_j}) ;
          b_i' :=        argmin          (δ_k) ;
                   k ∈ [w_left...w_right]
until b = b';
```

The computational cost per iteration is $O(n \cdot m^2 \cdot |\ell|)$, where m is the typical segment length and n is the typical (maximum) number of repeated apparitions of a linguistic label in ℓ

Experiments

The tests were intended, both to show the feasibility of the proposed method, and to compare the automatic segmentations with those carried out manually. The speech data consisted of 160 single-speaker utterances of the Spanish numbers from 0 to 1000, with possible repetitions. The average speaking-rate was 18 phonemes/second. The acoustic representations were sequences of 11-dimensional vectors of cepstral coefficients, at a rate of 66.66 vectors per second. Two kinds of linguistic units were adopted for segmentation. The first one consisted of rather large "lexemic" or "morphemic" units like 'θero', 'uno', 'dieθi', 'veinti', 'tre', 'nta', etc... The second type consisted of the set of Spanish phonemes that appears in the chosen data. The results are shown in Table 1.

		Number of iterations	Initial Dist. (Di)	Final Dist. (Df)	Improvements (%)		Statistics of Discrepancies (Average/Std.Dev.)	
					Di-Df / Di	Dm-Df / Dm	Initial vs. Final	Automatic vs. Manual
Lexemic Units	Linear initialization	11	20772	14354	30.9	16.5	5.95/6.12	1.59/2.43
	Manual initialization	6	Dm=17194	14221	17.3		1.49/2.23	
Phonemes	Linear initialization	12	79782	64833	18.7	6.8	1.83/2.05	0.92/1.24

Table 1. Experimental Results (Distortion of the manual phonetic segmentation = 69557)

For the lexemic segmentation, both an arbitrary (linear) segmentation and a manual one were used for initialization, while only a linear segmentation of the previous automatic lexemic segments was used to initialize the phonetic segmentation. The results shown are the initial and final distortions, the percentage of initial-to-final improvement, and the percentage of distortion reduction with respect to the manual segmentations. The statistics of mark displacements, from their initial to their final positions, are also presented, as well as the statistics of the discrepancies between these final automatic boundaries and the manual ones.

It is worth noting that automatic segmentation always leads to smaller distortion, even if manual initialization is used to start the procedure. On the other hand, the discrepancies between the position of automatic and manual boundary marks are small, on the average – less than one frame for the phonetic segmentation.

Figure 2 shows a segmentation of the sentence /tresθientosθinkuentaitres/ (353). For the lexemic units, the automatic and manual segmentations are fairly similar. The only discrepancies appear in those parts of the signal in which the exact segmentation point is rather ambiguous; e.g. between the strong and weak fricatives of 'tre*sθ*ientos'. In other parts, where boundaries are clear both segmentations are in perfect agreement. Similar behaviour is observed at the phonetic level even though, owing to the high speaking rate, the segmentation into so small units tends to be a rather challenging task, both for the automatic and manual procedures.

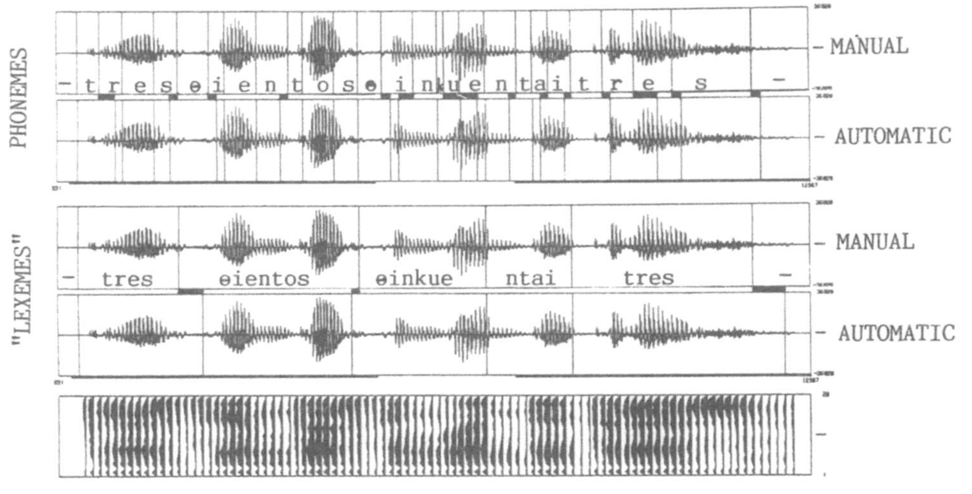

Figure 2. A sample of manual and automatic segmentations into lexemes and phonemes

Discussion and Conclusions

The distortion of fully automatic segmentation is always smaller than that of manual segmentation. Furthermore, the automatic method can also be used to improve significantly the distortion of an initial manual segmentation. On the other hand, the average discrepancy between manual and automatic boundaries is small and, what is more interesting, these small discrepancies are in fact what contribute to the smaller distortion in the automatic case. From a qualitative point of view, the automatic results are considered appropriate in general and, as shown by the numerical results, they are indeed more homogeneous than the manual ones. From all these results it can be expected that automatically obtained *Model-Free Linguistically Constrained* acoustic segments would lead to a better modelling of the corresponding Linguistic Units.

References

1 E. Vidal and A. Marzal: "A Review and New Approaches for Automatic Segmentation of Speech Signals". EUSIPCO 90 Proc. 1990.

2 M. Okochi and T. Shakai: "Trapezoidal DP Matching with Time Reversibility". ICASSP 82 Proc. 1982.

Segmentation of speech based upon a linear model of the effects of coarticulation

P.J.Dix and G.Bloothooft

Research Institute for Language and Speech, University of Utrecht
Trans 10, 3512 JK Utrecht, The Netherlands

Abstract: Most acoustic-phonetic, knowledge based approaches to speech recognition are based on the phonemic hypothesis. According to this hypothesis it is possible to divide a speech utterance into a sequence of phonemes, each phoneme corresponding to a phonetic event or, as for instance in the case of plosives and diphthong, to several phonetic events. The purpose of speech segmentation algorithms is to automatically carry out a division of an utterance into a sequence of events. Due to coarticulation, neighboring events are often poorly separated in time and this makes segmentation a difficult task. In this article we will describe a speech segmentation algorithm which is based on a linear model of the effects of coarticulation. The model assumes that for a suitable set of speech parameters the inter-segment transition between two neighboring articulatory events can be described by a linear combination of two speech vectors.

Keywords: Temporal Decomposition, Segmentation

1. Introduction

Speech signals are generated by a slowly moving articulatory system. These movements take place towards and away from articulatory targets, resulting in a sequence of phonetic events. Since the articulatory system is rather slow, movements towards and away from adjacent targets overlap in time. As a result of this overlap, adjacent targets can be poorly separated in time, which makes speech segmentation a difficult task. In 1983 Atal [1] devised a method for efficient speech coding, called temporal decomposition (TD), which models this overlap by decomposing a time sequence of speech parameters into a series of time-overlapping interpolation functions and an associated series of data vectors. We have used TD as a starting point for a segmentation algorithm based on a model for coarticulation effects.

TD approximates a sequence of speech parameters $\mathbf{y}(n)$ by means of a series of time-overlapping so called phi functions $\phi_k(n)$ and an associated series of so called target vectors $\mathbf{a}_k$:

NATO ASI Series, Vol. F 75
Speech Recognition and Understanding. Recent Advances
Edited by P. Laface and R. De Mori

$$\mathbf{y}(n) \approx \mathbf{y}^*(n) \equiv \sum_{k=1}^{K} \mathbf{a}_k \, \phi_k(n) \quad 1 \le n \le N \tag{1}$$

In this formula N equals the number of frames in the given utterance and $\mathbf{y}^*(n)$ denotes the approximation of the observed speech parameter vector $\mathbf{y}(n)$. We can interpret this equation as a description of a speech utterance in terms of a sequence of K articulatory targets in which the temporal evolution of the k-th event is described by phi function $\phi_k(n)$ and in which the k-th articulatory target is described by target vector $\mathbf{a}_k$; the coarticulation is modelled by the overlap of neighboring phi functions.

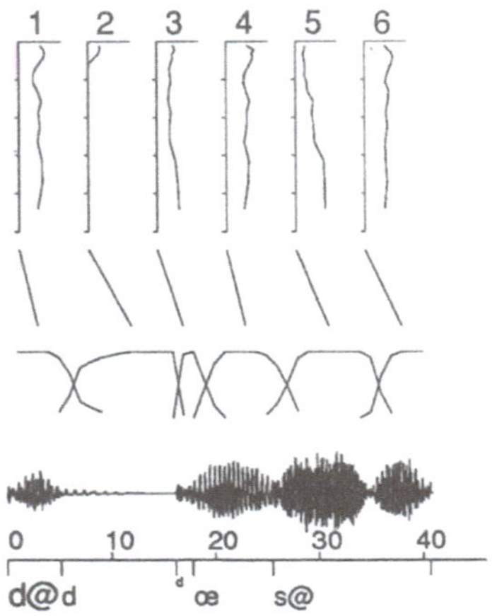

Fig. 1 An example of the decomposition into phi functions and target vectors. The overlapping curves depict the phi functions. The target vectors, in this case filterbank output spectra, are drawn at the top

2. Temporal decomposition and segmentation

Intrinsically, TD is a kind of segmentation algorithm: the phi functions can be considered as segments and the target vectors can be viewed as a weighted mean data vector of the associated segment. Using TD for speech segmentation purposes poses several problems, however. For instance, it is possible to have three or more phi functions that overlap at the same instant of time. This makes it difficult to devise an automatic procedure to determine segment boundaries. Even when there are only two overlapping phi functions that can readily be associated with two speech events, an obvious rule such as taking the point of intersection of two phi functions would often yield an unsatisfactory boundary. A further complication is that quite often the algorithm yields a phi function with an unacceptable shape. These problems make it difficult to use rules which are based on the shape of phi functions. To overcome these difficulties we have devised a modified algorithm, which yields segment boundaries directly, without constructing the phi functions as an intermediate step.

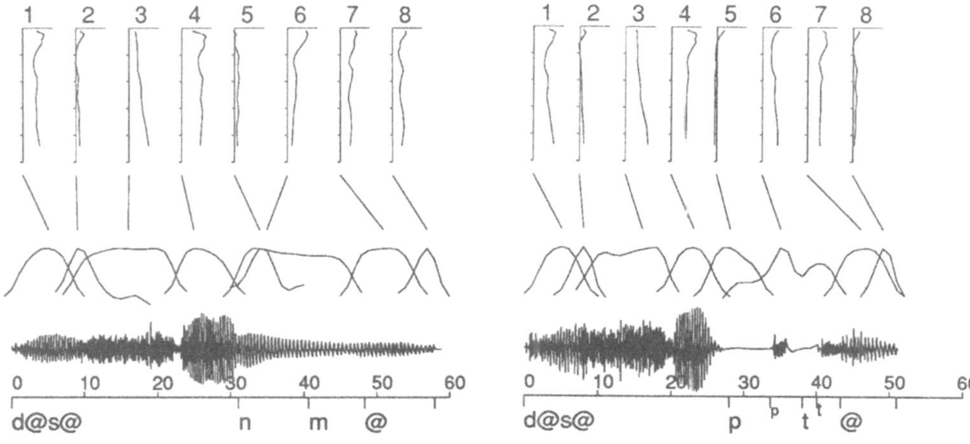

Fig. 2 Some problems associated with an interpretation of phi functions as segments. Phi function nr 5 in the left picture is entirely contained within phi function nr 6. In the right picture, phi function nr 6 has an unacceptable shape

Following Atal, we assume that we can approximate the observed speech parameters $\mathbf{y}(n)$ by a linear combination of target vectors and phi functions. We assume that at any instant of time there are only two phi functions, which are adjacent in time, are non-zero. In this case the movement from articulatory event k towards articulatory event k+1 can be approximated by:

$$\mathbf{y}(n) \approx \phi_k(n)\mathbf{a}_k + \phi_{k+1}(n)\mathbf{a}_{k+1}, \quad 1 \leq n \leq M \tag{2}$$

This equation is approximately valid in the time interval [1,M] during which there is a movement from articulatory event k towards articulatory event k+1. In (2) the speech parameters $\mathbf{y}(n)$ are approximated as a linear combination of phi functions and, by inverting this equation, it follows that the phi functions can be approximated as a linear combination of the speech parameters $\mathbf{y}(n)$. However, in order to be able to unambiguously invert this equation, the number of independent vectors $\mathbf{y}(n)$ must equal the number of phi functions in the analysis window, in this case two. A principal component analysis procedure (PCA) is used to determine the two most important dimensions of the linear subspace generated by $\mathbf{y}(n)$. If $\mathbf{u}_1$ and $\mathbf{u}_2$ are the first two eigenvectors yielded by the PCA procedure, we now approximate $\phi_k(n)$ as a linear combination of these two vectors:

$$\phi_k(n) \approx w_1 u_1(n) + w_2 u_2(n), \quad 1 \leq n \leq M \tag{3}$$

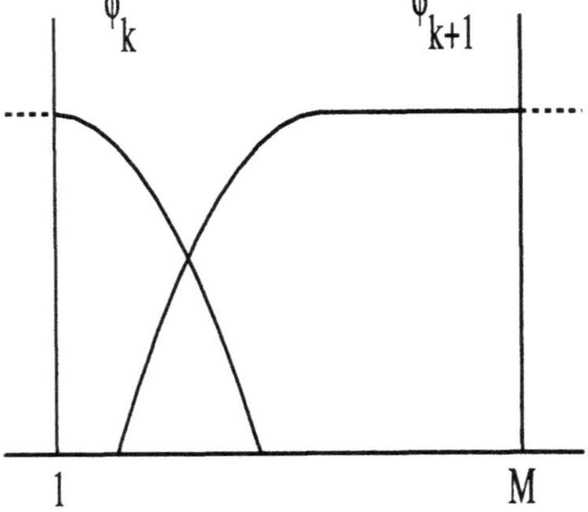

Fig. 3 Transition from event k towards event k+1 in analysis window [1,M]. $\phi_k(n)$ has a high level in the left part of the analysis window and a low level in the right part of the window

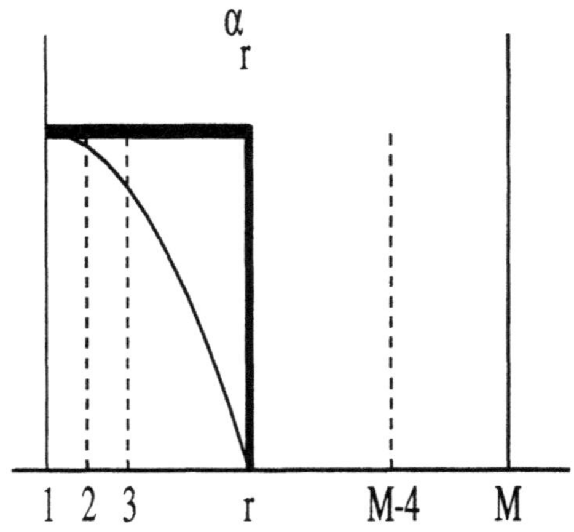

Fig. 4 Every value of r in the interval [1,M-4] is a candidate for a boundary between event k and event k+1. Every r defines a rectangular shaped function $\alpha(r)$. These functions $\alpha(r)$ are matched with phi functions

In order to determine the coefficients w_1 and w_2, a boundary condition must be imposed on the shape of $\phi_k(n)$. Since phi functions describe the temporal evolution of speech events, they are non-zero only in a limited time interval. So, if there is a transition in the time interval [1,M], we expect $\phi_k(n)$ to have a high level in the left part of the analysis window and a low level in the right part of the analysis window. This boundary condition on $\phi_k(n)$ is used to determine the coefficients w_1 and w_2. The following expression is maximized as a function of w_1 and w_2 for each value of r in the interval [1,M-4]:

$$\sum_{n=1}^{r} \phi_{kr}(n)^2 \Big/ \sum_{n=1}^{M} \phi_{kr}(n)^2 \qquad (4)$$

In this way a phi function is constructed which is prominent in the first part [1,r] of the analysis window. Each value of r gives us a phi function $\phi_{kr}(n)$ and an associated rectangular shaped function α_r given by:

$$\alpha_r(n)=1 \quad \text{for } 1 \le n \le r \qquad (5)$$
$$\alpha_r(n)=0 \quad \text{elsewhere.}$$

We will use these rectangular shaped functions as stylizations of $\phi_{kr}(n)$ and take their right borders as candidates for a segment boundary between the speech events k and k+1. The best candidate for the boundary is determined by choosing the best matching pair of phi function $\phi_{kr}(n)$ and α_r . As a measure for the resemblance between $\phi_{kr}(n)$ and α_r, we determine the cosine of the angle between these two M-dimensional vectors:

$$\cos(\phi_{kr},\alpha_r) = < \phi_{kr},\alpha_r > / (||\phi_{kr}||*||\alpha_r||) \tag{6}$$

If $\cos(\phi_{kr},\alpha_r)$ has a maximum value for r less than M-4, that value of r is chosen as the position of the boundary. In this case r is subsequently used as starting point for the search for a boundary between the next two events. If the maximum is obtained for r equal to M-4 , it may be possible to find a better choice for the location of the boundary to the right of the position M-4. In that case the whole analysis window is shifted two frames to the right and the search for a boundary between event k and event k+1 starts all over again in this shifted analysis window. The parameter M is associated with the duration of transitions. We have found it useful to use two different lengths for the analysis window, one for long transitions and one for short transitions. For long transitions we take $M=M_{max}$ equal to 12 and for short transitions we take $M=M_{min}$ equal to 8. For both values of M the values $\cos(\phi_{kr},\alpha_r)$ are computed and for matching values of r we simply take the biggest of the two. The analysis window is shifted to the right only if $\cos(\phi_{kr},\alpha_r)$ has a maximum value for r equal to M_{max} - 4 .

3. Performance Results

To test the segmentation algorithm, we used a Dutch test corpus consisting of phonotactically allowed CV and CC combinations. In order to reduce the number of possible combinations we selected one or two representatives from the broad phonetical classes plosives, fricatives, liquids, glides and nasals. The CV set consisted of nonsense words which had the form /d@CVs@/. In this set there were 81 words, which were pronounced by one male and one female speaker. The CC set contained 33 words with format /d@s@CC@/. The words in this set were pronounced by one male and one female speaker too.

In table 1 and 2 the relations between phoneme class and number of segments is given. From a phonetic point of view there is an oversegmentation, especially in the case of vowels. From an acoustic point of view however, some over-segmentation is in many cases quite acceptable, since there often is an considerable intrinsic spectral change in these vowels. In our view, the one to one correspondence between phonemes and segments is a bit over-stressed in segmentation algorithms: the real goal is to get rid of transitional phenomena.

Number	0	1	2	3	4	N
Short vowels		82	12	7	10	60
Long vowels		19	50	27	9	48
Diphthongs		2	35	52	11	54
Plosives	2	4	93	4		56
Liquids	2	81	17			52
Nasals		93	7			28
Glides		92	8			26

Table 1 Percentages of number of segments per phoneme in the CV data set. The far right column gives the total number of phonemes for each class

Number	0	1	2	3	4	N
Plosives	3	26	56	15		34
Fricatives		63	25	13		24
Liquids	10	80	8	3		40
Nasals		88	13			24
Glides	30	60	10			10

Table 2 Percentages of number of segments per phoneme in the CC data set. The far right column gives the total number of phonemes for each class

4. Summary and conclusions

We have described a segmentation algorithm which is based on a linear model of the effects of coarticulation. The test results showed a good initial segmentation of the speech material. From a phonetic point of view there is in general some oversegmentation, which is however quite acceptable from a acoustic point of view.

References

1. Atal, B.S. (1983): 'Efficient coding of LPC parameters by temporal decomposition', Proc. ICASSP 83, Boston, pp 81-84.

2. Van Dijk-Kappers, A.M.L. and Marcus, S.M. (1987): 'Temporal decomposition of speech', IPO Annual Progress Report 22, 41-50.

3. Van Dijk-Kappers, A.M.L. (1989): 'Temporal decomposition of speech and its relation to phonetic information', PhD-thesis Technical University of Eindhoven.

4. Gerbrands, J.J. (1981) On the relationships between SVD, KLT and PCA, Pattern Recognition, 14, 375-381.

5. Niranjan, M. and Fallside, F. (1989): 'Temporal decomposition: a framework for enhanced speech recognition', Proc. ICASSP 89, pp 655-658

Subject Index

NATO ASI Series F

Including Special Programmes on Sensory Systems for Robotic Control (ROB) and on Advanced Educational Technology (AET)

Vol. 1: Issues in Acoustic Signal – Image Processing and Recognition. Edited by C. H. Chen. VIII, 333 pages. 1983.

Vol. 2: Image Sequence Processing and Dynamic Scene Analysis. Edited by T. S. Huang. IX, 749 pages. 1983.

Vol. 3: Electronic Systems Effectiveness and Life Cycle Costing. Edited by J. K. Skwirzynski. XVII, 732 pages. 1983.

Vol. 4: Pictorial Data Analysis. Edited by R. M. Haralick. VIII, 468 pages. 1983.

Vol. 5: International Calibration Study of Traffic Conflict Techniques. Edited by E. Asmussen. VII, 229 pages. 1984.

Vol. 6: Information Technology and the Computer Network. Edited by K. G. Beauchamp. VIII, 271 pages. 1984.

Vol. 7: High-Speed Computation. Edited by J. S. Kowalik. IX, 441 pages. 1984.

Vol. 8: Program Transformation and Programming Environments. Report on a Workshop directed by F. L. Bauer and H. Remus. Edited by P. Pepper. XIV, 378 pages. 1984.

Vol. 9: Computer Aided Analysis and Optimization of Mechanical System Dynamics. Edited by E. J. Haug. XXII, 700 pages. 1984.

Vol. 10: Simulation and Model-Based Methodologies: An Integrative View. Edited by T. I. Ören, B. P. Zeigler, M. S. Elzas. XIII, 651 pages. 1984.

Vol. 11: Robotics and Artificial Intelligence. Edited by M. Brady, L A. Gerhardt, H. F. Davidson. XVII, 693 pages. 1984.

Vol. 12: Combinatorial Algorithms on Words. Edited by A. Apostolico, Z. Galil. VIII, 361 pages. 1985.

Vol. 13: Logics and Models of Concurrent Systems. Edited by K. R. Apt. VIII, 498 pages. 1985.

Vol. 14: Control Flow and Data Flow: Concepts of Distributed Programming. Edited by M. Broy. VIII, 525 pages. 1985.

Vol. 15: Computational Mathematical Programming. Edited by K. Schittkowski. VIII, 451 pages. 1985.

Vol. 16: New Systems and Architectures for Automatic Speech Recognition and Synthesis. Edited by R. De Mori, C.Y. Suen. XIII, 630 pages. 1985.

Vol. 17: Fundamental Algorithms for Computer Graphics. Edited by R.A. Earnshaw. XVI, 1042 pages. 1985.

Vol. 18: Computer Architectures for Spatially Distributed Data. Edited by H. Freeman and G.G. Pieroni. VIII, 391 pages. 1985.

Vol. 19: Pictorial Information Systems in Medicine. Edited by K. H. Höhne. XII, 525 pages. 1986.

Vol. 20: Disordered Systems and Biological Organization. Edited by E. Bienenstock, F. Fogelman Soulié, G. Weisbuch. XXI, 405 pages. 1986.

Vol. 21: Intelligent Decision Support in Process Environments. Edited by E. Hollnagel, G. Mancini, D.D. Woods. XV, 524 pages. 1986.

NATO ASI Series F

Vol. 22: Software System Design Methods. The Challenge of Advanced Computing Technology. Edited by J. K. Skwirzynski. XIII, 747 pages. 1986.

Vol. 23: Designing Computer-Based Learning Materials. Edited by H. Weinstock and A. Bork. IX, 285 pages. 1986.

Vol. 24: Database Machines. Modern Trends and Applications. Edited by A. K. Sood and A. H. Qureshi. VIII, 570 pages. 1986.

Vol. 25: Pyramidal Systems for Computer Vision. Edited by V. Cantoni and S. Levialdi. VIII, 392 pages. 1986. *(ROB)*

Vol. 26: Modelling and Analysis in Arms Control. Edited by R. Avenhaus, R. K. Huber and J. D. Kettelle. VIII, 488 pages. 1986.

Vol. 27: Computer Aided Optimal Design: Structural and Mechanical Systems. Edited by C. A. Mota Soares. XIII, 1029 pages. 1987.

Vol. 28: Distributed Operating Systems. Theory und Practice. Edited by Y. Paker, J.-P. Banatre and M. Bozyiğit. X, 379 pages. 1987.

Vol. 29: Languages for Sensor-Based Control in Robotics. Edited by U. Rembold and K. Hörmann. IX, 625 pages. 1987. *(ROB)*

Vol. 30: Pattern Recognition Theory and Applications. Edited by P. A. Devijver and J. Kittler. XI, 543 pages. 1987.

Vol. 31: Decision Support Systems: Theory and Application. Edited by C. W. Holsapple and A. B. Whinston. X, 500 pages. 1987.

Vol. 32: Information Systems: Failure Analysis. Edited by J. A. Wise and A. Debons. XV, 338 pages. 1987.

Vol. 33: Machine Intelligence and Knowledge Engineering for Robotic Applications. Edited by A. K. C. Wong and A. Pugh. XIV, 486 pages. 1987. *(ROB)*

Vol. 34: Modelling, Robustness and Sensitivity Reduction in Control Systems. Edited by R. F. Curtain. IX, 492 pages. 1987.

Vol. 35: Expert Judgment and Expert Systems. Edited by J. L. Mumpower, L. D. Phillips, O. Renn and V. R. R. Uppuluri. VIII, 361 pages. 1987.

Vol. 36: Logic of Programming and Calculi of Discrete Design. Edited by M. Broy. VII, 415 pages. 1987.

Vol. 37: Dynamics of Infinite Dimensional Systems. Edited by S.-N. Chow and J. K. Hale. IX, 514 pages. 1987.

Vol. 38: Flow Control of Congested Networks. Edited by A. R. Odoni, L. Bianco and G. Szegö. XII, 355 pages. 1987.

Vol. 39: Mathematics and Computer Science in Medical Imaging. Edited by M. A. Viergever and A. Todd-Pokropek. VIII, 546 pages. 1988.

Vol. 40: Theoretical Foundations of Computer Graphics and CAD. Edited by R. A. Earnshaw. XX, 1246 pages. 1988.

Vol. 41: Neural Computers. Edited by R. Eckmiller and Ch. v. d. Malsburg. XIII, 566 pages. 1988.

NATO ASI Series F

Vol. 42: Real-Time Object Measurement and Classification. Edited by A. K. Jain. VIII, 407 pages. 1988. *(ROB)*

Vol. 43: Sensors and Sensory Systems for Advanced Robots. Edited by P. Dario. XI, 597 pages. 1988. *(ROB)*

Vol. 44: Signal Processing and Pattern Recognition in Nondestructive Evaluation of Materials. Edited by C. H. Chen. VIII, 344 pages. 1988. *(ROB)*

Vol. 45: Syntactic and Structural Pattern Recognition. Edited by G. Ferraté, T. Pavlidis, A. Sanfeliu and H. Bunke. XVI, 467 pages. 1988. *(ROB)*

Vol. 46: Recent Advances in Speech Understanding and Dialog Systems. Edited by H. Niemann, M. Lang and G. Sagerer. X, 521 pages. 1988.

Vol. 47: Advanced Compuling Concepts and Techniques in Control Engineering. Edited by M. J. Denham and A. J. Laub. XI, 518 pages. 1988.

Vol. 48: Mathematical Models for Decision Support. Edited by G. Mitra. IX, 762 pages. 1988.

Vol. 49: Computer Integrated Manufacturing. Edited by I. B. Turksen. VIII, 568 pages. 1988.

Vol. 50: CAD Based Programming for Sensory Robots. Edited by B. Ravani. IX, 565 pages. 1988. *(ROB)*

Vol. 51: Algorithms and Model Formulations in Mathematical Programming. Edited by S. W. Wallace. IX, 190 pages. 1989.

Vol. 52: Sensor Devices and Systems for Robotics. Edited by A. Casals. IX, 362 pages. 1989. *(ROB)*

Vol. 53: Advanced Information Technologies for Industrial Material Flow Systems. Edited by S. Y. Nof and C. L. Moodie. IX, 710 pages. 1989.

Vol. 54: A Reappraisal of the Efficiency of Financial Markets. Edited by R. M. C. Guimarães, B. G. Kingsman and S. J. Taylor. X, 804 pages. 1989.

Vol. 55: Constructive Methods in Computing Science. Edited by M. Broy. VII, 478 pages. 1989.

Vol. 56: Multiple Criteria Decision Making and Risk Analysis Using Microcomputers. Edited by B. Karpak and S. Zionts. VII, 399 pages. 1989.

Vol. 57: Kinematics and Dynamic Issues in Sensor Based Control. Edited by G. E. Taylor. XI, 456 pages. 1990. *(ROB)*

Vol. 58: Highly Redundant Sensing in Robotic Systems. Edited by J. T. Tou and J. G. Balchen. X, 322 pages. 1990. *(ROB)*

Vol. 59: Superconducting Electronics. Edited by H. Weinstock and M. Nisenoff. X, 441 pages. 1989.

Vol. 60: 3D Imaging in Medicine. Algorithms, Systems, Applications. Edited by K. H. Höhne, H. Fuchs and S. M. Pizer. IX, 460 pages. 1990.

Vol. 61: Knowledge, Data and Computer-Assisted Decisions. Edited by M. Schader and W. Gaul. VIII, 421 pages. 1990.

Vol. 62: Supercomputing. Edited by J. S. Kowalik. X, 425 pages. 1990.

Vol. 63: Traditional and Non-Traditional Robotic Sensors. Edited by T. C. Henderson. VIII, 468 pages. 1990. *(ROB)*

Vol. 64: Sensory Robotics for the Handling of Limp Materials. Edited by P. M. Taylor. IX, 343 pages. 1990. *(ROB)*

Vol. 65: Mapping and Spatial Modelling for Navigation. Edited by L. F. Pau. VIII, 357 pages. 1990. *(ROB)*

NATO ASI Series F

Vol. 66: Sensor-Based Robots: Algorithms and Architectures. Edited by C.S.G. Lee. X, 285 pages. 1991. *(ROB)*

Vol. 67: Designing Hypermedia for Learning. Edited by D.H. Jonassen and H. Mandl. XXV, 457 pages. 1990. *(AET)*

Vol. 68: Neurocomputing. Algorithms, Architectures and Applications. Edited by F. Fogelman Soulié and J. Hérault. XI, 455 pages. 1990.

Vol. 69: Real-Time Integration Methods for Mechanical System Simulation. Edited by E.J. Haug and R.C. Deyo. VIII, 352 pages. 1991.

Vol. 70: Numerical Linear Algebra, Digital Signal Processing and Parallel Algorithms. Edited by G.H. Golub and P. Van Dooren. XIII, 729 pages. 1991.

Vol. 71: Expert Systems and Robotics. Edited by T. Jordanides and B. Torby. XII, 744 pages. 1991.

Vol. 72: High-Capacity Local and Metropolitan Area Networks. Architecture and Performance Issues. Edited by G. Pujolle. X, 536 pages. 1991.

Vol. 73: Automation and Systems Issues in Air Traffic Control. Edited by J.A. Wise, V.D. Hopkin and M.L. Smith. XIX, 594 pages. 1991.

Vol. 74: Picture Archiving and Communication Systems (PACS) in Medicine. Edited by H.K. Huang, O. Ratib, A.R. Bakker and G. Witte. XI, 438 pages. 1991.

Vol. 75: Speech Recognition and Understanding. Recent Advances, Trends and Applications. Edited by P. Laface and Renato De Mori. XI, 559 pages. 1992.

Vol. 76: Multisensor Fusion for Computer Vision. Edited by J.K. Aggarwal. *(ROB)*

Vol. 77: Computer Algorithms for Solving Linear Algebraic Equations. The State of the Art. Edited by E. Spedicato. VIII, 352 pages. 1991.

Vol. 78: Integrating Advanced Technology into Technology Education. Edited by M. Hacker, A. Gordon, M. de Vries. VIII, 185 pages. 1991. *(AET)*

Vol. 79: Logic, Algebra, and Computation. Edited by F.L. Bauer. VII, 485 pages. 1991.

MIX
Papier aus verantwortungsvollen Quellen
Paper from responsible sources
FSC® C105338

If you have any concerns about our products,
you can contact us on
ProductSafety@springernature.com

In case Publisher is established outside the EU,
the EU authorized representative is:
Springer Nature Customer Service Center GmbH
Europaplatz 3, 69115 Heidelberg, Germany

Printed by Libri Plureos GmbH
in Hamburg, Germany